ADVANCES IN
QUANTUM CHEMISTRY

VOLUME 30

ADVANCES IN
QUANTUM CHEMISTRY
MODERN TRENDS IN ATOMIC PHYSICS

EDITOR-IN-CHIEF

PER-OLOV LÖWDIN

PROFESSOR EMERITUS

DEPARTMENT OF QUANTUM CHEMISTRY
UPPSALA UNIVERSITY
UPPSALA, SWEDEN

AND

QUANTUM THEORY PROJECT
UNIVERSITY OF FLORIDA
GAINESVILLE, FLORIDA

EDITORS

JOHN R. SABIN
MICHAEL C. ZERNER

QUANTUM THEORY PROJECT
UNIVERSITY OF FLORIDA
GAINESVILLE, FLORIDA

ERKKI BRÄNDAS

DEPARTMENT OF QUANTUM CHEMISTRY
UPPSALA UNIVERSITY
UPPSALA, SWEDEN

GUEST EDITORS

DAG HANSTORP

DEPARTMENT OF PHYSICS
CHALMERS UNIVERSITY OF TECHNOLOGY
GÖTEBORG UNIVERSITY
GÖTEBORG, SWEDEN

HANS PERSSON

DEPARTMENT OF PHYSICS
CHALMERS UNIVERSITY OF TECHNOLOGY
GÖTEBORG UNIVERSITY
GÖTEBORG, SWEDEN

VOLUME 30

ACADEMIC PRESS

San Diego London Boston New York Sydney Tokyo Toronto

Academic Press Rapid Manuscript Reproduction

This book is printed on acid-free paper. ♾

Academic Press
a division of Harcourt Brace & Company
525 B Street, Suite 1900, San Diego, California 92101-4495, USA
http://www.apnet.com

Academic Press Limited
24-28 Oval Road, London NW1 7DX, UK
http://www.hbuk.co.uk/ap/

International Standard Book Number: 0-12-034830-6

PRINTED IN THE UNITED STATES OF AMERICA
98 99 00 01 02 03 QW 9 8 7 6 5 4 3 2 1

Contents

Contributors ix
Preface xiii

Four Decades of Atomic Physics: A Review of Ingvar Lindgren's Career until Now 1
Carl Nordling

Spectroscopy with Highly Coherent Radiation 5
Norman F. Ramsey

Atoms in Optical Lattices: Cooling, Trapping, and Squeezing Atoms with Light 19
William D. Phillips

Coherent Quantum State Manipulation of Trapped Atomic Ions 41
D. J. Wineland, C. Monroe, D. M. Meekhof, B. E. King, D. Leibfried, W. M. Itano, J. C. Bergquist, D. Berkeland, J. J. Bollinger, and J. Miller

Resonance Fluorescence of a Single Ion 65
J. T. Höffges, H. W. Baldauf, T. Eichler, S. R. Helmfrid, and H. Walther

Quantum Electrodynamics and the Fundamental Constants 77
Peter Mohr

High Precision Spectroscopy of Positronium and Muonium 99
Vernon W. Hughes

Quantum Electrodynamics of Strong Fields: Status and Perspectives 125
Gerhard Soff, T. Beier, M. Greiner, H. Persson, and G. Plunien

State-Specific Multi-reference Coupled Cluster Formulations: Two Paradigms 163
Uttam Sinha Mahapatra, Barnali Datta, Barun Bandyopadhyay, and Debashis Mukherjee

Correlations in the Vacuum 195
Walter Greiner

High-Power Lasers and Their Applications 209
S. Svanberg

A Periodic Table in Three Dimensions: A Sightseeing Tour in the Nanometer World 235
Arne Rosén

Relativistic and Dynamic Contributions in Ion–Atom and Ion–Solid Collisions 273
B. Fricke, J. Anton, K. Schulze, W.-D. Sepp, and P. Kürpick

Atomic Radiative Lifetimes Measured by Pulsed Laser Spectroscopy in the UV/VUV Spectral Region 283
U. Berzinsh and S. Svanberg

A Theoretical and Experimental Study of the Quartet System of B I 301
Yaming Zou, Tomas Brage, and Indrek Martinson

Two Electron Dynamics in Photodetachment 311
D. Hanstorp, G. Haeffler, A. E. Klinkmüller, U. Ljungblad, U. Berzinsh, I. Yu. Kiyan, and D. J. Pegg

Collinear Laser Spectroscopy on Radioactive Isotopes 335
J. R. Persson

Four Decades of Hyperfine Anomalies 343
Martin G. H. Gustavsson and Ann-Marie Mårtensson-Pendrill

Spins and Moments of Nuclei far from Stability Determined by On-Line Atomic-Beam Techniques 361
Curt Ekström

Regularization Corrections to the Partial-Wave Renomalization Procedure 379
Hans Persson, Sten Salomonson, and Per Sunnergren

The Reference State Coulomb-Breit QED Corrections for the Few-Electron Highly Charged Ions 393
L. N. Labzowsky and M. A. Tokman

Some Aspects on the Bloch-Lindgren Equation and a Comparison with the Partitioning Technique 415
Per-Olov Löwdin

What Is the Force between Electrons? 433
J. Sucher

Atomic Physics and the Laser Metrology of Time and Length 445
L. R. Pendrill and L. Robertsson

Optical Trapping of Absorbing Particles 469
H. Rubinsztein-Dunlop, T. A. Nieminen, M. E. J. Friese, and N. R. Heckenberg

Index 493

Contributors

Numbers in parentheses indicate the pages on which the authors' contributions begin.

J. Anton (273), Fachbereich Physik, Universität Kassel, D-34209 Kassel, Germany

H. W. Baldauf (65), Max-Planck-Institut für Quantenoptik and Sektion Physik der Universität, München, 85748 Garching, Germany

Barun Bandyopadhyay (163), Department of Physical Chemistry, Indian Association for the Cultivation of Science, Calcutta 700032, India

T. Beier (125), Institut für Theoretische Physik, Technische Universität Dresden, D-01062 Dresden, Germany

J. C. Bergquist (41), Ion Storage Group, Time and Frequency Division, NIST, Boulder, Colorado 80303

D. Berkeland (41), Ion Storage Group, Time and Frequency Division, NIST, Boulder, Colorado 80303

U. Berzinsh (283, 311), Department of Physics, Lund Institute of Technology, S-221 00 Lund, Sweden

J. J. Bollinger (41), Ion Storage Group, Time and Frequency Division, NIST, Boulder, Colorado 80303

Tomas Brage (301), Department of Physics, University of Lund, S-223 62 Lund, Sweden

Barnali Datta (163), Department of Physical Chemistry, Indian Association for the Cultivation of Science, Calcutta 700032, India

T. Eichler (65), Max-Planck-Institut für Quantenoptik and Sektion Physik der Universität, München, 85748 Garching, Germany

Curt Ekström (361), The Svedberg Laboratory, Box 533, S-751 21 Uppsala, Sweden

B. Fricke (273), Fachbereich Physik, Universität Kassel, D-34209 Kassel, Germany

M. E. J. Friese (469), Department of Physics, The University of Queensland, Brisbane, Queensland 4072, Australia

M. Greiner (125), Institut für Theoretische Physik, Technische Universität Dresden, D-01062 Dresden, Germany

Walter Greiner (195), Theoretische Physik, Universität Frankfurt, D-60054, Frankfurt/Main, Germany

Martin G. H. Gustavsson (343), Chalmers University of Technology, Göteborg University, Department of Experimental Physics, Atomic and Molecular Physics Groups, Fysikgränd 3, S-412 96 Göteborg, Sweden

G. Haeffler (311), Department of Physics, Göteborg University and Chalmers University of Technology, S-412 96, Göteborg, Sweden

Dag Hanstorp (311), Department of Physics, Göteborg University and Chalmers University of Technology, S-412 96 Göteborg, Sweden

N. R. Heckenberg (469), Department of Physics, The University of Queensland, Brisbane, Queensland 4072, Australia

S. R. Helmfrid (65), Max-Planck-Institut für Quantenoptik and Sektion Physik der Universität, München, 85748 Garching, Germany

J. T. Höffges (65), Max-Planck-Institut für Quantenoptik and Sektion Physik der Universität, München, 85748 Garching, Germany

Vernon W. Hughes (99), Yale University, Physics Department, New Haven, Connecticut 06520

W. M. Itano (41), NIST, Boulder, Colorado 80303

B. E. King (41), NIST, Boulder, Colorado 80303

I. Yu. Kiyan (311), Department of Physics, Göteborg University and Chalmers University of Technology, S-412 96, Göteborg, Sweden

A. E. Klinkmüller (311), Department of Physics, Göteborg University and Chalmers University of Technology, S-412 96 Göteborg, Sweden

P. Kürpick (273), Fachbereich Physik, Universität Kassel, D-34209 Kassel, Germany

Leonti Labzowsky (393), St. Petersburg State University, Department of Physics, St. Petersburg, Russia

D. Leibfried (41), Ion Storage Group, NIST, Boulder, Colorado 80303

U. Ljungblad (311), Department of Physics, Göteborg University and Chalmers University of Technology, S-412 96 Göteborg, Sweden

Per-Olov Löwdin (415), Department of Quantum Chemistry, Uppsala Universitet, S-75120 Uppsala, Sweden; and Florida Quantum Theory Project, Department of Physics and Chemistry, University of Florida, Gainesville, Florida 32611-8435

Uttam Sinha Mahapatra (163), Department of Physical Chemistry, Indian Association for the Cultivation of Science, Calcutta 700032, India

Ann-Marie Mårtensson-Pendrill (343), Department of Physics, Göteborg University, and Chalmers University of Technology, S-412 96 Göteborg, Sweden

Indrek Martinson (301), Department of Physics, University of Lund, S-223 62 Lund, Sweden

D. M. Meekhof (41), NIST, Boulder, Colorado 80303

J. Miller (41), NIST, Boulder, Colorado 80303

Peter J. Mohr (77), National Institute of Standards and Technology, Gaithersburg, Maryland 20899-0001

C. Monroe (41), NIST, Boulder, Colorado 80303

D. Mukherjee (163), Department of Physical Chemistry, Indian Association for the Cultivation of Science, Calcutta 700032, India

T. A. Nieminen (469), Department of Physics, The University of Queensland, Brisbane, Queensland 4072, Australia

Carl Nording (1), Department of Physics, Uppsala University, S-751 21 Uppsala, Sweden

D. J. Pegg (311), Department of Physics, Göteborg University, and Chalmers University of Technology, S-412 96 Göteborg, Sweden

L. R. Pendrill (445), Swedish National Testing & Research Institute, S-501 14 Borås, Sweden

J. R. Persson (335), School of Physics and Space Research, The University of Birmingham, Edgbaston, Birmingham B15 2TT, United Kingdom

Hans Persson (125, 379), Department of Physics, Göteborg University, and Chalmers University of Technology, S-412 96 Göteborg, Sweden

William D. Phillips (19), National Institute of Standards and Technology, Gaithersburg, Maryland 20899

G. Plunien (125), Institut für Theoretische Physik, Technische Universität Dresden, D-01062 Dresden, Germany

Norman F. Ramsey (5), Lyman Physics Laboratory, Harvard University, Cambridge, Massachusetts 02138

L. Robertsson (445), Swedish National Testing & Research Institute, S-501 14 Borås, Sweden

Arne Rosén (235), Department of Physics, Göteborg University, and Chalmers University of Technology, S-412 96 Göteborg, Sweden

H. Rubinsztein-Dunlop (469), Department of Physics, The University of Queensland, Brisbane, Queensland 4072, Australia

Sten Salomonson (379), Department of Physics, Göteborg University and Chalmers University of Technology, S-412 96 Göteborg, Sweden

K. Schulze (273), Fachbereich Physik, Universität Kassel, D-34209 Kassel, Germany

W.-D. Sepp (273), Fachbereich Physik, Universität Kassel, D-34209 Kassel, Germany

Gerhard Soff (125), Institut für Theoretische Physik, Technische Universität Dresden, D-01062 Dresden, Germany

Joe Sucher (433), Center for Theoretical Physics and Department of Physics, University of Maryland, College Park, Maryland 20742

Per Sunnergren (379), Department of Physics, Göteborg University and Chalmers University of Technology, S-412 96 Göteborg, Sweden

Sune Svanberg (209, 283), Department of Physics, Lund Institute of Technology, S-221 00 Lund, Sweden

M. A. Tokman (393), St. Petersburg State University, Department of Physics, St. Petersburg, Russia
H. Walther (65), Max-Planck-Institut für Quantenoptik and Sektion Physik der Universität, München, 85748 Garching, Germany
D. J. Wineland (41), NIST, Boulder, Colorado 80303
Yaming Zou (301), Department of Physics, University of Lund, S-223 62 Lund, Sweden

Preface

The symposium Modern Trends in Atomic Physics was held on May 24–25, 1995, at the Hjortviken Conference Centre just outside Göteborg in Sweden. It was organized in honor of Professor Ingvar Lindgren on the occasion of his 65th birthday, which also coincided with the date of his retirement. He had been appointed professor in the Department of Physics, Göteborg University/Chalmers University of Technology, 30 years earlier. During this time he achieved a national and international reputation as a leading scientist. Many of Ingvar's scientific friends and all of his past and present students were invited to the symposium. In total, 90 people attended.

This book contains the text of 11 invited talks presented at the symposium. Unfortunately, the limited time of the symposium made it impossible to allow many leading scientists to give oral presentations. They were instead invited to submit written papers for publication in this proceeding. In total, 13 such invited papers are presented. The first paper is written by Ingvar's long-time colleague and friend, Professor Carl Nordling from Uppsala University, who gives a personal presentation of Ingvar's career. The scientific papers start with that of Nobel laureate Norman F. Ramsey from Harvard University, who gives a historical review of the early development of experimental studies using coherent radiation. The volume then covers many areas of experimental and theoretical atomic physics, including both historical reviews and papers covering very recent results.

The Swedish Academy of Sciences and Göteborg University are acknowledged for the financial support that made the conference possible. M. Sc. Martin G. H. Gustavsson is kindly acknowledged for his work on the organizing committee and for his assistance in the preparation of this proceeding.

At this time we thank Ingvar for his support over the years, and wish him good luck in his future work as Director General of the Foundation for Strategic Research.

It is also a pleasure to recognize one of the invited speakers, Dr. William D. Phillips, who was awarded the 1997 Nobel Prize in Physics for "development of methods to cool and trap atoms with laser light."

DAG HANSTORP
HANS PERSSON

Four decades of atomic physics
A review of Ingvar Lindgren's career until now

by Carl Nordling
Department of Physics
University of Uppsala

Ingvar graduated from high-school in Uppsala in 1950. For further education he went to the Royal Institute of Technology in Stockholm, from which he graduated in 1954. In the same year his physics professor, Kai Siegbahn, was appointed a professor of physics at Uppsala university. Siegbahn had observed Ingvar's unusual talents for physics and invited Ingvar to join him as a graduate student.

In this capacity Ingvar immediately became involved in atomic physics research. Or was it rather nuclear physics research? The title of one of his first scientific publications was "Atomic beam resonance apparatus with six-pole magnets for radioactive isotopes", a title that suggests an involvement of both atomic and nuclear physics. The effective solid angle of his instrument was an order of magnitude larger than could be obtained with the conventional method, so Ingvar could produce resonance curves and determine hfs interaction constants and nuclear spins and magnetic moments for radioactive atoms that had not previously been within reach.

Obviously, at that time Ingvar was doing experimental physics and designing new instruments for his experiments. And he has continued to work as an experimentalist and supervise experimental work in atomic beam resonance spectroscopy, laser spectroscopy and environmentally oriented applications, but theoretical work has become an increasingly large part of his scientific activity. Indeed, so much so that in a selective list of his publications that I have obtained, only theoretical publications are mentioned! Also, the nuclear physics has to a large extent given way to atomic physics in his research.

In May of 1959 Ingvar took his PhD with highest marks (marks are not given anymore). I presented my thesis on the following day, and we had a common dinner party to celebrate both events. A few years later, in 1966, Ingvar moved to Göteborg as a professor of physics at Chalmers Institute of Technology and Göteborg University. He soon became head of the physics department (1967-70) and dean of the faculty (1968-70, 75-79 and 88-93). More than 60 students have studied för the PhD exam under his supervision. With the symposium on Modern

ADVANCES IN QUANTUM CHEMISTRY, VOLUME 30

0065-3276/98 $25.00

Trends in Atomic Physics we have now celebrated his promotion to the level of emeritus professor.

During the late 1950:s and first part of the 60:s Ingvar made major contributions to our knowledge of spins and electromagnetic moments of nuclear ground and excited states. In the 1965 edition of the "Bible" of nuclear spectroscopy, "Alpha-, Beta- and Gamma-ray Spectroscopy", Editor K.Siegbahn, he contributed an exhaustive table of nuclear spins and moments, obtained with various techniques.

Experiments had now shown that the single-particle model was unable to satisfactorily explain the atomic hyperfine interaction. Many-body effects had to be taken into account. So, after the transfer to Göteborg Ingvar became more and more involved in theoretical work, in particular electronic many-body perturbation theory.

He extended the Rayleigh-Schrödinger perturbation formalism to the case of a model space that is not necessarily degenerate [Journal of Physics B **7**, 2441-70, 1974]. He presented a new approach to the diagrammatic formulation of many-body perturbation theory for open-shell systems [International Journal of Quantum Chemistry **S12**, 33-58, 1978]. He applied many-body perturbation theory in the coupled-cluster formulation to perform very accurate calculations on the 2^2S and 2^2P states of the lithium atom [Physical Review A **31**, 1273-86, 1985]. In a review article with D. Mukherjee he took a close look into the various aspects of the linked-cluster theorem for open-shell systems, using as little constraints on the starting functions as possible - and practicable [Physics Reports **151**, 93-127, 1987]. A rigorous method for calculating the first-order self-energy in a model potential, using partial-wave renormalization, was presented by Ingvar and several coworkers in 1993 [Physical Review A **47**, R4555-58, 1993]. A numerical scheme for evaluating the part of the one-photon vacuum-polarization effect that is not accounted for by the Uehling potential was published in the same year [Physical Review A **48**, 2772-78, 1993], and in 1995 a complete, numerical calculation of the effect of exchange of two virtual photons between the electrons in the ground states of heliumlike systems was published [Physical Review A **51**, 1167-95, 1995].

I have mentioned here only a small selection of theoretical papers on Ingvar's publication list, which contains about 150 titles in total. A few weeks ago I received a preprint of his invited talk at the 1996 ICAP conference where Ingvar and coworkers review QED effects in heavy, highly charged ions, in particular the Lamb shift for one-, two-, and three-electron ions [Proc. 15th Int. Conf. on Atomic Physics, World Scientific 1996], and he has published at least four more papers this year.

Let us for a moment descend from these lofty heights and land in more pedestrian theoretical territory. The year is 1967. The ESCA people in Uppsala, under the leadership of Kai Siegbahn, were then busy writing the first ESCA book. Most of what we wanted to report in this book was of an experimental nature, but theory had to be considered and applied to our data on electron binding energies and their chemical shifts. Ingvar, with his solid theoretical expertness, was consulted on how to calculate wave functions, energy levels and chemical shifts in the atomic core. As a result of these discussions the theoretical level of the book was raised considerably and Ingvar became a co-author of the book.

There are five more books on Ingvar's list of publications. His course in atomic physics for graduate students in the 1960:s was the first of its kind in Uppsala, and resulted in a book. Together with Erik B. Karlsson he also gave a course in tensor operators, again resulting in a book. At that time students on the undergraduate level wanted to have course books in the Swedish language. A set of five books, covering the whole field of physics, and written i Swedish, was produced. Ingvar was co-author of the quantum physics part. Soon thereafter the wind had changed direction and the students demanded coursebooks written in the English language... A book on atomic many-body theory for graduate students, with J Morrison as co-author, was published in 1982 (with a second edition 1986) and the proceedings of the eighth ICAP in Göteborg 1982 had Ingvar Lindgren, Arne Rosén and Sune Svanberg as editors.

To this impressive list I would also like to add Ingvar's involvement in the Swedish National Encyclopedia which was completed in 1996 and contains twenty heavy-weight volumes. In this enterprise Ingvar has acted both on the physics expert panel and as an author, see for example the title-words ”atom” and ”kvantelektrodynamik” (quantum electrodynamics). About one-hundred thousand subscribers can now explore the world of atomic physics with the help of Ingvar's articles in the Encyclopedia.

For fourteen years Ingvar was a member of the Nobel Committee for physics of the Royal Swedish Academy of Sciences and chaired the Committee for three years at the end of this period (1989-91). He served on the Swedish Natural Science Research Council (NFR) 1971-77 and was entrusted in 1991 with the task of evaluating the Swedish National Research Facilities. Academic memberships and awards, university committees and national commitments include an APS fellowship, a membership of the Swedish Academy for Engineering Sciences (1995), a von Humboldt Research award (1993) and many others. He has been a member of many scientific conference committees, the main organizer of two Nobel

Symposia (*Many-Body Theory of Atomic Systems,* 1979, and *Heavy-ion Spectroscopy and QED Effects in Atoms,* 1992) and one ICAP conference (Göteborg 1982). Ingvar has been an invited speaker and session chairman at innumerable scientific conferences all over the world.

Since 1994 Ingvar is managing director and scientific leader of the *Swedish Foundation for Strategic Research*, based in Stockholm. In this new capacity he sits on top of a respectable amount of money, about eight billion Swedish kronor or more than one billion US dollars. The primary goal of the foundation is to improve the scientific and technological competence in Sweden in areas of relevance for the competitiveness of our country.

Finally, let me return to atomic physics, which has been the main theme of Ingvar's research over the years. It is obvious that atomic physics has experienced a renaissance during the last few decades as a testing ground for fundamental physical laws and theoretical advances. Ingvar has made, and is still making important contributions to this development, which Norman Ramsey described at the Hjortviken symposium as "an exciting and highly productive period in atomic physics". We all look forward to Ingvar's continued participation in this development.

SPECTROSCOPY WITH HIGHLY COHERENT RADIATION

Norman F. Ramsey

Lyman Physics Laboratory
Harvard University
Cambridge, MA 02138

ABSTRACT

The origins and history of Spectroscopy with coherent radiation is reviewed from the earliest molecular beam magnetic resonance experiments up to the development of the laser.

INTRODUCTION

It is an honor to speak at this celebration of Ingvar Lindgren and his many contributions to experimental and theoretical atomic physics. We should also celebrate his having started so many young students into productive scientific careers.

Ingvar asked me as the first speaker to provide some historical perspective by discussing my personal recollections of some of the advances in atomic physics during the beginning of the present era of atomic physics, but there have been far too many to cover the entire field in my allowed half hour. Therefore, I am limiting myself to the origins of spectroscopy with highly coherent radiation, which is the feature that most distinguishes atomic physics during the past 60 years from all previous periods. Before spectroscopy with radio frequency and microwave radiations, all spectroscopy was with radiation sources of low coherency whereas present atomic physics research is dominated by the use of highly coherent radiation from radio frequency oscillators, microwave sources and lasers.

MOLECULAR BEAM MAGNETIC RESONANCE METHOD

In the summer of 1937 I had the good fortune to start working for my Ph. D. with I. I. Rabi at Columbia. Up to that time there had

ADVANCES IN QUANTUM CHEMISTRY, VOLUME 30

0065-3276/98 $25.00

been only one successful spectroscopy experiment with highly coherent radiation. Cleeton and Williams [1] in 1934 had observed with microwaves a single broad resonance due to the tunneling transition in NH_3 as shown in Fig. 1, but the resonance was single and so broad that its width exceeded its frequency. There were no follow-up experiments and there were no other microwave spectroscopy experiments for the next eleven years. In 1936 Gorter [3] had tried unsuccessfully to observe resonant heating of a crystal in a magnetic field when exposed to an oscillating magnetic field at the nuclear Larmor precession frequency.

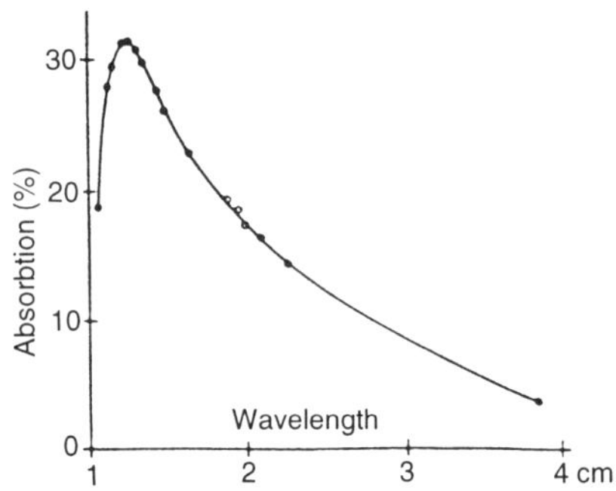

Fig. 1. Absorption band of NH_3 plotted on a wave-length scale. Only a single broad resonance band is observed with the width of the band being approximately equal to wave-length at maximum absorption [1].

Early in 1937 Rabi developed what proved later to be the fundamental exact resonance formula for radio frequency induced transitions between two energy levels, but initially he did not recognize the sharp resonance character of the solution. In September of 1937, Rabi [2], following a visit by Gorter [2,3], invented the molecular beam magnetic resonance method for measuring nuclear magnetic moments, as shown schematically in Fig. 2. Immediately two of Rabi's research groups, including mine, started simultaneously to modify their molecular beam apparatuses for resonance measurements. As expected Rabi, Millman, Kusch and Zacharias (RMKZ) [2] obtained the first resonances, since they were working with the easily detected LiCl whereas Kellogg, Rabi, Ramsey and Zacharias (KRRZ) [4] were working with the difficult, but more interesting isotopes of hydrogen. The first nuclear magnetic resonance is shown in Fig. 3. Soon thereafter we obtained our first resonance with H_2 but were shocked by the extremely broad resonance which looked more like noise over a wide frequency range than like the expected resonance that had been seen in LiCl. However we did obtain resonances of the expected shape with HD and from these determined the magnetic moments of the proton and deuteron, which was our original objective.

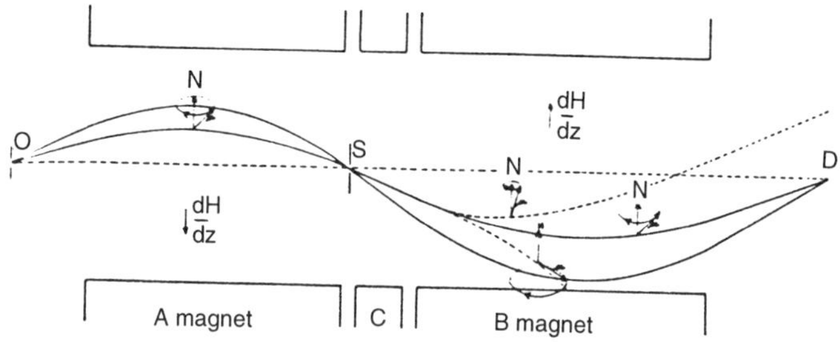

Fig. 2. Schematic diagram [2] showing the principles of the first molecular beam magnetic resonance experiments. The two solid curves indicate two paths of molecules having different orientations that are not changed during passage through the apparatus. The two dashed curves in the region of the B magnet indicate two paths of molecules whose orientation has been changed in the C region so the refocusing is lost due to the change in the component along the direction of the magneti field.[2].

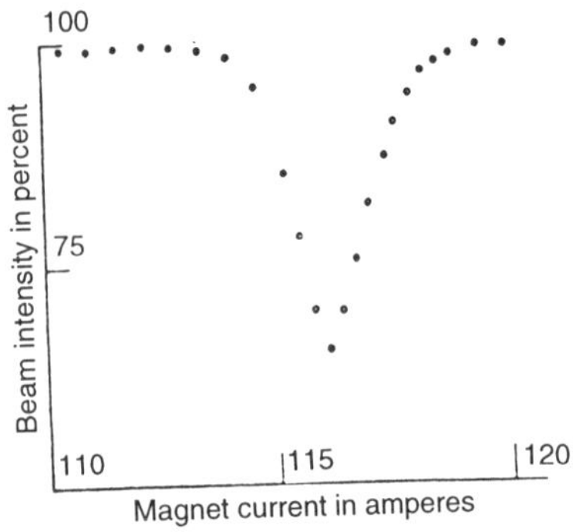

Fig.3. First observed nuclear magnetic resonance [2]. The refocused beam intensity is plotted for various values of the magnetic field. One ampere corresponds to about 1.84×10^{-4} T. The frequency of the oscillating field was held constant at 3.518×10^6 Hz [4].

At that time Columbia had the rule that a Ph.D. thesis had to be published with many copies and a single author. This meant in a collaborative experiment it was necessary to find some uninteresting topic to assign to the student for his thesis. Since we thought the peculiar resonance shape was probably an apparatus effect, I was given the thesis topic of exploring the H_2 and D_2 resonances and trying to find out why they were so peculiar. I spent the summer of 1938 in these studies while Rabi was a visiting professor at Stanford and while Zacharias and Kellogg prepared improvements in our apparatus. I found that we had been using much too much radio frequency current. When I lowered the current, the confusing broad single resonance for H_2, became six distinct resonances as shown in Fig. 4. Up to then we had assumed internal interactions within the molecule were negligible compared to the interaction of the proton magnetic moment with the external magnetic field, but it then became clear that we were actually observing a radio frequency spectrum with the displacements being due to the magnetic interactions of one proton with another and with magnetic field due to the molecular rotation.

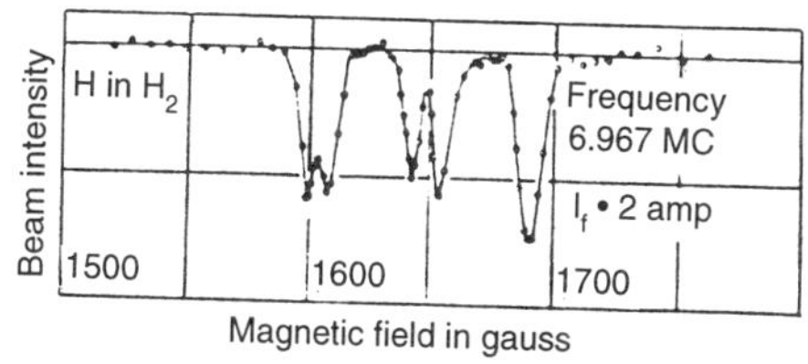

Fig. 4. Radio frequency spectrum of H_2 in the vicinity of the proton resonance frequency [4] in first clearly observed multiple line spectra with coherent radiation. The resonance frequencies are primarily determined by the interaction of the proton magnetic moment with the external magnetic field, but the states of different m_I and m_J are displaced relative to each other by the different values of the nuclear spin-spin and spin rotational interaction energies [4].

My studies of D_2 turned up another surprise. Although there was a principal peak that corresponded to one of the HD resonances, there was also a persistent 10% background, which with weaker oscillating magnetic fields began to look like overlapping resonances. To make the D_2 curves even more interesting the spread of these resonances was far greater than could be accounted for by the magnetic interactions that accounted for the H_2 spectrum so well. It was clear that an additional interaction was required and the data could be well fitted by an assumed deuteron quadrupole moment, which implied the existence of a new and previously unsuspected tensor force between a neutron and a proton. At this point it became clear, with four of us having cooperated in construction of the apparatus, that the studies of H_2 and D_2 were no longer uninteresting problems to assign exclusively to one student, so we all joined in developing an improved apparatus and published together our discoveries and the improved resonances shown in Figs. 5 and 6. In the H_2 paper we used for the first time the subtitle "Radio frequency Spectroscopy". I took for my thesis the study of the radio frequency spectra associated with the rotational magnetic moments of H_2 , D_2 and HD. I was incredibly lucky to write the first

Ph. D. thesis using magnetic resonance and to participate in the first discovery of the deuteron quadrupole moment and the first publication on radio frequency spectroscopy.

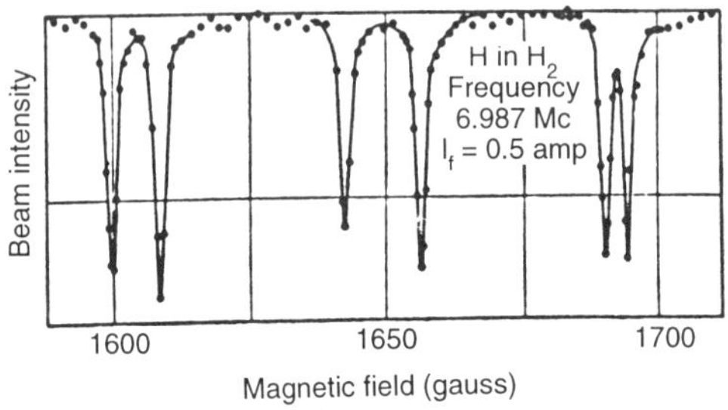

Fig. 5. Radio frequency spectrum of H_2 in the vicinity of the proton resonance frequency with improved apparatus[4].

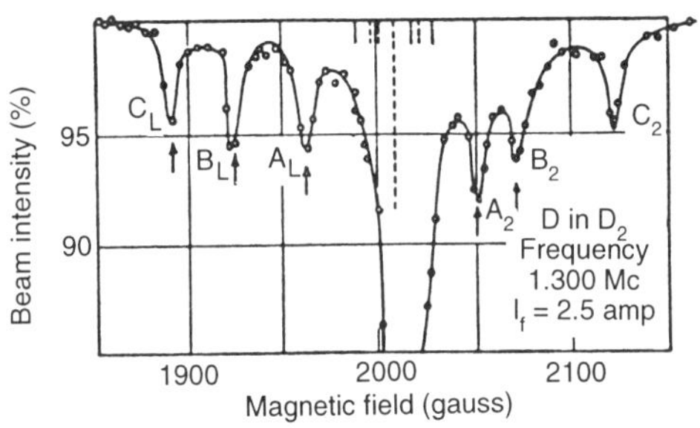

Fig. 6. Radio frequency spectrum of D_2 in the vicinity of the deuteron resonance frequency [4], Dotted lines indicate the spectrum predicted on the assumption of no deuteron electric quadrupole moment.

The first molecular beam magnetic resonance experiments were with $^1\Sigma$ molecules, but in 1940 Kusch, Millman and Rabi [5,6] extended the method to paramagnetic atoms and including

$\Delta F = \pm 1$ transitions of atoms. For such transitions, the relative orientations of the nuclear and electronic magnetic moments are changed, so the resonance frequencies are primarily determined by internal properties of the atom.

From 1938 to 1946, Rabi's laboratory was the only place in the world doing spectroscopy with coherent radiation. Although we recognized the radiation was coherent and calculated transition probabilities accordingly, we made little explicit use of the coherency; we merely needed oscillations at radio frequencies and normal radio oscillators produce coherent radiation whereas gas discharge tubes do not. Rabi and his associates during this time measured many nuclear and atomic magnetic moments as well as atomic hyperfine structure and were greatly helped by excellent resident theorists including Lamb and Schwinger. This exciting work was brought to an abrupt end by World War II when Rabi and many of his associates went to the MIT Radiation Laboratory to work on radar. Kellogg and Lamb stayed on at Columbia directing the Columbia Radiation Laboratory developing 1 cm. wave length magnetrons.

Following WWII, Rabi and I returned to Columbia, Lamb remained there and we were later joined by Kusch. Research resumed quickly. Harold K. Hughes [7] succeeded in making the first resonance apparatus with electric oscillating and deflecting fields and used it to study CsF. Nierenberg revived the old alkali resonance apparatus to study atomic cesium for his thesis.

During the war Rabi and I decided that we should test fundamental quantum mechanics by measuring the atomic hyperfine separation of hydrogen, which could be accurately calculated theoretically. When I returned to Columbia after the war, I found pieces of our old hydrogen apparatus in the Columbia attic and started Nafe and Nelson [8] assembling it for the experiment. At the same time Lamb and Retherford [9] independently began measuring the fine structure of atomic hydrogen in the $n = 2$ state. Both experiments disagreed with theory and both were reported at the famous 1947 Shelter Island Conference.

Although the hyperfine experiment was published first, the Lamb fine structure result attracted the greatest attention at the meeting because it was too big to ignore and it could be interpreted more easily.

The Lamb experiment was the primary stimulus leading to the development of a non-relativistic Quantum Electrodynamics (QED) by Kramers, Bethe, Schwinger, Weisskopf and others. The Lamb shift was interpreted as the effect of the vacuum on the electron.

Although no explanation of Nafe, Nelson and Rabi's [8] anomalous hyperfine interaction arose at the Shelter Island conference, an explanation did emerge later in the same year with an even more profound effect on the development of QED. The hyperfine anomaly inspired Schwinger [10] to develop his relativistic QED. Shortly after the Shelter Island conference Breit [10] noted that the anomalous hyperfine interaction could be explained by assuming the electron had an anomalous moment. Schwinger realized that he could explain the anomalous hyperfine interaction on a more fundamental basis by developing a relativistic QED. Since he knew this would be difficult, he did not wish to do so unless he was sure of the experimental evidence; I can clearly remember being vigorously cross examined by Julian in the summer of 1947 as to whether I was convinced that the hyperfine anomaly was real. Schwinger then seriously began developing his theory. While his work was in progress but before its publication [10] Kusch and Foley found more direct evidence [11] for the anomalous magnet moment of the electron by measuring the g_J ratios of the $^2P_{3/2}$ and $^2P_{1/2}$ states of Gallium. The announcement in early 1948 of Schwinger's successful relativistic QED, involving mass and charge renormalization, created a great sensation among physicists. This theory with subsequent developments served as the model for future quantum field theories including QCD, electroweak theories and efforts to achieve grand unification.

RADIO FREQUENCY AND MICROWAVE SPECTROSCOPY

Following the war, other new and important methods for radio frequency and microwave spectroscopy were introduced. In 1946 Purcell, Torrey and Pound [12] and Bloch, Hansen and Packard [13] at radio frequencies developed the nuclear magnetic resonance absorption and induction methods that later came to be known as NMR and provided the basis NMR chemical analysis and for magnetic resonance imaging (MRI).

During World War II, there were enormous advances in microwave technology at radar development laboratories, especially in

Britain and the United States, and these were well described in a series of M. I. T. Radiation Laboratory publications. As a result the long dormant field of microwave spectroscopy exploded. Prior to 1946 the only publication on microwave spectroscopy had been the Cleeton and Williams 1934 paper, but in 1946 there were important publications from different laboratories by the following authors [14,15]: Bleaney, Penrose, Beringer, Townes, Dicke, Lamb, Becker, Autler, Strandberg, Dailey, Khyl, Van Vleck, Wilson, Dakin, Good, Coles, Hershberger, Lamont, Watson, Roberts, Beers, Hill, Merritt and Walter. In 1947 there were more than 60 publications including papers by authors with reports the previous year and by Gordy, Jen and others. Although most microwave spectroscopy experiments utilized molecular electric dipole transitions, paramagnetic resonance experiments were also carried out in solids and a few molecules beginning with Zavoisky's first experiment in 1946 and followed by Bleaney, Beringer and many others.

SUCCESSIVE COHERENT PULSES AND EXPLICIT USES OF COHERENCY

Although the radio frequency and microwave radiations by their methods of production were intrinsically coherent and calculations of transition probabilities depended on the coherency, there was little effort to make explicit use of the coherency. The first explicit uses were those in my 1949 and subsequent papers [15,16] on the methods of separated and successive oscillatory fields. In particular I pointed out that, if the atoms or molecules being studied are subjected to a short radio frequency pulse of duration τ, followed by a time T with no radiation and followed finally by a second phase coherent pulse of duration τ, the observed resonances are narrower and can be changed in shape from resonance absorption to dispersion by adjusting the relative phases of the two coherent pulses. The method averages over field inhomogeneities in the intermediate region, eliminates first order Doppler shifts and is adaptable to molecular beam resonances at much higher frequencies since the wavelength of the radiation need only be less than dimensions of the regions in which short pulses are applied, not the long drift region which gives rise to the sharp resonances. In subsequent publications [16] I generalized the method to arbitrary numbers of successive oscillatory regions with arbitrary phases and amplitudes. Since the expressions for the transition amplitudes for a single region are exact, so also are the solutions by successively applying the transformations. The separated oscillatory field method (often called the Ramsey Method) is used in the Cs atomic beam clocks developed in 1954 and subsequent years in by Zacharias, Perry, Essen and many others [15]. The unit of time, the second, is now defined in terms of such clocks.

In 1950, Hahn [17] described his ingenious spin echo NMR techniques in which he first used a radio frequency pulse to flip nuclear spins by 90°, allowed the signal to disappear by field inhomogeneities or T_2 relaxation and then τ seconds after the first pulse he applied a 180° pulse so that the signal reappeared τ seconds later as a nuclear induction "spin echo". The spin echo technique has been extensively used to measure nuclear magnetic resonance phenomena, particularly relaxation times. Subsequently a variety of pulse shapes and phases have been used by have been used in NMR, in Fourier transform spectroscopy and in Magnetic Resonance Imaging (MRI) by Hahn, Slichter, Ernst, Anderson, Pines, Waugh, Warren, Lauterbuhr and many others to produce patterns and spectra of desired shapes.

Most of the early radio frequency and microwave spectroscopy could be well treated by theories that were quantum mechanical in the determination of atomic energy levels but used classical electromagnetism in the calculations of transition probabilities and coherency effects. For example, in classical electromagnetism the effective dipole moment of N different dipoles closer together than one wavelength and excited coherently by an electromagnetic field is proportional to N so the radiation intensity, proportional to the square of the effective electric dipole moment, is proportional to N^2. This relation was used from the beginning of radio frequency spectroscopy but in 1956 was formalized and extended to spontaneous radiation processes in Dicke's quantum mechanical theory [18] of "super radiant" states. In 1963 Glauber [18] and others developed the quantum theory of optical coherence.

OPTICAL PUMPING

In 1949 Bitter [15] showed the possibility of studying nuclear properties in optically excited states and Kastler and Brossel [15] the following year developed the double resonance method and optical pumping to increase the population of certain states, so that stimulated emission is not nearly canceled by absorption.

FOCUSING AND TRAPPING OF ATOMS AND IONS

In the 1950's there were several independent developments that greatly affected future spectroscopy even though they were not directly concerned with radiative transitions.

In 1949 and the following two years, Vauthier [15], Friedberg and Paul [15] and Korsunskii [15] independently suggested the use of inhomogeneous magnetic fields to produce space focusing of molecular beams. Friedberg and Paul showed that a hexapole magnet provided the desired focusing for magnetic dipoles of fixed magnitude while Bennewitz and Paul [15] showed that a qudrupole magnet was best for a magnetic moment whose strength was proportional to the field.

In 1951 Kantrowitz and Grey [15] first proposed the use of higher pressure aerodynamic jet sources in molecular beams. Jet sources provide lower beam temperatures, greater beam intensities and increased populations of lower energy molecular rotational states.

In 1953 Dicke [18] proposed the use of an inert buffer gas such as helium to reduce the first order Doppler width of a radiating system by partially averaging the velocities when the times between collisions are short compared to the radiative lifetime.

In 1956, Ramsey [19] proposed doing radio frequency and microwave spectroscopy on atoms with long lived excited states stored in a trap consisting of an evacuated bottle with suitably coated walls. Kleppner, Ramsey and Fjelstadt [15,19] in 1958 observed the hyperfine resonance of Cs atoms which had entered the trap after having been state selected and subjected to the first of two successive oscillatory fields and then after storage in the trap emerging through the second oscillatory field and state analyzer. The success of this experiment showed that the hyperfine state was not destroyed by the wall collisions and led to the development [15] of the highly stable atomic hydrogen maser in which hydrogen atoms are stored in an evacuated teflon coated quartz bulb for more than a second, during which time each atom makes more than 10,000 wall collisions. The averaging of the velocity by the frequent wall collisions effectively eliminates the first order Doppler shift and broadening. The long storage time produces very narrow resonances and high stability.

By Earnshaw's theorem, it is impossible to make a trap for ions static electric fields, but Penning in 1936 developed a stable ion trap that was a combination of a static uniform magnetic field and inhomogeneous electrostatic fields. In 1958 Paul and his associates made a successful electric quadrupole ion trap without a magnetic field by having the electric field oscillate sinusoidally at radio frequencies.

In 1962 Dehmelt [20] reported the first success in studying radio frequency spectra of the $^4He^+$ ions in a static Penning trap to reduce first order Doppler shifts by velocity averaging. Although the intrinsic widths of the spectrum from the uncertainty principle are narrow due to the long trapping time, the resonances were broadened by second order Doppler shifts due to the 1 eV kinetic energies of the trapped ion. With the later invention of laser cooling this problem is eliminated and ion traps are now effective for precision measurements and atomic clocks.

MASERS

In 1951, Purcell, Pound and Ramsey [21] did some NMR experiments with inverted populations of the nuclear spin systems in LiF and noted that the spin systems were at negative absolute temperatures and that they were intrinsic amplifiers rather than absorbers. The first suggestions to use systems with inverted populations as practical amplifiers or oscillators were made independently in the early 1950's by Townes [15,22], Weber [15] and Bassov and Prokhorov [15]. The first such amplifier was a molecular beam apparatus operating on the NH_3 inversion states and built by Gordon, Zeiger and Townes [22] and was called a microwave amplifier by stimulated emission of radiation (MASER).

Originally it was thought that atomic hyperfine transition frequencies and atomic magnetic moments were too small for atomic hyperfine masers to be possible, but Ramsey and Kleppner pointed out that by storing atoms in a suitably coated bottle [15] coherency could be maintained for several seconds and the hyperfine resonance would be made so narrow that maser oscillations would take place. In 1961 Goldenberg, Kleppner and Ramsey [15] made an atomic hydrogen maser and found it had extremely high stability.

LASERS

When masers were being developed it was thought that the methods could not be extended to optical frequencies because of the short wave length of visible light. Either the resonant region would have to be less than a wavelength or its effectiveness would be diminished by phase differences throughout the region. In 1958, Schalow and Townes [21] pointed out that this argument was incorrect since the laser radiation time is so short that the atoms moved less than a wavelength during the radiation time. They provided a valid theoretical analysis of the requirements for a laser. Maiman, in 1961 made the first successful laser with ruby and in 1961 Ali Javan made the first atomic maser using a mixture of mixture of helium and neon gases,

CONCLUSIONS

Due to lack of time I have had omit many atomic experiments, all theory and all physics after 1961. I particularly regret the omission of atomic theory since Ingvar Lindgren, in whose honor we are now meeting, has made so many important contributions to this field. I hope that subsequent speakers will make up for some of these deficiencies.

Although I have had to omit many important topics for lack of time, it should be clear that the period from 1934 to 1961 was a an exciting and highly productive time in atomic physics. But the same was also true for the period from 1961 to the present and there is every reason to expect this excitement will continue into the future.

REFERENCES

1. C. E. Cleeton and N. H. Williams, *Phys. Rev..*, 1934, **45**, 234.
2. I. I. Rabi, *Phys. Rev..*, 1937, **51**, 652 and . I. Rabi, J. R. Zacharias, S. Millman and P. Kusch, *Phys. Rev..*, 1938, **53**, 318 and 1939, **55**, 526.
3. C. J. Gorter, *Physica*, 1936, **3**, 503 and 995 and *Physics Today*, 1967, **20**, 76 (Jan.)

4. J. M. B. Kellogg, I. I. Rabi, N. F. Ramsey and J. R. Zacharias, *Phys. Rev..*, 1939, **55**, 729; 1939, **57**, 728; and 1940, **57**, 677
5. P. Kusch, S. Millman and I. I. Rabi, *Phys. Rev..*, 1940, **57**, 765.
6. S. Millman and P. Kusch, *Phys. Rev..*, 1940, **57**, 438.
7. H. K. Hughes, *Phys. Rev..*, 1949, **72**, 314
8. J. E. Nafe, E. B. Nelson and I. I. Rabi, *Phys. Rev..*, 1947, **71**, 914 and 1948, **73**, 718
9. W. E. Lamb and R. C. Retherford, *Phys. Rev.*, 1947, **72**, 241.
10. G. Breit, *Phys. Rev.* 1947, **71**, 984; J. Schwinger, *Phys. Rev..*, 1948, **73**, 416 and 1949, **76**, 790.
11. P. Kusch and H. M. Foley, *Phys. Rev..*, 1947, **72**, 1256 and *Phys. Rev..*, 1948, **74**, 250.
12. E. M. Purcell, H. G. Torrey and R. V. Pound, *Phys. Rev..*, 1946, **69**, 37.
13. F. Bloch, W. Hansen and M. E. Packard, *Phys. Rev..*, 1946, **69**, 127.
14. C. H. Townes and A. L. Schawlow, *Microwave Spectroscopy*, 1955, McGraw-Hill, New York.
15. N. F. Ramsey, *Molecular Beams*, 1956 and 1990, Oxford University Press, Oxford, England. and *Journal of Research of National Bureau of Standards*, 1983, **88**, 301. The book and the historical review give extensive references to the original publications.
16. N. F. Ramsey, *Phys. Rev..*, 1949,**75**, 1326 and **76**, 996; N. F. Ramsey, *Phys. Rev..*, 1958, **109**, 822 and *Rev. Mod. Phys.*, 1990, **62**, 541.
17. E. L. Hahn, *Phys. Rev.*, 1950, **77**, 297 and **80**, 580.
18. R. H. Dicke, *Phys. Rev..*, 1953, **89**, 472; R. J. Glauber, *Phys. Rev..*, 1963, **130**, 2529.
19. Ramsey, *Rev. Scient. Instr.*, 1957, **28**, 57; D. Kleppner, N. F. Ramsey and P. Fjelstadt. *Phys. Rev. Lett.*, 1958, **1**, 232.
20. H. G. Dehmelt, *Phys. Rev.*, 1959, **109**, 381.
21. R. V. Pound, *Phys. Rev.*, 1951, **81**, 156; N. F. Ramsey and R. V. Pound, *Phys. Rev.*, 1951, **81** 278;. E. M. Purcell and R. V. Pound, *Phys. Rev.*, 1951, **81**, 279; and N. F. Ramsey, *Phys. Rev.*, 1956, **103**, 20.
22. J. P. Gordon, H. J. Zeiger and C. H. Townes, *Phys. Rev.*, 1954, **95**, 282 and 1955, **99**, 1264.
23. A. L. Schalow and C. H. Townes, *Phys. Rev.*, 1958, **112**, 1940.
24. T. H. Maiman, *Nature*, 1960, **187**, 493.
25. A. Javan,W. Bennett and W. Herriott, *Phys. Rev. Lett.* 1969, **6**, 106.

ATOMS IN OPTICAL LATTICES:
Cooling, Trapping and Squeezing Atoms with Light

William D. Phillips
National Institute of Standards and Technology
Gaithersburg, MD 20899 USA

ABSTRACT: Radiative forces on atoms can be used both to cool the atoms to temperatures on the order of microkelvins and to trap them in a periodic array of microscopic potential wells formed by the interference of multiple laser beams, i.e., an optical lattice. The quantum motion of such lattice-trapped atoms can be studied by spectroscopic techniques. Atoms trapped in this way may be further manipulated so as to be cooled by adiabatic expansion, localized by sudden compression or driven into mechanical oscillations.

INTRODUCTION: This paper is a summary of a talk I presented at Modern Trends in Atomic Physics, a symposium in honor of Ingvar Lindgren, held in Göteborg on 24-25 May 1996. Laser cooling [1, 2, 3] is certainly one of the modern trends in atomic physics. It is one of the fastest growing areas of atomic physics research and its techniques were instrumental in the recent observation of Bose-Einstein condensation [4, 5]. This paper gives a brief summary of some of the progress we have made in laser cooling of atoms, and suggests that there are still things to be learned in this maturing field.

One of the many delights of the Symposium was an old photograph of Ingvar Lindgren in front of a black board covered with material of surprising relevance to the topics under discussion at the symposium. The editors of this special issue have kindly reproduced that photograph in this volume. Just as several speakers referred to it during their talks, I will refer to it in this paper, as "the famous picture."

DOPPLER COOLING: The first ideas for laser cooling of atoms [6] and ions [7] were published in 1975. The basic idea of what is now called Doppler cooling can be understood by referring to the famous picture. Here we see a typical atomic resonance curve, which we can take to show the rate at which an atom

ADVANCES IN QUANTUM CHEMISTRY, VOLUME 30

0065-3276/98 $25.00

absorbs photons from a laser beam as a function of the laser's frequency.

Let us consider such an atom to be irradiated with equal intensity, counterpropagating laser beams tuned slightly to the low frequency (red) side of the resonance curve. If the atom is at rest, it will absorb photons at equal rates (somewhat less than the maximum, resonant rate) from each beam. Since the re-radiated fluorescence photons are emitted in opposite directions with equal probability, the atom at rest feels no net average force from the photon absorption and emission. If the atom moves along the laser beams' direction, the Doppler shift makes the laser beam into which it is moving appear to be bluer and therefore closer to resonance. The atom preferentially absorbs light from this laser beam and feels a force opposing its motion. The effect of this damping is to slow or cool the atom.

The discreteness and randomness of the photon absorption and emission process causes an effective heating against which the cooling is balanced. Equilibrium leads to a thermal energy as low as the energy width of the atomic resonance [8]. The lowest temperature possible is the Doppler cooling limit of $k_B T_{Dopp} = \hbar\Gamma/2$, and is achieved for a detuning from resonance of $-\Gamma/2$ and low laser intensity. For a resonance linewidth of $\Gamma = 2\pi \times 5$ MHz (as in laser cooling of cesium), this is about 120 μK.

The same argument applies to a three dimensional configuration of three orthogonal pairs of counterpropagating laser beams. These provide damping of the atomic velocity in three dimensions. The damping is similar to that of a particle moving in a viscous fluid, so this laser cooling configuration is usually referred to as "optical molasses" [9, 10, 11], molasses being a very viscous, sugary fluid. The predicted Doppler cooling limit in 3D is the same as in 1D. Initial experiments with optical molasses yielded temperatures at about the Doppler cooling limit, although with large uncertainties [9, 12]

SUB-DOPPLER COOLING: In 1988 the NIST-Gaithersburg group made careful measurements of the temperature of atoms laser cooled in optical molasses, and found temperatures significantly below the Doppler cooling limit [13]. The initial measurements on laser cooled sodium atoms gave temperatures of about 40 μK, about six times lower than the predicted lower limit of 240 μK.

Later measurements [10] on sodium found temperatures as low as 25 μK, and later measurements on cesium at the Ecole Normale Supérieure in Paris found temperatures as low as 2.5 μK, about 50 times lower than the predicted 120 μK limit [14].

This striking departure of the experimental results from the predictions of theory was unexpected and surprising. The theory was simple and compelling, but it did treat the laser cooled atoms as two-level systems. In fact, the alkali atoms used in laser cooling experiments have multiple levels. In particular, the electronic ground state displays both hyperfine structure and Zeeman structure, with multiple quantum states being degenerate in the absence of an external field. An example of this kind of structure, and the lifting of the degeneracy by a magnetic field, can be found in the famous picture.

While it was well known that the alkalis were not two-level atoms as supposed by the theory of Doppler cooling, it was generally believed that this did not affect the cooling limit, since the limit depended on the lifetime of the excited state and this was the same for all the degenerate sublevels. Experiments showing that the temperature for sub-Doppler cooling depended on the magnetic field and on the polarization of the laser beams suggested, however, that the magnetic Zeeman sublevels were involved in the cooling process; a linear dependence of temperature on laser intensity suggested a cooling mechanism quite different from that of Doppler cooling, which predicted that the temperature would be independent of laser intensity, in the limit of low intensity, below saturation [13, 15].

A new theory of laser cooling was soon advanced [16, 17, 18] that made explicit use of the multilevel character of the atoms as well as the gradient of polarization of the light. Following ref. [17] we can understand the basic idea of the new theory by reference to fig. 1. Here we presume that an atom with two normally degenerate Zeeman sublevels (ground state J = 1/2, m = ±1/2) moves in a light field formed by two counterpropagating laser beams with orthogonal linear polarization (the lin-perp-lin configuration). The light couples the atom to a J =3/2 optically excited state. The interference of the two orthogonally polarized beams results in "fringes" of polarization such that the light changes from σ^+ circular polarization to σ^- circular polarization as one moves a distance of one fourth of the optical wavelength. The degeneracy of the sublevels is lifted by the action of the light, which shifts the two sublevels differently. This light shift is the second-order perturbation of the energy levels due to the optical field. It is also known as the ac Stark effect. Because the strength of the atom's interaction with the light depends on the light's polarization and on the internal state of the atom, the light shift changes with the atom's position in the light field, and is different for the two different atomic states.

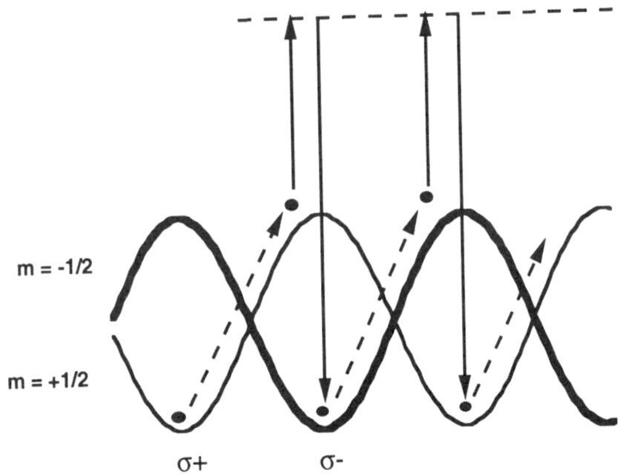

Fig. 1. Schematic representation of Sisyphus cooling. An atom with a J = 1/2 ground state in a 1D lin-perp-lin laser field has two energy levels that are split differently by the light shift as a function of position. Atoms are optically pumped into the lower energy level at places where the light polarization is circular. Moving atoms climb the potential hill created by the light shift, losing kinetic energy. Near the top of the hill, they are optically pumped to the other state, and are again at the bottom of a hill. The process efficiently reduces the kinetic energy.

Now consider what happens when a atom is at a position in space where the polarization is circular σ^+, that is, with positive angular momentum of the light along the axis of quantization. An atom at that place will be optically pumped into the state of highest projection of the atomic angular momentum along the quantization axis, m = +1/2. This is also the state that has the lowest energy at that point in space, when the laser detuning is red of resonance, and corresponds to the bottom of a potential well for m = +1/2 atoms, the potential energy being the light shift of the levels.

As an atom in the m = +1/2 level moves, it must climb out of the potential well, converting kinetic to potential energy and slowing down. As the atom moves away from the position where the light is purely σ^+, there is a greater tendency for optical pumping to put the atom into the other, m = -1/2 level. In the simplest picture of the cooling process, the atom climbs to the top of the potential well, at which point the polarization of the light is σ^-. The atom is then optically excited and decays into the m = -1/2 level, which corresponds to the bottom of another potential well. This process, known as Sisyphus cooling, repeats, efficiently removing kinetic energy from the atom, and converting the potential energy into photon energy.

Of course, the actual Sisyphus cooling cycle is somewhat more complicated than this simple picture. In general, an atom will not optically pump to the other potential well exactly at the top of the potential well. Faster atoms climb higher before pumping, and so, for atoms slow enough that they do not go up and down the potential hills before pumping, more energetic atoms lose energy faster. If the light intensity is higher, the atoms have a higher hill to climb and can lose more energy, but the higher intensity makes them optically pump faster, so they climb a smaller fraction of the hill. The two effects of higher potential and faster pumping cancel so that an atom moving past the wells at constant velocity experiences a damping force that is independent of the laser intensity, at least for intensities that are not too large or small [19].

The highly effective nature of Sisyphus cooling brings the atoms to much a lower temperature than does Doppler cooling. The temperature achieved corresponds to a thermal energy about one tenth of the energy depth of the potential wells. Also, both the simple theory [17] and the more complete theories [19, 20] predict a temperature that, to a good approximation, varies linearly with the potential energy well depth when that depth is not too small (too close to the recoil energy). This prediction has been verified repeatedly [10, 14, 21], and in the work presented here. This is one example of the success of the Sisyphus picture, which has guided much of the thinking about laser cooling.

OPTICAL LATTICES: Considering Fig. 1, and the fact that the thermal energy of the atoms cooled by the Sisyphus process is about a tenth of the energy depth of the pictured potential wells, one sees that the atoms will become trapped in the potential wells. This is still true when one uses level schemes that are more complicated than J = 1/2 in the ground state. In spite of the atoms being trapped the temperature is linearly dependent on the depth of the well, just as predicted by the simple theory in which atoms are cooled as they move at a quasi-constant velocity across many wells.

Because atoms are trapped in a periodic array of potential wells (formed by the interference of the beams used for laser cooling) such a situation is referred to as an optical lattice. We do not use the term "crystal" because in a crystal the atoms are self-organized due to the interactions between the atoms. In an optical lattice, the organization is imposed by the external light field and does not depend on interactions between atoms. In fact, in most optical lattices, only a small fraction of the lattice sites are occupied by atoms and the effect of atom-atom interactions is negligible.

The first optical lattices to be recognized as such were realized in one dimension [22, 23]. These first experiments were distinguished from earlier 3D experiments with optical molasses by the fact that the interference pattern was well defined and trapped atoms at locations where optical pumping tended to keep them in the trapped state. A 3D optical molasses is usually formed from 6 laser beams, 3 counterpropagating pairs (1D standing waves). Such an arrangement produces an interference pattern that depends on the relative time phase of the three standing waves. By contrast, a single pair of counterpropagating beams produces an interference pattern that does not depend on the relative time phase of the two beams. Only the position of the standing wave pattern depends on that phase. Later, it was realized [24] that using four laser beams in a 3D configuration results in the same phase-insensitivity of the interference pattern. Only the position of the pattern changes as the relative phase changes. But in the 3D, 6-beam experiments, the phases and the interference pattern varied randomly with the acoustic vibrations of the optical mounts, and atoms trapped in some location with some polarization might soon find themselves subject to a very different polarization and potential. While it was possible to detect the trapping of atoms in such a situation

[25] it was unsatisfactory for detailed studies or for observation of the quantized motion of atoms trapped in an optical lattice.

The first 1D optical lattices used the lin-perp-lin configuration of Fig. 1. Here, the atoms are trapped at the locations of pure circular polarization, and optical pumping puts the atoms into the extreme m_F state, $m_F = \pm F$, the state that is most strongly trapped. Most photon scattering processes therefore keep the atom in the same m-state and keep the atom trapped. In such a situation an atom can make many oscillations in its potential well before it escapes or otherwise has its oscillatory motion disturbed. Quantum mechanically speaking, there exist distinct and well resolved vibrational levels, with coherence between levels maintained for longer than an oscillation period. In this situation it is possible to see the quantized motion of the trapped atoms. In experiments in Paris [22] this was seen in the absorption spectrum of a probe beam passing through the optical lattice. Features spaced at the vibrational energy separation (typically 100 kHz) showed the quantum level structure.

In experiments at NIST, we instead observed the emission spectrum of the trapped atoms. It may seem remarkable, at first, that we resolve features much smaller than 100 kHz when the natural linewidth of the optical transition is 6 MHz (for Rb atoms.) Let us consider the spectrum emitted by a 2-level atom irradiated by a single-frequency laser beam. For weak excitation, the radiated spectrum is simply the spectrum of the laser: the atom absorbs light, radiates it while returning to the original ground state, and the elastic process results in the radiated photons having the same spectrum as the incident light, except for possible Doppler shifts. When the excitation is strong, energy can be shared between radiated photons and the spectrum of emitted light contains the original narrow feature for elastic scattering plus a three-peaked spectrum with features having widths on the order of the natural linewidth [26]. The central feature of this "Mollow triplet" is centered at the frequency of the laser and the sidebands are spaced at the Rabi frequency (the strength of the atom-field coupling). Our experiments see only the "elastic" part of this spectrum, and hence are not limited at all in their resolution by the natural linewidth. These experiments resolve spectral features about 1000 times narrower than the natural linewidth. Similar experiments in Munich have recently resolved the elastic feature in the spectrum of a single ion [27], a result also reported at this symposium.

The spectrum of lattice-trapped atoms is recorded using a heterodyne technique. Light fluoresced by the trapped atoms is combined with light (frequency shifted by a modulator) from the laser forming the lattice. When the beams mix on a photodiode they create a beat signal at the difference frequency between the fluorescence and the frequency-shifted laser. The power spectrum of the photocurrent is identical to the fluorescence power spectrum, but centered at radio frequency. This heterodyne technique is not sensitive to the frequency jitter of the laser because the jitter is common between the fluorescence and the laser, which acts as a local oscillator.

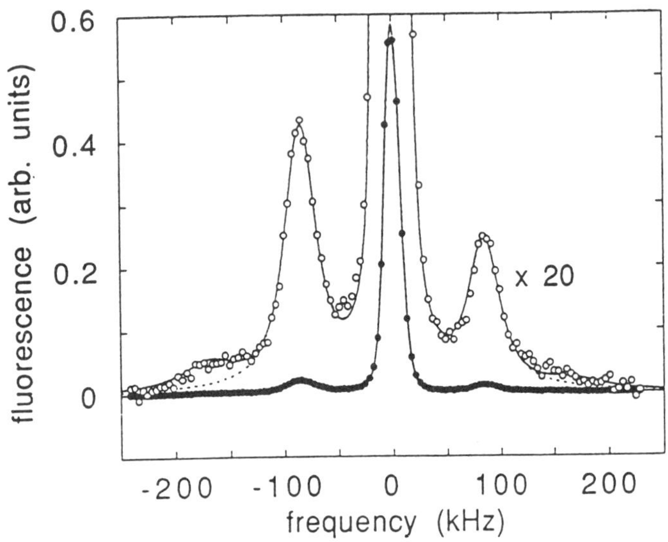

Fig. 2. Emission spectrum of ^{85}Rb in a 1D optical lattice. The solid curve is a fit including the second sidebands, while the dotted curve does not include them.

Figure 2 shows a typical spectrum obtained in this fashion [23] from Rb atoms in a 1D lattice. The central peak corresponds to a process where lattice laser light is absorbed by an atom in a given vibrational state and the atom then radiates, returning to the same vibrational state. This elastic scattering is therefore at the laser frequency (zero frequency in the figure). The sidebands correspond to processes where the atom returns to a different vibrational state from the initial one. The lower frequency sideband results from the final vibrational level being higher than the initial one while the higher frequency sideband results from the final state being lower than the initial one. Much of the width of the spectral features in this case was due to residual Doppler broadening.

Spectra such as those of Fig. 2 contain considerable information about the atomic motion. The spacing of the sidebands gives the vibrational frequency for the trapped atoms. The intensity of the sidebands relative to the central feature indicates the spread of the atomic position distribution about the center of the potential well (lattice site). When the atoms are well localized at their lattice sites, processes that change the vibrational state are strongly suppressed. This is the Lamb-Dicke effect, which also results in the elimination of Doppler broadening for radiators that are confined to a region small compared to the wavelength of the radiation [28]. The phenomenon is also very similar to the Mössbauer effect of recoilless emission. It accounts for the smallness of the first sidebands relative to the carrier and the near absence of second sidebands. The asymmetry of the sidebands, in which the lower frequency sideband is stronger that the high frequency one, is related to the temperature of the trapped atoms. The population of atoms in lower vibrational levels is greater than that of higher levels by a Boltzmann factor. The higher relative population of the lower levels is responsible for the greater relative intensity of the lower sideband.

Through an analysis of spectra like that of Fig. 2 we have confirmed that the vibrational frequency of the atoms varies as the square root of the potential depth, as expected for harmonic motion. We have verified that the temperature of the trapped atoms is linearly dependent on the well depth, just as predicted by the simple, non-trapping theory. Furthermore, a first-principles, fully quantum calculation of the spectrum [29] has reproduced all the details of the experimental spectrum.

The study of optical lattices has been extended from 1D to 2D and 3D using the phase-insensitive configurations mentioned above, and also by stabilization of the relative phases of the beams [30]. The spectra of such lattices has been studied by both absorption and emission techniques [24, 31, 32, 33, 34], as well as by photon correlation techniques [35, 36]. There have also been studies of lattices tuned far from resonance, using non-spectroscopic methods [37, 38] .

The degree of confinement or localization of atoms at their lattice sites can be a small fraction of a wavelength of the laser light, as small as λ/18 in 1D lattices. This opens the opportunity of "trading" some of that localization for lower temperature by performing adiabatic expansion of the atoms within the lattice. The adiabatic expansion is performed by turning down the intensity of lattice laser light with a characteristic time long compared to the vibrational frequency. Figure 3 shows the evolution of the temperature of Cs atoms during such an expansion in a 3D lattice. The temperature does not go as low as one would expect for expansion in a single harmonic potential well because of the periodicity of the optical lattices. A band structure analysis yields predictions in good agreement with the observed temperatures. In that 3D experiment we attained temperatures as low as 700 nK for cesium atoms [39], at that time a record low 3D kinetic temperature, but one that was soon superseded by evaporative cooling [40] and Bose-Einstein condensation [4].

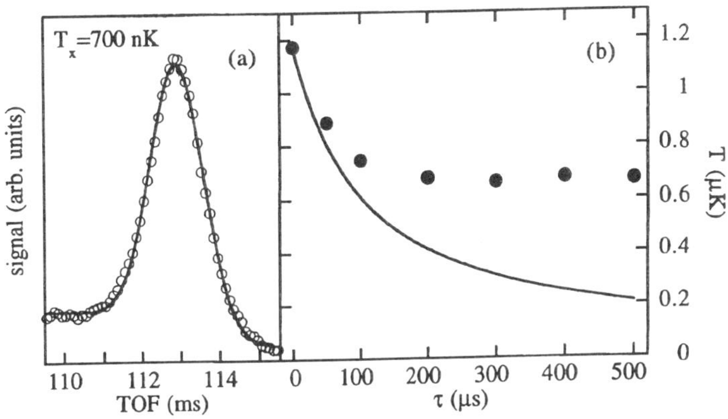

Fig. 3a. Time of flight spectrum for 700 nK atoms. Fig 3b. Time evolution of the temperature of Cs atoms during adiabatic expansion (points). The solid curve is the evolution expected if the atoms were expanding in a weakening harmonic oscillator potential.

BRAGG SCATTERING FROM OPTICAL LATTICES: Just as x-rays may be Bragg reflected from crystal lattices where both the lattice spacings and the x-ray wavelength are of the order of tenths of nanometers, light may be scattered from optical lattices where the lattice spacing and optical wavelength are on the order of hundreds of nanometers. The conditions for scattering are the same: the scattering is elastic and the change in wavevector upon scattering is equal to a reciprocal lattice vector. This is equivalent to specular reflection from a set of parallel planes of atoms along with the condition of constructive interference of the scattering from the various planes. We have studied Bragg scattering from Cs atoms in both 1D and 3D optical lattices [41] . A schematic of the 3D experimental arrangement is shown in Fig. 4. Bragg scattering from a lattice of Rb atoms has been observed in Munich [42] .

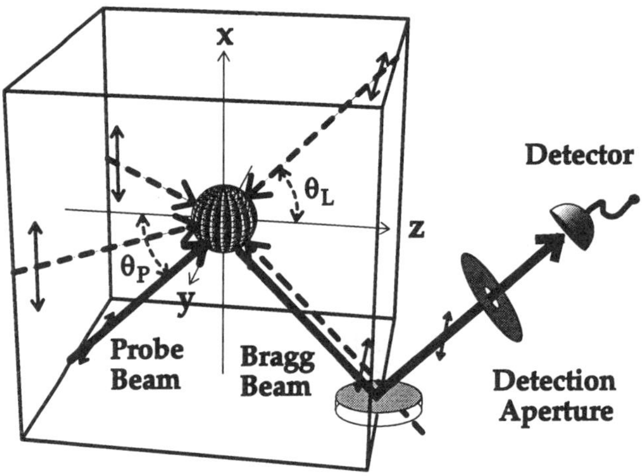

Fig. 4. Schematic representation of the Bragg scattering experiment, including the laser configuration. The angles θ_L and θ_P are about 45°. The probe and Bragg reflected beams are counterpropagating to lattice laser beams.

Our typical optical lattice has about 1000 lattice planes. We introduce the probe beam along the appropriate direction and tune it near the atomic resonance frequency, the same resonance near which the lattice laser beams are tuned. In order to avoid the possibility of 4-wave-mixing in which a lattice laser beam is reflected from a grating formed by the interference of the probe beam with another lattice laser beam, we extinguish the lattice laser beams just before probing. Under typical conditions we find that the scattered light in the Bragg direction is about 10^5 times more intense than would be the case for scattering from a disordered gas of the same density. The angular width for this intense Bragg scattering is about one milliradian [41].

One of the important features of Bragg scattering is that the strength of the scattered beam depends on how well the atoms are localized at their ideal lattice locations. If the atoms are all located very close to the lattice positions defined by the interference pattern, they will all nearly satisfy the conditions for constructive interference, and the scattering will be strong.

If the atoms are not well localized (because of, for example, their thermal motion) the light scattered from them will not interfere constructively and the scattering in the Bragg direction will be reduced. This reduction is proportional to the Debye-Waller factor, well known in x-ray diffraction [43]: $\exp(-K^2/\Delta\xi^2)$ where K is the difference in the incident and scattered wavevectors and $\Delta\xi^2$ is the mean squared spread (along the K direction) in atomic position about the center of the potential wells. The magnitude of wavevector of the incident and scattered light is $k = 2\pi/\lambda$. By varying the delay between the time the lattice laser beams are extinguished and the time the probe beam for Bragg scattering is introduced, we can see the thermal spreading of the released atoms as the intensity of Bragg scattering is reduced by the Debye-Waller factor [41].

Because Bragg scattering provides such a sensitive measure of the localization of the atoms, we use it to study the time evolution of the atomic center-of-mass motion. In one such study we have determined the rate of localization or cooling of atoms in an optical lattice. We start with a disordered array of atoms, a gas that has been cooled in optical molasses and released. Then we turn on the optical lattice and observe the Bragg signal after the lattice has been on for varying times. From the Bragg signal, the expression for the Debye-Waller factor, and from the known steady state localization [34] we can determine the time evolution of the localization. We find that for a wide range of parameters the localization decays exponentially to its steady state value [44]. The rate of this exponential decay is found to be proportional to the photon scattering rate of atoms trapped at the lattice sites. In 1D experiments with Cs atoms, the localization rate is about 30 times slower than the scattering rate and in 3D it is about 200 times slower. This dramatic difference in the 1D and 3D rates is not understood. Some results in 1D and 3D are shown in Figs. 5 and 6.

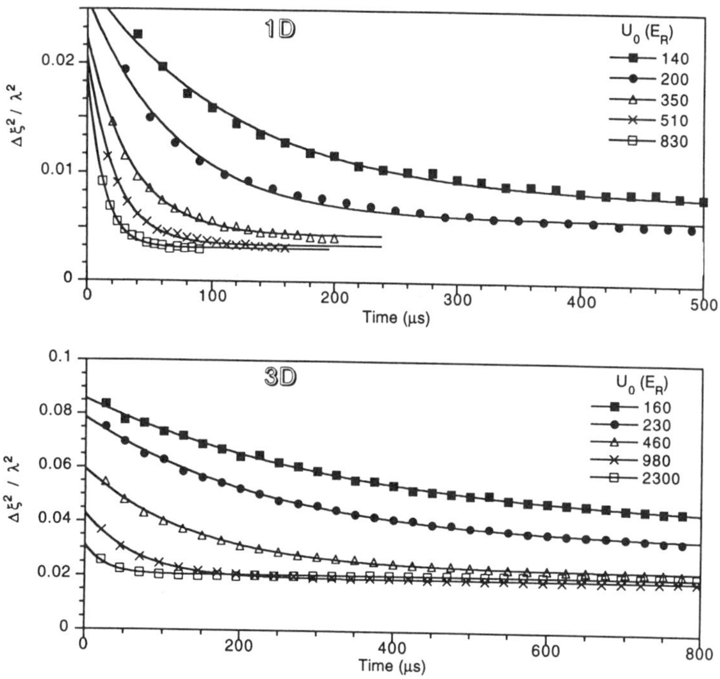

Fig. 5. Time evolution of the mean squared position spread, in units of the optical wavelength, for 1D and 3D lattices, for various values of the potential well depth at a laser detuning of 5 linewidths. The potential depths Uo are in units of the recoil energy $\hbar^2 k^2/2m$.

Considering the virial theorem as applied to a harmonic oscillator, for a gas in quasi-equilibrium we expect a close relationship between the localization and the temperature. The mean squared position spread (proportional to the potential energy) would be proportional to the mean squared momentum spread (proportional to the temperature). This does not strictly apply in our case because our potential is anharmonic and the momentum distribution is not truly Maxwellian. 1D quantum Monte Carlo wavefunction simulations [45] have shown that when we take the temperature from the full width at half maximum (FWHM) of the momentum distribution, assuming it to be Gaussian, the temperature and the localization decay at essentially the same rate [44]. If we take instead the mean square momentum to determine the temperature, the decay rates are not equal, but the temperature is still found to decay at a rate proportional to the photon scattering rate. The theoretical calculations of the evolution of the localization in 1D are in agreement with the 1D experiments. We have not performed 3D calculations.

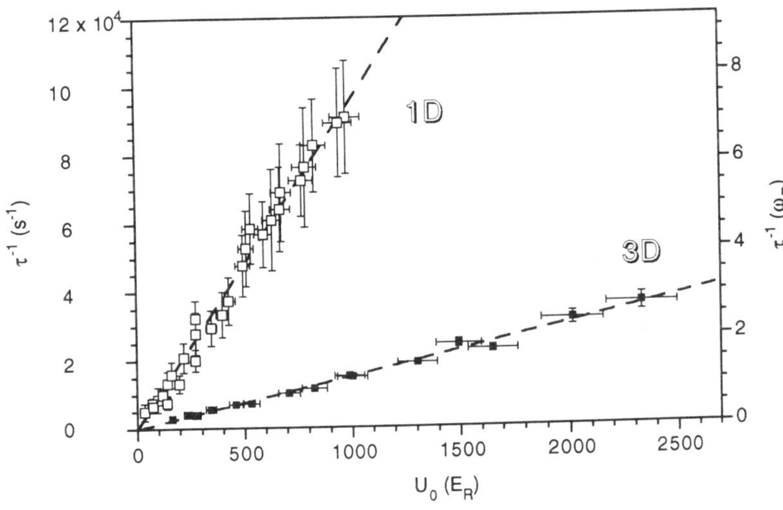

Fig. 6. Measured localization rates τ^{-1} in units of s^{-1} and in units of $\omega_R = E_R/\hbar = 2\pi \times 2.07$ kHz. Rates are plotted against the potential depth at a constant lattice laser detuning, which depth is proportional to the laser intensity. Measurements at various detunings show that the localization rates are proportional to the photon scattering rate.

On examination, this proportionality of the localization (or cooling) rate and the photon scattering rate is rather surprising. The simple Sisyphus model of laser cooling [17], where the atom moves at constant velocity, predicts a cooling rate independent of laser intensity (and thus independent of photon scattering rate) for constant laser detuning. Of course, our situation is far from the constant velocity one, but difficulties remain. The usual way of describing how equilibrium is reached in laser cooling is as a balance between a cooling rate and a heating rate. The fact that there is exponential damping of the temperature suggests that there is a rate of damping of the thermal energy that is proportional to the energy. The rate of exponential approach to equilibrium (which we take to be the cooling rate) is measured to be proportional to the photon scattering rate. Furthermore, there is good reason to believe that the heating rate is proportional to the laser intensity at constant laser detuning, that is, to the photon scattering rate. This idea is supported by the fact that the rate of inelastic photon scattering processes (those in which the vibrational state of the atom changes) is indicated by the strength of the sidebands in spectra like Fig. 2. We have found that the relative strength of these sidebands does not change with laser intensity at constant detuning, so the ratio of inelastic to elastic processes remains the same. This suggests that the rate of inelastic processes, and thus the heating rate, is proportional to the total photon scattering rate.

Now consider the consequence of a cooling rate and a heating rate both proportional to the photon scattering rate. In equilibrium governed by the balance of heating and cooling the equilibrium temperature is proportional to the ratio of the heating and cooling rates. For cooling and heating both proportional to laser intensity (i.e., to the scattering rate, for constant detuning) this would lead to a temperature independent of intensity (as in the case of Doppler cooling [10].) But we have confirmed many times that the temperature is proportional to the intensity at constant detuning (it is in fact proportional to the potential well depth) [23, 34]. This implies that the traditional ways of looking at laser cooling are inadequate to explain the results. In particular, it is likely that the idea of balance between heating and cooling determining the temperature needs to re-examined. It may be that ideas relating to relative rates in and out of quantized trap levels need to be invoked instead [46].

In spite of these apparent difficulties, the quantum Monte Carlo calculations of cooling rates in 1D agree very well with the observations. This indicates that the basic physical interactions are all understood. What is lacking is a good physical picture of the cooling process, such as the Sisyphus picture that has guided the development of laser cooling so well until now.

DRIVEN MOTION OF ATOMS IN AN OPTICAL LATTICE: Just as Bragg scattering can detect the time evolution of the localization of atoms as they cool naturally in an optical lattice, it can detect changes in the localization caused by deliberate changes in the lattice potential. Consider an optical lattice in steady state, with the atoms reasonably well localized near the lattice sites. Now let the intensity of the lattice laser beams, and therefore the depth of the potential wells, increase suddenly. Under the influence of this compression the atoms will accelerate toward the centers of the potential wells, and oscillate through the centers, producing a breathing motion at twice the trap vibrational frequency. This breathing motion represents an oscillating value of the mean squared position spread and should result in an oscillating Bragg scattered intensity.

Fig. 7 shows the results of such an experiment [47] in a 3D Cs optical lattice. The initial decrease in the mean squared spread follows the sudden increase in the lattice potential depth, with the oscillations occurring at twice the trap vibration frequency. The oscillations are damped mainly by the anharmonicity of the trapping potential. Classically, atoms oscillating with different amplitudes have different vibrational frequencies, and so get out of phase with each other. Over a longer period of time, the average value of the mean squared spread increases, indicating a heating as the atoms equilibrate at the higher temperature characteristic of the deeper potential wells. At equilibrium the higher temperature in the deeper potential well results in a $\Delta\xi^2$ that is essentially independent of the potential depth [34] . Therefore, the value to which $\Delta\xi^2$ increases in Fig. 7 is very close to the value it had before the sudden compression.

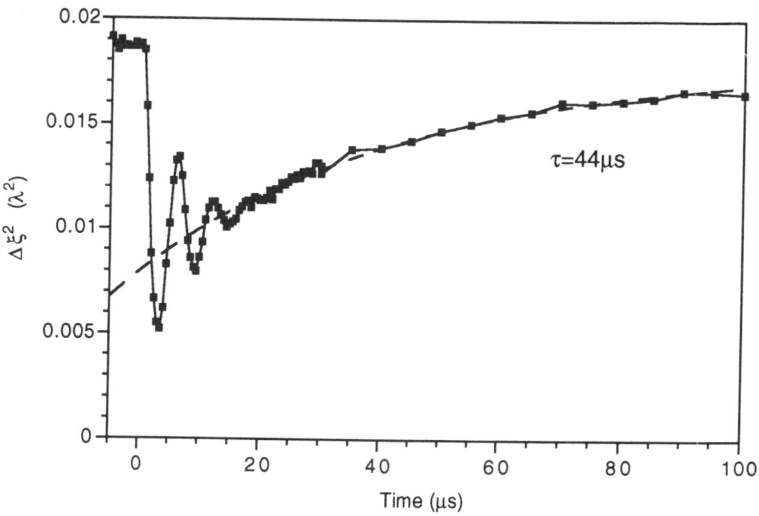

Fig. 7. Oscillations in the mean squared position spread induced by a sudden increase in the 3D optical lattice potential at time = 0. The slow increase in the spread after the oscillations are damped has a 44 μs time constant.

In addition to such motion driven by a sudden increase in the lattice potential, we have used parametric driving of the atomic motion. By modulating the intensity of the lattice at twice the trap vibrational frequency, we drive a breathing motion of the atoms. Observation of Bragg scattering at various phases of the parametric drive shows this breathing motion. During some parts of the drive cycle the atomic position spread is less than the spread with no driving at all, indicating a sort of "squeezing" of the motion of the atoms, although we have not observed squeezing below the standard quantum limit corresponding to the position spread of the ground vibrational state.

CONCLUSIONS: Optical lattices have allowed a series of new measurements on the trapping and cooling of atoms in laser fields. We have measured the rates of cooling and observed driven motion and squeezing of the position spread of atoms trapped in the lattices. In some cases the results conflict with the usual images of laser cooling, although where calculations have been possible, they have been in good agreement with the experiments.

ACKNOWLEDGMENTS: I wish to thank all the members of the Laser Cooling and Trapping Group at NIST-Gaithersburg who have been involved in the study of optical lattices: G. Birkl, I. Deutsch, M. Gatzke, C. Gerz, P. S. Jessen, A. Kastberg, P. D. Lett, G. Raithel, S. L. Rolston, R. J. C. Spreeuw, and C. I. Westbrook. I also want to remember and acknowledge Richard N. Watts, who died in November 1996 at the age of 39. His pioneering work on laser cooling with diode lasers, sub-Doppler laser cooling and localization of laser cooled atoms prepared the way for all the subsequent work on optical lattices.

The work reported here was supported in part by the U. S. Office of Naval Research and by NSF grant no. PHY 9312572.

REFERENCES

1. See for example the special issue on Mechanical Effects of Light, J. Opt. Soc. Am. B **2**, 1706 (1985).

2. See, for example, the special issue on laser cooling, J. Opt. Soc. Am. B **6**, (1989).

3. E. Arimondo, W. D. Phillips and F. Strumia, Eds., *Laser Manipulation of Atoms and Ions* (North Holland, Amsterdam, 1992).

4. M. H. Anderson, J. R. Ensher, M. R. Matthews, C. E. Wieman and E. A. Cornell, Science **269**, 198 (1995).

5. K. B. Davis, et al., Phys. Rev. Lett. **75**, 3969 (1995).

6. T. Hänsch and A. Schawlow, Opt. Commun. **13**, 68 (1975).

7. D. Wineland and H. Dehmelt, Bull. Am. Phys. Soc. **20**, 637 (1975).

8. D. Wineland and W. Itano, Phys. Rev. A **20**, 1521 (1979).

9. S. Chu, L. Hollberg, J. Bjorkholm, A. Cable and A. Ashkin, Phys. Rev. Lett. **55**, 48 (1985).

10. P. D. Lett, et al., J. Opt. Soc. Am. B **6**, 2084 (1989).

11. W. D. Phillips, J. Prodan and H. Metcalf, J. Opt. Soc. Am. B **2**, 1751 (1985).

12. D. Sesko, C. G. Fan and C. E. Wieman, J. Opt. Soc. Am B **5**, 1225 (1988).

13. P. D. Lett, et al., Phys. Rev. Lett. **61**, 169 (1988).

14. C. Salomon, J. Dalibard, W. D. Phillips, A. Clairon and S. Guellati, Europhys. Lett. **12**, 683 (1990).

15. W. D. Phillips, et al., in *Atomic Physics 11* S. Haroche, J. C. Gay, G. Grynberg, Eds. (World Scientific, Singapore, 1989) p. 633.

16. J. Dalibard, et al., in *Atomic Physics 11* S. Haroche, J. C. Gay, G. Grynberg, Eds. (World Scientific, Singapore, 1989) p. 199.

17. J. Dalibard and C. Cohen-Tannoudji, J. Opt. Soc. Am. B **6**, 2023 (1989).

18. P. J. Ungar, D. S. Weiss, E. Riis and S. Chu, J. Opt. Soc. Am. B **6**, 2058 (1989).

19. Y. Castin, J. Dalibard and C. Cohen-Tannoudji, in *Light induced kinetic effects on atoms, ions and molecules* L. Moi, S. Gozzini, C. Gabbanini, E. arimondo, F. Stumia, Eds. (ETS Editrice, Pisa, 1991) p. 5.

20. Y. Castin and K. Mølmer, Phys. Rev. Lett. **74**, 3772 (1995).

21. C. Gerz, et al., Europhys. Lett. **21**, 661 (1993).

22. P. Verkerk, et al., Phys. Rev. Lett. **68**, 3861 (1992).

23. P. S. Jessen, et al., Phys. Rev. Lett. **69**, 49 (1992).

24. G. Grynberg, B. Lounis, P. Verkerk, J.-Y. Courtois and C. Salomon, Phys. Rev. Lett. **70**, 2249 (1993).

25. C. I. Westbrook, et al., Phys. Rev. Lett. **65**, 33 (1990).

26. B. R. Mollow, Phys. Rev. **188**, 188 (1969).

27. J. T. Höffges, H. W. Baldauf, T. Eichler, S. R. Helmfrid and H. Walther, Optics Communications **133**, 170 (1997).

28. R. H. Dicke, Phys. Rev. **89**, 472 (1953).

29. P. Marte, R. Dum, R. Taïeb, P. Lett and P. Zoller, Phys. Rev. Lett. **71**, 1335 (1993).

30. A. Hemmerich and T. W. Hansch, Phys. Rev. Lett. **70**, 410 (1993).

31. P. Verkerk, et al., Europhys. Lett. **26**, 171 (1994).

32. A. Hemmerich and T. W. Hänsch, Phys. Rev. **A48**, R1753 (1993).

33. A. Hemmerich, C. Zimmermann and T. W. Hänsch, Phys. Rev. Lett. **72**, 625 (1994).

34. M. Gatzke, et al., to be published , (1996).

35. C. Jurczak, et al., Opt. Comm. **115**, 480 (1995).

36. C. Jurczak, et al., Phys. Rev. Lett. **77**, 1727 (1996).

37. P. S. Jessen, Bull. Am. Phys. Soc. **41**, 1088 (1996).

38. B. Anderson, T. Gustavson and a. M. Kasevich, Phys. Rev. A **53**, R3727 (1996).

39. A. Kastberg, W. D. Phillips, S. L. Rolston, R. J. C. Spreeuw and P. S. Jessen, Phys. Rev. Lett. **74**, 1542 (1995).

40. W. Petrich, M. H. Anderson, J. R. Ensher and E. A. Cornell, Phys. Rev. Lett. **74**, 3352 (1995).

41. G. Birkl, M. Gatzke, I. H. Deutsch, S. L. Rolston and W. D. Phillips, Phys. Rev. Lett. **75**, 2823 (1995).

42. M. Weidemüller, A. Hemmerich, A. Görlitz, T. Esslinger and T. Hänsch, Phys. Rev. Lett. **75**, 4583 (1995).

43. N. W. Ashcroft and N. D. Mermin, *Solid State Physics* (Saunders, Philadelphia, 1976).

44. G. Raithel, G. Birkl, A. Kastberg, W. D. Phillips and S. L. Rolston, to be published in Phys. Rev. Lett. , (1996).

45. J. Dalibard, Y. Castin and K. Mølmer, Phys. Rev. Lett. **68**, 580 (1992).

46. Y. Castin, Doctoral Dissertation, Ecole Normale Supérieure (1992).

47. G. Raithel, G. Birkl, W. D. Phillips and S. L. Rolston, to be published , (1996).

Coherent Quantum State Manipulation Of Trapped Atomic Ions[†]

By D.J. Wineland, C. Monroe, D.M. Meekhof, B.E. King,
D. Leibfried, W.M. Itano, J.C. Bergquist,
D. Berkeland, J.J. Bollinger, J. Miller

Ion Storage Group, Time and Frequency Division
NIST, Boulder, CO, 80303, USA

Abstract:

A single, laser-cooled, trapped $^9Be^+$ ion is used to investigate methods of coherent quantum-state synthesis. We create and characterize nonclassical states of motion including "Schrödinger-cat" states. A fundamental quantum logic gate is realized using the quantized motion and internal states as qubits. We explore some of the applications for, and problems in realizing, quantum computation based on multiple trapped ions.

1. INTRODUCTION

Currently, a major theme in atomic, molecular, and optical physics is coherent control of quantum states. This theme is manifested in a number of topics such as atom interferometry, Bose-Einstein condensation and the atom laser, cavity QED, quantum computation, quantum-state engineering, wavepacket dynamics, and coherent control of chemical reactions.

Here, we report related trapped-ion research at NIST on (1) the study of the dynamics of a two-level atomic system coupled to harmonic atomic motion, (2) the creation and characterization of nonclassical states of motion such as "Schrödinger-cat" superposition states, and (3) quantum logic for the generation of highly entangled states and for the investigation of scaling in a quantum computer.

2. TRAPPED ATOMIC IONS

Because of their overall charge, atomic or molecular ions can be confined by particular arrangements of electromagnetic fields for relatively long periods of times (hours or longer) with relatively small perturbations to their internal energy level structure. For studies of ions at low kinetic energy (< 1 eV), two types of trap are commonly used - the Penning and Paul traps. The Penning trap uses a combination of static electric and magnetic fields for confinement, and the Paul, or rf, trap confines ions primarily through ponderomotive fields generated by inhomogeneous oscillating electric fields. The operation of these traps is discussed in various reviews (for example, see Refs. 1 - 3) and in a recent book [4]. For brevity, we discuss one trap configuration which is useful for the topics discussed in this paper.

In Fig. 1 we show schematically a "linear" Paul trap. This trap is based on the one described in Ref. [5], which is derived from the original design of Paul [6]. It is basically a quadrupole mass filter which is plugged at the ends with static electric potentials. An oscillating potential $V_0\cos(\Omega_T t)$ is applied between diagonally opposite rods, which are fixed in a quadrupolar configuration, as indicated in Fig. 1. We assume that the rod segments along the z (horizontal) direction are coupled together with capacitors (not shown), so the rf potential is constant as a function of z. Near the axis of the trap this creates an oscillating potential of the form

$$V \approx \frac{V_o}{2}\cos(\Omega_T t)\left(1 + \frac{x^2 - y^2}{R^2} \right), \tag{1}$$

where R is equal to the distance from the axis to the surface of the electrode [7].

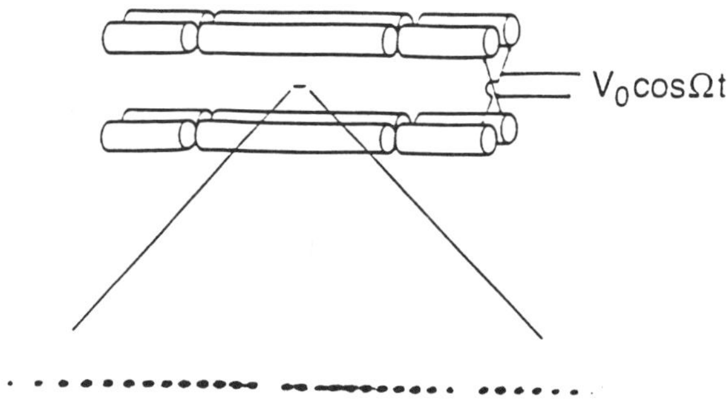

Fig. 1. The upper part of the figure shows a schematic diagram of the electrode configuration for a linear Paul-RF trap (rod spacing ≈ 1 mm). The lower part of the figure shows an image of a string of $^{199}Hg^+$ ions, illuminated with 194 nm radiation, taken with a UV-sensitive, photon-counting imaging tube. The spacing between adjacent ions is approximately 10 μm. The "gaps" in the string are occupied by impurity ions, most likely other isotopes of Hg^+, which do not fluoresce because the frequencies of their resonant transitions do not coincide with those of the 194 nm $^2S_{½} \rightarrow {}^2P_{½}$ transition of $^{199}Hg^+$.

This gives rise to a (harmonic) ponderomotive potential in the x-y direction which, for a single ion (or the center-of-mass (COM) motion of a group of ions), yields the motion

$$\begin{aligned} x(t) &\approx X(1+(q_x/2)\cos\Omega t)\cos(\omega_x t + \phi_x), \\ y(t) &\approx Y(1+(q_y/2)\cos\Omega t)\cos(\omega_y t + \phi_y), \end{aligned} \tag{2}$$

where $q_x = -q_y \equiv 2qV_o/(mR^2\Omega_T^2) = 2\sqrt{2}\omega_x/\Omega_T$, $\omega_x = \omega_y = \omega = qV_o/(\sqrt{2}mR^2\Omega_T)$, and q and m are the ion mass and charge. Equations (2) are valid for the typical conditions that $|q_x|, |q_y| << 1$. In this approximation, if we neglect the micromotion (the $(q_i/2)\cos\Omega t$ terms in Eq. (2)), the ion behaves as if it were confined in a pseudopotential Φ_p given by

$$q\Phi_p = \frac{1}{2} m \omega_i^2 (x^2 + y^2) , \tag{3}$$

in which the ion oscillates with secular frequency ω_i. To provide confinement along the z direction, static potentials are applied to the end segments of the rods relative to the central segments. Near the center of the trap, this provides a static harmonic well along z. This necessarily weakens the well in the radial direction [5], but we will assume that the binding in the radial direction is large compared to that along z, so this effect is small. Figure 1 also shows an image of a "string" of $^{199}Hg^+$ ions which are confined near the z axis of the trap described in Ref. 8. This string of ions can be thought of as a linear molecule or a super-molecule composed of the atomic ions bound to a much heavier atom, which is the trap.

When this kind of trap is installed in a high-vacuum apparatus (pressure < 10^{-8} Pa), ions can be confined for days with minimal perturbations to their internal structure. Collisions with background gas are infrequent (the mean time between collisions is more than a few minutes). Even though the ions interact strongly through their mutual Coulomb interaction, the fact that the ions are localized necessarily means that $\langle \vec{E} \rangle = 0$ at the positions of the ions; therefore electric field perturbations occur only in second order. These perturbations are typically very small [9]. Magnetic field perturbations are important; however, if we operate at fields where the energy separation between levels is at an extremum with respect to field, the coherence time for superposition states of the two levels can be very long. For example, in a $^9Be^+$ (Penning trap) experiment operating in a field of 0.82 T, a coherence time between hyperfine levels exceeding 10 minutes was observed [10]. As described below, we will be interested in coherently exciting the quantized modes of the ion motion in the trap. Not surprisingly, the coupling to the environment is relatively strong because of the ions' charge. One measure of the decoherence rate due to environmental coupling is the rate of transitions between the ions' quantized oscillator levels. Transition times from the zero-point energy level (achieved from laser cooling) to the first excited motional state have been measured for single $^{198}Hg^+$ ions to be about 0.15 s [11] and for single $^9Be^+$ ions to be about 1 ms [12].

Ion trap experiments can achieve high detection efficiency of the ion's internal states. Unit detection efficiency has been achieved in previous experiments on "quantum jumps" [13] where the internal state of the ion is indicated by light scattering (or lack thereof), correlated with the ion's internal state.

In the experiments we discuss here, a single $^9Be^+$ ion was stored in a conventional Paul trap [1-3] where the ion's confinement is characterized by secular oscillation frequencies $(\omega_x, \omega_y, \omega_z)/2\pi \approx (11, 19, 29)$ MHz along the three principal axes of the trap [14]. The future goal of the work will be able to extend these experiments to the case of multiple ions in a linear trap (Fig. 1).

3. ENTANGLED QUANTUM STATES

An entangled quantum state is one where the wave function of the overall system cannot be written as a product of the wave functions of the subsystems. In this case, a measurement on one of the subsystems will affect the state of the other subsystems. For example, consider a two-level atom bound in a 1-D harmonic well. Suppose we can create the state

$$\Psi = \frac{1}{\sqrt{2}}(|\downarrow\rangle|n\rangle + e^{i\phi}|\uparrow\rangle|n'\rangle), \tag{4}$$

where the kets $|\downarrow\rangle$ and $|\uparrow\rangle$ denote the two internal eigenstates of the atom (here, we use the spin-½ analog to a two-level system: $\sigma_z|\uparrow\rangle = +|\uparrow\rangle$, etc.), the second ket denotes a harmonic oscillator eigenstate $|n\rangle$, and ϕ is a (controlled) phase factor. This state is entangled because, if we measure the motional eigenstate of the atom and find it to be in level n, then it must also be found in the $\downarrow$ internal state if we measure σ_z. Similarly, if we find the atom in the n' motional state, it must be found in the $\uparrow$ internal state. Such correlations are central to "EPR" experiments [15]. Another interesting state is the state for N two-level atoms

$$\Psi = \frac{1}{\sqrt{2}}(|\downarrow\rangle_1|\downarrow\rangle_2...|\downarrow\rangle_N + e^{i\phi}|\uparrow\rangle_1|\uparrow\rangle_2...|\uparrow\rangle_N), \tag{5}$$

where the subscript i (= 1, 2, ..., N) denotes the ith atom. This state is "maximally entangled" in the sense that a measurement of σ_z on any atom automatically determines the value of σ_z of all other atoms.

4. LASER COOLING TO THE ZERO-POINT STATE

As a starting point for the experiments discussed in this paper, we want to prepare the $^9Be^+$ ion in a particular internal state and in the lowest quantized level of vibrational motion in the trap - the zero-point state. The ion can be optically pumped into a particular internal state using polarized light. Preparation in the zero-point state of motion is achieved with laser cooling.

We cool a single, trapped $^9Be^+$ ion to near the zero-point energy using resolved-sideband laser cooling [16] which can be explained as follows: Consider a two-level atom characterized by a resonant (optical) transition frequency ω_0 and radiative linewidth γ. For simplicity, we assume the atom is confined by a 1-D harmonic well of vibration frequency $\omega_x \gg \gamma$. If a laser beam (frequency ω) is incident along the direction of the atomic motion, the bound atom's absorption spectrum is composed of a "carrier" at frequency ω_0 and resolved (since $\omega_x \gg \gamma$) frequency-modulation sidebands that are spaced by ω_x and are generated from the Doppler effect (like vibrational substructure in a molecular optical spectrum). Cooling occurs if the laser is tuned to a lower sideband, for example, at $\omega = \omega_0 - \omega_x$. In this case, photons of energy $\hbar(\omega_0-\omega_x)$ are absorbed, and spontaneously emitted photons of average energy $\hbar\omega_0$ return the atom to its initial internal state (assuming $\hbar\omega_x$ is much greater than the photon recoil energy for the bound atom). This reduces the atom's kinetic energy by $\hbar\omega_x$ per scattering event. Cooling proceeds until the atom's mean vibrational quantum number in the harmonic well is given by $\langle n \rangle_{min} \simeq (\gamma/2\omega_x)^2 \ll 1$. This kind of laser cooling can be regarded as cooling through anti-Stokes spontaneous Raman scattering or "inverse-Stokes" scattering [16]. Experimentally, we find it convenient to use two-photon stimulated Raman transitions for cooling (as described below), but the basic idea for cooling is essentially the same as for single-photon transitions described here.

5. JAYNES-CUMMINGS-TYPE COUPLING BETWEEN INTERNAL AND MOTIONAL STATES

To achieve both laser cooling and entanglement, we need to provide a coupling between internal and motional quantum states. This can be achieved with the application of inhomogeneous (classical) electromagnetic fields. For example, consider an atom confined in a 1-D harmonic potential. The atom's dipole moment $\boldsymbol{\mu}$ is assumed to couple to an electric field E(x,t) through the Hamiltonian

$$H_I = -\mu E(x,t) = -\mu\left[E(x=0,t) + \frac{\partial E}{\partial x}x + \frac{1}{2}\frac{\partial^2 E}{\partial x^2}x^2 + \cdots\right]. \quad (6)$$

We have $\mu \propto \sigma_+ + \sigma_-$, where σ_+ and σ_- are the raising and lowering operators for the internal levels (in the spin-½ analog). In Eq. (6), the position **x** is an operator representing the position of the atom, which we write as $\mathbf{x} = x_0(\mathbf{a} + \mathbf{a}^\dagger)$, where **a** and $\mathbf{a}^\dagger$ are the usual harmonic oscillator lowering and raising operators, and x_0 is the rms spread of the n=0 zero-point state of motion. As a simple example, suppose the field is static and the motional oscillation frequency ω_x of the atom is equal to the resonance frequency ω_0 of the internal state transition. In its reference frame, the atom experiences an oscillating field due to the motion through the inhomogeneous field. Since $\omega_x = \omega_0$, this field resonantly drives transitions between the internal states. If the extent of the atom's motion is small enough that we need only consider the first two terms in Eq. (6), H_I can be approximated as $\hbar\Omega(\sigma_+\mathbf{a} + \sigma_-\mathbf{a}^\dagger)$ (in the interaction frame and using the rotating wave approximation), where Ω is a proportionality constant. This Hamiltonian is equivalent to the Jaynes-Cummings Hamiltonian of cavity-QED [17] which describes the coupling between a two-level atom and a single mode of the radiation field. This analogy has been pointed out in various papers [18-21]; for a review, see Ref. 22 and further references in Ref. 21.

5.1 Realization of a Jaynes-Cummings-type coupling for a trapped $^9Be^+$ ion

The relevant energy-level structure of $^9Be^+$ is summarized in Fig. 2. Because the ion is trapped, the internal $^9Be^+$ electronic states must include the ladder of external harmonic oscillator levels of energy $E_n = \hbar\omega(n+½)$, where we have considered only the x-dimension of the oscillator ($\omega_i \equiv \omega_x$) and its associated quantum number n ≡ $n_x \in (0, 1, 2, \ldots)$. The two internal levels of interest are the $^2S_{1/2}$ ground state hyperfine levels $|F=2, m_F=2\rangle$ (denoted by $|\downarrow\rangle$) and $|F=1, m_F=1\rangle$ (denoted by $|\uparrow\rangle$), which are separated in frequency by $\omega_0/2\pi \approx 1.25$ GHz. The other Zeeman levels (not shown) are resolved from the $|\downarrow\rangle$ and $|\uparrow\rangle$ states by the application of a ≈0.2 mT magnetic field [12,21].

Strong field gradients can be obtained with laser fields (e^{ikx} factor in $E_{laser} \propto e^{i(kx-\omega t)}$). In our experiment, the field corresponding to that in Eq. (6) is provided by two laser fields (R1 and R2 of Fig. 2a) which drive stimulated-Raman transitions between the levels of interest. The use of stimulated-Raman transitions

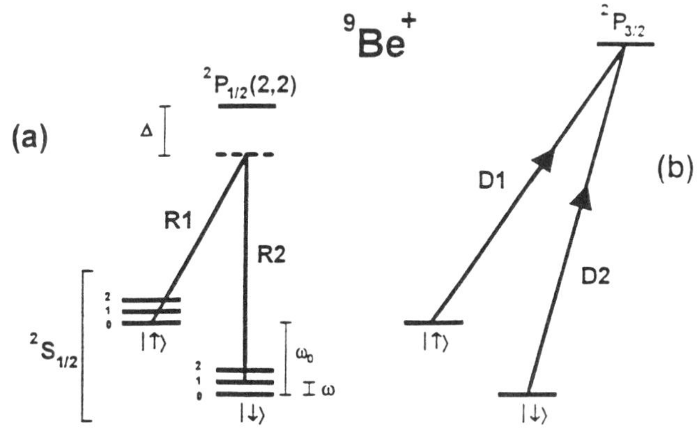

Fig. 2. (a) Electronic (internal) and motional (external) energy levels (not to scale) of the trapped $^9Be^+$ ion, coupled by indicated laser beams R1 and R2. The difference frequency of the Raman beams R1 and R2 is set near $\omega_0/2\pi \approx 1.250$ GHz, providing a two-photon coupling between the $^2S_{1/2}(F=2, m_F=2)$ and $^2S_{1/2}(F=1, m_F=1)$ hyperfine ground states (denoted by $|\downarrow\rangle$ and $|\uparrow\rangle$ respectively). The motional energy levels are depicted by a ladder of vibrational states separated in frequency by the trap frequency $\omega/2\pi \sim 11$ MHz. The Raman beams are detuned $\Delta/2\pi \approx -12$ GHz from the $^2P_{1/2}(F=2, m_F=2)$ excited state. As shown, the Raman beams are tuned to the red sideband (see text).
(b) Detection of the internal state is accomplished by illuminating the ion with σ^+-polarized "detection" beam D2, which drives the cycling $^2S_{1/2}(F=2, m_F=2) \rightarrow {}^2P_{3/2}(F=3, m_F=3)$ transition, and observing the scattered fluorescence. The vibrational structure is omitted from (b) since it is not resolved. Beam D1, also σ^+ polarized, provides spontaneous recycling from the $|\uparrow\rangle$ to $|\downarrow\rangle$ state.

has some technical advantages, but is formally equivalent to driving a narrow single-photon transition. Raman transitions between the $|\downarrow\rangle$ and $|\uparrow\rangle$ states can be driven by tuning the difference frequency of R1 and R2 to be near ω_0. The two Raman beams ($\lambda \sim 313$ nm) are generated from a single laser source and an acousto-optic modulator, allowing excellent stability of their relative frequency and phase. Both beams are detuned $\Delta/2\pi \approx -12$ GHz from the excited $^2P_{1/2}$ electronic state (radiative linewidth $\gamma/2\pi \approx 19.4$ MHz), and the polarizations are set to couple through the $^2P_{1/2}(F=2, m_F=2)$ level (the next nearest levels are the $^2P_{3/2}$ states which are over 200 GHz away and can be neglected). Because $\Delta >> \gamma$, the excited ^2P state can be adiabatically eliminated in a theoretical description, resulting in a coupling between the two ground states which exhibits a linewidth inversely proportional to the interaction time. When R1 and R2 are applied to the ion with wavevector difference $\delta\vec{k} = \vec{k}_1 - \vec{k}_2$ along the x direction, the effective coupling

Hamiltonian in the rotating-wave approximation is given by

$$H_I = g\left(\sigma_+ e^{i\eta(a^\dagger + a) - i\delta t} + \sigma_- e^{-i\eta(a^\dagger + a) + i\delta t}\right). \tag{7}$$

The coupling strength g depends on Δ and the intensity of the laser beams, $\eta = |\delta\vec{k}|x_0 \simeq 0.2$ is the Lamb-Dicke parameter, $x_0 = (\hbar/2m\omega_x)^{1/2} \approx 7$ nm, and δ is the difference between the relative frequency of the two Raman beams and ω_0. Setting $\delta\vec{k}$ to be parallel to the x axis of the trap yields almost no coupling between the internal states and motion in the y and z directions.

If $\delta = \omega_x(n'-n)$, transitions are resonantly driven between the levels $|\downarrow,n\rangle$ and $|\uparrow,n'\rangle$ at a rate $\Omega_{n,n'} \propto g$ which is dependent on n and n' [16,21,23] (analogous to Franck-Condon factors). Starting from the $|\downarrow\rangle|n\rangle$ state, application of a "Rabi π" pulse coherently transfers the ion to the $|\uparrow\rangle|n'\rangle$ state; this corresponds to applying the Raman beams for a duration τ such that $2\Omega_{n,n'}\tau = \pi$. If we apply the Raman beams for half of this time, we create the entangled state of Eq. (4). Here, we will assume the ion is confined in the Lamb-Dicke limit $(|\delta\vec{k}|\langle x^2\rangle^{1/2} \ll 1)$ and will consider three transitions. The carrier, at $\delta = 0$, drives transitions between states $|\downarrow,n\rangle \leftrightarrow |\uparrow,n\rangle$ with Rabi frequency $\Omega_{n,n} = g$. The first red sideband, corresponding to $\delta = -\omega_x$, drives transitions between states $|\downarrow,n\rangle \leftrightarrow |\uparrow,n-1\rangle$ with Rabi frequency $\Omega_{n,n-1} = g\eta\sqrt{n}$. This coupling is analogous to the case in cavity QED [17] where energy is coherently exchanged between the internal state and single-photon radiation field. The first blue sideband, at $\delta = +\omega_x$, drives transitions between states $|\downarrow,n\rangle \leftrightarrow |\uparrow,n+1\rangle$ with Rabi frequency $\Omega_{n,n+1} = g\eta(n+1)^{1/2}$.

Laser cooling to the $|\downarrow\rangle|n=0\rangle$ state is accomplished in two stages. We first use first Doppler cooling [16] to cool the ion to a limit given by $\langle n\rangle \simeq 1$ for our experimental conditions. We then apply sideband laser cooling using stimulated Raman transitions [12] to achieve $\langle n\rangle \simeq 0$. For sideband laser cooling, Rabi π pulses on the first red sideband $(|\downarrow\rangle|n\rangle \rightarrow |\uparrow\rangle|n-1\rangle)$ are alternated with repumping cycles using nearly resonant radiation (beam D1 of Fig. 2b) - which results (most probably) in transitions $|\uparrow\rangle|n\rangle \rightarrow |\downarrow\rangle|n\rangle$. These steps are repeated (typically 5 times) until the ion resides in the $|\downarrow\rangle|0\rangle$ state with high probability (> 0.9).

As described below, from the $|\downarrow\rangle|0\rangle$ state we are able to coherently create states of the form $|\downarrow\rangle\Psi(x)$, where the motional state $\Psi(x) = \Sigma_n C_n \exp(-in\omega_x t)|n\rangle$ and the C_n are complex. One way we can analyze the motional state created is as follows [21]: The Raman beams are pulsed on for a time τ, and the probability $P_\downarrow(\tau)$ that the ion is in the $|\downarrow\rangle$ internal state is measured. The experiment is repeated for a range of τ values. When the Raman beams are tuned to the first blue sideband, the expected signal is [21]

$$P_{\downarrow}(\tau) = \frac{1}{2}\left(1 + \sum_{n=0}^{\infty} P_n \cos(2\Omega_{n,n+1}\tau) e^{-\gamma_n \tau}\right), \tag{8}$$

where $P_n \equiv |C_n|^2$ is the probability of finding the ion in state n and γ_n are experimentally determined decay constants. The internal state $|\downarrow\rangle$ is detected by applying nearly resonant σ^+-polarized laser radiation (beam D2, Fig. 2b) between the $|\downarrow\rangle$ and $^2P_{3/2}(F=3,\ m_F=3)$ energy levels. Because this is a cycling transition, detection efficiency is near 1 [12,13,21]. The measured signal $P_{\downarrow}(\tau)$ can be inverted (Fourier cosine transform), allowing the extraction of the probability distribution of vibrational state occupation P_n. This signal does not show the phase coherences (phase factors of the C_n), which must be verified separately [21,24]. The most complete characterization is achieved with a state reconstruction technique [25] discussed below.

6. CREATION OF COHERENT AND SCHRÖDINGER-CAT STATES OF MOTION

We have created and analyzed thermal, Fock, squeezed, coherent, Schrödinger-cat states, and other superpositions of Fock states [21,24,25]; here we briefly describe the creation and measurement of coherent and Schrödinger-cat states [21,24]. We note that a scheme recently proposed for producing arbitrary states of the electromagnetic field [26] should be directly applicable to the ion case for producing arbitrary states of motion.

A coherent state of motion

$$\Psi(x) = |\alpha\rangle \equiv \exp(-|\alpha|^2/2) \sum_{n=0}^{\infty} \frac{\alpha^n}{\sqrt{n!}} |n\rangle, \tag{9}$$

corresponds to a displaced zero-point wave packet oscillating in the potential well with amplitude $2|\alpha|x_0$. From Eq. (8), $P_{\downarrow}(\tau)$ for a coherent state will undergo quantum collapses and revivals [27,28]. These revivals are a purely quantum effect due to the discrete energy levels and the narrow distribution of states [27,28].

We have produced coherent states of ion motion from the $|\downarrow\rangle|0\rangle$ state by applying either a resonant (frequency ω_x) classical driving field or a "moving standing wave" of laser radiation which resonantly drives the ion motion through the optical dipole force [21,24]. In Fig. 3, we show a measurement of $P_{\downarrow}(\tau)$ after creation of a coherent state of motion, exhibiting the expected collapse and revival signature. (For comparison, see the cavity-QED experiment of Ref. 29.) Eq. (8) is fitted to these data assuming a Poissonian distribution, allowing only $\langle n\rangle$ to vary. The inset shows the results of a separate analysis, which yield the probabilities of the Fock-state components, extracted by applying a Fourier cosine transform to $P_\downarrow(\tau)$ at the known frequencies as described above. These amplitudes display the expected Poissonian dependence on n.

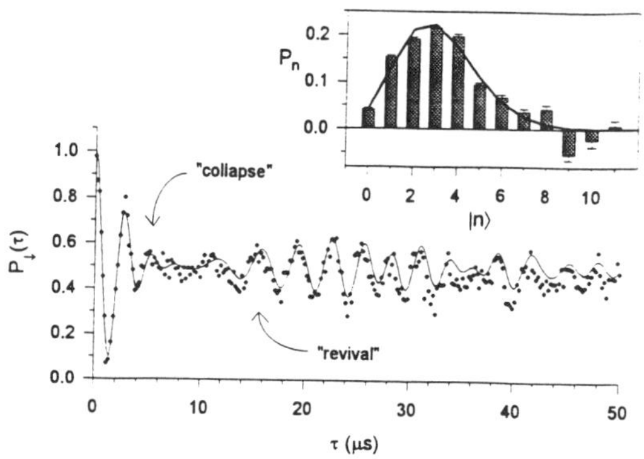

Fig. 3. $P_\downarrow(\tau)$ for a coherent state driven by the first blue sideband interaction, showing collapse and revival behavior. The data are fitted to a coherent state distribution, yielding $\langle n\rangle = 3.1(1)$. The inset shows the results of inverting the time-domain data by employing a Fourier cosine transform at the known Rabi frequencies $\Omega_{n,n+1}$, fitted to a Poissonian distribution, yielding $\langle n\rangle = 2.9(1)$. Each data point represents an average of ≈4000 measurements, or 1 s of integration. (from Ref. 21)

A Schrödinger-cat state is taken to be a coherent superposition of classical-like motional states. In Schrödinger's original thought experiment [30], he described how we could, in principle, entangle a superposition state of an atom with a macroscopic-scale superposition of a live and dead cat. In our experiment [24], we construct an analogous state, on a smaller scale, with a single atom. We create the state

$$\Psi = \frac{1}{\sqrt{2}}\left(|\downarrow\rangle|\alpha_1\rangle + e^{i\phi}|\uparrow\rangle|\alpha_2\rangle\right), \tag{10}$$

where $|\alpha_1\rangle$ and $|\alpha_2\rangle$ are coherent motional states (Eq. (9)) and ϕ is a (controlled) phase factor. The coherent states of the superposition are spatially separated by distances much greater than the size of the atom wavepacket which has an r.m.s. spread equal to x_0.

Analysis of this state is interesting from the point of view of the quantum measurement problem, an issue that has been debated since the inception of quantum theory by Einstein, Bohr, and others, and continues today [31]. One practical approach toward resolving this controversy is the introduction of quantum decoherence, or the environmentally induced reduction of quantum superpositions into classical statistical mixtures [32]. Decoherence provides a way to quantify the elusive boundary between classical and quantum worlds, and almost always precludes the existence of macroscopic Schrödinger-cat states, except for extremely short times. On the other hand, the creation of mesoscopic Schrödinger-cat states like that of Eq. (10) may allow controlled studies of quantum decoherence and the quantum-classical boundary. This problem is directly relevant to quantum computation, as we discuss below.

In our experiment, we create a Schrödinger-cat state of the single-ion $^9Be^+$ harmonic oscillator (Eq. (10)) with a sequence of laser pulses [24]. First, we create a state of the form $(|\downarrow\rangle + e^{i\xi}|\uparrow\rangle)|n=0\rangle/\sqrt{2}$ with a $\pi/2$ pulse on the Raman carrier transition (Sec. 5.1). To spatially separate the $|\downarrow\rangle$ and $|\uparrow\rangle$ components of the wave function, we apply a coherent excitation with an optical dipole force which, because of the polarization of the beams used to create the force, selectively excites the motion of only the $|\uparrow\rangle$ state. We then swap the $|\downarrow\rangle$ and $|\uparrow\rangle$ states with a carrier π pulse and reapply the dipole force with a different phase to create the state of Eq. (10). In principle, if we could make $|\alpha_{1,2}|$ large enough, we could design a detector which could directly detect the (distinguishable) position of the particle and correlate it with a spin measurement [33]. Instead, to analyze this state in our experiment, we apply an additional laser pulse to couple the internal states, and we measure the resulting interference of the distinct wavepackets. With this interferometer, we can establish the correlations inherent in Eq. (10), the separation of the wavepackets, <u>and</u> the phase coherence ϕ between components of the wavefunction. These experiments are described in Ref. 24. The interference signal should be very sensitive to decoherence. As the separation $|\alpha_1 - \alpha_2|$ is made larger, decoherence is expected to exponentially degrade the fringe contrast [32].

Other experiments generate Schrödinger cats in the same sense as in our experiment. Examples are atom interferometers [34] and superpositions of electron wavepackets in atoms [35] (also, see additional citations in Ref. 24). However, as opposed to these experiments, the harmonic oscillator cat states of Eq. (10) do not disperse in time. This lack of dispersion provides a simple visualization of the "cat" (for example, a marble rolling back and forth in a bowl which can be simultaneously at opposite extremes of motion) and should allow controlled studies of decoherence models. This has recently been realized in cavity

QED experiments [36].

7. COMPLETE RECONSTRUCTION OF THE MOTIONAL QUANTUM STATE

The controlled interaction of light and rf-electromagnetic fields with the trapped ion not only allows us to prepare very general states of motion, but also to determine these quantum mechanical states using novel techniques. Few experiments have succeeded in determining the density matrices or Wigner functions of quantum systems [37]. Here we present the theory and experimental demonstration of two distinct schemes to reconstruct both the density matrix in the number state basis and the Wigner function of the motional state of a single trapped atom. Other proposed methods are discussed in Refs. 38-41. Both of our measurement techniques rely on our ability to coherently displace the input state to several different locations in phase space. The technical realization of the displacements is described in the previous section in the context of the production of coherent states. After the coherent displacement we apply radiation on the first blue sideband for a time τ, which induces a resonant exchange between the motional and internal degrees of freedom. For each coherent displacement characterized by a complex number α, a time series of measurements of the population $P_{\downarrow}(\tau)$ is made by monitoring the fluorescence produced in driving the resonant dipole cycling transition just as in Eq. (8). The internal state at $\tau=0$ is always prepared to be $|\downarrow\rangle$ for the various input states, so the signal averaged over many measurements is

$$P_{\downarrow}(\tau,\alpha) = \frac{1}{2}\left(1 + \sum_{n=0}^{\infty} Q_n(\alpha)\cos(2\Omega_{n,n+1}\tau)e^{-\gamma_n\tau}\right). \tag{11}$$

Without the coherent displacement we would just produce the signal of Eq. (8) and would get the populations of the motional eigenstates only. However, since we repeat these measurements for several magnitudes and phases of the coherently displaced state, we are able to extract information about the off-diagonal elements of the density matrix and can also reconstruct the Wigner function [42] from the measured displaced populations $Q_n(\alpha)$. These can be found by a Fourier cosine transform and are linearly related to the elements of the motional density matrix in the number state basis (a detailed description of these relations is given in Ref. 25). These relations can be numerically inverted to reconstruct the density matrix of the motional state in the number state basis using a general linear least-squares method.

As pointed out by several authors, quasiprobability distribution functions F(α,s) [42,43] such as the Wigner function can be characterized by a displacement α and a parameter s. These functions have a particularly simple representation when expressed in populations of coherently displaced number states $Q_k(\alpha)$ [43].

$$F(\alpha,s) = \frac{1}{\pi}\sum_{n=0}^{\infty}[(s+1)/2]^n\sum_{k=0}^{n}(-1)^k\binom{n}{k}Q_k(\alpha). \tag{12}$$

For $s = -1$, the sum reduces to one term, and $F(\alpha,-1) = Q_o(\alpha)/\pi$ gives the value of the Q-quasi-probability distribution at the complex coordinate α [44]. The Wigner function is given by s = 0. F(α,0) = W(α) for every point α in the complex plane can be determined by the sum

$$W(\alpha) = \frac{2}{\pi}\left(\sum_{n=0}^{\infty}(-1)^n Q_n\right). \tag{13}$$

In our reconstruction, the sum is carried out only to a finite n_{max}, depending on the estimated magnitude of the states to reconstruct [25].

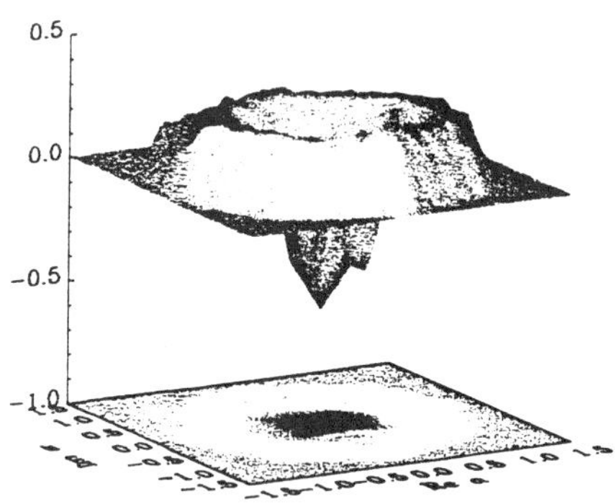

Fig. 4. Surface and contour plot of the Wigner function of an approximate |n=1⟩ number state.

In Fig. 4, we show the surface and contour plots of the Wigner function of an approximate $|n=1\rangle$ number state. The plotted points are the result of fitting a linear interpolation between the actual data points to a 0.1-by-0.1 grid. The octagonal shape is an artifact of the eight measured phases per radius. The white contour represents $W(\alpha)=0$. The negative values around the origin highlight the nonclassical character of this state.

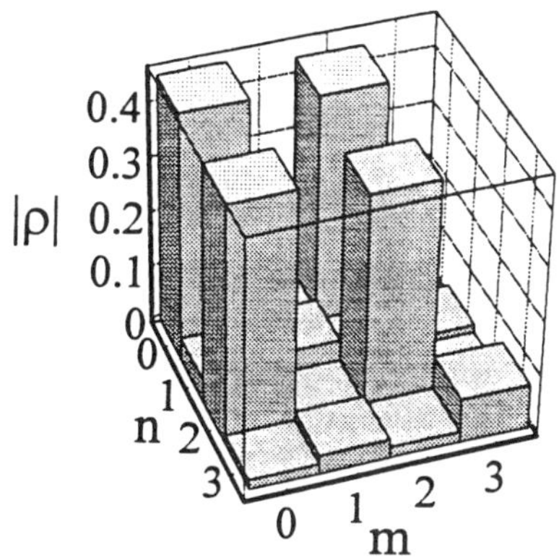

Fig. 5. Reconstructed density matrix amplitudes of an approximate $1/\sqrt{2}\,(|n=0\rangle - i\,|n=2\rangle)$ state. The amplitudes of the coherences indicate that the reconstructed density matrix is close to that of a pure state.

As an example of a reconstructed number state density matrix, we show in Fig. 5 our result for a coherent superposition of $|n=0\rangle$ and $|n=2\rangle$ number states. This state is ideally suited to demonstrate the sensitivity of the reconstruction to coherences. Our result indicates that the prepared motional states in our system are very close to pure states.

8. QUANTUM LOGIC

Quantum computation has received a great deal of attention recently because of the algorithm proposed by Peter Shor for efficient factorization [45,46]. Factorization has important implications for public-key data encryption where the security of these systems is due to the inability to efficiently factorize large numbers. Accomplishing quantum factorization will be formidable with any technology; however, other applications of quantum logic may be more tractable. One possibility is described in Sec. 9. Ignacio Cirac and Peter Zoller [47] proposed an attractive scheme for a quantum computer which would use a string of ions in a linear trap as "qubits," similar to what is shown in Fig. 1. This proposal has stimulated experimental efforts in several laboratories, including those at Innsbruck, Los Alamos National Laboratory, IBM, and NIST.

Each qubit in a quantum computer could be implemented by a two-level atomic system; for the ith qubit, we label these states $|\downarrow\rangle_i$ and $|\uparrow\rangle_i$ as above. In general, any quantum computation can be comprised of a series of single-bit rotations and two-bit controlled-NOT (CN) logic operations [46,48]. We are interested in implementing these two operations in a system of $^9Be^+$ ions. Single-bit rotations are straightforward and correspond to driving Raman carrier transitions (Sec. 5.1) for a controlled time. Such rotations have been achieved in many previous experiments. In the following, we describe the demonstration of a nontrivial CN logic gate with a single $^9Be^+$ ion [49].

8.1 *"Conditional dynamics" and a single-ion controlled-not logic gate*

The key to making a quantum logic gate is to provide conditional dynamics; that is, we desire to perform on one physical subsystem a unitary transformation which is conditioned upon the quantum state of another subsystem [46]. In the context of cavity QED, the required conditional dynamics at the quantum level has recently been demonstrated [50,51]. For trapped ions, conditional dynamics at the quantum level has been demonstrated in verifications of zero-point laser cooling where absorption on the red sideband depended on the motional quantum state of the ion [11,12]. Recently, we have demonstrated a CN logic gate; in this experiment, we also had the ability to prepare arbitrary input states to the gate (the "keyboard" operation of step (2a) below).

A two-bit quantum CN operation provides the transformation:

$$|\epsilon_1\rangle|\epsilon_2\rangle \rightarrow |\epsilon_1\rangle|\epsilon_1 \oplus \epsilon_2\rangle, \tag{14}$$

where $\epsilon_1, \epsilon_2 \in \{0,1\}$ and $\oplus$ is addition modulo 2. The (implicit) phase factor in the transformation is equal to 1. In this expression ϵ_1 is the called the control bit and ϵ_2 is the target bit. If $\epsilon_1 = 0$, the target bit remains unchanged; if $\epsilon_1 = 1$, the target bit flips. In the single-ion experiment of Ref. 49, the control bit is the quantized state of one mode of the ion's motion. If the motional state is $|n{=}0\rangle$, it is taken to be a $|\epsilon_1{=}0\rangle$ state; if the motional state is $|n{=}1\rangle$, it is taken to be a $|\epsilon_1{=}1\rangle$ state. The target states are two ground-hyperfine states of the ion, the $|\downarrow\rangle$ and $|\uparrow\rangle$ states of Sec. 5 with the identification here $|\downarrow\rangle \leftrightarrow |\epsilon_2{=}0\rangle$ and $|\uparrow\rangle \leftrightarrow |\epsilon_2{=}1\rangle$. Following the notation of Sec. 5, the CN operation is realized by applying three Raman laser pulses in succession:

(1a) A $\pi/2$ pulse is applied on the spin carrier transition. For a certain choice of initial phase, this corresponds to the operator $V^{1/2}(\pi/2)$ of Ref. 47.
(1b) A 2π pulse is applied on the first blue sideband transition between levels $|\uparrow\rangle$ and an auxiliary level $|aux\rangle$ in the ion (the $|F=2, M_F=0\rangle$ level in $^9Be^+$). This operator is analogous to the operator $U_n^{2,1}$ of Ref. 47. This operation provides the conditional dynamics for the controlled-not operation in that it changes the sign of the $|\uparrow\rangle|n=1\rangle$ component of the wavefunction but leaves the sign of the $|\uparrow\rangle|n=0\rangle$ component of the wavefunction unchanged.
(1c) A $\pi/2$ pulse is applied to the spin carrier transition with a 180° phase shift relative to step (1a). This corresponds to the operator $V^{1/2}(-\pi/2)$ of Ref. 47.

Steps (1a) and (1c) can be regarded as two resonant pulses (of opposite phase) in the Ramsey separated-field method of spectroscopy [52]. We can see that if step (b) is active (thereby changing the sign of the $|\uparrow\rangle|n=1\rangle$ component of the wave function) then a spin flip is produced by the Ramsey fields. If step (1b) is inactive, the net effect of the Ramsey fields is to leave the spin state unchanged. This CN operation can be incorporated to provide an overall CN operation between two ions in an ensemble of N ions if we choose the ion oscillator mode to be the center-of-mass (COM) mode of the ensemble. Specifically, to realize a controlled-not $C_{m,k}$ between two ions (m = control bit, k = target bit), we first assume the COM is prepared in the zero-point state. The initial state of the system is therefore given by

$$\Psi = \left(\sum_{M_1=\downarrow,\uparrow} \sum_{M_2=\downarrow,\uparrow} \cdots \sum_{M_N=\downarrow,\uparrow} C_{M_1,M_2,\ldots M_N} |M_1\rangle_1 |M_2\rangle_2 \ldots |M_N\rangle_N \right) |0\rangle . \tag{15}$$

$C_{m,k}$ can be accomplished with the following steps:

(2a) Apply a π pulse on the red sideband of ion m (the assumption is that ions can be addressed separately [47]). This accomplishes the mapping $(\alpha|\downarrow\rangle_m + \beta|\uparrow\rangle_m)|0\rangle \rightarrow |\downarrow\rangle_m(\alpha|0\rangle - e^{i\phi}\beta|1\rangle)$, and corresponds to the operator $U_m^{1,0}$ of Ref. 47. In our experiments, we prepare states of the form $(\alpha|\downarrow\rangle + \beta|\uparrow\rangle)|0\rangle$ using the carrier transition (Sec. 5.1). We can then implement the mapping $(\alpha|\downarrow\rangle + \beta|\uparrow\rangle)|0\rangle \rightarrow |\downarrow\rangle_m(\alpha|0\rangle - e^{i\phi}\beta|1\rangle)$ by applying a red sideband π pulse. This is the "keyboard" operation for preparation of arbitrary motional input states for the CN gate of steps 1a-1c above. Analogous mapping of internal state superpositions to motional state superpositions was demonstrated in Ref. 49.

(2b) Apply the CN operation (steps 1a-1c above) between the COM motion and ion k.

(2c) Repeat step (2a).

Overall, $\mathbf{C}_{m,k}$ provides the mappings $|\downarrow\rangle_m|\downarrow\rangle_k|0\rangle \rightarrow |\downarrow\rangle_m|\downarrow\rangle_k|0\rangle$, $|\downarrow\rangle_m|\uparrow\rangle_k|0\rangle \rightarrow |\downarrow\rangle_m|\uparrow\rangle_k|0\rangle$, $|\uparrow\rangle_m|\downarrow\rangle_k|0\rangle \rightarrow |\uparrow\rangle_m|\uparrow\rangle_k|0\rangle$,and $|\uparrow\rangle_m|\uparrow\rangle_k|0\rangle \rightarrow |\uparrow\rangle_m|\downarrow\rangle_k|0\rangle$ which is the desired logic of Eq. (14). Effectively, $\mathbf{C}_{m,k}$ works by mapping the internal state of ion m onto the COM motion, performing a CN between the motion and ion n, and then mapping the COM state back onto ion m. The resulting CN between ions m and k is not really different from the CN described by Cirac and Zoller, because the operations $V^{1/2}(\theta)$ and $U_m^{1,0}$ commute.

8.2 Quantum Registers and Schrödinger Cats

The state represented by Eq. (15) is of the same form as that of Eq. (10). Both involve entangled superpositions and both are subject to the destructive effects of decoherence. Creation of Schrödinger cats like Eq. (10) is particularly relevant to the ion-based quantum computer because the primary source of decoherence will probably be due to decoherence of the $|n=0,1\rangle$ motional states during the logic operations.

8.3 Perspective on ion quantum computation and quantum logic

To be useful for factorization, a quantum computer must be able to factorize a 200 digit or larger decimal number. This will require a few thousand ions and perhaps 10^9 or more elementary operations [46]. Therefore, given the current state of the art (one ion and about 10 operations before decoherence), we should be skeptical. Decoherence will be most decisive in determining the fate of quantum computation. Already, decoherence from spontaneous emission appears to limit the number of operations possible [53,54]. The experiments can be expected to improve dramatically, but we must hope for more efficient algorithms or ways to patch them (such as error correction schemes [55]) before large scale factoring is possible. Possibilities and limitations with the ion trap scheme are being explored [53,56-59].

Any quantum system that might be contemplated in quantum computation must be reproducible, stable, and well isolated from the environment. Quantum dots have the potential advantage of large-scale integration using microfabrication; however, at the present time, they suffer from lack of precise reproducibility and excessive decoherence. Trapped ions are reproducible and relatively immune to environmental perturbations - this is the reason they are candidates for advanced frequency standards [60]. In principle, higher information density could be achieved in ion traps by scaling down the size of the trap electrodes; however, we must then worry about excessive environmental coupling such as magnetic field perturbations caused by impurities and/or currents in the (nearby) trap electrodes. Similarly, electric field perturbations will also become important as the size of the trap electrodes becomes smaller. To avoid excessive coupling to the surrounding environment in any quantum logic scheme, the supporting matrix for quantum bits must therefore be reasonably separated from the qubits. In any scheme which uses atomic transitions, a scale size represented by present ion traps might be close to optimum.

Factorization, discrete logs, and certain other mathematical computations appear to be the hardest problems that quantum logic might be applied to. One of the applications for quantum computation that Richard Feynman originally had in mind was to simulate quantum mechanical calculations [61]. This idea is being explored again with new possibilities in mind [62]. Other potential applications might also be realized. At NIST, the original motivation for creation of entangled states was to fundamentally improve the signal-to-noise ratio in ion spectroscopy and frequency standards; we discuss this possibility in the next section.

9. QUANTUM LOGIC APPLIED TO SPECTROSCOPY

We conclude by discussing a possible application of quantum logic in the realm of atomic physics. This application has the advantage of being useful with a relatively small number of ions and logic operations.

Entangled atomic states can improve the quantum-limited signal-to-noise ratio in spectroscopy [19,63,64]. In spectroscopy experiments on N atoms, in which changes in atomic populations are detected, we can view the problem in the following way using the spin-½ analogy for two-level atoms. We assume that spectroscopy is performed by applying (classical) fields of frequency ω for a time T_R according to the Ramsey method of separated fields [52]. After applying these fields, we measure the final state populations. For example, we might measure the operator $\tilde{N}_\downarrow$ corresponding to the number of atoms in the $|\downarrow\rangle$ state. In the spin-½ analogy, this is equivalent to measuring the operator J_z, since $\tilde{N}_\downarrow = J\tilde{I} - J_z$ where $\tilde{I}$

is the identity operator.

If all sources of technical noise are eliminated, the signal-to-noise ratio (for repeated measurements) is fundamentally limited by the quantum fluctuations in the number of atoms which are observed to be in the $|\downarrow\rangle$ state. These fluctuations can be called quantum "projection" noise [65]. If spectroscopy is performed on N initially uncorrelated atoms (for example, $\Psi(t=0) = \Pi_i|\downarrow\rangle_i$), the imprecision in a determination of the frequency of the transition is limited by projection noise to $(\Delta\omega)_{meas.} = 1/(NT_R\tau)^{1/2}$ where $\tau >> T_R$ is the total averaging time. If the atoms can be initially prepared in entangled states, it is possible to achieve $(\Delta\omega)_{meas.} < 1/(NT_R\tau)^{1/2}$. Initial theoretical investigations [19,63] examined the use of correlated states which could achieve $(\Delta\omega)_{meas.} < 1/(NT_R\tau)^{1/2}$ when the population (J_z) was measured. More recent theoretical investigations [64] consider the initial state to be one where, after the first Ramsey pulse, the internal state is the maximally entangled state of Eq. (5). After applying the Ramsey fields, we measure the operator $\tilde{O} = \Pi_i\sigma_{zi}$ instead of J_z (or $\tilde{N}_\downarrow$). For unit detection efficiency, we can achieve $(\Delta\omega)_{meas.} = 1/(N^2T_R\tau)^{1/2}$ which is the maximum signal-to-noise ratio possible. The main reason for the reduced value of $(\Delta\omega)_{meas.}$ is that $\langle\tilde{O}\rangle = (-1)^N\cos[N(\omega - \omega_o)T_R]$; the derivative of this function with respect to applied frequency is N times larger than for the uncorrelated atom case. For an atomic clock where T_R is fixed by other constraints, this means that the time required to reach a certain measurement precision (stability) is reduced by a factor of N relative to the uncorrelated-atom case. In terms of quantum computation, this amounts to a computation of the function $\cos(N(\omega - \omega_o)T)$. Of course, this computation has special significance for the measurement of ω_o (an intrinsic computer parameter) but otherwise is much better suited for a classical computer!

Cirac and Zoller [47] have outlined a scheme for producing the state in Eq.(5) using quantum logic gates. Using the notation of Sec. 5.1, we would first prepare the atoms in the state $\Psi(t=0) = \Pi_i|\downarrow\rangle_i$ and then apply a $\pi/2$ rotation to ion 1 to create the state $\Psi = 2^{-1/2}(|\downarrow\rangle_1 + e^{i\phi}|\uparrow\rangle_1)|\downarrow\rangle_2|\downarrow\rangle_3...|\downarrow\rangle_N$. We then apply the CN gate of Eq. (14) sequentially between ion 1 and ions 2 through N to achieve the state of Eq. (5). An alternative method for generating this state, without the need of addressing individual ions, is described in Ref. 64.

10. ACKNOWLEDGMENTS

We gratefully acknowledge the support of the National Security Agency, the US Office of Naval Research, and the US Army Research Office. We thank E. Burt, J. Marquardt, D. Sullivan, and M. Young for helpful comments on the manuscript.

11. REFERENCES

† Contribution of NIST; not subject to US copyright.

1. H.G. Dehmelt, Adv. At. Mol. Phys. 3, 53 (1967) and 5, 109 (1969).
2. D.J. Wineland, Wayne M. Itano, and R.S. VanDyck, Jr., ibid. 19, 135 (1983).
3. R. C. Thompson, ibid. 31, 253 (1993).
4. *Ion Traps*, P.K. Ghosh (Clarendon Press, Oxford, 1995).
5. M. G. Raizen, J. M. Gilligan, J. C. Bergquist, W. M. Itano, and D. J. Wineland, *Phys. Rev.* A45, 6493 (1992).
6. J. Drees and W. Paul, Z. Phys. 180, 340 (1964).
7. Unless the rods conform to equipotentials of Eq. (1), this equation must be multiplied by a constant factor which, for simplicity, we take to be equal to 1; see for example Ref. [5].
8. M. E. Poitzsch, J. C. Bergquist, W. M. Itano, and D. J. Wineland, Rev. Sci. Instrum. 67, 129 (1996); J.D. Miller, D.J. Berkeland, F.C. Cruz, J.C. Bergquist, W.M. Itano, and D.J. Wineland, Proc. 1996 IEEE Int. Frequency Control Symp. (copies available from IEEE Service Center, 445 Hoes Lane, P.O. Box 1331, Piscataway, NJ 08855-1331) p. 1086.
9. see, for example: D. J. Wineland, J. C. Bergquist, J. J. Bollinger, W. M. Itano, D. J. Heinzen, S. L. Gilbert, C. H. Manney, and M. G. Raizen, IEEE Trans. on Ultrasonics, Ferroelectrics, and Frequency Control 37, 515 (1990).
10. J.J. Bollinger, J.D. Prestage, W.M. Itano, and D.J. Wineland, Phys. Rev. Lett. 54, 1000-1003 (1985); J. J. Bollinger, D. J. Heinzen, W. M. Itano, S. L. Gilbert, and D. J. Wineland, IEEE Trans. on Instrum. and Measurement 40, 126 (1991).
11. F. Diedrich, J.C. Bergquist, W. M. Itano, and D.J. Wineland, *Phys. Rev. Lett.* 62, 403 (1989).
12. C. Monroe, D. M. Meekhof, B. E. King, S. R. Jefferts, W. M. Itano, D. J. Wineland, and P. Gould, *Phys. Rev. Lett.* 75, 4011 (1995).
13. W. Nagourney, J. Sandberg, and H.G. Dehmelt, *Phys. Rev. Lett.* 56, 2797 (1986); Th. Sauter, R. Blatt, W. Neuhauser, and P.E. Toschek, *Phys. Rev. Lett.* 57, 1696 (1989); J.C. Bergquist, R.G. Hulet, W.M. Itano, and D.J. Wineland, *Phys. Rev. Lett.* 57, 1699 (1986).
14. S. R. Jefferts, C. Monroe, E. W. Bell, and D. J. Wineland, *Phys. Rev.* A51, 3112-3116 (1995).
15. A. Einstein, B. Podolsky, N. Rosen, *Phys. Rev.* 47, 777 (1935).

16. D.J. Wineland and W.M. Itano, Phys. Rev. A20, 1521 (1979).
17. *Cavity Quantum Electrodynamics*, ed. by P.R. Berman (Academic Press, Boston, 1994).
18. C.A. Blockley, D.F. Walls, and H. Risken, *Europhys. Lett.* 17, 509 (1992).
19. D. J. Wineland, J. J. Bollinger, W. M. Itano, F. L. Moore, and D. J. Heinzen, *Phys. Rev.* A46, R6797 (1992).
20. J.I. Cirac, R. Blatt, A.S. Parkins, and P. Zoller, *Phys. Rev. Lett.* 70, 762 (1993).
21. D.M. Meekhof, C. Monroe, B.E. King, W.M. Itano, and D.J. Wineland, *Phys. Rev. Lett.* 76, 1796 (1996); erratum, 77, 2346 (1996).
22. J.I. Cirac, A.S. Parkins, R. Blatt, P. Zoller, in *Adv. Atomic and Molecular Phys.*, to be published.
23. W. Vogel and R.L. de Matos Filho, Phys. Rev. A52, 4214 (1995).
24. C. Monroe, D. M. Meekhof, B. E. King, and D. J. Wineland, *Science* 272, 1131 (1996).
25. D. Leibfried, D.M. Meekhof, B.E. King, C. Monroe, W.M. Itano, and D.J. Wineland, Phys. Rev. Lett., to be published.
26. C.K. Law and J.H. Eberly, Phys. Rev. Lett. 76, 1055 (1996).
27. T. von Foerster, J. Phys. A: Math.Gen., 8, 95 (1975).
28. J.H. Eberly, N.B. Narozhny, and J.J. Sanchez-Mondragon, *Phys. Rev. Lett.* 44, 1323 (1980).
29. M. Brune, F. Schmidt-Kaler, A. Maali, J. Dreyer, E. Hagley, J.M. Raymond, and S. Haroche, Phys. Rev. Lett. 76, 1800 (1996).
30. E. Schrödinger, *Naturwissenschaften* 23, 807 (1935).
31. *Quantum Theory and Measurement*, ed. by J.A. Wheeler, W.H. Zurek (Princeton Univ. Press, Princeton, 1983).
32. W.H. Zurek, *Physics Today*, 44, 36 (1991).
33. J.F. Poyatos, J.I. Cirac, R. Blatt, P. Zoller, *Phys. Rev.* A, to be published.
34. O. Carnal and J. Mlynek, *Phys. Rev. Lett.* 66, 2689 (1991); D.W. Keith, C.R. Ekstrom, Q.A. Turchette, D.E. Pritchard, *Phys. Rev. Lett.*66, 2693 (1991); M. Kasevich and S. Chu, *Phys. Rev. Lett.* 67, 181 (1991); J. Lawall, S. Kulin, B. Saubamea, N. Bigelow, M. Leduc, and C. Cohen-Tannoudji, *Phys. Rev. Lett.* 75, 4194 (1995).
35. L.D. Noordam, D.I. Duncan, T.F. Gallagher, *Phys. Rev.* A45, 4734 (1992); R.R. Jones, C.S. Raman, S.W. Schumacher, P.H. Bucksbaum, *Phys. Rev. Lett.* 71, 2575 (1993); M.W. Noel and C.R. Stroud, Jr., *Phys. Rev. Lett.* 75, 1252 (1995); M.W. Noel and C.R. Stroud, Phys. Rev. Lett. 77, 1913 (1996).

36. M. Brune, E. Hagley, J. Dreyer, X. Maître, A. Maali, C. Wunderlich, J.M. Raimond, and S. Haroche, Phys. Rev. Lett., to be published.
37. J. R.Ashburn, R. A. Cline, P. J. M. van der Burgt, W. B. Westerveldt, and J. S. Risley, Phys. Rev. A **41**, 2407 (1990); D. T. Smithey, M. Beck, M. G. Raymer, and A.Faridani, Phys. Rev. Lett. **70**, 1244 (1993); G. Breitenbach, T. Müller, S. F. Pereira, J. Ph. Poizat, S. Schiller, and J. Mlynek, J. Opt. Soc. B **12**, 2304 (1995); T. J. Dunn, I. A. Walmsley, and S. Mukamel, Phys. Rev. Lett. **74**, 884 (1995); C. Kurtsiefer, J. Mlynek, Univ. Konstanz, private communication.
38. S. Wallentowitz and W. Vogel, Phys. Rev. Lett. 75, 2932 (1995).
39. J.F. Poyatos, R. Walser, J.I. Cirac, P. Zoller, and R. Blatt, Phys. Rev. A 53, R1966 (1966).
40. C. D'Helon and G.J. Milburn, Phys. Rev. A54, R25 (1996).
41. P.J. Bardroff, C. Leichtle, G. Schrade, and W.P. Schleich, Phys. Rev. Lett. 77, 2198 (1996); G. Schrade, P.J. Bardroff, R.J. Glauber, C. Leichtle, V. Yakovlev, and W.P. Schleich, Appl. Phys. B, to be published.
42. See, for example, V. Bužek and P.L. Knight, *Progress in Optics*, vol. XXXIV, 1 ((1995).
43. A.Royer, Phys. Rev. Lett. **52**, 1064 (1984); H. Moya-Cessa and P. L.Knight, Phys. Rev. A **48**, 2479 (1993); S. Wallentowitz and W. Vogel, Phys. Rev. A. **53**, 4528 (1996); K. Banaszek and K. Wodkiewicz, Phys.Rev. Lett. **76**, 4344 (1996).
44. J. F. Poyatos, R. Walser, J. I. Cirac, P. Zoller, and R. Blatt, Phys. Rev. A **53**, R1966 (1996).
45. P. Shor, Proc. 35th Ann. Symp. on the Foundations of Comp. Sci. (IEEE Comp. Soc. Press, NY, 1994). p. 124.
46. A. Ekert and R. Jozsa, *Rev. Mod. Phys.* 68, 733 (1996).
47. J.I. Cirac and P. Zoller, *Phys. Rev. Lett.* 74, 4091 (1995).
48. D.P. DiVincenzo, *Phys. Rev.* A51, 1051 (1995).
49. C. Monroe, D. M. Meekhof, B. E. King, W. M. Itano, and D. J. Wineland, *Phys. Rev. Lett.* 75, 4714 (1995).
50. M. Brune, P. Nussenzveig, F. Schmidt-Kaler, F. Bernardot, A. Maali, J.M. Raimond, and S. Haroche, *Phys. Rev. Lett.* 72, 3339 (1994).
51. Q. Turchette, C. Hood, W. Lange, H. Mabushi, H.J. Kimble, *Phys. Rev. Lett.* 75, 4710 (1995).
52. N.F. Ramsey, *Molecular Beams*, (Oxford University Press, London, 1963).
53. M.B. Plenio and P.L. Knight, Phys. Rev. A53, 2986 (1996).
54. The effects of spontaneous emission are significantly reduced if rf

transitions between hyperfine levels are induced with inhomogeneous rf fields (Sec. 3).

55. See, for example, D.P. DiVincenzo and P.W. Shor, Phys. Rev. Lett. 77, 3260 (1996) and references therein.
56. R.J. Hughes, D.F.V. James, E.H. Knill, R. Laflamme, and A.G. Petschek, Phys. Rev. Lett. 77, 3240 (1996).
57. A. Steane, Proc. Royal Soc. Lond. A, to be published.
58. A. Garg, Phys. Rev. Lett. 77, 964 (1996).
59. D.J Wineland, C. Monroe, D.M. Meekhof, B.E. King, and D. Leibfried, W.M. Itano, J.C. Bergquist, D. Berkeland, J.J. Bollinger, and J. Miller, *Atomic Physics 15*, Proc.15th Int. Conf. on Atomic Physics, Amsterdam, Aug., 1996, submitted.
60. see for example: *Proc., Fifth Symp. Freq. Standards and Metrology*, ed. by J.C. Bergquist, Woods Hole, MA, Oct. 1995 (World Scientific, Singapore, 1996).
61. R.P. Feynman, *Int. J. Theor. Phys.* 21, 467 (1982); *Opt. News* 11, 11 (1985); *Found. Phys.* 16, 507 (1986).
62. S. Lloyd, *Science* 273, 1073 (1996).
63. D. J. Wineland, J. J. Bollinger, W. M. Itano, and D. J. Heinzen, Phys. Rev. A50, 67-88 (1994).
64. J. J. Bollinger, D. J. Wineland, W. M. Itano, and D. J. Heinzen, Phys. Rev. A, to be published, Dec., 1996; J. J. Bollinger, D. J. Wineland, W. M. Itano, and D. J. Heinzen, *Proc., Fifth Symp. Freq. Standards and Metrology*, ed. by J.C. Bergquist, Woods Hole, MA, Oct. 1995 (World Scientific, Singapore, 1996), p. 107.
65. W. M. Itano, J. C. Bergquist, J. J. Bollinger, J. M. Gilligan, D. J. Heinzen, F. L. Moore, M. G. Raizen, and D. J. Wineland, *Phys. Rev.* A47, 3554 (1993).

RESONANCE FLUORESCENCE OF A SINGLE ION *

J. T. Höffges, H.W. Baldauf, T. Eichler, S.R. Helmfrid and H. Walther

Max-Planck-Institut für Quantenoptik and
Sektion Physik der Universität München
85748 Garching, Fed. Rep. of Germany

INTRODUCTION

Resonance fluorescence of atoms is a basic process in radiation-atom interactions, and has therefore always generated considerable interest. The methods of experimental investigation have changed continuously due to the availability of new experimental tools. A considerable step forward occurred when tunable and narrow band dye laser radiation became available. These laser sources are sufficiently intense to easily saturate an atomic transition. In addition, the lasers provide highly monochromatic light with coherence times much longer than typical natural lifetimes of excited atomic states. Excitation spectra with laser light using well collimated atomic beam lead to a width being practically the natural width of the resonance transition, therefore it became possible to investigate the frequency spectrum of the fluorescence radiation with high resolution. However, the spectrograph used to analyze the reemitted radiation was a Fabry-Perot interferometer, the resolution of which did reach the natural width of the atoms, but was insufficient to reach the laser linewidth, see e.g. (1,2). A considerable progress in this direction was achieved by investigating the fluorescence spectrum of ultra-cold atoms in an optical lattice in a heterodyne experiment (Jessen et al. (3)). In these measurements a linewidth of 1 kHz was achieved, however, the quantum aspects of the resonance fluorescence such as antibunched photon statistics cannot be investigated under these conditions since they wash out when more than one atom is involved.

Thus the ideal experiment requires a single atom to be investigated. Since some time it is known that ion traps allow to study the fluorescence from a single laser cooled particle practically at rest, thus providing the ideal case for the

*This paper is dedicated to Professor Ingvar Lindgren on the occasion of his 65th birthday

ADVANCES IN QUANTUM CHEMISTRY, VOLUME 30

0065-3276/98 $25.00

spectroscopic investigation of the resonance fluorescence. The other essential ingredient for achievement of high resolution is the measurement of the frequency spectrum by heterodyning the scattered radiation with laser light as demonstrated with many cold atoms. Such an optimal experiment with a single trapped Mg^+ ion is described in this paper. The measurement of the spectrum of the fluorescent radiation at low excitation intensities is presented. Furthermore, the photon correlation of the fluorescent light has been investigated under practically identical excitation conditions. The comparison of the two results shows a very interesting aspect of complementarity since the heterodyne measurement corresponds to a "wave" detection of the radiation whereas the measurement of the photon correlation is a "particle" detection scheme. It will be shown that under the same excitation conditions the wave detection provides the properties of a classical atom, i.e. a driven oscillator, whereas the particle or photon detection displays the quantum properties of the atom. Whether the atom displays classical or quantum properties thus depends on the method of observation.

The spectrum of the fluorescence radiation is given by the Fourier transform of the first order correlation function of the field operators, whereas the photon statistics and photon correlation is obtained from the second order correlation function. The corresponding operators do not commute, thus the respective observations are complementary. Present theory on the spectra of fluorescent radiation following monochromatic laser excitation can be summarized as follows: fluorescence radiation obtained with low incident intensity is also monochromatic owing to energy conservation. In this case, elastic scattering dominates the spectrum and thus one should measure a monochromatic line at the same frequency as the driving laser field. The atom stays in the ground state most of the time and absorption and emission must be considered as one process with the atom in principle behaving as a classical oscillator. This case was treated on the basis of a quantized field many years ago by Heitler (4). With increasing intensity upper and lower states become more strongly coupled leading to an inelastic component, which increases with the square of the intensity. At low intensities, the elastic part dominates since it depends linearly on the intensity. As the intensity of the exciting light increases, the atom spends more time in the upper state and the effect of the vacuum fluctuations comes into play through spontaneous emission. The inelastic component is added to the spectrum, and the elastic component goes through a maximum where the Rabi flopping frequency $\Omega = \Gamma/\sqrt{2}$ (Γ is the natural linewidth) and then disappears with growing Ω. The inelastic part of the spectrum gradually broadens as Ω increases and for $\Omega > \Gamma/2$ sidebands begin to appear. For a saturated atom, the form of the spectrum shows three well-separated Lorentzian peaks. The central peak has width Γ and the sidebands which are each displaced from the central peak by the Rabi frequency are broadened to $3\Gamma/2$. The ratio of the height of the central peak to the sidebands is 3:1. This spectrum was first calculated by Mollow (5). For other relevant papers see the review of Cresser et al. (2).

EXPERIMENTAL SETUP AND RESULTS

The experimental study of the problem requires, as mentioned above, a Doppler-free observation. In order to measure the frequency distribution, the fluorescent light has to be investigated by means of a high resolution spectrometer. The first experiments of this type were performed by Schuda et al. (6) and later by Walther et al. (7), Hartig et al. (7, 1) and Ezekiel et al. (8). In all these experiments, the excitation was performed by single-mode dye laser radiation, with the scattered radiation from a well collimated atomic beam observed and analyzed by Fabry-Perot interferometers. Experiments to investigate the elastic part of the resonance fluorescence giving a resolution better than the natural linewidth have been performed by Gibbs et al. (9) and Cresser et al. (2).

The first experiments which investigated antibunching in resonance fluorescence were also performed by means of laser-excited collimated atomic beams. The initial results obtained by Kimble, Dagenais, and Mandel (10) showed that the second-order correlation function $g^{(2)}(t)$ had a positive slope which is characteristic of photon antibunching. However, $g^{(2)}(0)$ was larger than $g^{(2)}(t)$ for $t \rightarrow \infty$ due to number fluctuations in the atomic beam and to the finite interaction time of the atoms (11, 12). Further refinement of the analysis of the experiment was provided by Dagenais and Mandel (11). Rateike et al. (13) used a longer interaction time for an experiment in which they measured the photon correlation at very low laser intensities (see Ref. (2) for a review). Later, photon antibunching was measured using a single trapped ion in an experiment which

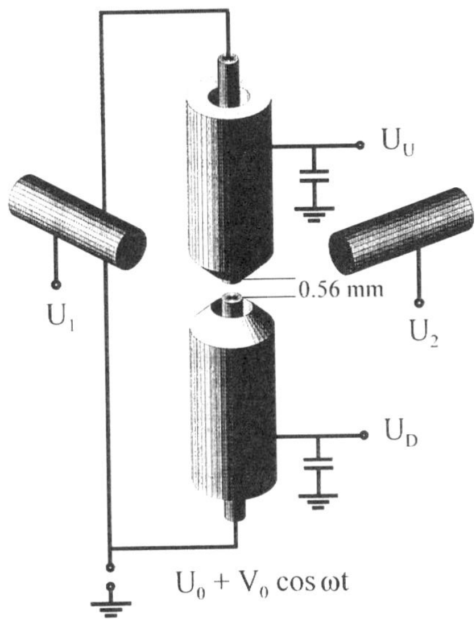

Fig. 1: Electrode configuration of the endcap trap. The open structure offers a large detection solid angle and good access for laser beams testing the micromotion of the ion. Micromotion is minimized by applying dc voltages: U_1, U_2, U_U, U_D.

avoids the disadvantages of atom number statistics and finite interaction time between atom and laser field (14).

As pointed out in many papers photon antibunching is a purely quantum phenomenon (see e.g. Refs. (2, 15). The fluorescence of a single ion displays the additional nonclassical property that the variance of the photon number is smaller than its mean value (i.e. it is sub-Poissonian). This is because the single ion can emit only a single photon and has to be re-excited before it can emit the next one which leads to photon emissions at almost equal time intervals. The sub-Poissonian statistics of the fluorescence of a single ion has been measured in a previous experiment (14) (see also Ref. (16) for comparison).

The trap used for the present experiment was a modified Paul-trap, called an endcap-trap (17) (see Fig. 1) which produces good confinement of the trapped ion. Therefore, the number of sidebands in the fluorescence spectrum, caused by the oscillatory motion of the laser cooled ion in the pseudopotential of the trap, is reduced. The trap consists of two solid copper-beryllium cylinders (diameter 0.5 mm) arranged co-linearly with a separation of 0.56 mm. These correspond to the cap electrodes of a traditional Paul trap, whereas the ring electrode is replaced by two hollow cylinders, one of which is concentric with each of the cylindrical endcaps. Their inner and outer diameters are 1 and 2 mm, respectively and they are electrically isolated from the cap electrodes. The fractional anharmonicity of this trap configuration, determined by the deviation of the real potential from the ideal quadrupole field is below 0.1 % (17). The trap is driven at a frequency of 24 MHz with typical secular frequencies in the xy-plane of approximately 4 MHz. This required a radio-frequency voltage with an amplitude on the order of 300 V to be applied between the cylinders and the endcaps, and with AC-grounding of the outer electrodes provided through a capacitor.

The measurements were performed using the $3^2S_{1/2}$ - $3^2P_{3/2}$ transition of the $^{24}Mg^+$-ion at a wavelength of 280 nm. The natural width of this transition is 42.7 MHz. The exciting laser light was produced by frequency doubling the light from a rhodamine 110 dye laser. The laser was tuned slightly below resonance in order to Doppler-cool the secular motion of the ion. All the measurements of the fluorescent radiation described in this paper were performed with this slight detuning.

For the experiment described here, it is important to have the trapped ion at rest as far as possible to minimize the light lost into motional sidebands. There are two reasons which may cause motion of the ion: the first one is the periodic oscillation of the ion within the harmonic pseudopotential of the trap and the second one is micromotion which is present when the ion is not positioned exactly at the saddle point of the trap potential. Such a displacement may be caused by a contact potential resulting, for example, by a coating of the electrodes by Mg produced when the atoms are evaporated during the loading procedure of the trap. Reduction of the residual micromotion can be achieved by adjusting the position of the ion with DC-electric fields generated by additional electrodes. For the present experiment they were arranged at an angle of 120° in a plane perpendicular to the symmetry axis of the trap electrodes. By applying auxiliary voltages (U_1 and U_2) to these electrodes and U_U and U_D to the outer trap electrodes (Fig. 1), the ion's position can be adjusted to settle at the saddle point of the trap potential.

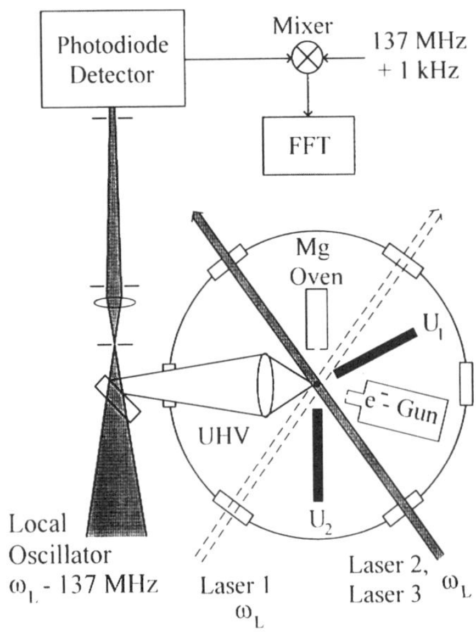

Fig. 2: Scheme of heterodyne detection. The trap is omitted in the figure with only two of the compensation electrodes shown. Laser 3 is directed at an angle of 22° with respect to the drawing plane and Laser 2.

The micromotion of the ion can be monitored using the periodic Doppler shift at the driving frequency of the trap which results in a periodic intensity modulation in the fluorescence intensity. This modulation can be measured by means of a transient recorder, triggered by the AC-voltage applied to the trap. There are three laser beams (lasers 1-3 in Fig. 2) passing through the trap in three different spatial directions which allow measurement of the three components of the micromotion separately. By adjusting the compensation voltages U_1 U_2, U_U and U_D the amplitude of the micromotion could be reduced to a value smaller than $\lambda/8$ in all spatial directions.

The amount of secular motion of the ion resulting from its finite kinetic energy cannot be tested by this method since the secular motion is not phase coupled to the trap voltage. However, the intensity modulation owing to thi motion can be seen in a periodic modulation of the photon correlation signal. For all measurements presented here, this amplitude was below $\lambda/5$. Thi correspnds to the ion having a temperature of approximately 1 mK and a mea vibrational occupation of n=5, which results in less than 50 % of the fluorescence energy being lost into the vibrational sidebands.

The heterodyne measurement is performed as follows. The dye laser excites the trapped ion with frequency ω_L while the fluorescence is observed in a direction of about 54° to the exciting laser beam (see Fig. 2). However, both the observation direction and the laser beam are in a plane perpendicular to the symmetry axis of the trap. Before reaching the ion, a fraction of this laser radiation is removed with a beamsplitter and then frequency shifted (by 137 MHz with an acousto-optic modulator (AOM)) to serve as the local oscillator. The local oscillator and fluorescence radiations are then overlapped and simultaneously focused onto the photodiode where the initial frequency mixing occurs. The frequency difference signal is amplified by a narrow band amplifier and then frequency down-converted to 1 kHz so that it could be analyzed by means of a fast Fourier analyzer (FFT). The intermediate frequency for this mixing of the signal was derived from the same frequency-stable synthesizer which was used to drive the accousto-optic modulator producing the sideband of the laser radiation so that any synthesizer fluctuations are canceled out.

An example of a heterodyne signal is displayed in Fig. 3, where $\Delta\omega$ is the frequency difference between the heterodyne signal and the driving frequency of the AOM. Frequency fluctuations of the laser beam cancel out and do not influence the linewidth because at low intensity the fluorescence radiation always follows the frequency of the exciting laser while the local oscillator is derived directly from the same laser beam. The residual linewidth results mainly from fluctuations in the optical path length of the local oscillator or of the fluorescent beam. Both beams pass through regular air and it was observed that a forced motion of the air increased the frequency width of the heterodyne signal. The frequency resolution of the FTT was 3.75 Hz for the particular measurement. The heterodyne measurements were performed at a saturation parameter $s = \frac{\Omega^2/2}{\Delta^2 + (\Gamma^2/4)}$ of 0.9, where Δ is the laser detuning. In this region, the elastic part of the fluorescent spectrum has a maximum (18).

The signal to noise ratio observed in the experiment is shot noise limited. The signal in Fig. 3 corresponds to a rate of the scattered photons of about 10^4 s^{-1} which is an upper limit since photons were lost from detection due to scattering into sidebands caused by the secular motion of the ion. In order to reduce

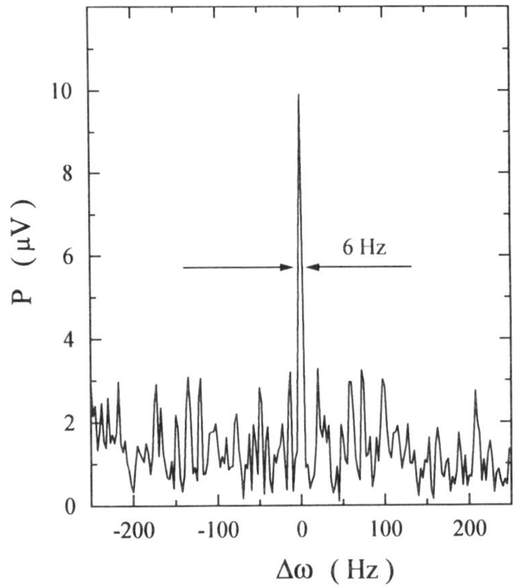

Fig. 3: Heterodyne spectrum of a single trapped $^{24}Mg^{+}$-ion for s = 0.9, Δ = -2.3 Γ, Ω = 3.2 Γ. Integration time: 267 ms.

this loss as much as possible, a small angle between the directions of observation and excitation was used.

Investigation of photon correlations employed the ordinary Hanbury-Brown and Twiss setup with two photomultipliers and a beam splitter. The setup was essentially the same as described in Ref. (14). The pulses from the photomultipliers (RCA C31034-A02) were amplified and discriminated by a constant fraction discriminator (EG&G model 584). The time delay t between the photomultiplier signals was converted by a time-to-amplitude converter into a voltage amplitude proportional to the time delay. A delay line of 100 ns in the stop channel allowed for the measurement of $g^{(2)}(t)$ for both positive and negative t in order to check the symmetry of the measured signal. The output of the time-to-amplitude converter was accumulated by a multichannel analyzer in pulse height analyzing mode. Three typical measurements, each at small values of s but with different detunings, are shown in Fig. 4. The detunings and Rabi frequencies were determined by fitting the formula given in Ref. (12) to the measurements. For small time delays (< 20 ns) the nonclassical antibunching

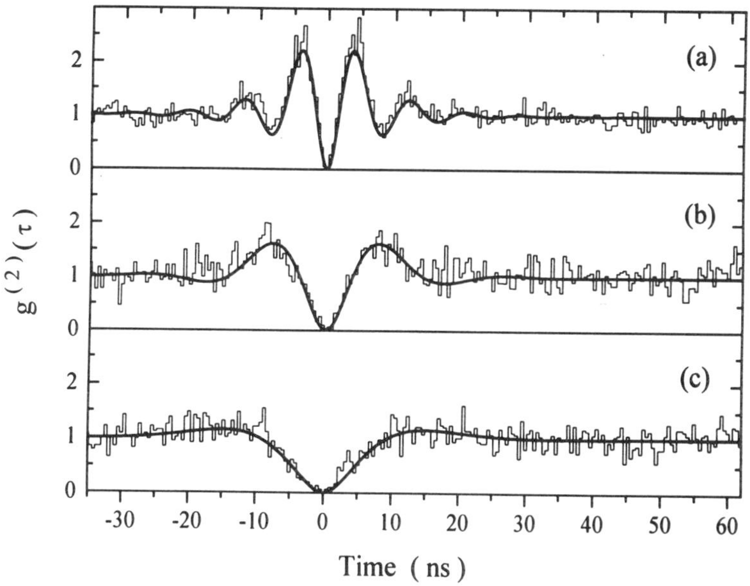

Fig. 4: Antibunching signals of a single $^{24}Mg^{+}$-ion. The integration time t_0 was limited by the storage time of the ion. (a), $\Delta = -2.3\ \Gamma$, $\Omega = 2.8\ \Gamma$. s = 0.7 and t_0 = 165 min. (b) $\Delta = -1.1\ \Gamma$, $\Omega = 1.0\ \Gamma$. s = 0.3 and t_0 = 95 min. (c) $\Delta = -0.5\ \Gamma$, $\Omega = 0.6\ \Gamma$. s = 0.4 and t_0 = 220 min. The solid line is a theoretical fit, see text for details.

effect is observed, superposed with Rabi oscillations which are damped out with a time constant corresponding to the lifetime of the excited state.

A measurement of $g^{(2)}(t)$ with an averaging time of hours and time delays up to 500 ns resulted in no visible micromotion effects when the compensation voltages U_1, U_2, U_U and U_D were correctly adjusted. Micromotion results in a periodic modulation of the photon correlation at the driving frequency of the trap (compare Ref. (14)). The stray-light counting rate was so low that there was no need to correct the measurement shown in Fig. 4 (b) for accidental counts. There was actually not a single count in the t = 0 channel within the integration time of 220 min.

CONCLUSIONS

In conclusion, we have presented the first high-resolution heterodyne measurement of the elastic peak in resonance fluorescence of a single ion. At identical experimental parameters we have also measured antibunching in the photon correlation of the scattered field. Together, both measurements show that, in the limit of weak excitation, the fluorescence light differs from the excitation radiation in the second-order correlation but not in the first order correlation. However, the elastic component of resonance fluorescence combines an extremely narrow frequency spectrum with antibunched photon statistics, which means that the fluorescence radiation is not second-order coherent as expected from a classical point of view. This apparent contradiction can be explained easily by taking into account the quantum nature of light, since first-order coherence does not imply second-order coherence for quantized fields (19). The heterodyne and the photon correlation measurement are complementary since they emphasize either the classical wave properties or the quantum properties of resonance fluorescence, respectively.

In a recent treatment of a quantized trapped particle (20) it was shown that a trapped ion in the vibrational ground state of the trap will also show the influence of the micromotion since the wavefunction distribution of the ion is pulsating at the trap frequency. This means that a trapped particle completely at rest will also scatter light into the micromotion sidebands. Investigation of the heterodyne spectrum at the sidebands may give the chance to confirm these findings. It is clear that such an experiment will not be easy since other methods are needed to verify that the ion is actually at rest at the saddle point of the potential .

ACKNOWLEDGEMENTS

We would like to thank Roy Glauber for many discussions in connection with his quantum treatment of a trapped particle. We also thank Girish S. Agarwal for many discussions.

REFERENCES

(1) Hartig, W., Rasmussen, W., Schieder, R., Walther, H. *Z. Physik* **1976,** *A278,* 205.

(2) Cresser, J. D., Häger, J., Leuchs, G., Rateike, F. M., Walther, H. *Topics in Current Physics* **1982,** *27,* 21.

(3) Jessen, P.S., Gerz, C., Lett, P.D., Philipps, W.D., Rolston, S.L., Spreuuw, R.J.C., and Westbrook, C.I. *Phys. Rev. Lett.* **1992,** *69,* 49.

(4) Heitler, W. In „The Quantum Theory of Radiation," Third Edition, Oxford University Press, 1954, p. 186.

(5) Mollow, B.R. *Phys. Rev.* **1969,** *188,* 1969.

(6) Schuda, F., Stroud, C., Jr., Hercher, M. *J. Phys.* **1974,** *B1,* L198.

(7) Walther, H. *Lecture Notes in Physics* **1975,** *43*, Springer, p. 358.

(8) Wu, F. Y., Grove, R. E., Ezekiel, S. *Phys. Rev. Lett.* **1975,** *35,* 1426; Grove, R.E., Wu, F. Y., Ezekiel, S. *Phys. Rev. Lett.* **1977,** *A 15*, 227.

(9) Gibbs H.M. and Venkatesan, T.N.C. *Opt. Comm.* **1976, 17**, 87.

(10) Kimble, H. J., Dagenais, M., and Mandel, L. *Phys. Rev. Lett.* **1977,** *39,* 691.

(11) Jakeman, E., Pike, E. R., Pusey, P. N., and Vaugham, J. M. *J. Phys.* **1977,** *A 10,* L257.

(12) Kimble, H. J., Dagenais, M., and Mandel, L. *Phys. Rev.* **1978,** *A 18,* 201; Dagenais, M., Mandel, L. *Phys. Rev.* **1978,** *A 18*, 2217.

(13) Rateike, F.-M., Leuchs, G., and Walther, H., results cited in Ref. (2).

(14) Diedrich, F., Walther, H. *Phys. Rev. Lett.* **1987,** *58,* 203.

(15) Walls, D.F. *Nature* **1979,** *280,* 451.

(16) Short, R. and Mandel, L. *Phys. Rev. Lett.* **1983,** *51,* 384, and in „Coherence and Quantum Optics", Mandel, L. and Wolf, E., Eds.; Plenum: New York, 1984. Vol. V, p. 671.

(17) Schrama, C. A., Peik, E., Smith, W. W., and Walther, H. *Opt. Comm.* **1993,** *101,* 32.

(18) Cohen-Tannoudji, C., Dupont-Roc, J., Grynberg, G. In „Atom-Photon Interactions", J. Wiley & Sons, Inc.: New York, 1992, 407-514

(19) Loudon, R. *Rep. Progr. Phys.* **1980,** *43,* 913.

(20) Glauber, R. In „Proceedings of the International School of Physics "Enrico Fermi", Course CXVIII Laser Manipulation of Atoms and Ions",. Arimondo, E., Phillips, W.D., Strumia, F., Eds.; North Holland 1992, p. 643

Quantum Electrodynamics and the Fundamental Constants

Peter Mohr
National Institute of Standards and Technology
Gaithersburg, MD 20899-0001, USA

Abstract

The values of the fundamental constants and the theory of quantum electrodynamics (QED) are closely coupled. This is evident from the fact that the constants appear as parameters in the theoretical expressions that describe the physical properties of particles and matter, and most of these theoretical expressions are derived from QED. In practice, values of the constants are determined by a consistent comparison of the relevant measurements and theoretical expressions involving those constants. Such a comparison is being carried out in order to provide CODATA recommended values of the constants for 1997. This review describes some of the advances that have been made since the last set of constants was recommended in 1986. As a result of these advances, there is a significant reduction in the uncertainty of a number of constants included in the set of 1997 recommended values.

1 Introduction

An adjustment of the fundamental constants is underway to provide the CODATA 1997 recommended values. The results of this work will be described in a report of the *CODATA Task Group on Fundamental Constants* written by Barry Taylor *(NIST)*, E. Richard Cohen *(Rockwell Int., retired)*, and the present author.

The CODATA Task Group consists of the following people:

F. Cabiati, Istituto Elettrotecnico Nazionale "Galileo Ferraris," Italy;

E. R. Cohen, Science Center, Rockwell International (retired), United States of America;

T. Endo, Electrotechnical Laboratory, Japan;

V. Kose, Physikalisch-Technische Bundesanstalt, Germany;

Liu Ruimin, National Institute of Metrology, China (People's Republic of);

B. A. Mamyrin, A. F. Ioffe Physical-Technical Institute, Russia;

P. J. Mohr, National Institute of Standards and Technology, United States of America;

B. W. Petley, National Physical Laboratory, United Kingdom (GB);

T. J. Quinn, Bureau International des Poids et Mesures;

B. N. Taylor (Chairman), National Institute of Standards and Technology, United States of America;

V. S. Tuninsky, Mendeleev Institute of Metrology, Russia;

W. Wöger, Physikalisch-Technische Bundesanstalt, Germany; and

B. M. Wood, National Research Council, Canada.

The acronym CODATA stands for the *Committee on Data for Science and Technology*, which was established in 1966 by the International Council of Scientific Unions (ICSU).

The purpose of the adjustment is to determine and recommend values of various fundamental constants such as the fine-structure constant, Rydberg constant, Avagadro constant, Planck constant, electron mass, muon mass, and many others, which provide the greatest consistency among the most critical experiments, based on relationships derived primarily from QED and condensed matter theory.

2 Background

A pioneering evaluation of a set of fundamental physical constants was done by Raymond Birge in 1929 [1]. It is of interest to note that this is the first article (vol. **1**, p. 1) in the journal that was to become the Reviews of Modern Physics. Adjustments done since 1950 include the following:

DuMond and Cohen (1951) [2],
Bearden and Watts (1951) [3, 4],
DuMond and Cohen (1953) [5],
DuMond and Cohen (1955) [6],
Bearden and Thomsen (1957) [7],
Cohen and DuMond (1965) [8],
Taylor, Parker, and Langenberg (1969) [9],
Cohen and Taylor (1973) [10], and
Cohen and Taylor (1987) [11].

References [2] and [8] were based on reports to the National Research Council (NRC) of the National Academy of Sciences (NAS), while the latest two provided CODATA recommended values.

3 1986 Adjustment

The most recent adjustment gives the current CODATA recommended values as reported by E. R. Cohen and B. N. Taylor, "The 1986 adjustment of the fundamental physical constants", *Rev. Mod. Phys.* **59**, 1121 (1987).

The values from this adjustment are available on the World Wide Web through the National Institute of Standards and Technology (NIST) Physics Laboratory home page:
http://physics.nist.gov/
via the "Physical Reference Data" page, or they can be found directly at:
http://physics.nist.gov/fundcon.
This Web site is undergoing a number of improvements and will provide the new 1997 values when they become available.

4 1990 Update

In 1990, the effect of new information on the fundamental constants was examined by B. N. Taylor and E. R. Cohen, "Recommended values of the fundamental physical constants: A status report", *J. Res. Natl. Inst. Stand. Technol.* **95**, 497 (1990).

As pointed out in that paper, there was a significant improvement in the data relevant to the constants between 1986 and 1990 due primarily to the following three new results:

- A value of the Planck constant obtained from a realization of the watt measured at the National Physical Laboratory (NPL), UK [12].

- A value of the fine-structure constant obtained from the magnetic moment anomaly of the electron measured at the University of Washington [13].
- A value of the molar gas constant obtained from a measurement of the speed of sound in argon carried out at NIST [14].

Although these results had the potential to impact significantly on the values of the fundamental constants, it was felt at that time that it was premature to recommend a new set of constants [15].

5 Recent Progress

Recent progress that will give improved values of three particular constants in the current adjustment is discussed in more detail in the following. The three constants directly affected are:

Planck constant from "watt balance" experiments,

fine-structure constant from the anomalous magnetic moment of the electron, and

Rydberg constant from spectroscopic measurements in the hydrogen atom.

The Planck constant can be obtained by comparing a watt of mechanical energy measured in SI units to a watt of electrical energy measured in terms of the Josephson constant ($K_{\rm J} = 2e/h$) and von Klitzing constant ($R_{\rm K} = h/e^2$) in the combination

$$K_{\rm J}^2 R_{\rm K} = \frac{4}{h} \, . \tag{1}$$

The fine-structure constant is obtained from the measured electron magnetic moment anomaly by solving the equation

$$a_{\rm e}(\text{th}) \equiv f(\alpha) = a_{\rm e}(\exp) \tag{2}$$

for α, where $f(\alpha)$ is the theoretical expression for the anomaly as a power series in α, and $a_{\rm e}(\exp)$ is the experimental value.

Similarly, the Rydberg constant is determined from the relation

$$\nu_{ij} = f_{ij} R_\infty \, , \tag{3}$$

where ν_{ij} is the observed frequency of transitions between states i and j in hydrogen, and f_{ij} is a calculated coefficient that depends on QED theory and is weakly dependent on the fine-structure constant, the electron-proton mass ratio, and the electric charge radius of the proton.

These three examples reflect various aspects of quantum electrodynamics theory. The electron anomalous magnetic moment follows from free-electron QED, the transition frequencies in hydrogen follow from bound-state QED, and, at least in principle, the relevant condensed matter theory follows from the equations of many-body QED.

6 Planck Constant

For this example, Planck's constant from a watt-balance experiment, we first review the relevant descriptions of the Josephson effect and the Quantized Hall effect.

The Josephson effect is the phenomenon in which two superconducting materials weakly coupled through a non-superconducting interface show a sharp increase in conductance of current flowing from one of the superconductors to the other if radiation of a suitable frequency is applied to the interface. The relation between the voltage difference between the two superconductors $U_{\rm J}$ and the frequency f of the applied radiation for which there is an increase in current is given by

$$2eU_{\rm J}(n) = nhf \qquad n = 1, 2, 3, \ldots , \tag{4}$$

where n is any (small in practice) positive integer. This may be expressed as

$$nf = K_{\rm J}\, U_{\rm J}(n) = \frac{2e}{h}\, U_{\rm J}(n) , \tag{5}$$

which defines the Josephson constant.

The Quantum Hall effect manifests itself in a two-dimensional electron gas in certain semiconductor structures. When such a structure is placed in a magnetic flux density normal to the plane of the electron gas, and a current I flows along its length, there is a quantized voltage $U_{\rm H}$ across the semiconductor in the direction perpendicular to the current given by

$$U_{\rm H}(i) = R_{\rm K}\, \frac{I}{i} = \frac{h}{e^2}\, \frac{I}{i} \qquad i = 1, 2, 3, \ldots , \tag{6}$$

where i takes on positive integer values. This defines the von Klitzing constant. The second relation on the right-hand-side of Eq. (6) is based on theory and is taken to be exact at the level of present-day precision.

The principle of the watt-balance experiment is the following: A horizontal circular coil of wire is suspended in a radial magnetic flux density from one side of a pully and balanced by a counterweight on the other side as shown in Fig. 1.

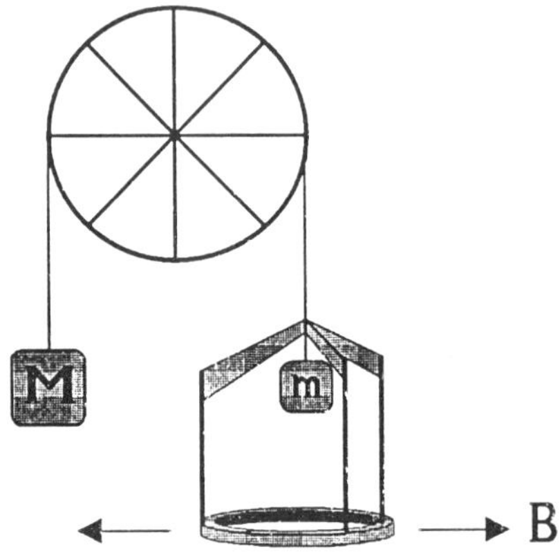

Fig. 1. Schematic figure of the watt-balance experiment.

In addition to the counterweight M on the left side in that figure there is another coil and radial magnetic flux density, not shown in the figure, to apply a force to control the motion of the wire over the pully. The measurement is divided into two parts.

Part 1: A current I is sent through the horizontal coil, shown in the figure, that produces a force that exactly matches the force produced by a standard mass m hung on the same side, $F_z = -mg$. The expression for the force in the z direction, in terms of the current density $\vec{J}$ and the magnetic flux density $\vec{B}$, is

$$F_z = \int d\vec{x}\ (\vec{J} \times \vec{B}) \cdot \hat{z} = I \int (d\vec{\ell} \times \vec{B}) \cdot \hat{z}\ , \tag{7}$$

where $d\vec{\ell}$ is an infinitesimal element of the wire in the coil. The current I is measured as a function of the vertical position of the coil.

Part 2: The coil is moved in the vertical direction with a constant velocity v, and the induced voltage across the coil U_v is measured. From classical electrodynamics the voltage is

$$U_v = \int d\vec{\ell} \cdot (\vec{v} \times \vec{B}) = -v \int (d\vec{\ell} \times \vec{B}) \cdot \hat{z}\ . \tag{8}$$

This equation assumes that $\vec{v} = v\hat{z}$ exactly, which may require a small correction.

In view of Eqs. (7) and (8), the dependence on the flux density $\vec{B}$ and the coil geometry $d\vec{\ell}$ can be eliminated by writing

$$F_z = -mg = -\frac{I}{v}\, U_v\ . \tag{9}$$

The current I is measured in terms of voltage and resistance given by the Josephson and Quantum Hall effects and the voltage U_v is measured in terms of the Josephson effect, so we have

$$\begin{aligned} mgv &= I\,U_v = a_1 \frac{U_{\mathrm{H}}}{R_{\mathrm{K}}} U_v = a_1 \frac{e^2}{h} a_2 \frac{hf_2}{2e} a_3 \frac{hf_3}{2e} \\ &= a_1\, a_2\, a_3\, f_2\, f_3\, \frac{h}{4}\,, \end{aligned} \tag{10}$$

or

$$h = \frac{4mgv}{a_1\, a_2\, a_3\, f_2\, f_3}\,, \tag{11}$$

where a_1, a_2, and a_3 are ratios from the voltage and resistance calibrations, and f_1 and f_2 are the frequencies employed in the Josephson effect calibrations. The mass m is a standard mass calibrated in terms of the kilogram, the local gravitational constant g is determined by an absolute gravimeter, and the velocity v is measured by means of laser interferometry. All of these quantities can thus be given in terms of SI (International System of) units, and so Planck's constant is thereby given in its SI unit.

Two measurements of h based on the general principles described here have been carried out. One done at the National Physical Laboratory (NPL) in 1988 [12] gave

$$h = 6.626\,068\,21(90) \times 10^{-34}\ \mathrm{J \cdot s}\,, \tag{12}$$

and the other done at the National Institute of Standards and Technology (NIST) in 1989 [16, 17] gave

$$h = 6.626\,070\,4(88) \times 10^{-34}\ \mathrm{J \cdot s}\,. \tag{13}$$

These values can be compared to the 1986 CODATA recommended value

$$h = 6.626\,075\,5(40) \times 10^{-34}\ \mathrm{J \cdot s}\,. \tag{14}$$

It is remarkable that a fundamental quantum mechanical constant is best measured with an apparatus whose operation is based on classical mechanics and classical electrodynamics.

7 Fine-Structure Constant

Deviation of the electron magnetic moment from the Dirac value $g_{\mathrm{e}}(\mathrm{Dirac}) = 2$ is given in terms of the magnetic moment anomaly a_{e} by

$$g_{\mathrm{e}} = 2(1 + a_{\mathrm{e}})\,. \tag{15}$$

Measurement of a_e by the University of Washington group [13] has yielded

$$a_{e^-}(\text{exp}) = 1\,159\,652\,188.4(4.3) \times 10^{-12}, \tag{16}$$

$$a_{e^+}(\text{exp}) = 1\,159\,652\,187.9(4.3) \times 10^{-12}. \tag{17}$$

The theoretical value of the anomaly is given as a sum of a QED contribution, a weak interaction contribution, and a hadronic contribution

$$a_e(\text{th}) = a_e(\text{QED}) + a_e(\text{weak}) + a_e(\text{had}) . \tag{18}$$

The QED theoretical value can be written as

$$\begin{aligned} a_e(\text{QED}) &= A_1 + A_2\,(m_e/m_\mu) + A_2\,(m_e/m_\tau) \\ &\quad + A_3\,(m_e/m_\mu, m_e/m_\tau) , \end{aligned} \tag{19}$$

where A_1 arises from Feynman diagrams with only electron lines, A_2 from diagrams with one heavier lepton (μ or τ), and A_3 from diagrams with both heavier leptons. Each of these contributions can be written as a power series in α:

$$\begin{aligned} A_i &= A_i^{(2)}\left(\frac{\alpha}{\pi}\right) + A_i^{(4)}\left(\frac{\alpha}{\pi}\right)^2 \\ &\quad + A_i^{(6)}\left(\frac{\alpha}{\pi}\right)^3 + A_i^{(8)}\left(\frac{\alpha}{\pi}\right)^4 + \cdots , \end{aligned} \tag{20}$$

where $i = 1, 2, 3$. For the electron diagrams, the first two terms are known to be [18]

$$A_1^{(2)} = \tfrac{1}{2} , \tag{21}$$

and [19, 20, 21, 22]

$$\begin{aligned} A_1^{(4)} &= \tfrac{197}{144} + \left(\tfrac{1}{2} - 3\ln 2\right)\zeta(2) + \tfrac{3}{4}\zeta(3) \\ &= -0.328\,478\,965 \cdots . \end{aligned} \tag{22}$$

The most recent results concern the higher-order terms. For the sixth-order term the numerical calculation has been improved [23, 24, 25], and there is a recently completed analytic result [26, 27]:

$$\begin{aligned} A_1^{(6)} &= \frac{100\,a_4}{3} - \frac{215\,\zeta(5)}{24} + \frac{83\,\pi^2\,\zeta(3)}{72} + \frac{139\,\zeta(3)}{18} \\ &\quad + \frac{25\,\ln^4 2}{18} - \frac{25\,\pi^2\,\ln^2 2}{18} - \frac{298\,\pi^2\,\ln 2}{9} \\ &\quad - \frac{239\,\pi^4}{2160} + \frac{17101\,\pi^2}{810} + \frac{28259}{5184} \\ &= 1.181\,241\,456 \cdots , \end{aligned} \tag{23}$$

where $a_4 = \sum_{n=1}^{\infty} 1/(2^n n^4)$. There has also been improvement in the numerical evaluation of the eighth-order coefficient, which consists of 891 Feynman diagrams, with the result [28]

$$A_1^{(8)} = -1.409\,2(384)\,. \tag{24}$$

There are additional corrections from muon and tau vacuum polarization loops [29, 30]

$$A_2(m_e/m_\mu) = 2.712 \times 10^{-12}\,, \tag{25}$$

$$A_2(m_e/m_\tau) = 0.009 \times 10^{-12}\,. \tag{26}$$

The hadronic and weak interaction contributions are [28, 31, 32]

$$a_e(\text{had}) = 1.648(41) \times 10^{-12}\,, \tag{27}$$

$$a_e(\text{weak}) = 0.030(1) \times 10^{-12}\,. \tag{28}$$

Taking these corrections into account, the value of the fine-structure constant for which the theoretical and experimental values of the electron anomalous magnetic moment are equal is

$$\alpha^{-1}(a_{e^-}) = 137.035\,999\,91(53)\,. \tag{29}$$

This value has the smallest assigned uncertainty among the various methods of determining α.

This value can be compared to the current CODATA 1986 recommended value, which is

$$\alpha^{-1}(\text{CO86}) = 137.035\,989\,5(61)\,. \tag{30}$$

The new uncertainty is 0.09 times the old uncertainty, and the value is shifted by about 1.7 times the earlier uncertainty.

8 Rydberg Constant

The Rydberg constant is evaluated by comparison of theory and experiment for energy levels in hydrogen. There have been advances in both experiment and theory for the transition frequencies. The most recent experiments calibrate the measured frequency by a chain of comparisons that link to the cesium atomic standard for the second which provides a significant improvement in accuracy compared to earlier methods. Results of these measurements are listed at the end of this section. The main emphasis in the rest of this section is on theoretical developments. There have been improvements in calculations of the binding effects in the lowest-order self-energy correction as well as a calculation of the first binding correction to the two-photon QED corrections. The rest of the theoretical contributions are summarized here as well.

The theoretical energy levels are determined to high accuracy by the Dirac eigenvalue, quantum electrodynamic effects such as the self energy and vacuum polarization, finite-nuclear-size corrections, and nuclear motion effects.

Dirac Eigenvalue: This is given exactly by

$$\begin{aligned} E_{\mathrm{D}} &= \left[1+\frac{(Z\alpha)^2}{(n-\delta)^2}\right]^{-1/2} m_{\mathrm{e}}c^2\,, \\ \delta &= |\kappa| - \left[\kappa^2-(Z\alpha)^2\right]^{1/2}\,. \end{aligned} \tag{31}$$

Relativistic Recoil: The reduced mass correction up to order $(Z\alpha)^4$ is [33, 34]

$$\begin{aligned} E_M &= Mc^2 + [f(n,j)-1]m_{\mathrm{r}}c^2 - [f(n,j)-1]^2\frac{m_{\mathrm{r}}^2c^2}{2M} \\ &\quad + \frac{1-\delta_{l0}}{\kappa(2l+1)}\frac{(Z\alpha)^4 m_{\mathrm{r}}^3c^2}{2n^3m_{\mathrm{N}}^2} + \cdots\,, \end{aligned} \tag{32}$$

where

$$f(n,j) = \left[1+\frac{(Z\alpha)^2}{(n-\delta)^2}\right]^{-1/2}\,, \tag{33}$$

$M = m_{\mathrm{e}} + m_{\mathrm{N}}$, and $m_{\mathrm{r}} = m_{\mathrm{e}}m_{\mathrm{N}}/(m_{\mathrm{e}}+m_{\mathrm{N}})$. Relativistic-recoil corrections to Eq. (32) are given by [35]

$$\begin{aligned} E_{\mathrm{S}} &= \frac{m_{\mathrm{r}}^3}{m_{\mathrm{e}}^2m_{\mathrm{N}}}\frac{(Z\alpha)^5}{\pi n^3}m_{\mathrm{e}}c^2 \\ &\quad \times\Big\{\tfrac{1}{3}\delta_{l0}\ln(Z\alpha)^{-2} - \tfrac{8}{3}\ln k_0(n,l) - \tfrac{1}{9}\delta_{l0} - \tfrac{7}{3}a_n \\ &\quad - \frac{2}{m_{\mathrm{N}}^2-m_{\mathrm{e}}^2}\delta_{l0}\left[m_{\mathrm{N}}^2\ln\left(\frac{m_{\mathrm{e}}}{m_{\mathrm{r}}}\right) - m_{\mathrm{e}}^2\ln\left(\frac{m_{\mathrm{N}}}{m_{\mathrm{r}}}\right)\right]\Big\}\,, \end{aligned} \tag{34}$$

where [36]

$$\begin{aligned} a_n &= -2\left[\ln\left(\frac{2}{n}\right) + \sum_{i=1}^{n}\frac{1}{i} + 1 - \frac{1}{2n}\right]\delta_{l0} \\ &\quad + \frac{1-\delta_{l0}}{l(l+1)(2l+1)}\,. \end{aligned} \tag{35}$$

The next-order correction is [37, 38, 39, 40]

$$E_{\rm R} = \frac{m_{\rm e}}{m_{\rm N}}\frac{(Z\alpha)^6}{n^3} m_{\rm e}c^2 D_{60} \,, \tag{36}$$

where

$$D_{60} = \begin{cases} 4\ln 2 - \frac{7}{2} & nS_{1/2}\,, \\ \frac{1}{3} & 2P_{1/2}\,, \\ \frac{1}{3} & 2P_{3/2}\,. \end{cases} \tag{37}$$

Self Energy: The lowest-order self energy is written as

$$E^{(2)}_{\rm SE} = \frac{\alpha}{\pi}\frac{(Z\alpha)^4}{n^3} F(Z\alpha)\, m_{\rm e}c^2 \,, \tag{38}$$

where

$$\begin{aligned} F(Z\alpha) &= A_{41}\ln(Z\alpha)^{-2} + A_{40} + A_{50}(Z\alpha) \\ &\quad + A_{62}(Z\alpha)^2\ln^2(Z\alpha)^{-2} + A_{61}(Z\alpha)^2\ln(Z\alpha)^{-2} \\ &\quad + G_{\rm SE}(Z\alpha)(Z\alpha)^2 \,, \end{aligned} \tag{39}$$

with [41]

$$\begin{aligned} A_{41} &= \tfrac{4}{3}\,\delta_{l0}\,, \\ A_{40} &= -\tfrac{4}{3}\ln k_0(n,l) + \tfrac{11}{18}\,\delta_{l0} - \frac{1}{2\kappa(2l+1)}\,, \\ A_{50} &= \left(\tfrac{139}{32} - 2\ln 2\right)\pi\,\delta_{l0}\,, \\ A_{62} &= -\delta_{l0}\,, \\ A_{61} &= \begin{cases} \frac{28}{3}\ln 2 - \frac{21}{20} & 1S_{1/2}\,, \\ \frac{16}{3}\ln 2 + \frac{67}{30} & 2S_{1/2}\,, \\ \frac{4}{3}\ln 2 + \frac{391}{80} & 4S_{1/2}\,, \\ \frac{103}{180} & 2P_{1/2}\,, \\ \frac{29}{90} & 2P_{3/2}\,. \end{cases} \end{aligned} \tag{40}$$

The Bethe logarithms that appear in Eq. (40) are given by [42]

$$\begin{aligned} \ln k_0(1,0) &= 2.984\,128\,555\cdots\,, \\ \ln k_0(2,0) &= 2.811\,769\,893\cdots\,, \\ \ln k_0(4,0) &= 2.749\,811\,840\cdots\,, \\ \ln k_0(2,1) &= -0.030\,016\,708\cdots\,. \end{aligned} \tag{41}$$

Table 1: The function $G_{\rm SE}(Z\alpha)$.

Z	$1S_{1/2}$	$2S_{1/2}$	$4S_{1/2}$	$2P_{1/2}$	$2P_{3/2}$
1	−30.31(4)	−31.23(4)	−30.9(2)	−0.98(4)	−0.49(2)
2	−29.79(4)	−30.68(4)	−30.4(2)	−0.96(4)	−0.48(2)

The function $G_{\rm SE}(Z\alpha)$ gives the higher-order contributions to the self energy. This function has been calculated precisely in the case $Z = 0$, $G_{\rm SE}(0) = A_{60}$, with the result [43]

$$A_{60} = \begin{cases} -30.928\,90(1) & 1S_{1/2} \\ -31.840\,47(1) & 2S_{1/2} \end{cases} . \tag{42}$$

For $Z = 1$ and 2 an extrapolation of a complete numerical calculation at higher Z gives the results in Table 1 [44, 45, 46, 47]. These results are shown in Fig. 2. It should be emphasized that the point shown at $Z = 0$ and the points with $Z \geq 1$ are the results of completely independent calculations. The point at $Z = 1$ includes the effect of higher-order binding corrections not contained in the power series result. Similar extrapolations, that take into account additional logarithmic terms, based on the same numerical calculations, have also been made [48, 49, 50].

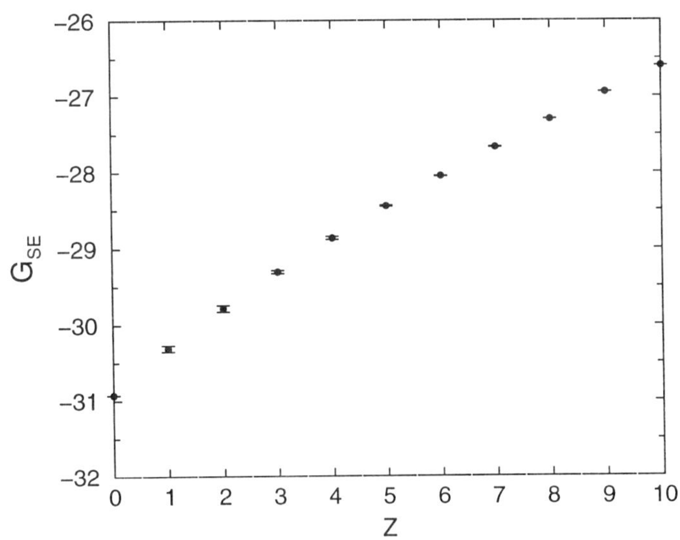

Fig. 2. Calculated values of $G_{\rm SE}(Z\alpha)$.

Table 2: The function $G_{\rm VP}^{(1)}(Z\alpha)$.

Z	$1S_{1/2}$	$2S_{1/2}$	$4S_{1/2}$	$2P_{1/2}$	$2P_{3/2}$
1	−0.6187	−0.8089	−0.8066	−0.0640	−0.0141
2	−0.6077	−0.7961	−0.7938	−0.0638	−0.0140

Vacuum Polarization: The vacuum polarization correction can be written as

$$E_{\rm VP}^{(2)} = \frac{\alpha}{\pi}\frac{(Z\alpha)^4}{n^3}H(Z\alpha)\, m_{\rm e}c^2 \,. \tag{43}$$

The function H can be divided into a Uehling potential part $H^{(1)}$ and the higher-order remainder $H^{(3)}+\cdots$, called the Wichmann-Kroll part

$$H(Z\alpha) = H^{(1)}(Z\alpha) + H^{(3)}(Z\alpha) + \cdots\,, \tag{44}$$

where the known lower-order terms are isolated as

$$\begin{aligned} H^{(1)}(Z\alpha) &= C_{40} + C_{50}(Z\alpha) + C_{61}(Z\alpha)^2\ln(Z\alpha)^{-2} \\ &\quad + G_{\rm VP}^{(1)}(Z\alpha)(Z\alpha)^2\,, \qquad (45)\\ H^{(3+)}(Z\alpha) &= G_{\rm VP}^{(3+)}(Z\alpha)(Z\alpha)^2\,, \qquad (46) \end{aligned}$$

with

$$\begin{aligned} C_{40} &= -\tfrac{4}{15}\,\delta_{l0}\,, \\ C_{50} &= \tfrac{5}{48}\pi\,\delta_{l0}\,, \\ C_{61} &= -\tfrac{2}{15}\,\delta_{l0}\,. \end{aligned} \tag{47}$$

The Uehling part, $G_{\rm VP}^{(1)}(Z\alpha)$, can be calculated numerically to any precision, with the result shown in Table 2 [51]. The Wichmann-Kroll part $G_{\rm VP}^{(3+)}(Z\alpha)$ is small, with the leading term given by [52, 53, 54]

$$G_{\rm VP}^{(3)}(0) = \left(\tfrac{19}{45} - \tfrac{1}{27}\pi^2\right)\delta_{l0}\,. \tag{48}$$

Higher-Order Radiative Corrections: Corrections from diagrams with two virtual photons are given by

$$E^{(4)} = \left(\frac{\alpha}{\pi}\right)^2\frac{(Z\alpha)^4}{n^3}m_{\rm e}c^2\left[B_{40} + B_{50}(Z\alpha) + \cdots\right]\,, \tag{49}$$

where [20, 21, 55, 56, 57, 58, 59]

$$
\begin{aligned}
B_{40} &= \left[2\pi^2 \ln 2 - \tfrac{49}{108}\pi^2 - \tfrac{6131}{1296} - 3\zeta(3)\right]\delta_{l0} \\
&\quad + \left[\tfrac{1}{2}\pi^2 \ln 2 - \tfrac{1}{12}\pi^2 - \tfrac{197}{144} - \tfrac{3}{4}\zeta(3)\right]\frac{1}{\kappa(2l+1)}\,,
\end{aligned} \tag{50}
$$

and [60, 61, 62, 63, 64, 65, 66, 67]

$$
B_{50} = -21.558(3)\delta_{l0}\,. \tag{51}
$$

The latter coefficient is the result of a series of recent calculations. The unexpectedly large value of B_{50}, $|B_{50}/B_{40}| \approx 40$ for S states, causes some theoretical predictions to shift beyond prior uncertainty estimates.

Reduced-Mass and Radiative-Recoil Corrections: The dominant reduced-mass correction is obtained by making the modification

$$
\begin{aligned}
&F(Z\alpha) + H(Z\alpha) \\
&\rightarrow \left(\frac{m_{\rm r}}{m_{\rm e}}\right)^3 \left\{\left[\frac{4}{3}\ln\left(\frac{m_{\rm e}}{m_{\rm r}(Z\alpha)^2}\right) - \frac{4}{15} + \frac{10}{9}\right]\delta_{l0}\right. \\
&\quad \left. - \frac{4}{3}\ln k_0(n,l)\right\} - \left(\frac{m_{\rm r}}{m_{\rm e}}\right)^2 \frac{1-\delta_{l0}}{2\kappa(2l+1)} \\
&\quad + \left(\frac{m_{\rm r}}{m_{\rm e}}\right)^3 \left(\frac{139}{32} + \frac{5}{48} - 2\ln 2\right)\delta_{l0}\pi(Z\alpha) \\
&\quad + \cdots\,,
\end{aligned} \tag{52}
$$

where $m_{\rm r}$ is the reduced mass defined above. Higher-order corrections, termed radiative recoil effects, of order

$$
\alpha \frac{(Z\alpha)^5 m_{\rm e}}{n^3 m_{\rm N}} m_{\rm e}c^2\,, \tag{53}
$$

are the subject of recent and current study [68, 69, 70].

Nuclear Finite Size: At low Z, the effect of the finite nuclear size on level energies is given by

$$
E_{\rm NS} = \frac{2}{3}\frac{(Z\alpha)^4}{n^3}\left(\frac{R}{\lambda\!\!\!^{-}_{\rm C}}\right)^2 \delta_{l0}\, m_{\rm e}c^2\,, \tag{54}
$$

where R is the root-mean-square (rms) charge radius of the nucleus, and $\lambda\!\!\!^{-}_{\rm C} = \alpha a_0$ is the Compton wavelength of the electron.

Comparison of Experiment and Theory: The validity of the theory discussed above is tested here by comparison with experiments. For the Lamb

Table 3: Theoretical contributions to the Lamb shift in hydrogen.

Contribution	Order (m_ec^2/h)	Value (MHz)
Self energy	$\alpha(Z\alpha)^4$	1085.817
Vacuum polarization	$\alpha(Z\alpha)^4$	−26.897
Fourth order	$\alpha^2(Z\alpha)^4$	0.064(4)
Reduced mass	$(m_e/m_N)\alpha(Z\alpha)^4$	−1.646(3)
Relativistic recoil	$(m_e/m_N)(Z\alpha)^5$	0.355
Nuclear size	$(Z\alpha)^4(R/\lambda\!\!\!^{-}_C)^2$	0.145(4)
Total		1057.839(6)

Table 4: Lamb shift in hydrogen.

Reference	$\mathcal{S}$ (MHz)
Newton Andrews & Unsworth [73]	1057.862(20)
Pal'chikov Sokolov & Yakovlev [74]	1057.8514(19)
Lundeen & Pipkin [75]	1057.845(9)
Hagley & Pipkin † [76]	1057.841(12)
Theory	1057.839(6)

† Indirect measurement

shift in hydrogen, the theoretical contributions are given in Table 3. In this table, $R = 0.862(12)$ fm [71]. An earlier measurement, $R = 0.805(11)$ [72], yields $\mathcal{S} = 1057.820(6)$ MHz.

Experimental values for the Lamb shift are listed in Table 4. The measurement labelled "indirect" refers to a measurment of the $2S_{1/2}{-}2P_{3/2}$ splitting that leads to the listed value via subtraction of a theoretical value for the fine-structure splitting.

Measurements of the 1S state radiative level shift are listed in Table 5. It is conventional to include in this "Lamb shift" all contributions to the energy level except the Dirac contribution in Eq. (31) and the leading reduced mass correction in Eq. (32).

It is also of interest to compare theory and experiment for the Lamb shift in He^+, as shown in Table 6. The theory of this quantity is strongly affected by the newly calculated coefficient in Eq. (51).

Recent determinations of the Rydberg constant have been made by groups working in Garching and in Paris. The deduced values are based on Eq. (3),

Table 5: $1S$ Lamb shift in hydrogen.

Reference	$L(1S)$ (MHz)
Nez *et al.* [77]	8172.815(70)
Weitz *et al.* [78]	8172.86(6)
Berkeland Hinds & Boshier [79]	8172.827(51)
Bourzeix *et al.* [80]	8172.798(46)
Theory	8172.80(5)

Table 6: Lamb shift in He^+.

Reference	$\mathcal{S}$ (MHz)
Dewey & Dunford [81]	14 042.0(1.2)
van Wijngaarden Kwela & Drake [82]	14 042.52(16)
Theory	14 041.3(2)

with theoretical estimates of f_{ij} by those authors based in large part on the discussion given above. The results are

$$R_\infty = 109\,737.315\,683\,6(18)\ \mathrm{cm}^{-1} \tag{55}$$

from measurement of the 2S-8S/8D transitions [83] and

$$R_\infty = 109\,737.315\,684\,9(30)\ \mathrm{cm}^{-1} \tag{56}$$

from measurement of the 1S-2S transition as well as Lamb shift measurements [84].

These results can be compared to the 1986 CODATA recommended value

$$R_\infty(\mathrm{CO86}) = 109\,737.315\,34(13)\ \mathrm{cm}^{-1}. \tag{57}$$

The uncertainty in Eq. (55) is 0.014 times the uncertainty of the CODATA value, and the difference between the current values and the CODATA value is 2.6 times the uncertainty of the latter value.

9 Future Adjustments

Beyond the 1997 adjustment of the fundamental constants, the possibility of more frequent adjustments is being considered. The World Wide Web as a resource for making the results available in a timely fashion makes this feasible, and some might argue mandatory. The present thinking is to have the new constants available in every-other odd-numbered year (every fourth year), or every odd-numbered year, if the availability of new data would seem to warrant it. With this in mind, the data handling procedures are being automated so that new information can be incorporated with no delay.

10 Bibliographic Database

A bibliographic database of publications relevant to the fundamental constants is being maintained at NIST by Barry N. Taylor and the present author. It presently contains about 1600 references and is available over the World Wide Web at:
http://physics.nist.gov/fundconbib.
Suggestions for papers to include are welcome and can be sent to:
constant@sed.nist.gov.

References

[1] R. T. Birge, Rev. Mod. Phys. **1**, 1 (1929).

[2] J. W. M. DuMond and E. R. Cohen, Phys. Rev. **82**, 555 (1951).

[3] J. A. Bearden and H. M. Watts, Phys. Rev. **81**, 73 (1951).

[4] J. A. Bearden and H. M. Watts, Phys. Rev. **81**, 160 (1951).

[5] J. W. M. DuMond and E. R. Cohen, Rev. Mod. Phys. **25**, 691 (1953).

[6] E. R. Cohen, J. W. M. DuMond, T. W. Layton, and J. S. Rollett, Rev. Mod. Phys. **27**, 363 (1955).

[7] J. A. Bearden and J. S. Thomsen, Nuovo Cimento Suppl. **5**, 267 (1957).

[8] E. R. Cohen and J. W. M. DuMond, Rev. Mod. Phys. **37**, 537 (1965).

[9] B. N. Taylor, W. H. Parker, and D. N. Langenberg, Rev. Mod. Phys. **41**, 375 (1969).

[10] E. R. Cohen and B. N. Taylor, J. Phys. Chem. Ref. Data **2**, 663 (1973).

[11] E. R. Cohen and B. N. Taylor, Rev. Mod. Phys. **59**, 1121 (1987).

[12] B. P. Kibble, I. A. Robinson, and J. H. Belliss, Metrologia **27**, 173 (1990).

[13] R. S. Van Dyck, Jr., P. B. Schwinberg, and H. G. Dehmelt, Phys. Rev. Lett. **59**, 26 (1987).

[14] M. R. Moldover *et al.*, J. Res. Natl. Bur. Stand. **93**, 85 (1988).

[15] B. N. Taylor and E. R. Cohen, J. Res. Natl. Inst. Stand. Technol. **95**, 497 (1990).

[16] P. T. Olsen *et al.*, IEEE Trans. Instrum. Meas. **38**, 238 (1989).

[17] B. N. Taylor, Metrologia **31**, 181 (1994).

[18] J. Schwinger, Phys. Rev. **73**, 416 (1948).

[19] R. Karplus and N. M. Kroll, Phys. Rev. **77**, 536 (1950).

[20] C. M. Sommerfield, Phys. Rev. **107**, 328 (1957).

[21] A. Petermann, Helv. Phys. Acta **30**, 407 (1957).

[22] G. S. Adkins, Phys. Rev. D **39**, 3798 (1989).

[23] T. Kinoshita and W. B. Lindquist, Phys. Rev. D **42**, 636 (1990).

[24] T. Kinoshita, IEEE Trans. Instrum. Meas. **44**, 498 (1995).

[25] T. Kinoshita, Phys. Rev. Lett. **75**, 4728 (1995).

[26] S. Laporta and E. Remiddi, Phys. Lett. B **265**, 182 (1991).

[27] S. Laporta and E. Remiddi, Phys. Lett. B **379**, 283 (1996).

[28] T. Kinoshita, Rep. Prog. Phys. **59**, 1459 (1996).

[29] B. E. Lautrup and E. de Rafael, Phys. Rev. **174**, 1835 (1968).

[30] S. Laporta and E. Remiddi, Phys. Lett. B **301**, 440 (1993).

[31] R. Jackiw and S. Weinberg, Phys. Rev. D **5**, 2396 (1972).

[32] A. Czarnecki, B. Krause, and W. J. Marciano, Phys. Rev. Lett. **76**, 3267 (1996).

[33] W. A. Barker and F. N. Glover, Phys. Rev. **99**, 317 (1955).

[34] J. R. Sapirstein and D. R. Yennie, in *Quantum Electrodynamics*, edited by T. Kinoshita (World Scientific, Singapore, 1990), Chap. 12, p. 560.

[35] E. E. Salpeter, Phys. Rev. **87**, 328 (1952).

[36] G. W. Erickson, J. Phys. Chem. Ref. Data **6**, 831 (1977).

[37] M. Doncheski, H. Grotch, and G. W. Erickson, Phys. Rev. A **43**, 2152 (1991).

[38] R. N. Fell, I. B. Khriplovich, A. I. Milstein, and A. S. Yelkhovsky, Phys. Lett. A **181**, 172 (1993).

[39] K. Pachucki and H. Grotch, Phys. Rev. A **51**, 1854 (1995).

[40] E. A. Golosov, A. S. Elkhovskiĭ, A. I. Mil'shteĭn, and I. B. Khriplovich, Zh. Eksp. Teor. Fiz. **107**, 393 (1995), [JETP **80**, 208 (1995)].

[41] G. W. Erickson and D. R. Yennie, Ann. Phys. (N.Y.) **35**, 271, 447 (1965).

[42] G. W. F. Drake and R. A. Swainson, Phys. Rev. A **41**, 1243 (1990).

[43] K. Pachucki, Ann. Phys. (N.Y.) **226**, 1 (1993).

[44] P. J. Mohr, Phys. Rev. Lett. **34**, 1050 (1975).

[45] P. J. Mohr, Phys. Rev. A **46**, 4421 (1992).

[46] P. J. Mohr and Y.-K. Kim, Phys. Rev. A **45**, 2727 (1992).

[47] P. J. Mohr, in *Atomic, Molecular, & Optical Physics Handbook*, edited by G. W. F. Drake (AIP, Woodbury, NY, 1996), Chap. 28, p. 341.

[48] S. G. Karshenboĭm, Zh. Eksp. Teor. Fiz. **106**, 414 (1994), [JETP **79**, 230 (1994)].

[49] S. G. Karshenboim, Yad. Fiz. **58**, 309 (1995), [Phys. At. Nucl. **58**, 262 (1995)].

[50] S. G. Karshenboim, Yad. Fiz. **58**, 707 (1995), [Phys. At. Nucl. **58**, 649 (1995)].

[51] P. J. Mohr, Phys. Rev. A **26**, 2338 (1982).

[52] E. H. Wichmann and N. M. Kroll, Phys. Rev. **101**, 843 (1956).

[53] P. J. Mohr, in *Beam-Foil Spectroscopy*, edited by I. A. Sellin and D. J. Pegg (Plenum Press, N.Y., 1976), Vol. 1, p. 89.

[54] P. J. Mohr, At. Data Nucl. Data Tables **29**, 453 (1983).

[55] B. E. Lautrup, A. Peterman, and E. de Rafael, Phys. Lett. B **31**, 577 (1970).

[56] R. Barbieri, J. A. Mignaco, and E. Remiddi, Lett. Nuovo Cimento **3**, 588 (1970).

[57] A. Peterman, Phys. Lett. B **35**, 325 (1971).

[58] M. Baranger, F. J. Dyson, and E. E. Salpeter, Phys. Rev. **88**, 680 (1952).

[59] G. Källén and A. Sabry, Dan. Mat. Fys. Medd. **29**, no. 17 (1955).

[60] M. I. Eides, H. Grotch, and D. A. Owen, Phys. Lett B **294**, 115 (1992).

[61] M. I. Eides and H. Grotch, Phys. Lett. B **301**, 127 (1993).

[62] M. I. Eides and H. Grotch, Phys. Lett. B **308**, 389 (1993).

[63] M. I. Eides, H. Grotch, and P. Pebler, Phys. Rev. A **50**, 144 (1994).

[64] K. Pachucki, Phys. Rev. A **48**, 2609 (1993).

[65] M. I. Eides, S. G. Karshenboim, and V. A. Shelyuto, Phys. Lett. B **312**, 358 (1993).

[66] K. Pachucki, Phys. Rev. Lett. **72**, 3154 (1994).

[67] M. I. Eides and V. A. Shelyuto, Pis'ma Zh. Eksp. Teor. Fiz. **61**, 465 (1995), [JETP Lett. **61**, 478 (1995)].

[68] G. Bhatt and H. Grotch, Phys. Rev. A **31**, 2794 (1985).

[69] G. Bhatt and H. Grotch, Phys. Rev. Lett. **58**, 471 (1987).

[70] G. C. Bhatt and H. Grotch, Ann. Phys. (N.Y.) **178**, 1 (1987).

[71] G. G. Simon, Ch. Schmitt, F. Borkowski, and V. H. Walther, Nucl. Phys. **A333**, 381 (1980).

[72] L. N. Hand, D. G. Miller, and R. Wilson, Rev. Mod. Phys. **35**, 335 (1963).

[73] G. Newton, D. A. Andrews, and P. J. Unsworth, Philos. Trans. R. Soc. London **290**, 373 (1979).

[74] V. G. Pal'chikov, Yu. L. Sokolov, and V. P. Yakovlev, Pis'ma Zh. Eksp. Teor. Fiz. **38**, 347 (1983), [JETP Lett. **38**, 418 (1983)].

[75] S. R. Lundeen and F. M. Pipkin, Metrologia **22**, 9 (1986).

[76] E. W. Hagley and F. M. Pipkin, Phys. Rev. Lett. **72**, 1172 (1994).

[77] F. Nez *et al.*, Europhys. Lett. **24**, 635 (1993).

[78] M. Weitz *et al.*, Phys. Rev. Lett. **72**, 328 (1994).

[79] D. J. Berkeland, E. A. Hinds, and M. G. Boshier, Phys. Rev. Lett. **75**, 2470 (1995).

[80] S. Bourzeix *et al.*, Phys. Rev. Lett. **76**, 384 (1996).

[81] M. S. Dewey and R. W. Dunford, Phys. Rev. Lett. **60**, 2014 (1988).

[82] A. van Wijngaarden, J. Kwela, and G. W. F. Drake, Phys. Rev. A **43**, 3325 (1991).

[83] S. Bourzeix *et al.*, in *Symposium on Frequency Standards and Metrology*, edited by J. C. Bergquist (World Scientific, Singapore, 1996), pp. 145–150.

[84] M. Weitz *et al.*, in *Symposium on Frequency Standards and Metrology*, edited by J. C. Bergquist (World Scientific, Singapore, 1996), pp. 137–144.

High Precision Spectroscopy of Positronium and Muonium

Vernon W. Hughes
Yale University, Physics Department
New Haven, CT 06520, U.S.A.

I had the pleasure, profit and honor of having Ingvar Lindgren spend his sabbatical year 1970-1 with us at Yale. In addition to doing his theoretical work on many-electron atoms, he joined our experiments to measure precisely the positronium ground state hyperfine structure interval and Zeeman effect and to search for its n=2 state. Hence I hope that the topic High Precision Spectroscopy of Positronium and Muonium is appropriate for my contribution to this volume on **"Modern Trends in Atomic Physics,"** in honor of Ingvar Lindgren.

Positronium and muonium are the simplest atoms composed of leptons and are of great importance for testing quantum electrodynamics and, more generally, the modern standard theory. In particular, these atoms are useful for studying bound state quantum electrodynamics and for determining the properties of the positron as antiparticle to the electron and of the muon as a "heavy" electron.

1. Positronium

Although the positron was discovered in 1932 during studies of cosmic rays with a bubble chamber(1), positronium (Ps) was not discovered until 1951 from studies of the behaviour of positrons from a radioactive source stopped in gases(2). The difference in the annihilation γ rays from the singlet $^1S_0$ state (2γ) and the triplet $^3S_1$ state (3γ) of Ps provided the key for its discovery.

The energy level diagram of the n=1 and n=2 states of positronium is shown in Fig. 1. By now the hyperfine structure interval (really of order fine structure) in the ground n=1 state (or the $^3S_1$ to $^1S_0$ interval), the fine structure intervals in the n=2 state, and the 1S-2S interval have been measured with high precision.

1.1 Hyperfine structure in n=1 state

The original measurement of the hyperfine structure interval $\Delta\nu$ in the n=1 state was done at MIT(3). More recently, two groups at Yale(4) and at Brandeis(5) have reported higher precision measurements of $\Delta\nu$. The experiments have not observed the $\Delta\nu$ transition directly but determine $\Delta\nu$ from the Zeeman transition shown in Fig. 2. The energies of the M=$\pm$1 states are independent of the magnetic field H because of the equality of the magnitudes of the magnetic moments of e^- and e^+ as required by CPT invariance. The energies of the M=0 triplet and M=0 singlet states have approximately a

ADVANCES IN QUANTUM CHEMISTRY, VOLUME 30

0065-3276/98 $25.00

quadratic dependence on H which involves $\Delta\nu$. The frequency of the indicated transition between M=±1 and M=0 triplet states is given by:

$$f_{01} = \frac{\Delta\nu}{2}\left[(1+x^2)^{1/2} - 1\right]$$

where $x = \frac{2\mu_B^e g'H}{h\Delta\nu}$ in which $g' = g_e(1 - \frac{5}{24}\alpha^2)$ is the g-value of the electron in positronium.

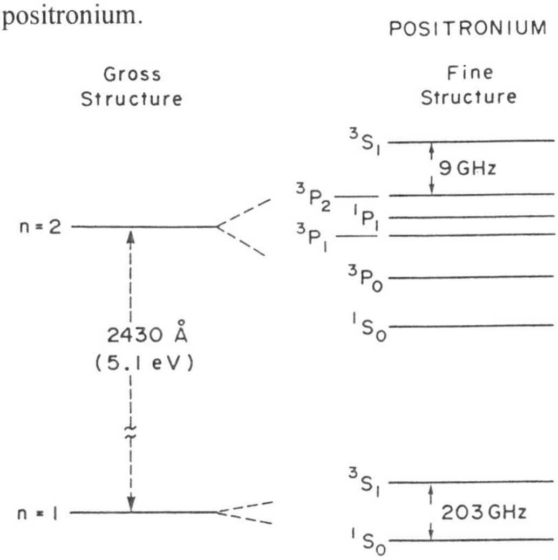

Figure 1. Energy levels of the n=1 and n=2 states of positronium.

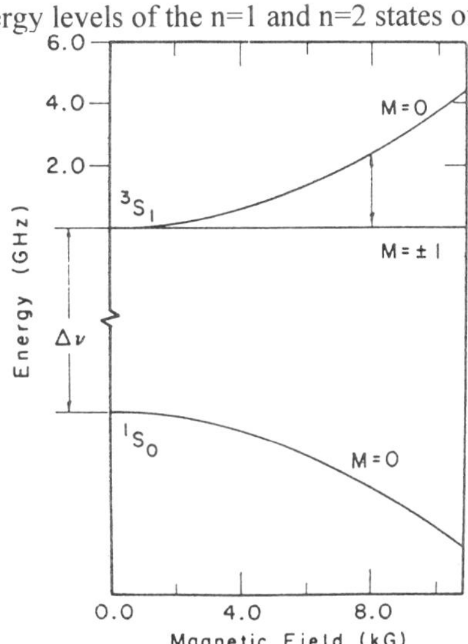

Figure 2. Zeeman energy levels of positronium in its ground state.

The principle of the resonance experiment rests on the unequal populations of the M=0 triplet and the M=±1 triplet states in a magnetic field due to the admixture of singlet state in the M=0 triplet state in a magnetic field H, and on the increase in two photon annihilation which results when transitions are induced from the M=±1 to the M=0 triplet states.

The experimental arrangement of the Yale experiment is shown in Fig. 3. A ^{22}Na positron source of about 15 mCi is placed inside a microwave cavity resonant in the TM_{110} mode at 2.323 GHz and filled with N_2 gas to a pressure between 0.25 and 3 atm. Eight NaI(Tl) detectors count in coincidence 0.5 MeV annihilation γ rays emitted at 180°. The magnetic field of about 8 kG is varied across the resonance line as indicated in Fig. 4. The signal is the increase in Ps(2γ) rate, and the linewidth is determined principally by the lifetime of the M=0 triplet state.

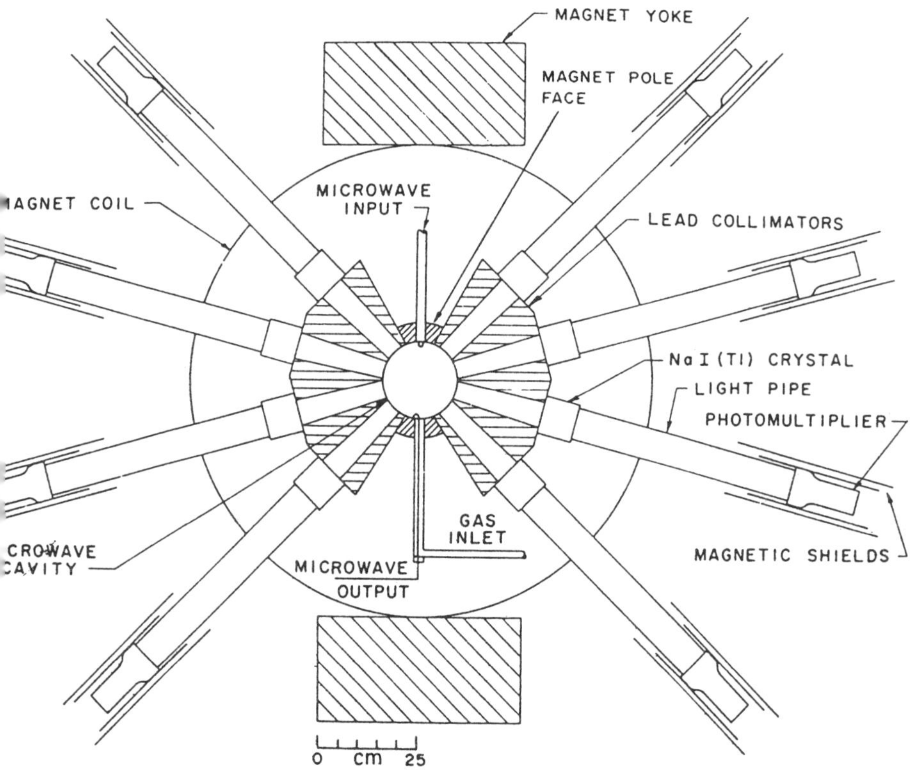

Figure 3. Schematic diagram of the experimental apparatus.

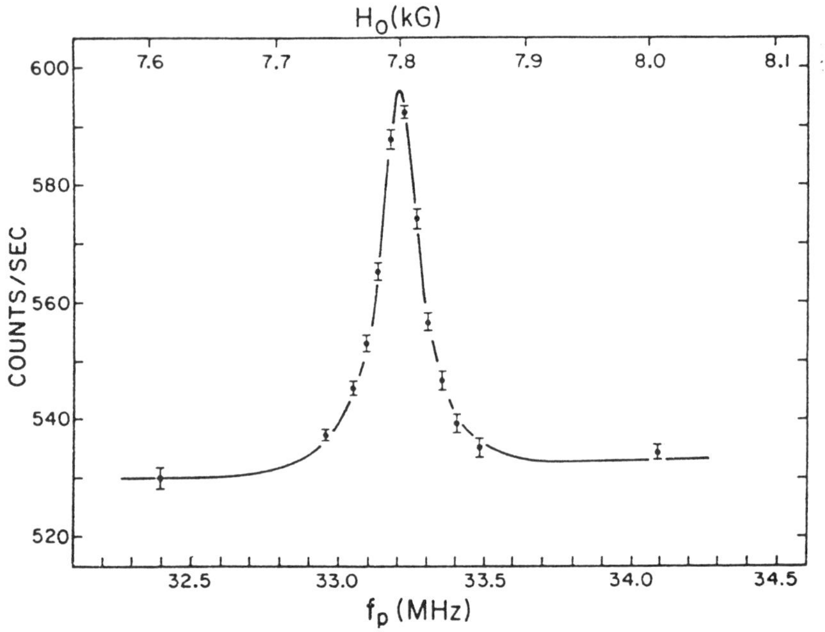

Figure 4. Typical observed resonance line, fit with the theoretical line shape. These data were taken with 5.5. lbs/in^2 (psi) of N_2.

Analysis of the resonance lines is based on the Hamiltonian for positronium in a magnetic field H_0+H_1 sin ωt in which H_0 is the static magnetic field and H_1 sin ωt is the time dependent microwave magnetic field of amplitude H_1 and angular frequency ω. The annihilation interaction has been treated by introducing decaying amplitudes for the states, and a correction has been made for the off-diagonal matrix elements of the annihilation interaction.

The values of Δν obtained from the Yale experiment and also that from the similar Brandeis experiment are:

$$\Delta\nu = 203\ 389.10 \pm 0.57 \pm 0.43 \text{ MHz } (3.6 \text{ ppm}) \quad (\text{Yale})$$

$$\Delta\nu = 203\ 387.5 \pm 1.6 \text{ MHz } (8 \text{ ppm}) \quad (\text{Brandeis})$$

in which the error of 1.6 MHz for the Brandeis Δν is the combined statistical and systematic error. These values are in good agreement and we combine them to obtain the current experimental value:

$$\Delta\nu_{exp} = 203\ 388.65 \pm 0.67 \text{ MHz } (3.3 \text{ ppm}).$$

The theoretical value for Δν has been calculated fully through order $\alpha^4 \ln\alpha^{-1} R_\infty$:

$$\Delta\nu = \alpha^2 R_\infty \left[\frac{2}{3} + \frac{1}{2} - \frac{\alpha}{\pi}\left(\ln 2 + \frac{16}{9} \right) + \frac{5}{12}\alpha^2 \ln\alpha^{-1} + K\alpha^2 \right]$$

Partial calculations have been made for K. The current theoretical value including all known contributions is(6)[1]

$$\Delta\nu_{th} = 203\ 404.5 \pm 10.0 \text{ MHz}$$

in which the dominant error is an estimate of the magnitude of uncalculated contributions to K. The experimental and theoretical values of $\Delta\nu$ are in reasonable agreement within the relatively large uncertainty assigned to $\Delta\nu_{th}$

$$\Delta\nu_{th} - \Delta\nu_{exp} = 15.8 \pm 10.0 \text{ MHz}.$$

As to the future, major efforts are underway (private communication from G. Adkins and J. Sapirstein) to calculate the full $\alpha^4 R_\infty$ term, and when this is accomplished the resulting theoretical uncertainty would be of order 1 ppm. This theoretical advance would justify an improved measurement of $\Delta\nu$.

A different and possibly feasible approach to an improved measurement of $\Delta\nu$ could be to study the direct $1^3S_1 \to 1^1S_0$ transition. The fractional linewidth in frequency of this transition is

$$\frac{\gamma_S}{2\pi\Delta\nu} \simeq 6 \times 10^{-3}$$ (in which γ_s is the annihilation rate of singlet positronium),

and indeed is the same as that of the transition studied between the $M=\pm1 \to M=0$ Zeeman levels in the triplet state. Hence determination of $\Delta\nu$ to 1 ppm requires determination of the center of the resonance line to 2 parts in 10^4. At the transition frequency of 203 GHz adequate microwave power sources of the EIO type are available. For a transition at magnetic field H=0 the microwave frequency must be swept through the resonance line while maintaining constant or known microwave power. Recently very promising data have been obtained in a precision measurement of $\Delta\nu$ for the n=1 state of muonium (see section 2.1 of this article) in which resonance lines were obtained by sweeping the microwave frequency with fixed H. The advantages of a measurement at H=0 obtained by sweeping the microwave frequency are that no knowledge of the magnetic field or of the Ps g-values is required, and no systematic errors associated with dependences of e^+ trajectories or detector efficiency on magnetic field are present.

An unusually high intensity cw e^+ radioactive source of ^{64}Cu (~100 Ci) was developed at BNL with their HEBR reactor(7) and was used to study e^+-atom scattering. This source would be very suitable for a measurement of Ps $\Delta\nu$.

A measurement involving line-narrowing by using long-lived Ps atoms could also be considered, as has been successfully done with muonium(8). A

---

[1] We use the same values of the fundamental constants used in Ref. 6, since use of more recent values does not change the value of $\Delta\nu_{th}$ significantly with respect to its estimated uncertainty.

pulsed high intensity, high repetition rate electron linac with energy in the range of 50 to 100 MeV could provide the e^+ source(9).

1.2 Fine structure and Lamb shift in n=2 state

Since the discovery of positronium, the fine structure and Lamb shift in the n=2 state were regarded as very interesting intervals to measure. During Lindgren's stay at Yale an attempt was made to produce the n=2 state by optical excitation from the n=1 state(10). The idea was to stop positrons in a gas (Ar) where n=1 state positronium would be formed in both the $^1S_0$ and $^3S_1$ states. Light at the 1S→2P transition wavelength of 2430 A would preferentially excite electric dipole transitions to the triplet states because of their longer lifetime. Positronium in the excited n=2 states would collide with Ar atoms and ^3P state Ps could be converted to ^1P Ps, thus increasing the fraction of Ps annihilating by 2 γ rays. The apparatus used is shown in Fig. 5. A statistically significant increase in 2γ annihilation was observed with a Sn arc which has a strong line at 2430 A, and this was interpreted as evidence for the production of n=2 state Ps.

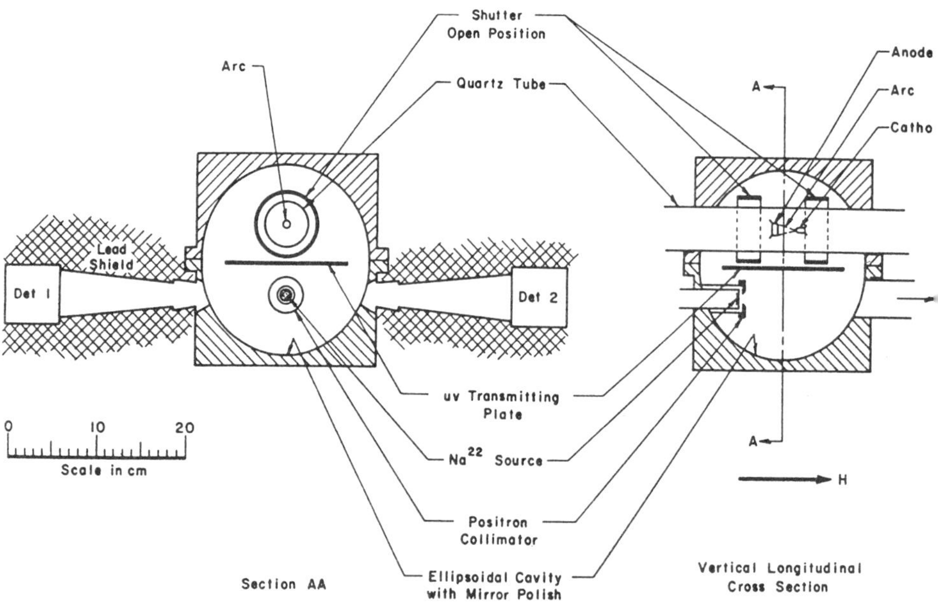

Figure 5. Schematic diagram of experimental arrangement to excite Ps from the n=1 to n=2 states.

However, the important breakthrough which has led to microwave spectroscopy measurements on the n=2 state was the discovery that Ps is formed when slow positrons collide with certain surfaces in vacuum(11). Indeed not only is n=1 state Ps produced but there is a small yield of 2S Ps-about 10^{-4} 2S Ps per incident e^+. With 2^3S_1 Ps available in vacuum, microwave transitions are induced to P states and the Lyman α 2430 A radiation is observed as the signal.

The setup for the first fine structure measurement is shown in Fig. 6(12). A slow e^+ beam strikes a wall of a microwave cavity. The microwave electric field induces transitions from 2^3S_1 to 2^3P_2. Lyman α photons are detected in coincidence with 1S annihilation γ rays as the microwave frequency is swept through the resonance line.

Subsequent experiments were done at Michigan(13). The most recent and accurate experiment, done at Mainz(14), employs a similar method and has measured additional intervals. The experimental setup is shown in Fig. 7, and in Fig. 8 a resonance line whose signal is the increase in Lyman-α counts.

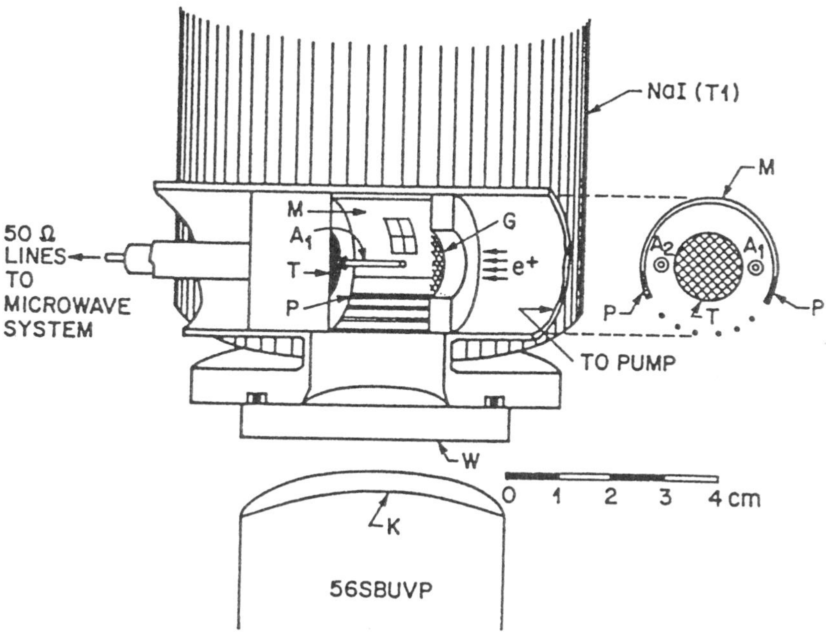

Figure 6: Positron target chamber and microwave cavity for measuring fine structure interval in the n=2 state of positronium. G, grid; T, copper target; M, mirror; W, window; K, CsTe photocathode of the uv photon detector; A, antennae; NaI(Tl), annihilation γ -ray detector.

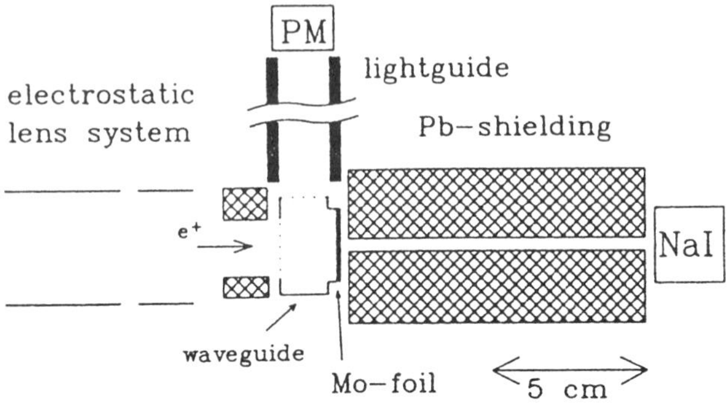

Figure 7: Interaction region of Ps atoms (n=2) and microwave radiation.

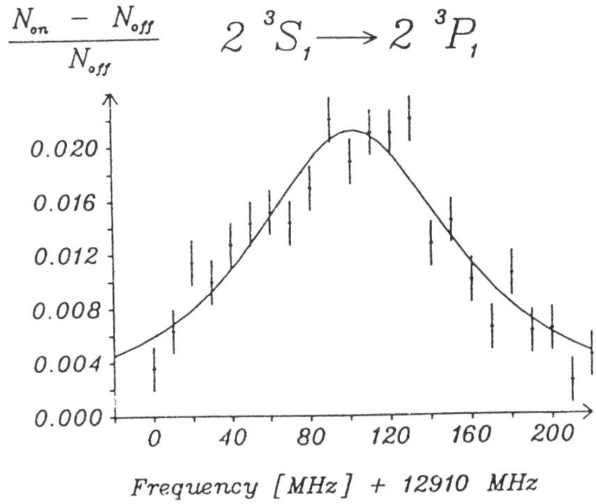

Figure 8: Example spectrum of a $2^2S_1 \rightarrow 2^3P_1$ transition. The center frequency is ν_1=13011.0±2.2 MHz. The experimental points are fitted by a Lorentzian line shape (microwave power = 4 mW/cm^2).

The results are given in Table I for the $2^3S_1 \rightarrow 2^3P_{0,1,2}$ transition frequencies. The first error is statistical and the second is systematic. The current theoretical values are also shown in Table I(6,15). The theoretical calculations have been evaluated through the term of order $m\alpha^6 \ln\alpha^{-1}$, but calculation of the $m\alpha^6$term has not been completed. The experimental and theoretical values are in good agreement.

Table I. Fine-structure transition in Ps (n=2). All frequencies are in MHz.

	Theory	Experiment
$2^3S_1 \rightarrow 2^3P_0$	18 496.98	18 499.65±1.20±4.00
$2^3S_1 \rightarrow 2^3P_1$	13 011.74	13 012.42±0.67±1.54
$2^3S_1 \rightarrow 2^3P_2$	8 626.09	8 624.38±0.54±1.40

1.3 The 1S-2S interval

The 1^3S_1-2^3S_1 interval in Ps is attractive for a precision measurement because its natural line width is determined solely by the annihilation rate of the triplet states and is small compared to the 1S→2S interval, so that the fractional natural linewidth is only $\frac{\gamma_t}{2\pi\nu} \sim 0.9 \times 10^{-9}$. As for the 1S-2S measurements in hydrogen the two-photon Doppler-free transition is used(16). Fig. 9 shows the scheme for Ps

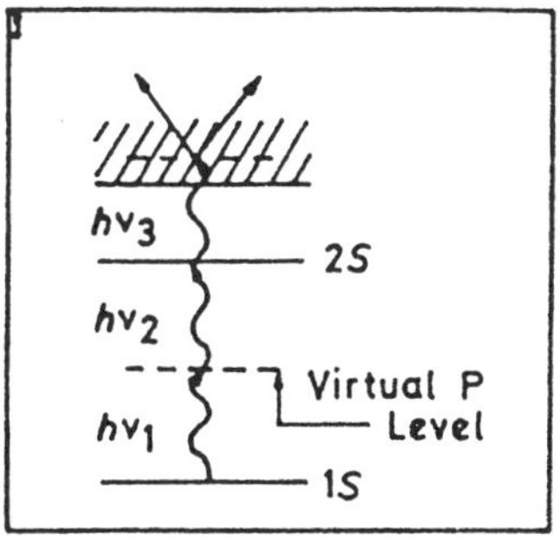

RESONANT 3 PHOTON IONIZATION

Figure 9: 1S-2S resonant three photon ionization of positronium.

where following the resonant two photon excitation from 1S to 2S, a third photon from the laser field photoionizes the atom and the e^+ is detected.

Since high laser power is necessary for the two photon transition, a pulsed laser is required and correspondingly a pulsed e^+ source is provided. Fig. 10 shows the experimental setup of the thermal positronium-laser beam interaction region. The pulsed e^+ source is formed starting from a ^{58}Co radioactive source and moderator. The slow e^+ are stored in a magnetic trap and then ejected in 10 ns pulses with an electric field. Ps was formed into vacuum when the e^+ struck a clean Al surface.

Two counter propagating 486 nm laser pulses excited the $1^3S_1 \rightarrow 2^3S_1$ transition. Ps atoms were ionized by the light and collected by an electron multiplier detector. Fig. 11 shows an observed resonance line with a simultaneously recorded Te_2 reference line and the frequency marker signal.

The experimental result is

$$\nu(1^3S_1\text{-}2^3S_1)_{exp} = 1\,233\,607\,218.9 \pm 10.7 \text{ MHz (12 ppb)}.$$

The error is a systematic error arising from ac Stark effect, second order Doppler shift, a frequency offset between the cw dye laser oscillator and the high-powered pulsed amplifier output and laser metrology. Substantial improvement in the accuracy of this measurement appears technically feasible.

The current theoretical value is(6,15)

$$\nu(1^2S_1\text{-}2^3S_1)_{th} = 1\,233\,607\,222.5 \pm 10 \text{ MHz}$$

where the uncertainty is an estimate of the value of the uncalculated α^4Ry terms. In view of the uncertainties, theoretical and experimental values are in reasonable agreement.

Two excellent reviews have been written on fundamental positronium research(17,18).

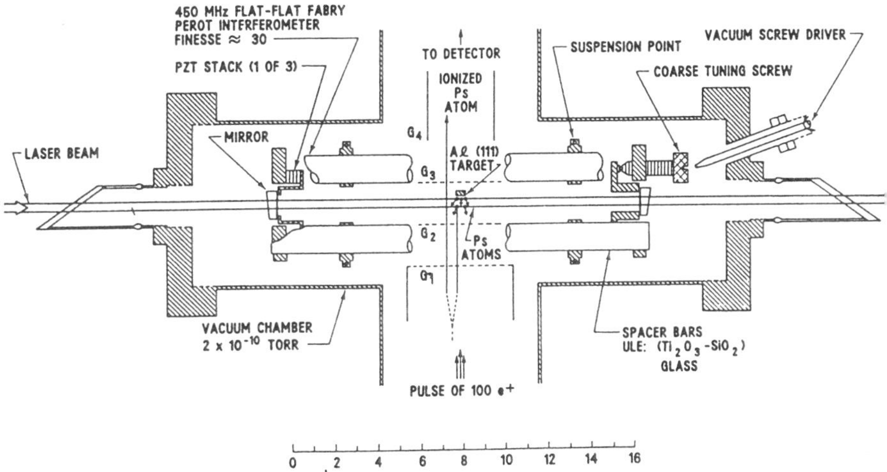

Figure 10: Thermal positronium-laser beam interaction region. Positronium is formed by a bunch of positrons that is stopped by a clean Al surface in ultrahigh vacuum. Positronium atoms thermally desorbed from the surface are ionized by the laser and the e^+ fragments are collected by a single particle detector. The laser pulse is narrowed in frequency by the Fabry-Perot interferometer.

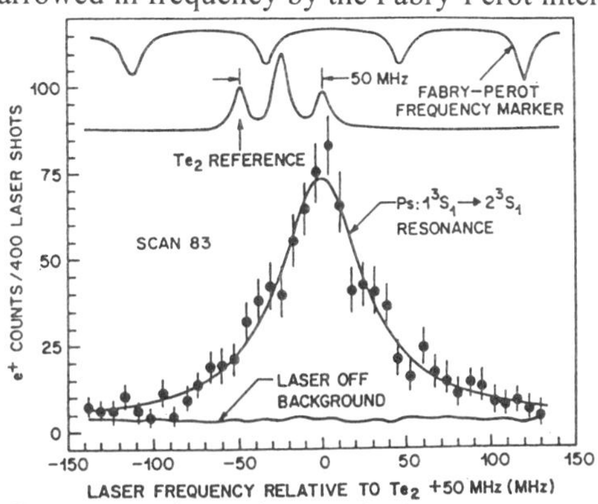

Figure 11: Resonant three-photon ionization of positronium due to $1^3S_1+3h\nu\rightarrow 2^3S_1+h\nu\rightarrow e^++e^-$. The Te_2 reference line has been split into three lines by acousto-optically modulating the cw dye laser at 50 MHz. For this scan, the line center was 25.9±2.7 MHz above the Te_2 line.

2. Muonium

The muon was discovered in cosmic radiation in 1937(19) and muonium was first observed in 1960(20) in a precession experiment based on parity nonconservation in the $\pi \rightarrow \mu \rightarrow e$ decay chain.

The energy level diagram of the n=1 and n=2 states of muonium is shown in Fig. 12 with the corresponding and very similar diagram for H for comparison. From the viewpoint of tests of fundamental theory muonium has the advantage of being free of hadronic effects, apart from the very small and well known modification of the photon propagator associated with virtual hadronic processes. The energy intervals shown can be measured for both H and M, but for M the measurements are more difficult, primarily because fewer atoms are available, and hence the precision of the measurements is lower at present.

2.1 Hyperfine structure and Zeeman effect in the n=1 state

The general method of the experiment to measure ground n=1 state energy levels is microwave magnetic resonance spectroscopy as applied to muonium(21,22). It relies on parity nonconservation in the $\pi^+ \rightarrow \mu^+ \nu_\mu$ decay to produce polarized μ^+ and in the $\mu^+ \rightarrow e^+ \bar{\nu}_\mu \nu_e$ decay to indicate the spin direction of μ^+. The most recent LAMPF completed experiment provides the present experimental values(23).

$\Delta\nu_{exp} = 4\ 463\ 302.88\ (16)$ kHz (36 ppb); $\mu_\mu/\mu_p = 3.183\ 346\ 1\ (11)$ (360 ppb)

The theoretical expression of $\Delta\nu$ can be written as(6):

$$\Delta\nu_{th} = \Delta\nu(\text{binding,rad}) + \Delta\nu(\text{recoil}) + \Delta\nu(\text{rad-recoil}) + \Delta\nu(\text{weak})$$

Except for the small term $\Delta\nu(\text{weak})$ coming from the weak interaction, and a small known contribution arising from hadronic contributions to the photon propagator, this expression arises solely from the electromagnetic interaction of two point-like leptons of different masses in their bound state. The present theoretical value is (24,25)

$$\Delta\nu_{th} = 4\ 463\ 302.38(1.34)(0.04)(0.17)\ \text{kHz}\ (0.3\ \text{ppm})$$

The principal error of 1.34 kHz arises from the uncertainty of μ_μ/μ_p. The second uncertainty arises from that in α based on the electron g-2 experiment(26), and the third is the estimate of the theoretical error in the latest QED calculation(24). Weak neutral current effects associated with Z exchange in the e - μ interaction contribute -0.065 kHz or 15 ppb and are included in $\Delta\nu_{th}$. The experimental and theoretical values for $\Delta\nu$ agree well:

$$\Delta\nu_{th} - \Delta\nu_{exp} = -0.50(1.4)\ \text{kHz:}\quad \frac{\Delta\nu_{th} - \Delta\nu_{exp}}{\Delta\nu_{exp}} = -(0.11 \pm 0.3)\ \text{ppm}.$$

The Breit-Rabi energy level diagram for ground state muonium is shown in Fig. 13. The history of the various determinations of $\Delta\nu$ and of μ_μ/μ_p is shown in Fig. 14.

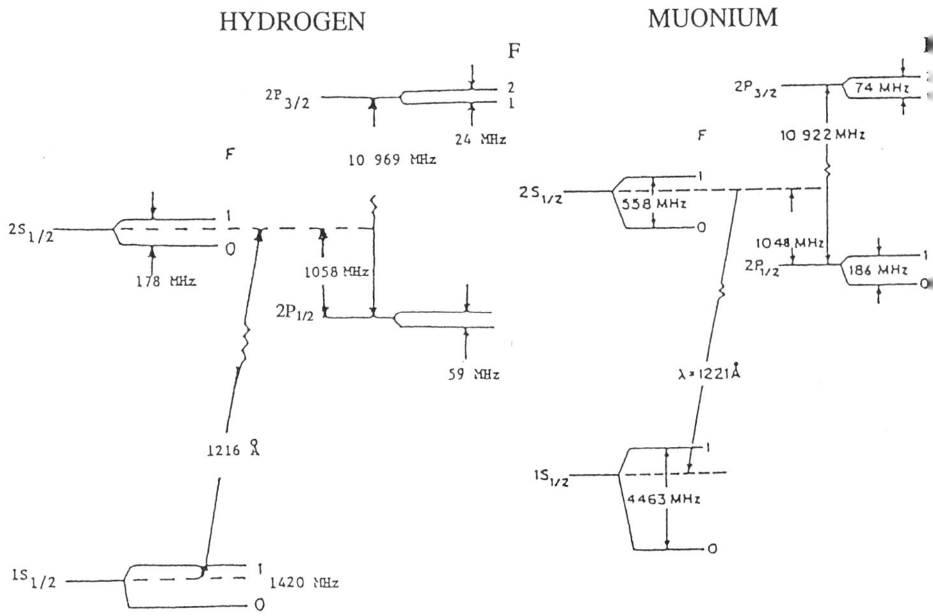

Figure 12: Energy levels in the n=1 and n=2 states.

The new experiment(27) improves on the earlier one(23) in several ways. First, the μ^+ beam intensity is now larger by a factor of 3 and is 1×10^7 μ^+/s, with a duty factory of 6 to 9% and has greater purity, achieved principally by use of an $\vec{E}x\vec{B}$ separator to reduce the e^+ background in the beam. Second, the magnetic field from a commercial Magnetic Resonance Imaging superconducting magnet system operating in persistent mode provides a field with a homogeneity of better than 1 ppm over the active region of the microwave cavity and with a stability of 0.01 ppm to 0.1 ppm/hr. Third, an electrostatic chopper in the muon beam line provides a muon beam with an on-period of 4 μs and an off-period of 10 μs, which allowed the observation of a resonance line from muonium atoms which had lived longer than the 2.2 μs mean lifetime τ_μ of muons. Such resonance lines can be narrower than the natural linewidth determined by τ_μ, and indeed narrower than a line obtained by the conventional method by factors of up to 3.

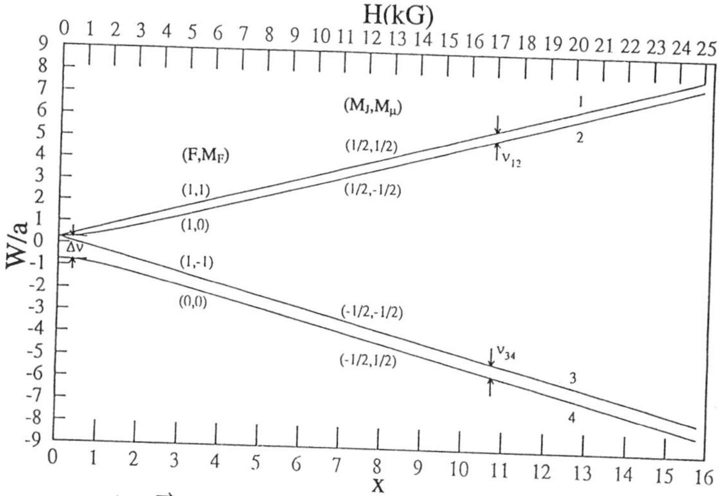

$$H = a\,\vec{I}\cdot\vec{J} + \mu_B^e g_J\,\vec{J}\cdot\vec{H} - \mu_B^\mu g'_\mu\,\vec{I}\cdot\vec{H} \qquad \Delta\nu = \nu_{12} + \nu_{34}$$

$$x = (g_J\mu_B^e + g'_\mu\mu_B^\mu)\,H/(h\Delta\nu) \qquad \nu_{34} - \nu_{12} = \frac{2\mu_B^\mu g'_\mu H}{h} + \Delta\nu\,[-x + (1+x^2)^{1/2}]$$

$$a/h = \Delta\nu = \Delta W/h \approx 4463\ \text{MHz}$$

$$\vec{F} = \vec{J} + \vec{I}$$

$$W_{F=\frac{1}{2}\pm\frac{1}{2},M_F} = -\frac{1}{4}\Delta W - \mu_B^\mu g'_\mu M_F H \pm \frac{1}{2}\Delta W\sqrt{1 + 2M_F x + x^2}$$

Figure 13. Breit-Rabi energy level diagram for muonium in its $1^2S_{1/2}$ ground state in a magnetic a field.

The experimental setup is shown in Fig. 15. The longitudinally polarized μ^+ beam of about 26 MeV/c is stopped in a microwave cavity contained in a pressure vessel filled with krypton at a pressure between 0.5 and 1.5 atm. The entire apparatus is in a region of a strong magnetic field of 1.7 T. Muonium is formed in an electron capture reaction between μ^+ and a Kr atom. The e^+ from μ^+ decay is detected in a scintillator telescope. Figure 16 shows the MRI solenoid and indicates the characteristics of this precision magnet which is operated in persistent mode. Resonance lines for transitions ν_{12} and ν_{34} are observed by sweeping the magnetic field with fixed microwave frequency.

The conditions and lineshapes for the conventional and "old muonium" resonances are given in Fig. 17. Observed resonance curves are shown in Fig. 18. It can be seen that a narrower resonance line is obtained for muonium atoms which have lived a longer time. Also wings develop for cases where older muonium atoms are observed, and larger signal amplitudes are obtained.

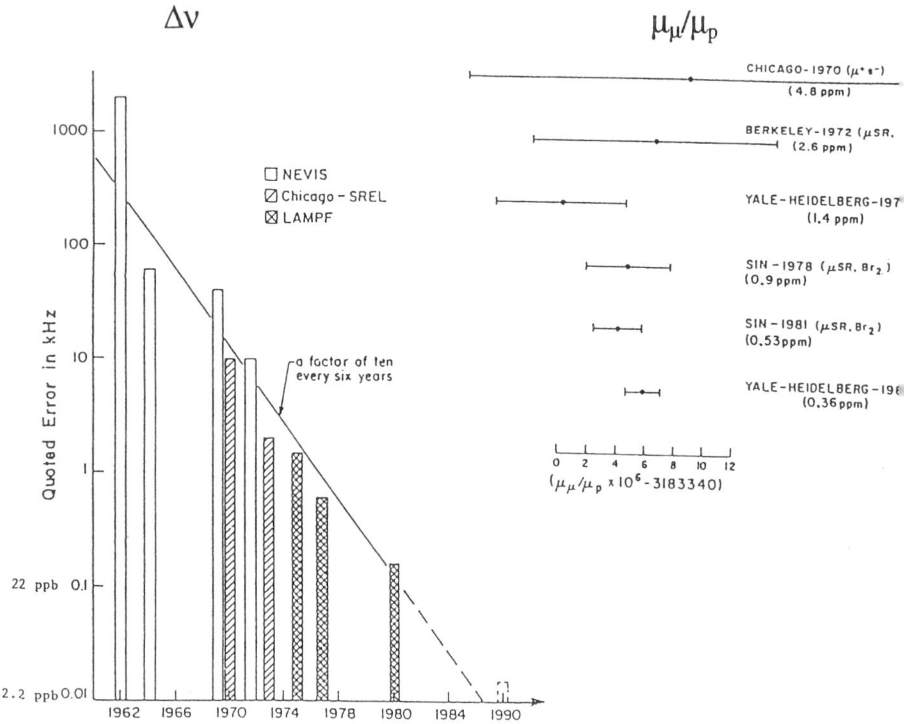

Figure 14. History of muonium $\Delta\nu$ and μ_μ/μ_p measurements.

During runs in 1994 and 1995 with a total beam time of about 3 mos., data were taken with stopping Kr gas pressures of 0.5, 1.0 and 1.5 atm. Some 2,000 sweeps of the resonance line were obtained by varying the microwave frequency with fixed magnetic field. A set of lines obtained in 1996 by varying the microwave frequency with fixed magnetic field is shown in Fig. 19. The relative microwave power was measured as the frequency was varied. About 500 such resonance lines were obtained in 1996. These data should provide an important check of 1994-95 data for systematic errors(28). When all of the data of the current experiment (1994-1996) have been analyzed, it is expected that a substantial increase in precision for $\Delta\nu$ and for μ_μ/μ_p will be obtained.

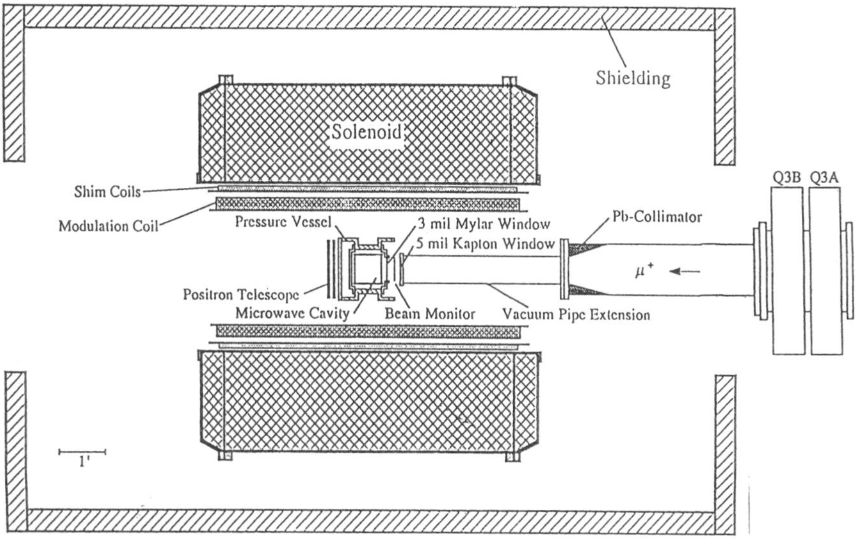

Figure 15. Experimental setup in 1994-1996.

Figure 16. LAMPF superconducting solenoid 2T MRI magnet.

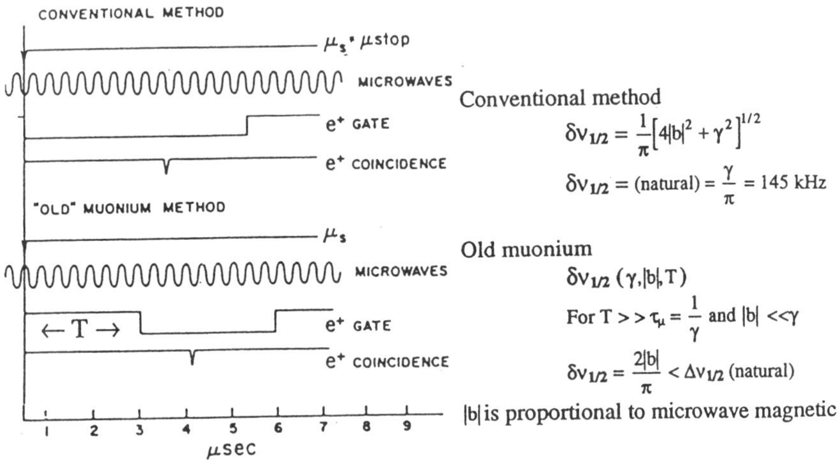

Figure 17. The conventional and old muonium methods of observing resonan lines and the associated linewidths.

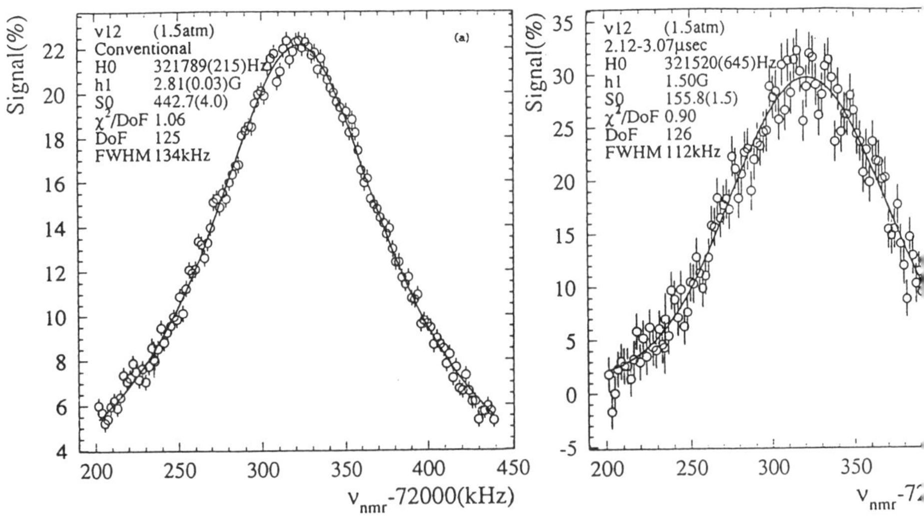

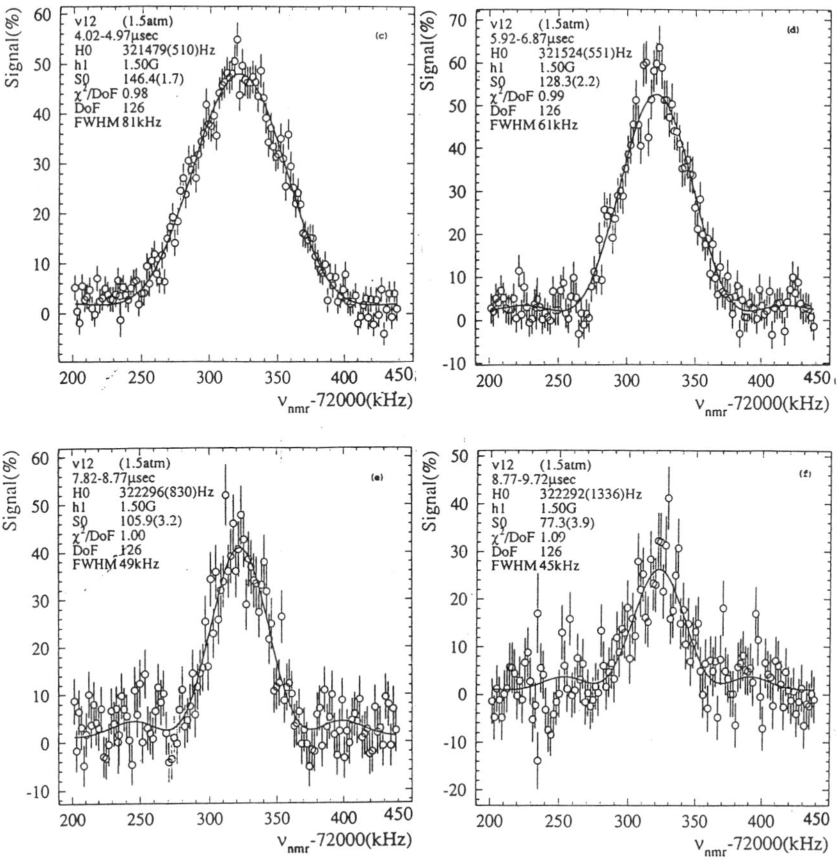

Figure 18. Conventional and old muonum resonance lines. Observed ν_{12} resonance lines obtained by a magnetic field scan with fixed microwave frequency for a 1.5 atm Kr target. The signal $= (N^{e^+}_{on} - N^{e^+}_{off}) / N^{e^+}_{off}$ [%] in which $N^{e^+}_{on}$ is the number of decay e^+ counts with microwaves on. The magnetic field is measured by the proton nmr frequency ν_{nmr}. Solid curves are fits to the theoretical line shape.

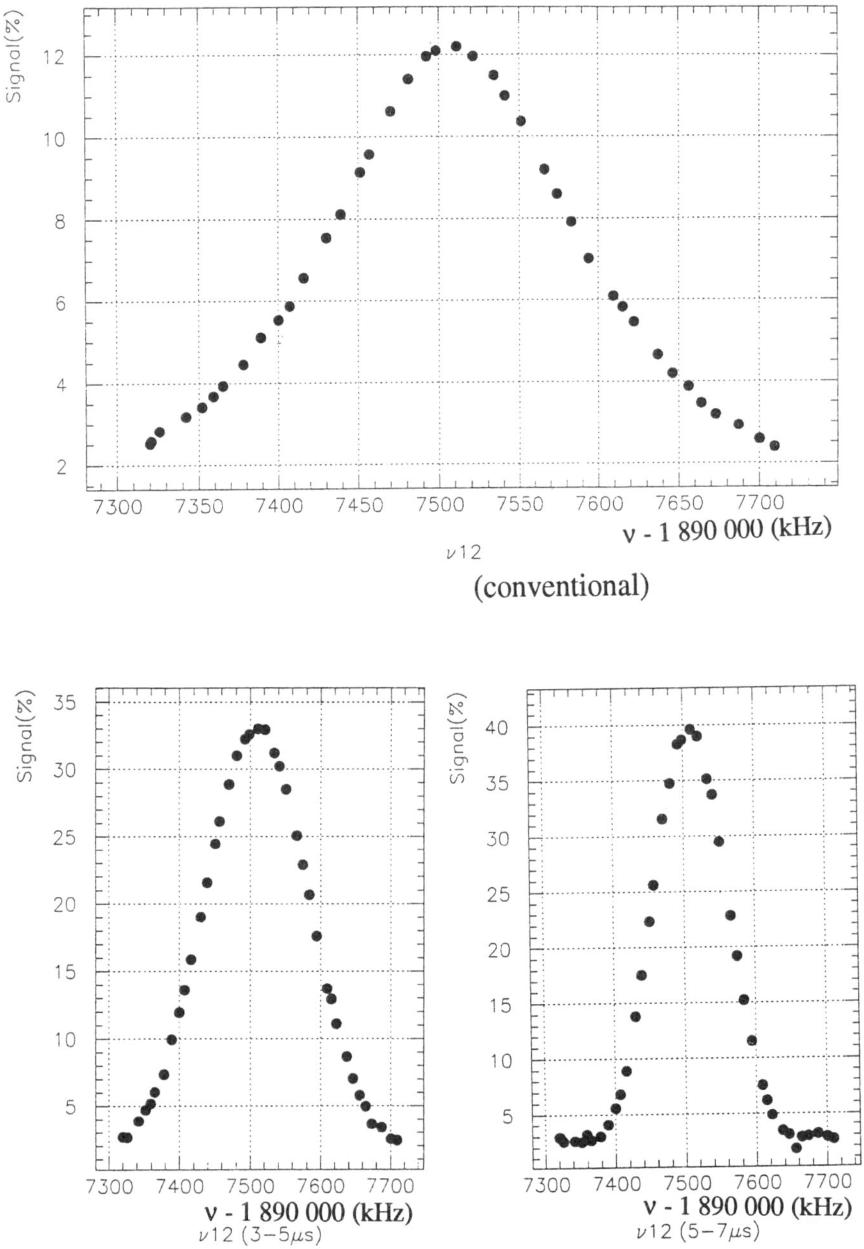
Signal(%)
ν - 1 890 000 (kHz)
ν12
(conventional)
Signal(%)
ν - 1 890 000 (kHz)
ν12 (3–5μs)
Signal(%)
ν - 1 890 000 (kHz)
ν12 (5–7μs)

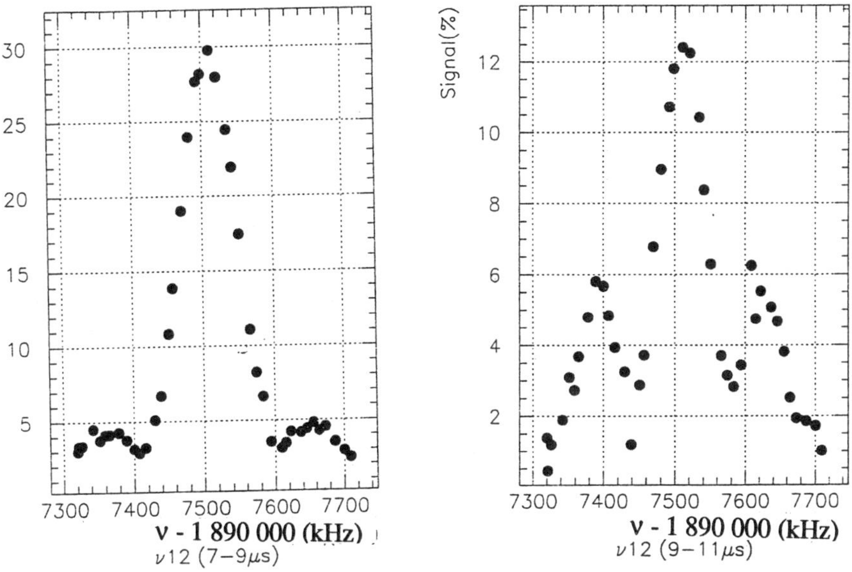

Figure 19. Conventional and old muonum resonance lines. Observed ν_{12} resonance lines obtained by a microwave frequency scan with a fixed magnetic field for a 1.5 atm Kr target. The signal $=(N_{on}^{e^+} - N_{off}^{e^+}) / N_{off}^{e^+}$ [%] in which $N_{on}^{e^+}$ is the number of decay e^+ counts with microwaves on.

The scientific importance of a more precise determination of $\Delta\nu$ and μ_μ/μ_p includes the following:

1. Precise test of QED for the two lepton bound state and of the behavior of the muon as a heavy electron.

2. Precise determination of the fundamental constants: muon magnetic moment and muon mass. Also a value of the fine structure constant α accurate to about 30 ppb could be obtained; this accuracy is comparable to that from condensed matter determinations.

3. Precise value of m_μ and μ_μ are very important for the BNL muon g-2 experiment(29) and in the determination of the muon neutrino mass m_{μ_ν} (30).

2.3 Lamb shift and fine structure in the n=2 state

The Lamb shift and the fine structure of the n=2 state of muonium has been measured by microwave spectroscopy in experiments at LAMPF(31) and TRIUMF(32) very similar to the classical measurement of the Lamb shift in hydrogen. However, the muonium experiments suffer from a dramatically smaller number of atoms--a factor of at least 10^{10}, and consequentially the accuracy achieved is much poorer.

The experimental setup for the LAMPF experiment is shown in Fig. 20(31). The muon beam is measured with a scintillation counter and then passes through an Al foil where electron capture leads to a small fraction of the beam being in the 2S metastable state. An rf transition to the 2P state is induced in the rf coaxial line and is observed as a decrease in the M(2S) beam as measured from the electric field quenching and accompanying Lyman α photon. A resonance curve is shown in Fig. 21 where the large statistical errors are evident. Table II shows values of the measured intervals, together with the theoretical values(6). Theory and experiment agree within the stated errors. The theoretical value of the Lamb shift in muonium unlike in hydrogen has no uncertainty associated with muon structure and hence is a better atom than H for testing QED. Recoil corrections are relatively larger than for H and are of order 10^{-3}.

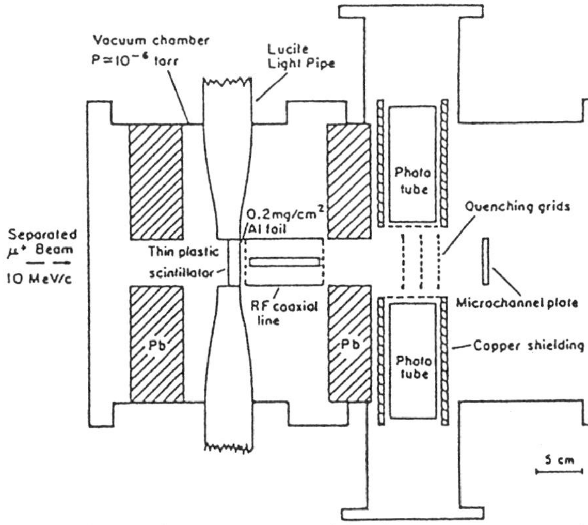

Figure 20: Experimental apparatus used in the LAMPF muonium Lamb shift experiment.

Table IIa. Theoretical value of muonium Lamb shift for n=2 state.

Correction	Order (mc^2)	Value (MHz)
Self-energy	$\alpha(Z\alpha)^4[\ln(Z\alpha)^{-2},1,Z\alpha,...]$	1085.812
Vacuum polarization	$\alpha(Z\alpha)^4(1,Z\alpha,...)$	-26.897
Fourth order	$\alpha^2(Z\alpha)^4$	0.102
Reduced mass	$\alpha(Z\alpha)^4(m_e/m_\mu)[\ln(Z\alpha)^{-2},1]$	-14.493
Relativistic recoil	$(Z\alpha)^5(m_{ee}/m_\mu)[\ln(Z\alpha)^{-2},1,Z\alpha]$	3.159
Higher-order recoil	$(Z\alpha)^4(m_e/m_\mu)^2$	-0.171
Radiative recoil	$\alpha(Z\alpha)^5 m_e/m_\mu$	-0.022
	Total	1047.490(300)

Table IIb. Experimental values of Lamb shift and fine structure.

	Experimental Values
$2^2S_{1/2}$-$2^2P_{1/2}$	(1054 ± 22) MHz, LAMPF
	$(1070\,^{+12}_{-15})$ MHz, TRIUM
$2^2P_{3/2}$-$2^2P_{1/2}$	$(10\,943\,^{+35}_{-30})$ MHz LAMPF

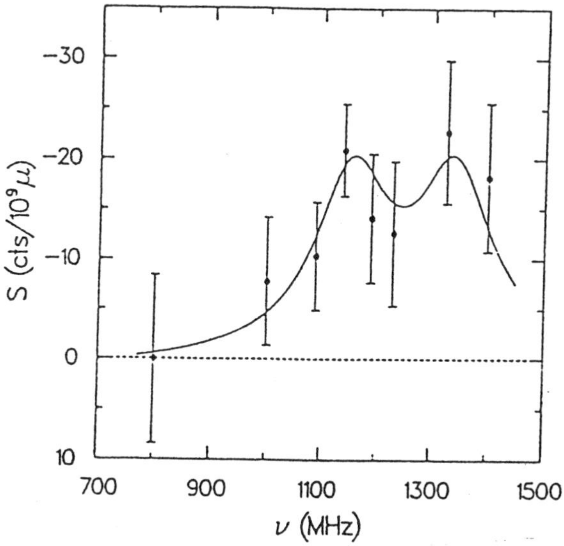

Figure 21: Microwave resonance data for n=2 Lamb shift measurement, and the best fit line shape.

2.3 The 1S-2S transition

The 1S-2S transition in muonium has also been measured by laser spectroscopy. The transition is induced by a two-photon Doppler-free process and detected through the subsequent photoionization of the 2S state in the laser field. The key to success in this experiment was the production of muonium into vacuum from the surface of heated W or of $Si0_2$ powder. The discovery experiment(33) was done at the KEK facility in Japan with a pulsed muon beam and an intense pulsed laser system. A subsequent experiment(34) done with the pulsed μ^+ beam at RAL and a similar pulsed laser has improved the signal substantially and has achieved a a precision of about 10^{-8} in the 1S→2S interval, thus determining the Lamb shift in the 1S state to about 1% accuracy (Fig. 22). The precision of this experiment should be greatly improved in a new experiment now underway at RAL. This experiment will provide a precise determination of the muon mass from the muonium-hydrogen isotope shift in the transition since this shift is primarily due to reduced mass.

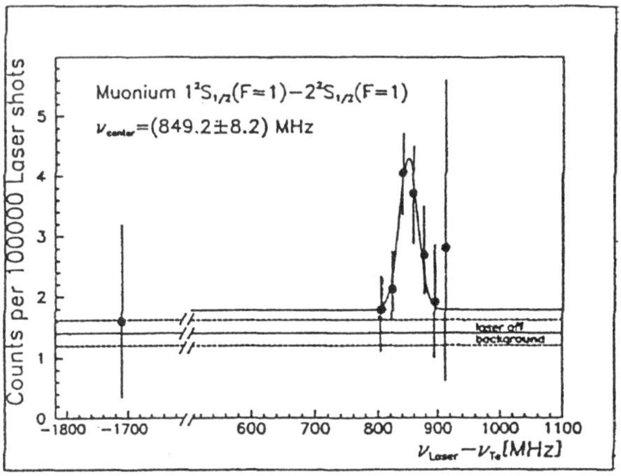

Transition Frequency

$$M(1S\text{-}2S)_{EXPT} = 2,455{,}529{,}002(33)(46)\ \text{MHz}$$
$$M(1S\text{-}2S)_{QED} = 2,455{,}528{,}934(3.6)\ \text{MHz}$$

Lamb shifts

$$[L_{1S} - L_{2S}]_{EXPT} = 6{,}988(33)(46)\ \text{MHz}$$
$$[L_{1S} - L_{2S}]_{QED} = 7{,}056.1(1.0)\ \text{MHz}$$

agrees with QED at the 0.8% level.

Figure 22: Observed 1S-2S transition in muonium by 2 photon laser spectroscopy.

References

(1) C.D. Anderson, Science **LXXVI**, 238 (1932); Phys. Rev. **XLIII**, 491 (1933).

(2) M. Deutsch, Phys. Rev. **82**, 455 (1951); M. Deutsch and E. Dulit, Phys. Rev. **84**, 601 (1951).

(3) M. Deutsch and S.C. Brown, Phys. Rev. **85**, 1047 (1952).

(4) E.R. Carlson, V.W. Hughes and I. Lindgren, Phys. Rev. **A15**, 241 (1977); P.O. Egan, V.W. Hughes and M.H. Yam, Phys. Rev. **A15**, 251 (1977); M.W. Ritter, P.O. Egan, V.W. Hughes and K.A. Woodle, Phys. Rev. **A30**, 1331 (1984).

(5) A.P. Mills and G.H. Bearman, Phys. Rev. Lett. **34**, 246 (1975); A.P. Mills, Phys. Rev. **A27**, 262 (1983).

(6) J.R. Sapirstein and D.R. Yennie, in **Quantum Electrodynamics** ed. by T. Kinoshita (World Scientific, Singapore, 1990) p. 560.

(7) K.G. Lynn, M. Weber, L.O. Roellig, A.P. Mills, Jr. and A.R. Moodenbaugh in **Atomic Physics with Positrons** ed. by J.W. Humberston and E.A. Armour (Plenum, New York, 1987) p. 161; M. Weber, A. Schwab, D. Becker and K.G. Lynn, Hyperfine Interactions **73**, 147 (1992).

(8) M.G. Boshier, S. Dhawan, X. Fei, V.W. Hughes, M. Janousch, K. Jungmann, W. Liu, C.R. Prigl, G. zu Putlitz, I. Reinhard, W. Schwarz, P.A. Souder, O. van Dyck, X. Wang, K.A. Woodle and Q. Xu, Phys. Rev. **A52**, 1948 (1995).

(9) R.H. Howell, R.A. Alverez, and M. Stanek, Appl. Phys. Lett. **40**, 751 (1982); R.H. Howell, R.A. Alverez, K.A. Woodle, S. Dhawan, P.O. Egan, V.W. Hughes and M.W. Ritter, IEEE Trans, Nucl. Sci. **NS-30**, 1438 (1983).

(10) S.L. Varghese, E.S. Ensberg, V.W. Hughes and I. Lindgren, Phys. Lett. **49A**, 415 (1974).

(11) K.F. Canter, A. Mills, Jr., and S. Berko, Phys. Rev. Lett. **33**, 7 (1974).

(12) A.P. Mills, Jr., S. Berko and K.F. Canter, Phys. Rev. Lett. **34**, 1541 (1975).

(13) S. Hatamian, R.S. Conti and A. Rich, Phys. Rev. Lett. **58**, 1833 (1987).

(14) D. Hagena, R. Ley, D. Weil, G. Werth, W. Arnold and H. Schneider, Phys. Rev. Lett. **71**, 2887 (1993).

(15) R.N. Fell, Phys. Rev. Lett. **68**, 25 (1992); R.N. Fell, Phys. Rev. A48, 2634 (1993).

(16) S. Chu and A.P. Mills, Jr., Phys. Rev. Lett. **48**, 1333 (1982), S. Chu, A.P. Mills, Jr. and J.L. Hall, Phys. Rev. Lett. **52**, 1689 (1984).

(17) A. Rich, Rev. Mod. Phys. **53**, 127 (1981).

(18) A.P. Mills, Jr. and S. Chu, in **Quantum Electrodynamics** ed. by T. Kinoshita, (World Scientific, Singapore, 1990) p.774.

(19) C.D. Anderson and S.H. Neddermeyer, Phys. Rev. **51**, 884 (1937); Phys. Rev. **54**, 88 (1938); C. Street and E. Stevenson, Phys. Rev. **51**, 1005 (1937).

(20) V.W. Hughes, D.W. McColm, K. Ziock, and R. Prepost, Phys. Rev. Lett. **5**, 63 (1960); V.W. Hughes, D.W. McColm, K. Ziock and R. Prepost, Phys. Rev. **A1**, 595 (1970).

(21) V.W. Hughes, Ann. Rev. Nucl. Sci. **16**, 445 (1966).

(22) V.W. Hughes and G. zu Putlitz, in **Quantum Electrodynamics** ed by T. Kinoshita, (World Scientific, Singapore, 1990) p. 822.

(23) F.G. Mariam , W. Beer, P.R. Bolton, P.O. Egan, C.J. Gardner, V.W. Hughes, D.C. Lu, P.A. Souder, H. Orth, J. Vetter, U. Moser and G. zu Putlitz, Phys. Rev. Lett. **49**, 993 (1982).

(24) T. Kinoshita and M. Nio, Phys. Rev. Lett. **72**, 3803 (1994); Phys. Rev. **D53**, 4909 (1996).

(25) S.G. Karshenboim, Z. Phys. **D36**, 11 (1996).

(26) T. Kinoshita, CLNS 96/1418 (1996).

(27) M.G. Boshier, D.E. Ciskowski, S. Dhawan, X. Fei, V.W. Hughes, M. Janousch, K. Jungmann, W. Liu, C. Pillai, R. Prigl, G. zu Putlitz, W. Schwarz, P.A. Souder, O. van Dyck, X. Wang, K.A. Woodle, and Q. Xu, **Proceedings of the International Workshop on Low Energy Muon Science - LEMS 93** LA-12698-C ed. by M. Leon (Los Alamos National Laboratory, 1994).

(28) M.G. Boshier, S. Dhawan, P. Egan, X. Fei, V.W. Hughes, M. Janousch, K. Jungmann, W. Liu, F. Mariam, C. Pillai, R. Prigl, G. zu Putlitz, I. Reinhard, W. Schwarz, O. van Dyck and K. Woodle, Abstract 15th International Conference on Atomic Physics, Amsterdam, (!996).

(29) V.W. Hughes in **A. Gift of Prophecy** ed by E.C.G. Sundarshan (World Scientific, Singapore, 1994) p. 222.

(30) Review of Particle Properties 1996, Phys. Rev. **D54**, 1 (1996).

(31) K.A. Woodle, A. Badertscher, V.W. Hughes, D.C. Lu, M. Ritter, M. Gladisch, H. Orth, G. zu Putlitz, M. Eckhause, J. Kane and F.G. Mariam, Phys.

Rev. **41**, 94 (1990); A. Badertscher, V.W. Hughes, D.C. Lu, M.W. Ritter, K.A. Woodle, M. Gladisch, H. Orth, G. zu Putlitz, M. Eckhause, J. Kane and F.G. Mariam, **Atomic Physics 9,** ed by R.S. Van Dyck, Jr. and E.N. Fortson (World Scientific, Singapore, (1985) p. 83; S.H. Kettell, H.E. Ahn, A. Badertscher, F.c. Chmely, M. Eckhause, P. Guss, V.W. Hughes, J.R. Kane, Y. Kuang, B.E. Matthias, J.-J. Mundinger, B. Ni, H. Orth, G. zu Putlitz, H.R. Schaefer, and K.A. Woodle, Bull. Am. Soc. **36**, 1258 (1991).

(32) C.J. Oram, J.M. Bailey, P.W. Schmor, C.A. Fry, R.F. Kiefl, J.B. Warren, G.M. Marshall and A. Olin, Phys. Rev. Lett. **52**, 910 (1984); C.J. Oram, in **Atomic Physics 9,** ed. by R.S. Van Dyck, Jr., and E.N. Fortson (World Scientific, Singapore, 1985) p. 75.

(33) S. Chu, A.P. Mills, A.G. Yodh, K. Nagamine, Y. Miyake and T. Kuga, Phys. Rev. Lett. **60**, 101 (1985); K. Sanzman, M.S. Free and S. Chu, Phys. Rev. **A39,** 6072 (1989).

(34) F.E. Maas. B. Braun, H. Geerds, K. Jungmann, B.E. Matthias, G. zu Putlitz, I. Reinhard, W. Schwarx, L. Willmann, L. Zhang, P.E.G. Baird, P.G.H. Sandars, G.S. Woodman, G.H. Eaton, P. Matousek, W.T. Toner, M. Towrie, J.R.M. Barr, A.I. Ferguson, M.A. Persaud, E. Riis, D. Berkeland, M.G. Boshier, and V.W. Hughes, Phys. Lett. **A187**, 247 -(1994).

Research supported in part by the U.S. Department of Energy.

Quantum electrodynamics of strong fields: Status and perspectives[‡]

G. Soff[†], T. Beier[†], M. Greiner[†], H. Persson[*], G. Plunien[†]

† Institut für Theoretische Physik
Technische Universität Dresden
Mommsenstr. 13, D-01062 Dresden, Germany

* Department of Physics
Chalmers University of Technology and the University of Göteborg
SE-412 96 Göteborg, Sweden

Abstract

We report on the current status of Lamb shift contributions in hydrogenlike heavy ions which have to be calculated non-perturbatively in $Z\alpha$. We subsequently outline the quantum electrodynamical corrections of first order in α, the effects resulting from nuclear mass and size, the quantum electrodynamical corrections of second order in α and the nuclear polarization effects. An excellent agreement with experimental data is found at the current level of precision but insufficient knowledge of nuclear parameters appears to be the major boundary to much more precise predictions. Additionally we also focus on the hyperfine splitting of the ground state in hydrogenlike ^{209}Bi, where again fair agreement between the quantum electrodynamical calculations and the experiment is found.

Introduction

The experimental success in obtaining precise spectroscopical data in few electron high-Z ions has considerably increased in recent years. As an example, the $2p_{1/2} - 2s_{1/2}$ Lamb shift of lithiumlike uranium (U^{89+}) has been measured to be 280.59 ± 0.09 eV at the Bevalac in Berkeley by Schweppe *et al.* [1]. In particular, the new ion source SuperEBIT in Livermore as well as the new relativistic heavy ion facilities SIS/ESR at GSI in Darmstadt provide new stimulating possibilities for spectroscopy. One-electron ions being the simplest systems have also been widely analysed. E.g., for gold a ground-state Lamb shift measurement of 202 ± 8 eV was reported by Beyer *et al.* in 1994 [2]. The increasing experimental precision for the same number in uranium is shown in Fig. 1, together with the most recent theoretical value. The impressive values of Stöhlker *et al.* [4] and Beyer *et al.* [6] are thought to become even more precise in the near future.

[‡]This contribution is dedicated to Prof. I. Lindgren on the occasion of his 65th birthday.

ADVANCES IN QUANTUM CHEMISTRY, VOLUME 30

0065-3276/98 $25.00

Other outstanding measurements in hydrogenlike heavy ions are the ground state hyperfine splitting of $^{209}Bi^{82+}$ [8] and the planned experiment on the g-factor of the bound electron [9]. The transition wavelength of the bismuth hyperfine splitting was found to be 243.87(4) nm which corresponds to an energy of 5.0840(8) eV.

Besides the gain in experimental skill, these experiments are particularly useful in several ways. First, they allow the investigation of bound electron states which are highly relativistic and strongly localized, therefore incorporating very high momentum components. Second, the strength of the electrical field caused by the nucleus is close to the value expected in critical systems which leads to spontaneous electron-positron pair creation [10]. For one-electron systems, we display the expectation value $\langle E \rangle$ for the lowest lying electron states in Fig. 2. The increase of more than six orders of magnitude from $Z = 1$ to $Z = 92$ is not only due to the increasing nuclear charge but also to the closer localization of the wave function as mentioned above.

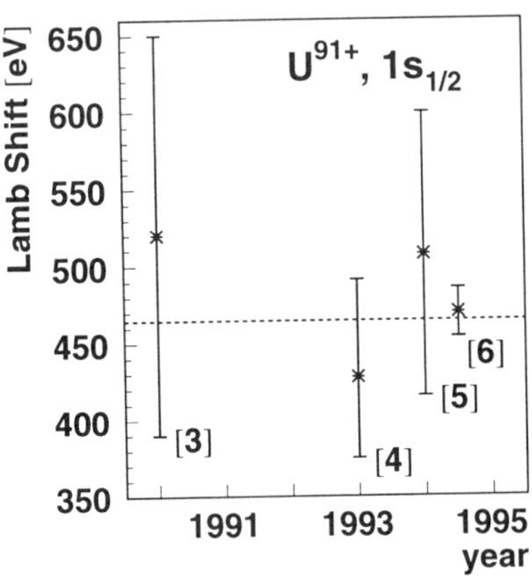

Fig. 1. Recent experimental results for the $1s_{1/2}$ Lamb shift in hydrogenlike uranium. [3] – [6] correspond to the references. The dashed line indicates the current theoretical prediction [7].

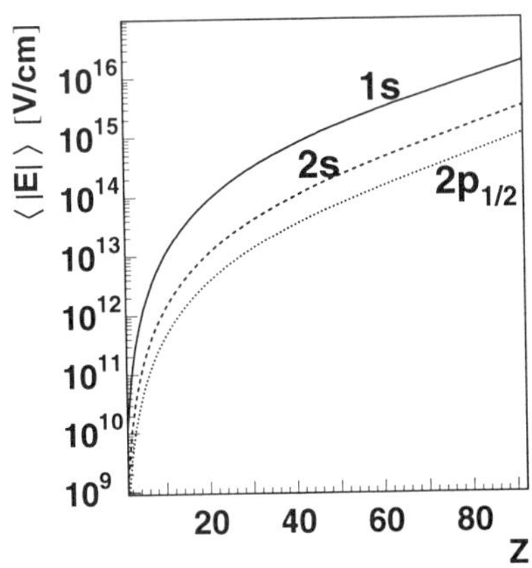

Fig. 2. Expectation value of the electric field strength for the lowest-lying bound states in hydrogenlike atoms with nuclear charge numbers Z.

The point we will mainly focus on in this contribution is the coupling constant $Z\alpha$ of the electromagnetic nuclear-electron interaction. As $\alpha \sim 1/137$, for light nuclei the coupling constant allows a perturbation expansion for all quantum electrodynamical processes under consideration. For heavier nuclei, this is no longer true, because $Z\alpha$ approaches unity as Z approaches 137. Already for uranium $Z\alpha = 0.67$. A perturbation expansion in $Z\alpha$ is therefore useless, and all quantum electrodynamical processes have to be calculated non-perturbatively, including all orders of $Z\alpha$.

Nowadays, powerful techniques for these needs have been developed, as we will discuss in the following.

The bound states of high-Z few-electron systems are not only particularly well suited for the study of QED in strong Coulomb fields. They are also most sensitive [11] to yet unknown possible nonlinear extensions of the QED-Lagrangian, leading to

$$\mathcal{L} = \mathcal{L}_{\text{QED}} + \mathcal{L}_{\text{cor}} \tag{1}$$

with

$$\mathcal{L}_{\text{cor}} = \lambda \left(\overline{\Psi}\Gamma\Psi\right)^n \quad . \tag{2}$$

Here, Γ represents a general vertex function and n is an integer different from one. A Lamb-shift analysis of high-Z atoms yields the most stringent boundary for the coupling constant λ. Up to now no experiment contradicts the assumption $\lambda = 0$.

The experimental success in spectroscopy and the mentioned characteristics of few-electron high-Z systems lead to a challenge for theoreticians to predict the spectroscopic data including all known corrections most precisely. In the following we will examine all corrections to the electron binding energy of the lowest lying states of heavy one-electron ions. The time independent Dirac equation (units chosen so that $\hbar = c = 1$, $\alpha = e^2$)

$$\left[-i\boldsymbol{\alpha}\cdot\boldsymbol{\nabla} + V(\boldsymbol{x}) + \beta\, m - E_{nlj}\right] \Psi_{nlj}(\boldsymbol{x}) = 0 \tag{3}$$

yields an energy eigenvalue of

$$E_{nlj} = m \Bigg/ \sqrt{1 + \left(\frac{Z\alpha}{n - (j+1/2) + \sqrt{(j+1/2)^2 - (Z\alpha)^2}}\right)^2} \tag{4}$$

for a one-electron state $|nlj\rangle$ with $j = l \pm 1/2$ in the field of a pointlike nucleus. In these equations, m denotes the electron rest mass.

The deviation of all corrections from this Dirac energy eigenvalue is widely known as Lamb shift of a state $|nlj\rangle$. This is an analogy to the "classical" Lamb shift, which is the energy difference between the $2s_{1/2}$ and the $2p_{1/2}$ states. However, by convention [12] the so-called "nonrelativistic reduced mass correction"

$$\Delta E_{\text{rm}\,nlj} = \frac{m}{m+M}\left(m - E_{nlj}\right) \tag{5}$$

where m and M are the masses of electron and nucleus, respectively, is not included in the term "Lamb shift" for one single state, and its value has to be given separately. For heavy ions the nonrelativistic reduced mass correction amounts only to a few tenth of an eV, whereas the experimental uncertainty is still more than 10 eV. But this might change in the very near future.

Self energy and vacuum polarization

The two most dominant corrections to the Dirac energy eigenvalue are the self energy and the vacuum polarization. The leading diagrams are shown in Fig. 3. They contain only one photon line and are thus of order α. Therefore they yield the significant contributions of both effects. The diagrams of order α^2 which contain two photon lines will be discussed later on.

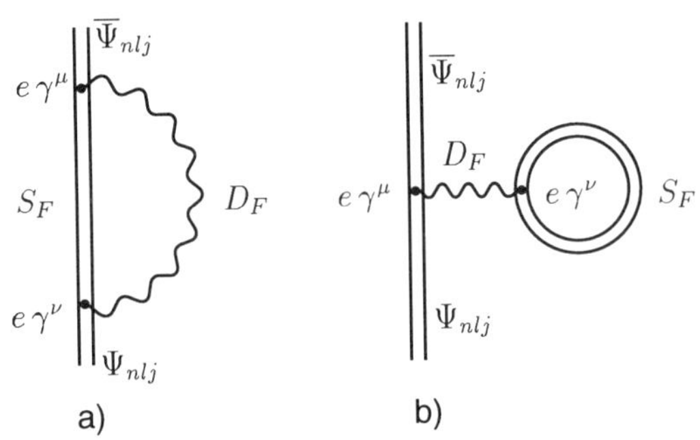

Fig. 3. Self energy and vacuum polarization diagrams of order α. Double lines indicate bound fermions and wavy lines stand for photons. The letters point to the accompanying expressions in the basic equations (6) and (22).

We start discussing the self energy (diagram Fig. 3 a)). Classically, self energy in QED means the emission and reabsorption of a photon by a fermion according to Heisenberg's uncertainty relation. While the photon is present, the fermion can be in any energy state without violating energy conservation. After reabsorption, however, the original state is present again. As long as the fermion is in the intermediate state, its interaction with the external field results in an energy shift of the original state.

According to Feynman rules of QED, this energy shift is given by

$$\begin{aligned} \Delta E_{\mathrm{SE}\,nlj} &= 4\pi\, i\, \alpha \int d(t_y - t_x) \int d^3x \int d^3y\ \times \\ &\times \overline{\Psi}_{nlj}(x)\, \gamma^\mu S_F(x,y)\, \gamma^\nu\, \Psi_{nlj}(y)\, g_{\mu\nu}\, D_F(x,y) \\ &- \delta m \int d^3x\, \overline{\Psi}_{nlj}(x)\, \Psi_{nlj}(x) \quad . \end{aligned} \tag{6}$$

D_F denotes the photon propagator given by

$$D_F\,(x,y) = \frac{-i}{(2\pi)^4} \int d^4k\, \frac{\exp\left[-ik\,(y-x)\right]}{k^2 + i\,\epsilon} \quad . \tag{7}$$

S_F represents the Feynman propagator in the field of the external potential $A^{\rm ex}$ of the nucleus and obeys the equation

$$\left[\gamma^\mu \left(p_\mu - eA^{\rm ex}_\mu(x)\right) - m\right] S_F(x,y) = \delta^4(x-y) \quad . \tag{8}$$

As is well known, the mass counter term in equation (6) accounts for the use of the observable instead of the "bare" mass of the electron. The latter is not observable due to the interaction of an electrical charge with its own radiation field which is always present. The magnitude of δm is fixed by

$$\lim_{Z\alpha\to 0} \Delta E_{{\rm SE}\,nlj}(Z\alpha) = 0 \quad , \tag{9}$$

as the energy of a free electron by definition is not shifted due to quantum electrodynamical effects.

Early relativistic calculations of the self energy were based on an expansion in $Z\alpha$ of the external field. In terms of Feynman diagrams, this means an expansion of the bound state propagator (the double line in Fig. 3) into free states (Fig. 4).

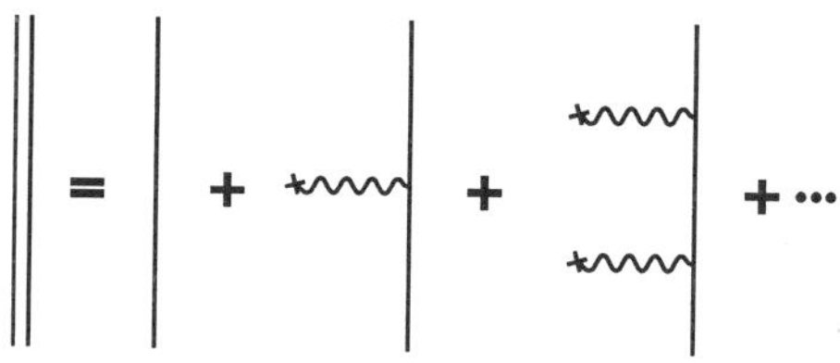

Fig. 4. Symbolical expansion of the bound propagator into free propagators yields any number of interactions with the external potential. Each interaction accounts for one power of $Z\alpha$.

The self energy level shift can be expressed as [13]

$$E_{\rm SE} = \frac{\alpha}{\pi}\frac{(Z\alpha)^4}{n^3} F(Z\alpha)\, mc^2 \,. \tag{10}$$

On evaluating $F(Z\alpha)$, the lower-order terms in the $Z\alpha$-expansion can be identified separately [13]:

$$\begin{aligned} F(Z\alpha) &= A_{41}\ln(Z\alpha)^{-2} + A_{40} + (Z\alpha)A_{50} \\ &\quad + (Z\alpha)^2 A_{62}\ln^2(Z\alpha)^{-2} + (Z\alpha)^2 A_{61}\ln(Z\alpha)^{-2} \\ &\quad + (Z\alpha)^2 G_{\rm SE}(Z\alpha)\,. \end{aligned} \tag{11}$$

The function $G_{\rm SE}(Z\alpha)$ in (11) includes all higher-order contributions to the self energy which are not given explicitely. The low-Z limit of this function $G_{\rm SE}(0) = A_{60}$ has been calculated by Pachucki [14]:

$$A_{60} = \begin{cases} -30.928\,90(1) & 1s_{1/2} \\ -31.840\,47(1) & 2s_{1/2} \end{cases} \,. \tag{12}$$

For larger Z, the expansion (11) is completely dominated by these higher-order contributions. In Fig. 5 we display the value of

$$\frac{\text{higher order self energy}}{\text{total self energy}} = \frac{[G_{\text{SE}}(Z\alpha) - G_{SE}(0)]\,(Z\alpha)^2}{F(Z\alpha)} \tag{13}$$

for the $1s_{1/2}$ state for hydrogenlike ions.

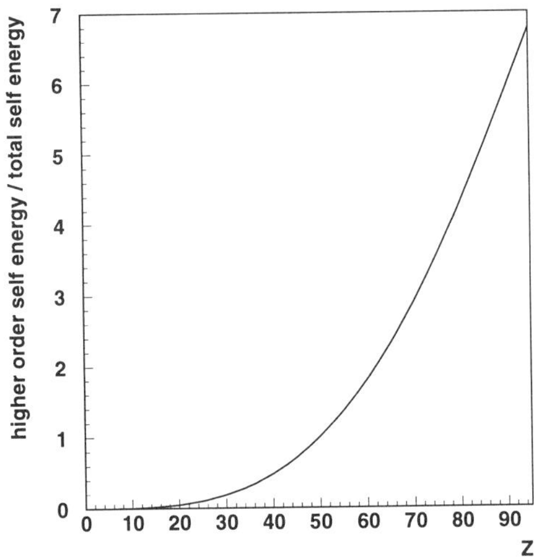

Fig. 5. Comparison of the exact self-energy calculation to the $Z\alpha$-expansion (11). Displayed is the value given by eq. (13).

From this presentation it is clear that any expansion in $Z\alpha$ is useless even in the region of intermediate Z and the calculation has to be performed to all orders of $Z\alpha$. Methods for these calculations were developed by decomposing the intermediate bound state into several terms but keeping all orders of $Z\alpha$. In the following discussion we employ the usual bound interaction ("Furry") picture of QED. The external field is considered as instantaneous Coulomb interaction according to an infinitely heavy atomic nucleus. Deviations from this assumption result only in minor corrections and will be discussed later on. The Furry picture results in a possible separation of time and space variables contrary to the covariant formulation of eqs. (6 – 8).

Using the relation

$$S_F(x,y) = \int_{C_F} \frac{dz}{2\pi i} \exp\left[-iz(t_y - t_x)\right] \mathcal{G}(\boldsymbol{x},\boldsymbol{y},z)\,\gamma^0 \tag{14}$$

and the coordinate space representation of the photon propagator,

$$D_{F\,\mu\nu}(x,y) = \frac{1}{2\pi i}\int dk_0 \left(-\frac{\exp\left[-a|\boldsymbol{y}-\boldsymbol{x}|\right]}{4\pi\,|\boldsymbol{y}-\boldsymbol{x}|}\right) \exp\left[-ik_0(t_y-t_x)\right] \ , \quad (15)$$

where

$$a = -i\left[k_0^2 + i\epsilon\right]^{1/2} \quad ; \quad \mathrm{Re}(a) > 0 \quad , \quad (16)$$

eq. (6) becomes

$$\begin{aligned} \Delta E_{\mathrm{SE}\,nlj} &= -\frac{i\,\alpha}{2\pi}\int d^3y \int d^3x \int_{C_F} dz\, \Psi^\dagger_{nlj}(\boldsymbol{y})\,\alpha_\mu\,\mathcal{G}(\boldsymbol{y},\boldsymbol{x},z) \\ &\quad \times \alpha^\mu \Psi_{nlj}(\boldsymbol{x})\,\frac{\exp\left[-b|\boldsymbol{y}-\boldsymbol{x}|\right]}{|\boldsymbol{y}-\boldsymbol{x}|} \\ &\quad -\delta m \int d^3x\,\overline{\Psi}_{nlj}(\boldsymbol{x})\,\Psi_{nlj}(\boldsymbol{x}) \quad . \end{aligned} \quad (17)$$

with

$$b = -i\left[\left(E_{nlj} - z\right)^2 + i\epsilon\right]^{1/2} \quad ; \quad \mathrm{Re}(b) > 0. \quad (18)$$

The Green's function $\mathcal{G}$ corresponds to the Dirac Hamiltonian $H = \boldsymbol{\alpha}\cdot\boldsymbol{p} + V(\boldsymbol{x}) + \beta m_e$ by

$$\mathcal{G} = \frac{1}{(H-z)} \quad . \quad (19)$$

The difficulty in evaluating (17) comes from large imaginary arguments of z, as

$$\lim_{z\to\pm i\infty} \exp\left[-(b+c)|\boldsymbol{y}-\boldsymbol{x}|\right] \to \exp\left[-2|z||\boldsymbol{y}-\boldsymbol{x}|\right] \quad , \quad (20)$$

where c is part of the electron Green's function, $c = (1-z^2)^{1/2}$, $\mathrm{Re}(c) > 0$. The integrand is exponentially damped except in the region $|\boldsymbol{y}-\boldsymbol{x}| \to 0$, which is the source of the singular terms. To handle these problems an expansion of $\mathcal{G}$ in terms of $|z|^{-1}$ is carried out. The leading terms are evaluated analytically, whereas the convergent remainder can be treated numerically without difficulty. The decomposition of $\mathcal{G}$ can symbolically be written as

$$\mathcal{G}(z) \to \mathcal{G}_A(z) + \left[\mathcal{G}(z) - \mathcal{G}_A(z)\right] \quad . \quad (21)$$

In the past, different $\mathcal{G}_A(z)$ were employed [15 – 19]. An elaborate comprehensive discussion of these various techniques is given by P. Mohr [20]. The most recent calculation following equation (6) was employed by Mohr and Soff [21] according to the technique presented by Indelicato and Mohr [22]. They obtained 355.0432 eV for the self energy shift of the $1s_{1/2}$ state in hydrogenlike uranium, and the most stringent limitation to this value is the incomplete knowledge of nuclear size and shape, as will be discussed below.

A somewhat different technique of renormalization was employed by Quiney and Grant [23, 24] and also by Lindgren, Persson, and coworkers [25, 26]. Whereas in most approaches the mass counterterm is treated separately, they use the same partial wave expansion for both terms in equation (6) and subtract the counterterm individually for each partial wave. This results in a proper convergence for each partial wave and the infinite terms move into the sum over the partial waves. We will not discuss this procedure in detail here.

The other principal radiative correction is the vacuum polarization (Fig. 3 b)). It describes the interaction of a fermion with virtual electron-positron pairs which can be thought present in the vacuum for short times without violating the energy-time uncertainty relation. If external fields are present, these virtual pairs are influenced and act like a polarizable medium. Therefore the Coulomb interaction of the nucleus with the electrons is modified which leads to an energy shift compared to the pure Coulomb potential energy eigenvalue.

From bound state QED [27] the energy shift corresponding to the total vacuum polarization is given by

$$\begin{aligned} \Delta E_{\mathrm{VP}\,nlj} &= 4\pi\, i\, \alpha \int d\,(t_y - t_x) \int d^3x \int d^3y \;\times \\ &\times \overline{\Psi}_{nlj}(x)\, \gamma^\mu\, \Psi_{nlj}(x)\, g_{\mu\nu}\, D_F(x,y)\, \mathrm{Tr}\,[\gamma^\nu\, S_F(y,y)] \end{aligned} \tag{22}$$

which can be written as

$$\Delta E_{\mathrm{VP}\,nlj} = \int d^3x\, \Psi^\dagger_{nlj}(\boldsymbol{x})\, \Psi_{nlj}(\boldsymbol{x})\, V_{\mathrm{VP}}(\boldsymbol{x}) \quad . \tag{23}$$

In this equation, $V_{\mathrm{VP}}(\boldsymbol{x})$ is an effective potential:

$$\begin{aligned} V_{\mathrm{VP}}(\boldsymbol{x}) &= 4\pi i\alpha \int d(t_1 - t_2) \int d^3y\, D_F(x,y)\;\; \mathrm{Tr}\,[\gamma_0\, S_F(y,y)] \\ &= \frac{i\,\alpha}{2\pi} \int d^3y\, \frac{1}{|\boldsymbol{x} - \boldsymbol{y}|} \int\limits_{C_F} d\,z\; \mathrm{Tr}\, \mathcal{G}(\boldsymbol{y}, \boldsymbol{y}, z) \quad . \end{aligned} \tag{24}$$

Here we again employed the Furry picture. Introducing the vacuum polarization charge density

$$\rho(\boldsymbol{x}) = \frac{e}{2\pi i} \int_{C_F} dz\, \mathrm{Tr}\, \mathcal{G}\,(\boldsymbol{x}, \boldsymbol{x}, z) \tag{25}$$

it simply follows

$$V_{\mathrm{VP}}(\boldsymbol{x}) = -e \int d^3y\, \frac{\rho(\boldsymbol{y})}{|\boldsymbol{x} - \boldsymbol{y}|} \quad . \tag{26}$$

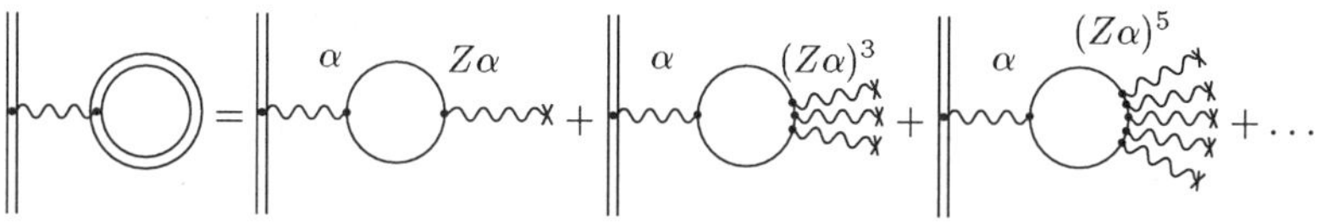

Fig. 6. Expansion of the vacuum polarization loop into different powers of $Z\alpha$. The first term of the expansion corresponds to the Uehling contribution, the remaining terms are known as Wichmann-Kroll contribution.

The expression for ρ still contains an infinite unrenormalized charge. To overcome this difficulty, again a separation into formal divergent and convergent terms is made. The bound fermion loop is expandend into powers of $Z\alpha$ (Fig. 6). Only loops with an even number of photon lines contribute according to the Furry theorem. From the expansion, only the first term is divergent. It is the major vacuum polarization contribution in systems with $Z\alpha \ll 1$ and its first evaluation including renormalization was done in 1935 by Uehling [28] at a time when no closed approach to Quantum Electrodynamics existed. A proper gauge-invariant formulation of the vacuum polarization was given by Schwinger 16 years later [29] after an invariant regularization scheme was developed by Pauli and Villars [30].

The separation of the loop also implies a separation of the corresponding potential (24) into the Uehling potential and the "Wichmann-Kroll" potential, as the higher orders of the $Z\alpha$ expansion were first considered by Wichmann and Kroll in 1956 [31].

The evaluation of the Uehling potential can be found in a number of articles [32 – 35]. For spherically symmetric extended charge densities ρ' (normalized to $4\pi \int_0^\infty dr' \, (r')^2 \, \rho'(r') = 1$), it yields

$$V_{\text{Uehl}}(r) = -\frac{2}{3}\,\alpha\, Z\alpha\,\frac{1}{r}\int_0^\infty dr' r' \rho'(r')\left[\chi\left(2\left|r - r'\right|\right) - \chi\left(2\left(r + r'\right)\right)\right] \,, \quad (27)$$

where $r = |\boldsymbol{x}|$ and

$$\chi(u) = \int_1^\infty ds \left(\frac{\exp\left[-su\right]}{s^2}\right)\left(1 + \frac{1}{2s^2}\right)\left(1 - \frac{1}{s^2}\right)^{1/2} \quad . \quad (28)$$

The Uehling potential is shown in Fig. 7. The plot displays also the asymptotic behaviour for $r \to \infty$, which reads

$$V_{\text{Uehl}}(r) \stackrel{r\to\infty}{=} \frac{\alpha}{4\sqrt{\pi}}\, Z\alpha\, \frac{\exp(-r)}{r^{5/2}} \quad . \quad (29)$$

In atoms, the dominating Uehling potential causes the vacuum polarization effect to be *attractive* instead of repulsive, as would be expected from classical polarization theory. The polarization of the vacuum can be imagined as in Fig. 8. This can be understood as a result of charge renormalization. The "bare" charge is unobservable and screened by a polarization cloud of opposite sign. At distances very close to the original charge the screening is weakened and therefore the net charge seems to be enhanced. This suggests an additional positive charge in the vicinity of the nucleus and an additional negative screening charge further away as a correction to the classical model of nuclear charge distribution.

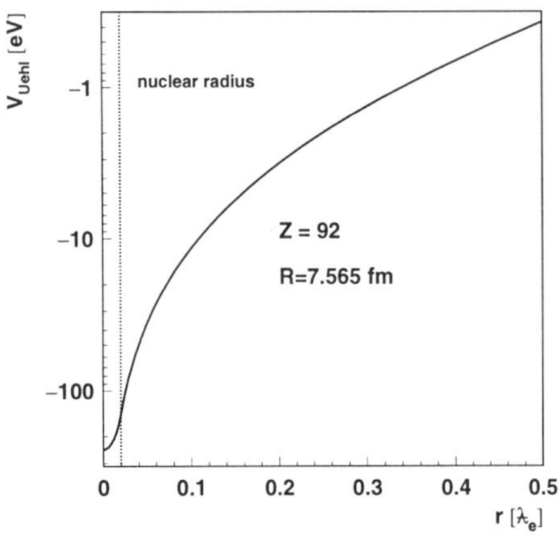

Fig. 7. Uehling potential of the uranium nucleus. The specified radius corresponds to $\langle r^2 \rangle^{1/2} = 5.86$ fm.

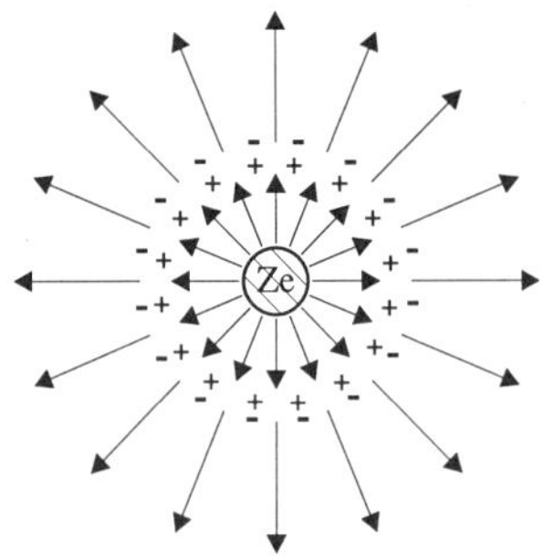

Fig. 8. The vacuum polarization charge around a central nucleus.

The higher order (Wichmann-Kroll) vacuum polarization charge densities show the "normal" opposite behaviour. In Fig. 9 we show the Wichmann-Kroll charge density according to its angular momentum decomposition, which allows the most accurate calculations. It was already proposed by Wichmann and Kroll [31]. Calculations were also done by Gyulassy [36], Neghabian [37], and Soff and Mohr [38] and Persson *et al.* [39]. All these calculations are based on a partial wave decomposition of the Greens function $\mathcal{G}$ [17, 18], which yields [38]

$$\begin{aligned}
\rho^{(\mathrm{WK})}(r) \quad = \quad & \frac{e}{2\pi^2} \int_0^\infty du \left(\sum_{\kappa=\pm 1}^{\pm\infty} |\kappa| \, \mathrm{Re} \left\{ \sum_{i=1}^{2} \mathcal{G}_\kappa^{ii}(r, r, iu) \right. \right. \\
& \left. \left. + \int_0^\infty dr' \, r'^2 \, V(r') \sum_{m,n=1}^{2} [\mathcal{F}_\kappa^{mn}(r, r', iu)]^2 \right\} \right) \\
& + \frac{e}{2\pi} \sum_{-m<E<0} |\kappa| \left\{ g_{nlj}^2(r) + f_{nlj}^2(r) \right\} \qquad (30)
\end{aligned}$$

for the Wichmann-Kroll vacuum polarization charge density $\rho^{(\mathrm{WK})}(r)$. In this equation, $\mathcal{F}^{ij}$ and $\mathcal{G}^{ij}$ denote the components of the free and bound electron radial Green function, and κ is the Dirac angular momentum quantum number ($|\kappa| = j + 1/2$). The last row in equation (30) accounts for bound state poles on the negative real axis in the complex z-plane. This sum only contributes for systems where the binding energy of a bound state electron exceeds the electron rest mass which is the case only in superheavy systems. In this row, g_{nlj} and f_{nlj} denote large and small components of the radial Dirac wave function of the corresponding state. The origin of this sum is mathematical. By a Wick rotation the imaginary axis is choosen as integration contour, and therefore poles lying left of it have to be considered separately. A detailed discussion of this procedure is provided by [38].

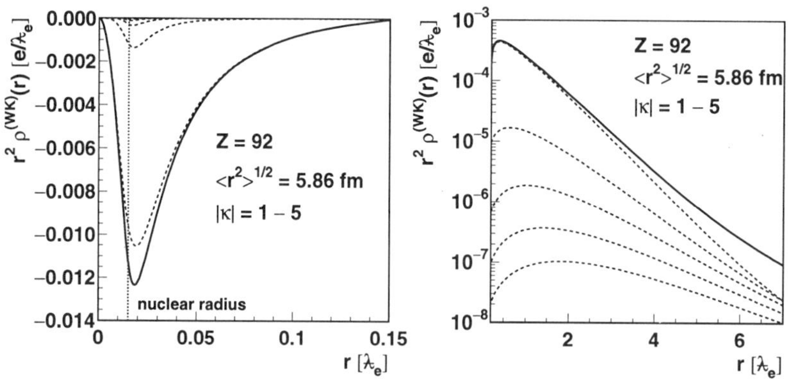

Fig. 9. Wichmann-Kroll vacuum-polarization charge density $r^2\,\rho^{(\mathrm{WK})}(r)$ and the individual components $r^2\rho_\kappa(r)$ for uranium plotted as a function of the radial coordinate r. The total density is shown as a solid line and the individual components are shown as dashed lines where the largest magnitude contribution corresponds to $|\kappa| = 1$ and the successively smaller magnitude contributions correspond to increasing values of $|\kappa|$. On the left the charge density is shown close to the nucleus on a linear scale. In this region the charge density is negative. On the right, the charge density is shown for regions far away from the nucleus where it has a positive sign.

The current calculations yield -88.60 eV for the vacuum polarization effect of the $1s_{1/2}$-state in uranium [39]. It is therefore of almost the same order of magnitude as the self energy in this heavy system contrary to light atoms, where the self energy alone is the dominating effect. Since vacuum polarization effects dominate in the region of the nucleus itself, the more localized wave functions of heavy ions are much stronger influenced than those of the lighter atoms. In Fig. 10 we display the self energy and vacuum polarization contributions of order α to the lamb shift together with the influence of the nuclear size which will be discussed later on.

An experimentally not yet proven effect is the r-dependance of the Wichmann-Kroll charge distribution. The potential corresponding to the $\alpha(Z\alpha)^3$ charge density (the second loop in the expansion of Fig. 6) is predicted to diminish as [32, 40]

$$V_{13}(r) \overset{r\to\infty}{\sim} \frac{\alpha\,(Z\alpha)^3}{\pi r}\frac{32}{225}\frac{1}{(2r)^4} \sim \frac{1}{r^5} \tag{31}$$

contrary to the exponential behaviour of the Uehling potential (29). For wave functions of Rydberg states with large $\langle r \rangle$, this implies a domination of the Wichmann-Kroll part over the Uehling part and thus a change in sign of the whole vacuum polarization effect. For uranium, the energy shifts are given by Soff and Mohr [41].

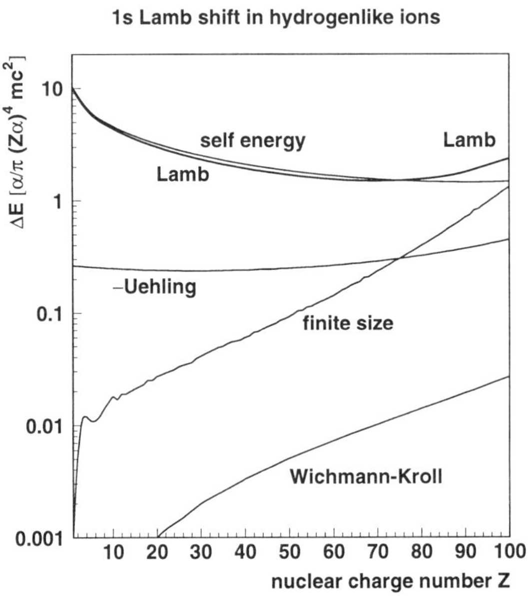

Fig. 10. Contributions of self energy, vacuum polarization, and finite size to the Lamb shift of the $1s_{1/2}$-state. For the energy values a dimensionless quantity similar to $F(Z\alpha)$ in eq. (10) is used. The vacuum polarization contribution is separated into Uehling and Wichmann-Kroll parts. The negative of the Uehling contribution is shown. For high Z, the finite nuclear size becomes a major contribution to the total Lamb shift. Also the Wichmann-Kroll part becomes more important for large Z.

These shifts amount to -1.67μeV for the Uehling part and 3.51μeV for the $\alpha(Z\alpha)^3$ Wichmann Kroll part in the $4f_{5/2}$-state which is the lowest lying state showing this feature. Unfortunately, this seems to be still beyond the experimental limit of precision but would be a particular proof of strong field QED in the future.

Nuclear mass and size

The effect of the nuclear mass was already mentioned in the introduction. In the Furry picture which is employed in the calculations of QED effects on bound electron states a static external field is assumed which corresponds to an infinitely heavy nucleus. In a non-relativistic approximation its finite mass is encountered by the reduced mass correction similar to the two-body problem in classical mechanics. In a relativistic treatment, however, this approach is oversimplyfied. Recently Artemyev *et al.* [42, 43] almost solved the whole problem by considering the nucleus as a simple Dirac particle with spin $1/2$, mass M and charge Ze. The interaction of the two Dirac particles electron and nucleus leads to a quasipotential equation in the center-of-mass system,

$$\left(E - \sqrt{\boldsymbol{p}^2 + M^2} - \boldsymbol{\alpha p} - \beta m\right) \Psi_{nlj}(\boldsymbol{p}) = \int V(E, \boldsymbol{p}, \boldsymbol{q})\, \Psi_{nlj}(\boldsymbol{q})\, d^3q \tag{32}$$

where V is a quasipotential defined by a Lippman-Schwinger equation. Here, $\boldsymbol{p}$ and $\boldsymbol{q}$ are the spatial coordinate and momentum of one of the particles in the center-of-mass frame, and M and m denote the masses of nucleus and electron, respectively. For details we refer to the original publications. The final result of this calculation is the total energy shift due to the finite nuclear mass (including also the non-relativistic part), which can be written as

$$\Delta E_{\mathrm{rec}\,nlj} = \Delta E^{(1)}_{nlj} + \Delta E^{(2)}_{nlj} \tag{33}$$

with

$$\Delta E^{(1)}_{nlj} = \frac{m}{M} \frac{(\alpha Z)^2}{2N^2} mc^2 \tag{34}$$

and

$$\Delta E^{(2)}_{nlj} = \frac{m}{M} \frac{(\alpha Z)^5}{\pi n^3} P_{nlj}(\alpha Z)\, mc^2 \quad . \tag{35}$$

Here we used

$$\begin{aligned} N^2 &= n^2 - 2(n - |\kappa|)(|\kappa| - \gamma) \quad , \\ \gamma &= \sqrt{\kappa^2 - (\alpha Z)^2} \quad , \\ \kappa &= (-1)^{j-\ell+\frac{1}{2}} \left(j + \frac{1}{2}\right) \quad . \end{aligned} \tag{36}$$

The function $P_{nlj}(\alpha Z)$ in (35) has been evaluated numerically in Refs. [42] and [43]. The tabulated values were interpolated with a B-spline routine to provide numerical values for any nuclear charge number Z. We display these values in Fig. 11 together with the nonrelativistic reduced mass correction which is included in ΔE_{rec}. For uranium, the total effect to the $1s_{1/2}$-state is 0.51 eV, including the nonrelativistic correction of 0.30 eV. Compared to the radiative effects discussed so far this might be thought tiny, but an experimental precision of better than 1 eV demands also for the proper calculation of Lamb shift contributions of this size of magnitude.

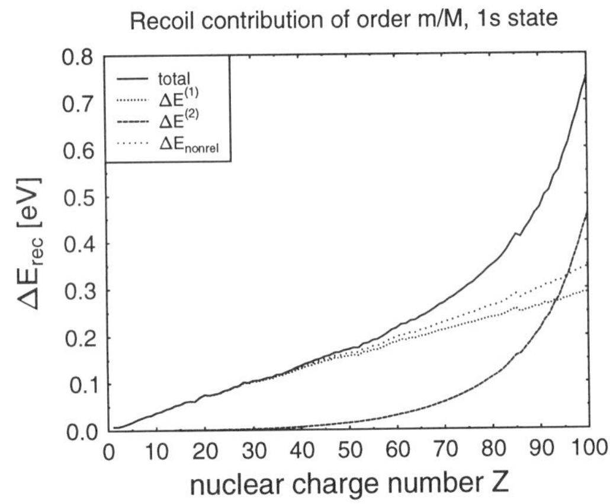

Fig. 11. Recoil contributions to the Lamb shift. The non-relativistic part is included in the sum of $\Delta E^{(1)} + \Delta E^{(2)}$.

We emphasize that the plots shown in Fig. 11 are an approximation obtained by considering the nucleus as a spin-$1/2$ particle. Nevertheless, these values may serve as a good approximation to the real effect and reduce the uncertainty of almost 100 % which was estimated for these values previously [44].

Contrary to the mass of the nucleus, its size influences the binding energy considerably in heavy ions (Fig. 10). In studying nuclear size effects nowadays always a spherically symmetric charge distribution of the nucleus is assumed which allows a separation of the Dirac equation and corresponding wave function into an angular part and a radial part similar to the point nucleus case. The radial Dirac equation then reads [45]

$$\begin{pmatrix} m + V(r) & -\frac{1}{r}\frac{d}{dr}r + \frac{\kappa}{r} \\ \frac{1}{r}\frac{d}{dr}r + \frac{\kappa}{r} & -m + V(r) \end{pmatrix} \begin{pmatrix} g_{nlj}(r) \\ f_{nlj}(r) \end{pmatrix} = E_{nlj} \begin{pmatrix} g_{nlj}(r) \\ f_{nlj}(r) \end{pmatrix} . \quad (37)$$

Solutions for E, g and f depend on $V(r)$, and by choosing an appropriate charge distribution $\rho(r)$ and solving the poisson equation which yields

$$V(r) = -4\pi\alpha \left\{ \frac{1}{r} \int_0^r \rho(r')\, r'^2 \, dr' + \int_r^\infty \rho(r')\, r' \, dr' \right\} \quad , \tag{38}$$

the effect of nuclear size and shape is obtained as difference between the point nucleus value E_{nlj} using $V(r) = -Z\alpha/r$ and that using a more realistic charge distribution. An often employed charge distribution for the nucleus is the two-parameter Fermi distribution given by

$$\rho(r) = \frac{N}{1 + \exp\{(r-c)/a\}} \tag{39}$$

where c denotes the half-density radius and a indicates the extension of the shell where the charge density decreases. The skin thickness $t = 4 \ln 3\, a$ is defined as the radial distance over which the charge density declines from 90% to 10% of its value at the origin. N is chosen to normalize ρ according to

$$4\pi \int_0^\infty \rho(r)\, r^2 \, dx = Z \quad . \tag{40}$$

A more simple distribution is obtained for $a \to 0$. In this case the nucleus is considered to be a homogeneous sphere with radius $R_0 = c$. The potential for this charge distribution reads

$$V_N(x) = \begin{cases} -\frac{3}{2}\frac{Z\alpha}{R_0} + \frac{1}{2}\frac{Z\alpha}{R_0^3} x^2 & \text{for } x < R_0 \\ -Z\alpha/x & \text{for } x \geq R_0 \quad . \end{cases} \tag{41}$$

In studying the effects of shape, the parameters R_0, t and c are varied so that a constant rms-radius $\langle r^2 \rangle^{1/2}$ is obtained for the charge distributions under comparison. Changing the rms-radius yields the size effects. The experimental values for each nucleus can be taken from a number of tables [46 – 50]. For uranium, the difference to the point nucleus binding energy value of the $1s_{1/2}$-state amounts to 198.82 eV. This number is obtained by assuming a Fermi distribution with $\langle r^2 \rangle^{1/2} = 5.860$ fm and $a = 0.5046$ fm [51, 52]. A 0.36 eV larger value is obtained when using the homogeneous-sphere distribution with the same $\langle r^2 \rangle^{1/2}$. As both models are only approximations to the real charge distribution in a nucleus, this difference can be considered as the uncertainty of the size effect itself, the more so, as most atoms do not possess spherical symmetric nuclei at all.

The large influence of the nuclear size on the binding energy arises the question whether the radiative corrections of first order in alpha are also influenced significantly by nuclear size effects. The extension of the nucleus influences all wave functions and therefore the propagators utilized in the previous section. For the self energy the total effect was already estimated by Mohr and Soff [21]. Due to the variation in the $\langle r^2 \rangle^{1/2}$-value of uranium published in the various tables listed above, they found it best to present a curve which allows for changes in this number.

We show this curve together with a similar one for the vacuum polarization in Fig. 12. Already a change of 2 % in the nuclear radius yields a change of ~ 0.1 eV in each of the radiative corrections. Due to the different sign of both contributions the net effect almost cancels.

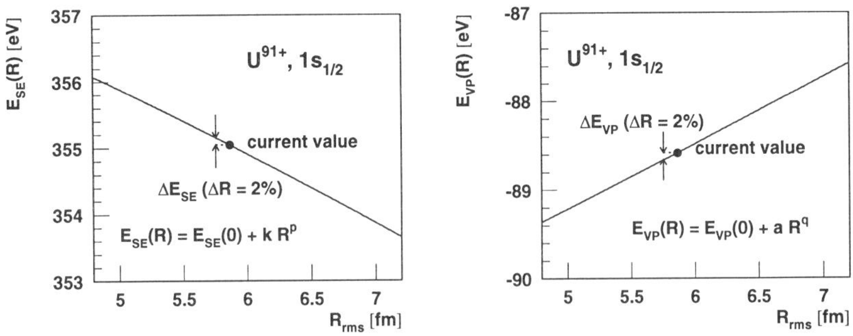

Fig. 12: Parametrization of self energy corrections (left) and vacuum polarization corrections (right) in uranium as function of the nuclear rms-radius $\langle r^2 \rangle^{1/2}$. The current value $\langle r^2 \rangle^{1/2} = 5.860$ fm [51] is indicated by •. The arrows denote the difference in the energy shift if an older value $\langle r^2 \rangle^{1/2} = 5.751$ fm [47] is utilized.

Higher order radiative corrections

Self energy and vacuum polarization of order α and the nuclear size account for the measured Lamb shift in hydrogenlike heavy ions at the current level of accuracy. Radiative corrections of the order α^2 contribute to the Lamb shift of the $1s_{1/2}$ state and amount to about 1 eV for uranium. Facing higher precision in experiments, these corrections have to be evaluated to yield a reliable Lamb shift calculation.

Fig. 13 shows the Feynman diagrams of all radiative corrections of order α^2, which implies two photon lines in each diagram. In this section, we will discuss the current status of the calculations for the ten diagrams displayed in Fig. 13.

The first row in Fig. 13 contains the diagrams consisting of two self energy loops. These diagrams are the ones most difficult to evaluate in a nonperturbative scheme including all orders in $Z\alpha$. For high-Z systems a renormalization scheme of the two-loop diagrams was recently suggested by Labzowsky and Mitrushenkov [53] which seems to be useful in combination with the partial-wave renormalization technique already mentioned. A major difficulty arises from the fact that the diagrams SESE a) – c) form a gauge invariant set only as a whole and not separately. Only in covariant gauges, it is possible to separate out the irreducible part of the SESE a) diagram, that is the diagram with all possible but not the reference state between the two self energy photon loops. Mitrushenkov *et al.* [54] obtained a numerical value of –0.971 eV for the $1s_{1/2}$ state of uranium. We emphasize that this value is by no means any approximation to the whole SESE contribution.

Next, we will consider the VPVP b) and c) diagrams. Similar to the vacuum polarization of order α, these contributions can be thought as the interaction of the bound electron with an additional potential where the interaction process is treated in first order perturbation theory. As for the vacuum polarization of order α, this requires proper renormalization. Both diagrams were evaluated first by Källén and Sabry [55] and are sometimes therefore named Källén-Sabry contribution. In their approach and in the calculations of all their successors an approximation was made using free fermion propagators in the vacuum polarization loops and considering these loops in the Uehling approximation only, which results in an expansion of the diagrams into those shown in Fig. 14.

The effective potential was derived by Blomqvist in 1972 [40] and generalized for finite charge distributions by Wayne Fullerton and Rinker [33]. It reads

$$V_{\rm KS}(r) = -(Z\alpha)\frac{\alpha^2}{\pi r}\int_0^\infty dr'\, r'\rho(r')\,[L_0(2|r-r'|) - L_0(2|r+r'|)]\ , \tag{42}$$

with

$$L_0(x) = \int^{-x} dy\, L_1(y)\ +\ \text{const.} \tag{43}$$

and

$$\begin{aligned} L_1(x) \ &= \ \int_1^\infty dt\, e^{-xt}\left\{\left(-\frac{13}{54t^2} - \frac{7}{108t^4} - \frac{2}{9t^6}\right)\sqrt{t^2-1}\right. \\ &+ \ \left(\frac{44}{9t} - \frac{2}{3t^3} - \frac{5}{4t^5} - \frac{2}{9t^7}\right)\ln\left[t+\sqrt{t^2-1}\right] \\ &+ \ \left(-\frac{4}{3t^2} - \frac{2}{3t^4}\right)\sqrt{t^2-1}\,\ln\left[8t(t^2-1)\right] \\ &+ \ \left(\frac{8}{3t} - \frac{2}{3t^5}\right)\int_t^\infty \mathrm{d}y\left(\frac{3y^2-1}{y(y^2-1)}\ln\left[y+\sqrt{y^2-1}\right]\right. \\ &- \ \left.\left.\frac{1}{\sqrt{y^2-1}}\ln\left[8y(y^2-1)\right]\right)\right\}\ . \end{aligned} \tag{44}$$

Considering point nuclei this energy shift was calculated by Beier and Soff [56]. Schneider *et al.* presented an improved calculation for extended nuclear charge distributions [57], which results in –0.72 eV for the $1s_{1/2}$-state of uranium. Currently research is in progress to overcome the Uehling approximation in this contribution.

For the remaining VPVP a) correction, values for a few Z were presented by Manakov and Nekipelov [58]. We present our method of calculation here. Instead of evaluating explicitely the corresponding energy shift,

$$\begin{aligned}\Delta E_{\mathrm{VPVPa}\,nlj} &= -e^4 \int \mathrm{d}^3x'\,\mathrm{d}^3x\,\overline{\Psi}_{nlj}(\boldsymbol{x}') \Big\{ \gamma^\mu \int \mathrm{d}^3x_2\, D_{\mu\alpha}(\boldsymbol{x}' - \boldsymbol{x}_2, 0) \\ &\times \int \frac{\mathrm{d}E_2}{2\pi}\, \mathrm{Tr}\,[\gamma^\alpha S_{\mathrm{F}}(\boldsymbol{x}_2, \boldsymbol{x}_2, E_2)] \Big\}\, S_{\mathrm{F}}(\boldsymbol{x}', \boldsymbol{x}, E_A) \\ &\times \Big\{ \gamma^\nu \int \mathrm{d}^3x_1\, D_{\nu\beta}(\boldsymbol{x} - \boldsymbol{x}_1, 0) \\ &\times \int \frac{\mathrm{d}E_1}{2\pi}\, \mathrm{Tr}\,\Big[\gamma^\beta S_{\mathrm{F}}(\boldsymbol{x}_1, \boldsymbol{x}_1, E_1)\Big] \Big\}\, \Psi_n lj(\boldsymbol{x}) \quad , \end{aligned} \tag{45}$$

and dealing with renormalization problems, the results of the vacuum polarization calculation of order α can be utilized. We insert the vacuum polarization potential V_{VP} given by Eq. (24) additionally into the Dirac equation, which yields

$$[-i\boldsymbol{\alpha}\cdot\boldsymbol{\nabla} + V(\boldsymbol{x}) + V_{\mathrm{VP}}(\boldsymbol{x}) + \beta m - E_{\mathrm{VP}\,nlj}]\,\Psi_{\mathrm{VP}\,nlj} = 0 \quad . \tag{46}$$

The subscripts VP indicate the changes in the energy eigenvalue and in the wave functions which are a consequence of this procedure. The difference of the energy eigenvalue $E_{\mathrm{VP}\,nlj}$ and that of the ordinary bound state problem, E_{nlj}, can be thought to be due to the vacuum polarization ladder graphs of any number of loops attached to the bound electron propagator (Fig. 15). Subtracting also the dominating first order contribution $\Delta E_{\mathrm{VP}\,nlj}$ (22) yields only the energy shift due to the remaining two-and-more loop graphs. The effect as a whole is small and results in 0.22 eV for the $1s_{1/2}$-state in uranium compared to –88.6 eV of the one-loop graph already discussed. This procedure is therefore considered being a good approximation for the effect of the graph VPVP a) alone. To summarize, one obtains

$$\Delta E_{\mathrm{VPVPa}\,nlj} = E_{\mathrm{VP}\,nlj} - E_{nlj} - \Delta E_{\mathrm{VP}\,nlj}\,. \tag{47}$$

We illustrate this calculation by Fig. 16.

A somewhat similar approach can also be used for the mixed self energy – vacuum polarization diagrams of Fig. 13. The detailed evaluation of these graphs is presented by Lindgren *et al.* [59] and we report only the result of their calculation here, which for the $1s_{1/2}$-state of uranium yields 1.12 eV in the Uehling approximation (no Wichmann-Kroll vacuum polarization potential included in the Dirac equation) and 1.14 eV by taking into account the Wichmann-Kroll potential also [7].

The remaining S(VP)E contribution has been evaluated in the Uehling approximation by Mallampalli and Sapirstein [60] and Persson *et al.* [7]. This effective self-energy diagram requires charge renormalization as well as mass renormalization. At present only the contribution of the Uehling term of the effective photon

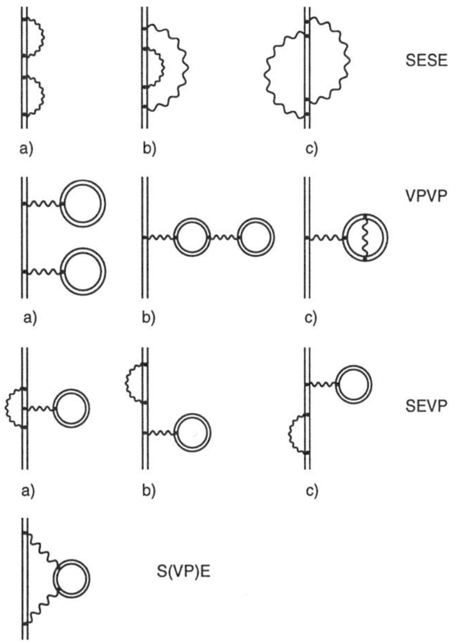

Fig. 13. QED diagrams of order α^2 (two photon lines) contributing to the Lamb shift in hydrogenlike systems. SE implies "self energy", VP "vacuum polarization". The nomenclature of the single graphs is therefore evident.

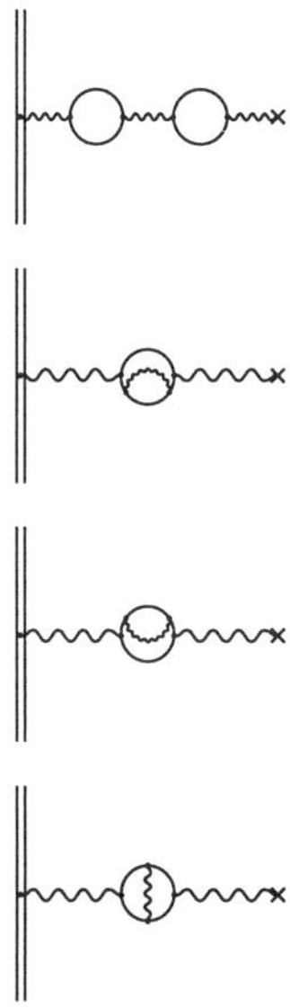

Fig. 14. Diagrams contributing to the Källén-Sabry potential. These diagrams represent only the Uehling approximation of the full diagrams VPVP b) and c).

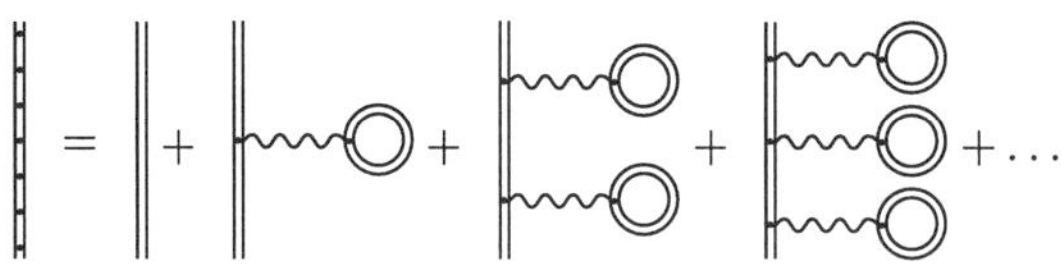

Fig. 15. The inclusion of the vacuum polarization potential into the Dirac equation results in an electron propagator dressed by vacuum polarization loops.

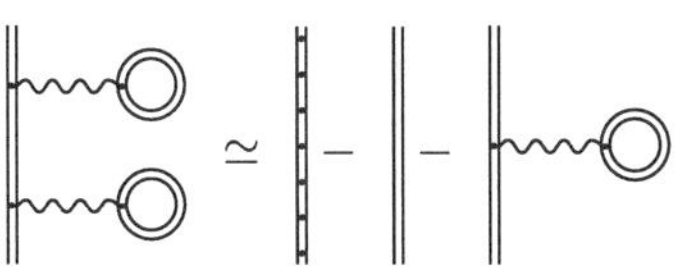

Fig. 16. Symbolical evaluation of the VPVP a) contribution (see Fig. 13) in terms of Feynman diagrams.

Table 1: Comparison of $Z\alpha$-expansion values to numerical calculcations including all orders in $Z\alpha$ for α^2-order QED contributions to the Lamb shift of the $1s_{1/2}$-state in lead and uranium. The Källén-Sabry contribution VPVP b) c) and the S(VP)E contribution are considered in the Uehling approximation only.

Z		VPVP a)	VPVP b) c)	SEVP	S(VP)E
82	numerical	-0.083 eV	-0.393 eV	0.534 eV	0.068 eV
	$Z\alpha$-exp.	-0.013 eV	-0.047 eV	0.406 eV	0.002 eV
92	numerical	-0.217 eV	-0.716 eV	1.139 eV	0.130 eV
	$Z\alpha$-exp.	-0.024 eV	-0.015 eV	0.722 eV	-0.007 eV

propagator but to all orders in $Z\alpha$ in the external electron line has been numerically evaluated. Mallampalli and Sapirstein [60] demonstrate that numerical techniques developed for high-precision calculations of the first-order self energy can be generalized and applied successfully to the case of the S(VP)E correction. The main reason for the feasibility is that the photon propagator keeps its simple analytic structure when it is modified by an Uehling loop. For the same reason the partial wave renormalization approach, which was already mentioned, also works in the case of the self energy - vacuum polarization correction [7]. Again we are not going to provide the details of the calculation and cite only the numerical result which yields 0.13 eV for the $1s_{1/2}$-state of uranium. To overcome the Uehling approximation, efforts are currently undertaken.

Up to now, we did not discuss any expansions in $Z\alpha$ series for the diagrams including vacuum polarisation contributions. For the $1s_{1/2}$-state of hydrogenlike lead and uranium we present in Table 1 a direct comparison between these values and our direct numerical calculations [7]. From this presentation it is evident that a $Z\alpha$ expansion also for the quantum electrodynamical corrections of order α^2 is completely meaningless. Reliable calculations for high-Z systems can only be performed by utilizing non-perturbative approaches.

Internal nuclear degrees of freedom

Besides the mass and the size of the nucleus also its internal degrees of freedom influence the binding energy. In the models we consider here the protons and neutrons in the nucleus act like an additional polarizable medium where virtual excitations may occur. Recent determinations of nuclear charge radii of highly-charged ions trapped in an EBIT indicate that experimental tests of the nuclear polarization might soon be possible by choosing isotopic systems where shifts of the nuclear mean-square-radius $\delta\langle r^2\rangle$ are very precisely known [61]. A relativistic field theoretical approach based on the concept of effective photon propagators has been developed in Refs. [62 – 64] where a systematic study of the nuclear polarization contributions in heavy even-A nuclei due to collective nuclear excitations has been presented. This formalism allows for the extension to single-particle excitations of a valence proton in odd nuclei [64]. In Ref. [65] a simple analytical formula for the nuclear polarization shift of energy levels in multicharged ions has been derived. It also turned out that previous numerical calculations have significantly overestimated the nuclear polarization effects in heavy ions. The relativistic consideration [63] indicates, that the contribution of the negative-frequency part to the nuclear polarization correction is also significant. We should emphazise that the expression for the energy shift obtained within the relativistic framework is by no means a straightforward generalization of ordinary second-order perturbation theory. A naive extension of the summation over the negative part of the Dirac spectrum in the corresponding energy formula incorporates this vacuum contribution with the wrong sign. However, in Refs. [62 – 65] a factor $(2\pi)^{-1}$ has been omitted in the formula for the nuclear polarization shift, and consequently all numbers listed there should be divided by 2π.

Nuclear polarization corrections for hydrogenlike ${}^{208}_{82}\mathrm{Pb}$ and ${}^{238}_{92}\mathrm{U}$ ions have been reconsidered recently [66]. Recalculations were performed within two different numerical approaches: the direct numerical integration [62 – 64] and the B-spline method [67]. Besides a comparison with analytical evaluations [65], an enlarged set of nuclear excitations has been taken into account in these calculations.

Fig. 17. Effective self energy diagram S(NP)E (left) and effective vacuum polarization diagram NP-VP (right) corresponding to the nuclear polarization. The diagrams marked "a" (hatched balls) show the polarization insertion into the photon propagator whereas those marked "b" and "c" show the actual interaction between electron and nucleus. The shaded lines indicate the nucleus with internal degrees of freedom.

Nuclear polarization gives rise to a modified photon interaction which contains nuclear polarization insertions. Performing standard perturbation theory for bound state QED, one is thus lead to an additional effective self energy diagram (Fig. 17, left) as well as to a diagram similar to vacuum polarization (Fig. 17, right). The energy shifts of bound electrons caused by these diagrams can be written as the contribution of an effective (first-order) self energy

$$\begin{aligned}\Delta E_{\mathrm{S(NP)E}\,nlj} &= i\,e^2 \int d^3x\, d^3x'\, \Psi^{+}_{nlj}(\boldsymbol{x}) \\ &\int \frac{dE}{2\pi}\, S_{\mathrm{F}}(\boldsymbol{x},\boldsymbol{x}',E)\, \tilde{\mathcal{D}}_{00}(\boldsymbol{x},\boldsymbol{x}',E_{nlj}-E)\gamma^0\Psi_{nlj}(\boldsymbol{x}') \quad (48)\end{aligned}$$

and a combined vacuum polarization - nuclear polarization correction

$$\Delta E_{\mathrm{VP-NP}nlj} = -i\,e^2 \int d^3x\, d^3x'\, \Psi^+_{nlj}(\boldsymbol{x})\, \Psi_{nlj}(\boldsymbol{x})\, \tilde{\mathcal{D}}_{00}(\boldsymbol{x},\boldsymbol{x}',0) \int \frac{dE}{2\pi}\, \mathrm{Tr}\left[\gamma^0\, S_{\mathrm{F}}(\boldsymbol{x}',\boldsymbol{x}',\,E)\right] . \tag{49}$$

Taking into account electric nuclear transitions only, gives rise to a longitudinal part $\tilde{\mathcal{D}}_{00}$ of the modified photon propagator, which reads in Coulomb gauge

$$\tilde{\mathcal{D}}_{00}(\boldsymbol{x},\boldsymbol{x}',E) = \sum_{LM} B(\mathrm{E}L;L\to 0)\, \frac{2E_L}{E^2 - E_L^2 + i\eta} \times F_L(|\boldsymbol{x}|)\, F_L(|\boldsymbol{x}'|)\, Y_{LM}(\hat{\boldsymbol{x}})\, Y^*_{LM}(\hat{\boldsymbol{x}}\,') \quad . \tag{50}$$

Here, $\hat{\boldsymbol{x}}$ denote the angles, E_L denotes the energy of a collective nuclear excitation with multipolarity L and corresponding reduced transition strength $B(\mathrm{E}L)$. The symbol $L \to 0$ indicates transitions to the ground state. The characteristic radial dependence is described by the functions

$$\begin{aligned} F_0(r) &= \frac{5\sqrt{\pi}}{2R_0^3}\left[1-\left(\frac{r}{R_0}\right)^2\right]\Theta(R_0-r), \quad L=0 \\ F_L(r) &= \frac{4\pi}{(2L+1)R_0^L} \\ &\times \left[\frac{r^L}{R_0^{L+1}}\,\Theta(R_0-r) + \frac{R_0^L}{r^{L+1}}\,\Theta(r-R_0)\right], \quad L\geq 1 \quad . \end{aligned} \tag{51}$$

The Y_{LM} denote the well-known spherical harmonics.

We have evaluated the energy shifts (48) by means of two different numerical methods. First, we performed a direct numerical integration over intermediate electron states. As in earlier calculations [62 – 64] the contributions arising from the bound electron states have not been taken into account. The corresponding results for the energy shifts caused by various collective nuclear excitations are given in the columns (a) of Tables 2 and 3. For the $1s_{1/2}$-state of some actinide nuclei, we also display the corresponding values in Figure 18. Secondly, we generated the complete spectrum of the Dirac equation using the B-spline method [67]. The effects of finite nuclear size were taken into account in both approaches. The calculations carried out in the framework of the B-spline method are listed in columns (b) of Tables 2 and 3. The polarization shifts obtained by the analytical formula derived in ref. [65] are also given in columns (c). The negative sign of the correction implies that it increases electron binding energies.

The corresponding experimental nuclear parameters are taken from Ref. [68] in the case of the $^{208}_{82}\mathrm{Pb}$ nucleus and from Ref. [69] for the $^{238}_{92}\mathrm{U}$. It should be noted that the difference between the results obtained within the different methods is smaller

Table 2: Energy shifts of 1s and 2s electrons in hydrogen-like $^{208}_{82}$Pb due to nuclear polarization effects. Upper half: The contributions from low-lying nuclear states are calculated using experimental values for excitation energies E_L and electric transition strengths $B(EL)$ [68]. Lower half: The resulting contributions from the isoscalar and isovector giant resonance states. The values for E_L and for $B(EL)$ are estimated according to Nefiodov *et al.* [66]. Column (a) corresponds to direct numerical integration; (b) – numerical calculations in the framework of the B-spline method; (c) – analytical evaluations according to [65]. Note that $\lambda = 2$ for monopole excitations and $\lambda = L$ in all other cases; b signifies barn.

E_L (MeV)	L	$B(EL)$ (e^2b^λ)	$-\Delta E^L_{1s}$ (meV)			$-\Delta E^L_{2s}$ (meV)		
			a	b	c	a	b	c
2.615	3^-	0.0870	0.8	0.8	0.8	0.1	0.2	0.1
4.085	2^+	0.0636	1.3	1.4	1.4	0.2	0.2	0.2
4.842	1^-	0.0005	0.0	0.1		0.0	0.0	
5.240	3^-	0.0186	0.2	0.2	0.2	0.0	0.0	0.0
5.292	1^-	0.0007	0.1	0.1		0.0	0.0	
5.512	1^-	0.0013	0.1	0.2		0.0	0.0	
6.193	2^+	0.0101	0.2	0.2	0.2	0.0	0.1	0.0
7.064	1^-	0.0005	0.0	0.1		0.0	0.0	
7.332	1^-	0.0007	0.1	0.1		0.0	0.0	
Total low-lying states			2.8	3.2		0.3	0.5	
10.522	2^+	0.154	2.7	2.9	3.5	0.5	0.5	0.6
10.522	3^-	0.117	1.0	1.0	1.1	0.2	0.2	0.2
13.328	1^-	0.185	17.3	17.5		3.0	3.0	
14.029	0^+	0.577	2.1	3.2		0.4	0.6	
22.447	2^+	0.111	1.6	1.6	2.5	0.3	0.3	0.4
22.447	3^-	0.084	0.6	0.6	0.8	0.1	0.1	0.1
28.058	0^+	0.444	1.2	1.8		0.2	0.3	
Total giant resonances			26.5	28.6		4.7	5.0	
Total shift			29.3	31.8		5.0	5.5	

than the inherent uncertainties of our calculations arising from the contribution of giant resonances. Quantities that can be deduced from experiment are the photoabsorption cross section and the spreading width. Therefore the $B(EL)$-values for giant resonances are estimated by means of well-known phenomenological energy-weighted sum rules. For a detailed discussion we refer to Ref. [66].

As can be deduced from the tables, the contributions of giant resonances turn out to be dominant for the lead nucleus and also significant for uranium. In the latter case the major contribution to the energy shifts arises due to the virtual excitation

Table 3: Energy shifts of K- and L-shell electrons in hydrogen-like $^{238}_{92}$U due to various collective excitations. Upper half: The contributions from low-lying nuclear states are calculated using experimental energies and transition probabilities [69]. Lower half: The contributions from giant resonance states. Excitation energies and corresponding reduced electric transition strengths are again estimated based on empirical formulae. Notations are the same as in Table 2.

E_L (MeV)	L	$B(EL)$ (e^2b^λ)	$-\Delta E^L_{1s}$ (meV)			$-\Delta E^L_{2s}$ (meV)			$-\Delta E^L_{2p_{1/2}}$ (meV)		
			a	b	c	a	b	c	a	b	c
0.0449	2^+	2.460	125.2	131.1	125.4	23.5	25.1	23.9	2.7	2.9	2.7
0.732	3^-	0.082	1.6	1.7	1.6	0.3	0.3	0.3	0.0	0.0	0.0
0.967	2^+	0.003	0.2	0.2	0.2	0.0	0.0	0.0	0.0	0.0	0.0
0.998	3^-	0.029	0.5	0.6	0.6	0.1	0.1	0.1	0.0	0.0	0.0
1.037	2^+	0.013	0.6	0.6	0.6	0.1	0.1	0.1	0.0	0.0	0.0
1.060	2^+	0.026	1.3	1.3	1.3	0.2	0.3	0.3	0.0	0.0	0.0
1.170	3^-	0.023	0.4	0.5	0.4	0.1	0.1	0.1	0.0	0.0	0.0
1.224	2^+	0.004	0.2	0.2	0.2	0.0	0.0	0.0	0.0	0.0	0.0
Total low-lying states			130.0	136.2	130.3	24.3	26.0	24.8	2.7	2.9	2.7
10.149	2^+	0.192	7.2	7.7	9.8	1.4	1.5	1.9	0.2	0.2	0.2
10.149	3^-	0.159	2.6	2.8	3.0	0.5	0.5	0.6	0.1	0.1	0.1
12.856	1^-	0.218	42.4	45.4		8.1	8.7		1.0	1.0	
13.533	0^+	0.721	6.0	9.6		1.1	1.9		0.1	0.2	
21.652	2^+	0.143	4.2	4.5	7.3	0.8	0.9	1.4	0.1	0.1	0.2
21.652	3^-	0.119	1.7	1.7	2.3	0.3	0.3	0.4	0.0	0.0	0.0
27.065	0^+	0.572	3.5	5.5		0.7	1.1		0.0	0.1	
Total giant resonances			67.6	77.2		12.9	14.9		1.5	1.7	
Total shift			197.6	213.4		37.2	40.9		4.2	4.6	

of the 2^+ rotational state within the ground-state band. The main reason for that is the very large value for the reduced transition strength. The numerical values for the energy shifts obtained by the direct summation and the B-spline method respectively deviate in the second digit in the case of low-lying nuclear states, while the deviations become more significant in the case of giant resonances. This might indicate that in the latter case the neglection of the contribution of the discrete Dirac spectrum should be avoided. However, we emphasize that numerical errors are well under control while nuclear parameters characterising the various nuclear excitations employed in the calculations remain as the major source for uncertainties. The total nuclear polarization shift of the ground state in the $^{238}_{92}$U ion is found to be about –0.2 eV. Note that this value has slightly changed compared to prior publications [7] as more states have been included in the present calculation. This alteration, however, gives an estimate of the uncertainty of the total nuclear polarization effect. While the shifts play only a minor role in $^{208}_{82}$Pb

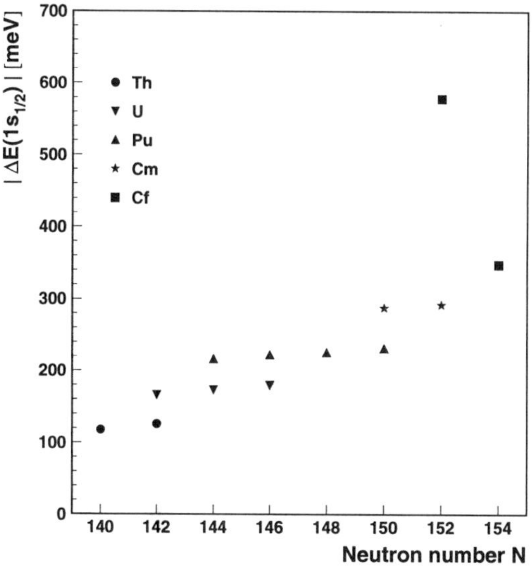

Fig. 18. Energy shift caused by the effective nuclear polarization self energy insertion (48) for $1s_{1/2}$ states of even-N even-A actinide nuclei.

this effect is essential for the determination of the natural limitations of present high-precision calculations and measurements of the Lamb shift in heavy systems. The contribution from the combined vacuum polarization - nuclear polarization correction $\Delta E_{\mathrm{VP-NP}}$ has been analysed recently for the hydrogenlike lead system [70]. For the ground state, we have obtained $\Delta E_{\mathrm{VP-NP}\,1s_{1/2}} = 0.023$ eV essentially arising from collective monopole vibrations of this particular nucleus. This value has nearly the same magnitude but opposite sign as the $\Delta E_{\mathrm{S(NP)E}\,1s_{1/2}}$ contribution of the same nucleus, which amounts to -0.029 eV [70]. This may underline that lead is a very promising candidate for tests of QED at utmost precision as the net effect of nuclear polarisation almost cancels. From the view point of high-precision QED-calculations, the effect of nuclear polarization introduces systematic errors mainly due to uncertainties of nuclear parameters employed and the restriction to a finite set of nuclear excitations. Accordingly, nuclear polarization represents a natural limitation of all more exact predictions as well as for measurements of the Lamb shift.

The status of the $1s_{1/2}$ Lamb shift calculations in hydrogenlike atoms

In the previous sections we have collected all corrections which contribute to the Lamb shift in hydrogenlike ions. For the radiative corrections from QED it was

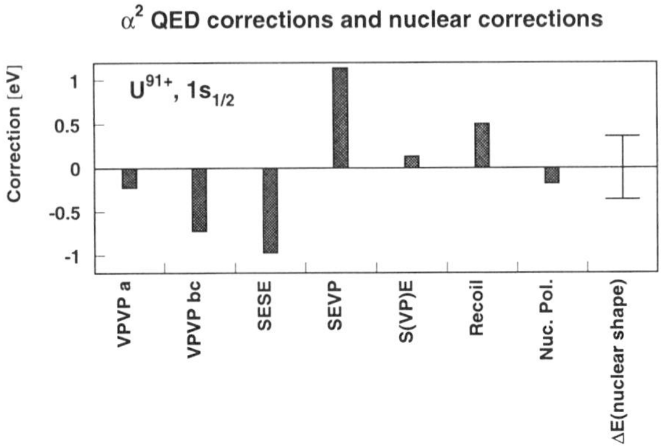

Fig. 19. Magnitude of QED contributions to the Lamb shift of order α^2 and contributions from nuclear properties which are of the same size. Note that the SESE value can serve only as an order-of-magnitude estimate as discussed previously.

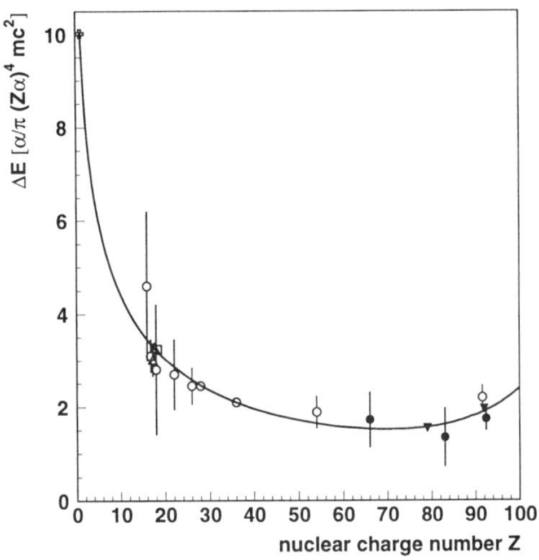

Fig. 20. Comparision of experimental measured values for the Lamb shift in hydrogenlike ions and the theoretical predicition. The energy shift is presented in the dimensionless unit $F(Z\alpha)$ similar to Fig. 10.

Table 4. Contributions to the $1s_{1/2}$ Lamb shift in hydrogenlike Pb and U. Values are given in eV. Parts of the SESE graphs are not yet calculated. The different contributions are obtained as discussed above. For the finite size correction we also quote the uncertainty obtained by the estimate comparing Fermi distribution and homogeneous sphere.

	Pb^{81+}	U^{91+}
Binding energy E_B for point nucleus:	-101581.37	-132279.96
Correction		
finite size	66.60 (14)	198.82 (36)
VP (order α)	-48.41	-88.60
SE (order α)	226.33	355.05
SESEa (irr)	-0.45	-0.97
SESEa (red) + SESE b,c	(?)	(?)
VPVPa	-0.08	-0.22
VPVPb,c	-0.39	-0.72
SEVP a,b,c	0.53	1.14
S(VP)E	0.07	0.13
Recoil	0.37	0.51
Nuclear pol.	-0.01	-0.20
Sum of corrections	284.56	464.94
Total binding energy	-101296.81	-131815.02
Reduced mass	0.27	0.30
Lamb Shift (Theory)	284.29	464.64
Lamb Shift (Exp)		470 ± 16

found that nowadays methods exist to evaluate almost all of them non-perturbatively. Besides these radiative corrections to the point nucleus binding energy, also the mass of the nucleus, its size, its shape and its internal degrees of freedom influence the binding energy. Whereas the size correction is rather large in heavy ions and of the same magnitude as the radiative corrections of order α, the uncertainty caused by the insufficiently known shape and polarizability are of almost the same magnitude as the QED corrections of second order in α. For the $1s_{1/2}$-state of uranium we plot this comparision in Fig. 19. Table 4 displays all known corrections to the point nucleus binding energy for the ground states of hydrogenlike ^{208}Pb and ^{238}U. In particular, the nuclear polarization contribution in lead is much smaller than that in uranium. Also the shape uncertainty in the almost ball shaped nucleus of ^{208}Pb is smaller than that in uranium and the presented uncertainty may well serve as a crude upper estimation. In Fig. 20 we present a number of experimental values for the Lamb shift over the whole range of Z compared to the theoretical prediction performed as discussed above. Note that no experiment contradicts the calculations. Therefore we can state that our knowledge of the theory of QED up to now is not contradicted even by the very strong electromagnetic fields which occur in heavy hydrogenlike ions.

The hyperfine structure splitting for $^{209}\mathrm{Bi}^{82+}$

The high relative accuracy of the experimentally measured hyperfine splitting of 5.0840(8) eV in hydrogenlike ^{209}Bi [8] implies a testing scenario of QED corrections not only under the presence of a strong electric but also a strong magnetic field. Lamb shift measurements prove the laws of quantum electrodynamics in strong Coulomb fields only.

The relativistic interaction Hamiltonian of the interaction of a bound electron with the nucleus is given by

$$\hat{H}_{\mathrm{int}} = e\boldsymbol{\alpha} \cdot \boldsymbol{A}(\boldsymbol{x}) \,, \tag{52}$$

where $\boldsymbol{A}(\boldsymbol{x})$ indicates the nuclear magnetic vector potential. In the magnetic dipole approximation, this nuclear magnetic field is given by

$$\boldsymbol{A}(\boldsymbol{x}) = \frac{\boldsymbol{m} \times \boldsymbol{x}}{4\pi|\boldsymbol{x}|^3} = g_I \mu_N \frac{\boldsymbol{I} \times \boldsymbol{x}}{4\pi|\boldsymbol{x}|^3} \,, \tag{53}$$

where g_I is the anomalous magnetic moment of the nucleus, μ_N denotes Bohr's nuclear magneton, and $\boldsymbol{I}$ is the nuclear spin operator. In first order perturbation theory, the energy splitting between the ground-state hyperfine-structure levels was already calculated by Breit in 1930 [71]

$$\begin{aligned} \Delta E^1(F_2 - F_1) &= -\frac{e g_I \mu_N}{4\pi}\frac{2}{3}\left[F_2(F_2+1) - F_1(F_1+1)\right] \\ &\quad \times 2\int \mathrm{d}r\, g_{nlj}(r) f_{nlj}(r) \,, \end{aligned} \tag{54}$$

where F denotes the total spin. The superscript "1" is used for "first-order".

Consideration of an extended nuclear charge distribution leads to a modification of (54) which can be expressed in terms of the Breit-Rosenthal parameter ε [72] [73]

$$\Delta E^1(\text{extended nucleus}) = (1-\varepsilon)\Delta E^1(\text{point nucleus}) \,. \tag{55}$$

This effect was first examined in detail by Rosenberg *et al.* [74] and in a more refined way by Blundell *et al.* [75] and also by Shabaev [76]. In total, a value of $\Delta E^1 = 5.1917(10)$ eV is obtained considering an extended bismuth nucleus with a two-parameter Fermi charge distribution of $c = 6.75(7)$ fm and $a = 0.468(39)$ fm which results in $\langle r^2 \rangle^{1/2} = 5.519$ fm [77] [78].

An extended nuclear magnetization distribution also causes deviations from the simple formula (54). This is known as Bohr-Weisskopf effect. The prediction of its numerical value strongly depends on the model which is used for the nuclear magnetization distribution. Recently, a number of approaches to the problem were made [78 – 80]. In all calculations one proton of the bismuth nucleus is considered as a valence proton moving in the field of the double magic ^{208}Pb core. The most elaborated model of Tomaselli *et al.* [80] also includes virtual excitations of the remaining core produced by this valence particle. They obtain a value of $\Delta E_{\rm BW} = -0.107(7)$ eV for the ground state hyperfine splitting in $^{209}\mathrm{Bi}^{82+}$.

The major point we focus on in this contribution are the quantum electrodynamical corrections to the measured effects. Again these consist of vacuum polarization and self energy. The vacuum polarization calculations were performed by Schneider *et al.* [81]. Restricting the calculation to single magnetic coupling three different contributions occur which are depicted in Fig. 21. We denote the shift caused by diagrams (a) and (b) $\Delta E_{\rm EL}$ and that caused by diagram (c) $\Delta E_{\rm ML}$. $\Delta E_{\rm EL}$ is treated in a similar manner as some of the vacuum polarization corrections of second order in α in the Lamb shift corrections. By including the additional Uehling and Wichmann-Kroll potentials in the Dirac equation wave functions $\Psi_{\mathrm{Uehl}\,nlj}$ and $\Psi_{\mathrm{VP}\,nlj}$ are generated which are used then to evaluate the first-order perturbation theory contribution (54). The difference

$$\Delta E_{\rm EL,\,Uehl} = \Delta E^1_{\rm Uehl}(\text{extended nucleus}) - \Delta E^1(\text{extended nucleus}) \qquad (56)$$

yields the Uehling contribution to $\Delta E_{\rm EL}$, which amounts to 0.0260 eV here. A similar calculation results in –0.0007 eV for the Wichmann-Kroll contribution.

For the magnetic–loop correction the energy splitting in the Uehling approximation is given by

$$\begin{aligned}\Delta E_{\mathrm{ML}\,nlj\,\mathrm{Uehl}} &= \frac{2\kappa}{4k^2-1} e g_I \mu_N \left[F(F+1) - I(I+1) - j(j+1)\right] \\ &\times \left(2\frac{2}{3}\frac{\alpha}{\pi} \int\limits_1^\infty dz \sqrt{1-\frac{1}{z^2}} \left(1+\frac{1}{2z^2}\right)\frac{1}{z} \right. \\ &\times \left. \int\limits_0^\infty dr\, g_{nlj}(r) f_{nlj}(r)\, e^{-2mrz}\left[2mrz+1\right] \right)\Bigg|_{(F_1)-(F_2)} \quad ,(57)\end{aligned}$$

where I denotes the total spin of the nucleus. The numerical power of our calculations is clearly demonstrated by Fig. 22 which shows the energy shift obtained by (57) compared to that obtained by the leading term in a nonrelativistic $Z\alpha$ expansion [81]. For the ground state of hydrogenlike ^{209}Bi the numerical value of $\Delta E_{\rm ML}$ amounts to 0.0093 eV. For a more detailed treatment of the influence of vacuum polarization to the hyperfine splitting we refer to [81].

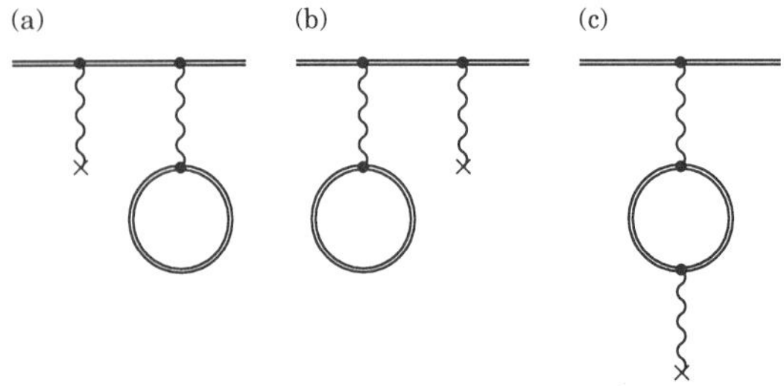

Fig. 21. The vacuum–polarization–like Feynman diagrams for the hyperfine–structure splitting. (a) and (b) are the wave–function corrections and (c) is the magnetic–loop modification. The cross × signifies the interaction with the nuclear magnetization distribution.

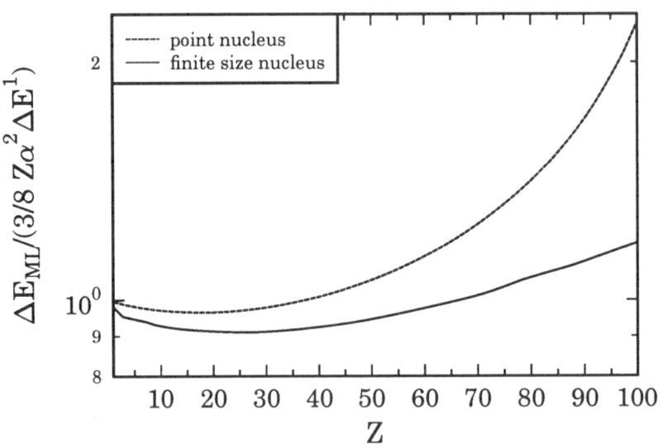

Fig. 22. The magnetic loop splitting energy E_{ML} of the ground state, normalized to $\frac{3}{8}\alpha(Z\alpha)\Delta E^{1.\mathrm{ord}}$, which is the leading term in a non-relativistic expansion. The dashed line refers to a relativistic point nucleus calculation whereas the solid line indicates the corresponding value for extended nuclei. The difference between both charge distribution models is notable. Note the logarithmic scale of the ordinate where 10^0 indicates 1.

The self energy diagrams which contribute to the hyperfine splitting energy are depicted in Fig. 23. They give rise to three different contributions which are further on named wave function modification contribution, binding energy correction and vertex correction. The wave function modification term includes the non-degenerate part of diagrams (a) and (b) in Fig. 23 (i.e., without the reference state between the magnetic interaction and the self energy loop) and reads

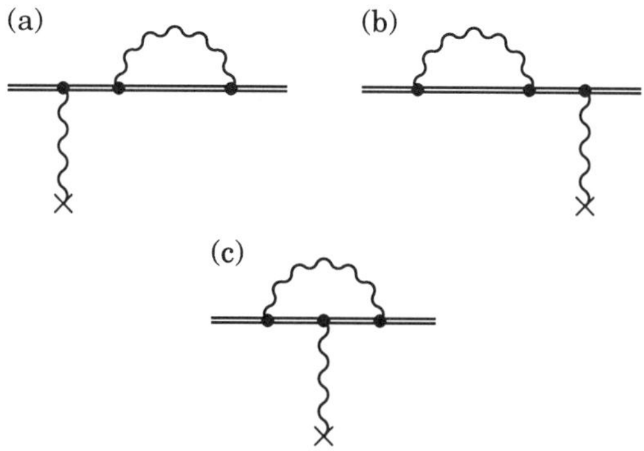

Fig. 23. The self–energy–like Feynman diagrams for the hyperfine–structure splitting in second order in e and first order in the external vector field. (a) and (b) are again wave–function corrections and (c) is the modification due to the perturbation of the propagator, the exact vertex correction.

$$E_{\mathrm{WFC}\,nlj} = -\frac{\alpha}{\pi}\sum_{p=0}^{\infty}(2p+1)\int \mathrm{d}k\, k \sum_q (-1)^q \sum_{t\neq nlj}\sum_u \frac{\langle nlj|e\boldsymbol{\alpha}\cdot\boldsymbol{A}(\boldsymbol{x}_3)|t\rangle\langle t|\alpha_\mu j_p(kr_2)\mathcal{C}_q^{[p]}|u\rangle\langle u|j_p(kr_1)\mathcal{C}_{-q}^{[p]}\alpha^\mu|nlj\rangle}{(E_{nlj}-E_t)(E_{nlj}-E_u-\mathrm{sign}(E_u)k)}\,. \tag{58}$$

In this equation, $\mathcal{C}_m^{[l]} = \sqrt{4\pi/(2l+1)}\mathcal{Y}_m^{[l]}(\vartheta,\varphi)$ and $\mathcal{Y}_m^{[l]}(\vartheta,\varphi) = i^{-l}\langle\vartheta,\varphi|lm\rangle$. This is the convention used in Ref. [82].

The binding energy correction is obtained from the degenerate part of the same diagrams,

$$E_{\mathrm{BEC}\,nlj} = \frac{\alpha}{\pi}\sum_{p=0}^{\infty}(2p+1)\int \mathrm{d}k\, k \sum_q (-1)^q \langle nlj|e\boldsymbol{\alpha}\cdot\boldsymbol{A}(\boldsymbol{x})|nlj\rangle \times\sum_u \frac{\langle nlj|\alpha_\mu j_p(kr_2)\mathcal{C}_q^{[p]}|u\rangle\langle u|j_p(kr_1)\mathcal{C}_{-q}^{[p]}\alpha^\mu|nlj\rangle}{(E_{nlj}-E_u-\mathrm{sign}(E_u)k)^2}\,. \tag{59}$$

Diagram Fig. 23 (c) corresponds to the vertex correction and reads

$$E_{\mathrm{VC}\,nlj} = -\frac{\alpha}{\pi}\sum_{p=0}^{\infty}(2p+1)\int \mathrm{d}k\, k \sum_q (-1)^q \times\sum_{t,u}\frac{\langle nlj|\alpha_\mu j_p(kr_3)\mathcal{C}_q^{[p]}|u\rangle\langle u|e\boldsymbol{\alpha}\cdot\boldsymbol{A}(\boldsymbol{x}_2)|t\rangle\langle t|j_p(kr_1)\mathcal{C}_{-q}^{[p]}\alpha^\mu|nlj\rangle}{(E_{nlj}-E_u-\mathrm{sign}(E_u)k)(E_{nlj}-E_t-\mathrm{sign}(E_t)k)}\,U\,, \tag{60}$$

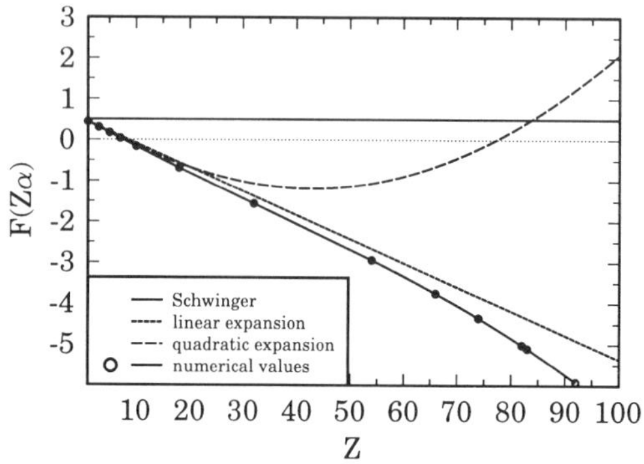

Fig. 24. Graphical comparison of our new numerical values (full line with circles) and the old $Z\alpha$–expansion results displayed as $F(Z\alpha)$. We plotted the Schwinger result (full line), the expansion up to terms proportional to $(Z\alpha)$ (short dashed line, second line from bottom), and the expansion up to terms proportional to $(Z\alpha)^2$ (long dashed line). The numerical results accidently are close to the linear expansion also for high Z.

Table 5: Hyperfine structure splitting contributions to the ground state of $^{209}Bi^{82+}$

rms radius: $\langle r^2 \rangle^{1/2}$ =5.519 fm
magnetic moment: $\mu = 4.1106\ \mu_B$

	ΔE [eV]
First order	5.1917
Bohr-Weisskopf	–0.107(7)
Vacuum polarization	
Uehling-like loop correction	0.0093
Uehling corr. of wave function	0.0260
Wichmann-Kroll corr. of wave f.	–0.0007
Self energy	–0.06144
Sum of QED corrections	–0.02684
Total	5.058(8)
Experiment	5.0840(8)

where a function U was introduced to account for the pole–structure in the complex z–plane,

$$U = 1 + (\text{sign}(E_u) - \text{sign}(E_t))\frac{k}{E_u - E_t} \ . \tag{61}$$

In equations (58), (59), and (60), the indices 1 – 3 distinguish between the different spatial coordinates of integration. For evaluation of these formulae and proper handling of divergencies we refer to [83]. For the hyperfine splitting in hydrogenlike ^{209}Bi, we obtain a self energy contribution of –0.06144(1) eV.

We emphasize that our result is obtained by a nonperturbative approach without any expansion in $Z\alpha$. To illustrate the breakdown of any $Z\alpha$ expansion in the case of high-Z systems, we compare our numerical results to an expansion up to $(Z\alpha)^2$, which is given by [84 – 88]

$$\begin{aligned} F(Z\alpha) &= \frac{1}{2} - (Z\alpha)\left[\frac{13}{4} - \ln 2\right]\pi \\ &+ (Z\alpha)^2\left[15.10 + \left[\frac{37}{72} - \frac{8}{3}\ln 2\right] ln(Z\alpha) - \frac{8}{3}\ln^2(Z\alpha)\right] \ . \end{aligned} \tag{62}$$

The function F is defined via

$$\Delta E_{\text{HFS SE}} = \frac{\alpha}{\pi}\Delta E_{\text{HFS}}^{\text{1. ord.}} F(Z\alpha) \ . \tag{63}$$

Fig. 24 displays the $Z\alpha$-expansion F in different approximations and our corresponding numerical values. In Table 5 we list all contributions to the hyperfine splitting energy. Surprisingly, the experimental value is almost exactly that obtained without any QED corrections. Adding the QED corrections we get a discrepancy of 0.026 eV between theory and experiment. This is probably due to the yet insufficient handling of the extended magnetic moment distribution in the bismuth nucleus which requires still more theoretical effort. We emphasize that the magnitude of the QED corrections is of almost the same size as the Bohr-Weisskopf contribution and can by no means be neglected in any experiment considering heavy hydrogenlike ions.

Final remarks

In the preceeding sections we have shown the current status of quantum electrodynamical and related calculations in heavy hydrogenlike systems with its unique strong electric and magnetic fields. From our present experimental knowledge there is no contradiction to the theory of quantum electrodynamics as we use it nowadays. However, an increasing experimental precision may still point to deviations in the theory since in Lamb shift calculations the predictions are still at least one order of magnitude more precise than the best experimental values. On the other hand we have also shown the limitations of theoretical predictions in the systems under consideration. Our limited experimental and theoretical knowledge of the nuclear parameters still limits the possibility to test more elaborated quantum electrodynamical calculations, as we see very clearly in the case of the hyperfine structure splitting. But in general quantum electrodynamics in strong fields is still a very stimulating area for both experimentalists and theoreticians to probe nature by the most precise physical theory presently known.

Acknowledgements

G. S., T. B., M. G., and G. P. thank the GSI, the BMBF, and the DFG for financial support. H. P. gratefully acknowledges support by the Alexander-von-Humboldt foundation. We are all grateful to M. G. H. Gustavsson, C. R. Hofmann, L. Labzowsky, P. Mohr, S. Salomonson, P. Sunnergren, and especially to I. Lindgren for countless stimulating and fruitful discussions.

References

[1] J. Schweppe, A. Belkacem, L. Blumenfeld, N. Claytor, B. Feinberg, H. Gould, V. E. Costram, L. Levy, S. Misawa, J. R. Mowat, and M. H. Prior, Phys. Rev. Lett. **66**, (1991), 1434.

[2] H. F. Beyer, D. Liesen, F. Bosch, K. D. Finlayson, M. Jung, O. Kleppner, R. Moshammer, K. Beckert, H. Eickhoff, B. Franzke, F. Nolden, P. Spädtker, M. Steck, G. Menzel, R. D. Deslattes, Phys. Lett. **A184**, (1994), 435.

[3] J. P. Briand, P. Chevallier, P. Indelicato, K. P. Ziock, and D. D. Dietrich, Phys. Rev. Lett. **65**, (1990), 2761.

[4] Th. Stöhlker, P. H. Mokler, K. Beckert, F. Bosch, H. Eickhoff, B. Franzke, M. Jung, T. Kandler, O. Klepper, C. Kozhuharov, R. Moshammer, F. Nolden, H. Reich, P. Rymuza, P. Spädtke, and M. Steck, Phys. Rev. Lett. **71**, (1993), 2184.

[5] J. H. Lupton, D. D. Dietrich, C. J. Hailey, R. E. Stewart, and K. P. Ziock, Phys. Rev. **A 50**, (1994), 2150.

[6] H. F. Beyer, IEEE Trans. Instrum. Meas. **44**, (1995), 510;
H. F. Beyer, G. Menzel, D. Liesen, A. Gallus, F. Bosch, R. Deslattes, P. Indelicato, Th. Stöhlker, O. Klepper, R. Moshammer, F. Nolden, H. Eickhoff, B. Franzke, and M. Steck, Z. Phys. **D 35**, (1995) 169.

[7] H. Persson, I. Lindgren, L. Labzowsky, T. Beier, G. Plunien, and G. Soff, Phys. Rev. **A 54**, (1996), 2805

[8] I. Klaft, S. Borneis, T. Engel, B. Fricke, R. Grieser, G. Huber, T. Kühl, D. Marx, R. Neumann, S. Schröder, P. Seelig, L. Völker, Phys. Rev. Lett. **73**, (1994) 2425.

[9] K. Hermanspahn, W. Quint, S. Stahl, M. Tönges, G. Bollen, H.-J. Kluge, R. Ley, R. Mann, and G. Werth, Acta Phys. Pol. **B 27**, (1996), 357.

[10] G. Soff, T. Beier, and C. Hofmann, Acta Phys. Pol. **B 27**, (1996), 369.

[11] D. C. Ionescu, J. Reinhardt, B. Müller, and W. Greiner, Phys. Rev. **A 38**, (1988), 616.

[12] J. R. Sapirstein and D. R. Yennie, in *Quantum Electrodynamics,* edited by T. Kinoshita (World Scientific, Singapore, 1990), p. 560.

[13] P. J. Mohr, *Tests of Fundamental Physics*, in Atomic Molecular, and Optical Handbook, ed. by G. W. F. Drake, (AIP, Woodbury, New York, 1996), p. 341.

[14] K. Pachucki, Ann. Phys. (N. Y.) **226**, (1993), 1.

[15] G. E. Brown, J. S. Langer, and G. W. Schaefer, Proc. Roy. Soc. (London) **A 251**, (1959), 92.

[16] A. M. Desiderio, W. R. Johnson, Phys. Rev. **A 3**, (1971), 1267.

[17] P. J. Mohr, Ann. Phys. (NY) **88**, (1974), 26.

[18] P. J. Mohr, Ann. Phys. (NY) **88**, (1974), 52.

[19] S. A. Blundell, N. J. Snyderman, Phys. Rev. **A 44**, (1991), 1427.

[20] P. J. Mohr, Physica Scripta **T 46**, (1993), 44.

[21] P. J. Mohr and G. Soff, Phys. Rev. Lett. **70**, (1993), 158.

[22] P. Indelicato, P. J. Mohr, Phys. Rev. **A 46**, (1992), 172.

[23] H. M. Quiney, I. P. Grant, Phys. Scr. **T 46**, (1993), 132.

[24] H. M. Quiney, I. P. Grant, J. Phys. **B 27**, (1994), L299.

[25] H. Persson, I. Lindgren, S. Salomonson, Physica Scripta **T 46**, (1993), 125.

[26] I. Lindgren, H. Persson, S. Salomonson, and A. Ynnerman, Phys. Rev. **A 47**, (1993), R4555.

[27] S. S. Schweber, *An Introduction to Relativistic Quantum Field Theory*, (Harper & Row, New York, 1961)

[28] E. A. Uehling, Phys. Rev. **48**, (1935), 55.

[29] J. Schwinger, Phys. Rev. **82**, (1951), 664.

[30] W. Pauli and F. Villars, Rev. Mod. Phys. **21**, 434.

[31] E. H. Wichmann and N. M. Kroll, Phys. Rev. **101**, 843.

[32] K.-N. Huang, Phys. Rev. **A 14**, (1976), 1311.

[33] L. Wayne Fullerton and G. A. Rinker, Jr., Phys. Rev. **A 13**, (1976), 1283.

[34] S. Klarsfeld, Phys. Lett. **B 66**, (1977), 86.

[35] D. J. Hylton, Phys. Rev. **A 32**, (1985), 1303.

[36] M. Gyulassy, Phys. Rev. Lett. **33**, (1974), 921.
M. Gyulassy, Phys. Rev. Lett. **32**, (1974), 1393.
M. Gyulassy, Nucl. Phys. **A 244**, (1975), 497.

[37] A. R. Neghabian, Phys. Rev. **A 27**, (1983), 2311.

[38] G. Soff and P. J. Mohr, Phys. Rev. **A 38**, (1988), 5066.

[39] H. Persson, I. Lindgren, S. Salomonson, P. Sunnergren, Phys. Rev. **A 48**, (1993), 2772.

[40] J. Blomqvist, Nucl. Phys. **B 48**, (1972), 95.

[41] G. Soff and P. J. Mohr, Phys. Rev. **A 40**, (1989), 2174.

[42] A. N. Artemyev, V. M. Shabaev, V. A. Yerokhin, Phys. Rev. **A 52**, (1995), 1884.

[43] A. N. Artemyev, V. M. Shabaev, V. A. Yerokhin, J. Phys. **B 28**, (1995), 5201.

[44] W. R. Johnson, G. Soff, At. Data and Nucl. Data Tables **33**, (1985), 405.

[45] M. E. Rose, *Relativistic Electron Theory*, (Wiley, New York, 1961)

[46] R. Engfer, H. Scheuwly, J. L. Vuilleumier, H. K. Walter, A. Zehnder, At. Data Nucl. Data Tables, **14**, 509 (1974)

[47] C. W. Jager, H. de Vries, C. de Vries, At. Data Nucl. Data Tables, **14**, (1974), 599.

[48] H. de Vries, C. W. de Jager, C. de Vries, At. Data Nucl. Data Tables, **36**, (1987), 495.

[49] E. G. Nadjakov, K. P. Marinova, Yu. P. Gangrsky, At. Data Nucl. Data Tables, (1994), 133.

[50] B. Nerlo-Pomorska, B. Mach, At. Data Nucl. Data Tables **60**, (1995), 287.

[51] D. Zumbro, E. B. Shera, Y. Tanaka, C. E. Bemis, Jr., R. A. Naumann, M. V. Hoehn, W. Reuter, R. M. Steffen, Phys. Rev. Lett. **53**, (1984), 1888.

[52] T. Franosch, G. Soff, Z. Phys. **D 18**, (1991), 219.

[53] L. N. Labzowsky and A. O. Mitrushenkov, Phys. Rev. **A 53**, (1996), 53.

[54] A. Mitrushenkov, L. Labzowsky, I. Lindgren, H. Persson, and S. Salomonson, Phys. Lett. **A200**, (1995), 51

[55] G. Källén, A. Sabry, Mat. Fys. Medd. Dan. Vid. Selsk. **29**, (1955), 17.

[56] T. Beier, G. Soff, Z. Phys. **D 8**, (1988), 129.

[57] S. M. Schneider, W. Greiner, G. Soff, J. Phys. **B 26**, (1993), L529

[58] N. L. Manakov and A. A. Nekipelov, in: Materials of the XXI Congress on Spectroscopy, Russia, Zvenigorod, 1995, p. 183 (in Russian).

[59] I. Lindgren, H. Persson, S. Salomonson, V. Karasiev, L. Labzowsky, A. Mitrushenkov, M. Tokman, J. Phys. B **26**, (1993), L 503.

[60] S. Mallampalli and J. Sapirstein, Phys. Rev. **A 54**, (1996), 2714.

[61] S. R. Elliott, P. Beiersdorfer, and M. H. Chen, Phys. Rev. Lett. **76**, (1996), 1031.

[62] G. Plunien, B. Müller, W. Greiner, and G. Soff, Phys. Rev. **A 39**, (1989), 5428.

[63] G. Plunien, B. Müller, W. Greiner, and G. Soff, Phys. Rev. **A 43**, (1991), 5853.

[64] G. Plunien and G. Soff, Phys. Rev. **A 51**, (1995), 1119; Phys. Rev. **A 53**, (1996), 4614.

[65] L. N. Labzowsky and A. V. Nefiodov, Phys. Lett. **A 188**, (1994), 371.

[66] A. V. Nefiodov, L. N. Labzowsky, G. Plunien, G. Soff, Phys. Lett. **A 222** (1996), 227.

[67] W. R. Johnson, S. A. Blundell, and J. Sapirstein, Phys. Rev. **A 37**, (1988), 307.

[68] M. J. Martin, Nucl. Data Sheets **47**, (1986), 797.

[69] E. N. Shurshikov, Nucl. Data Sheets **53**, (1988), 601.

[70] L. N. Labzowsky, A. V. Nefiodov, G. Plunien, T. Beier, G. Soff, J. Phys. **B 29**, (1996), 3841

[71] G. Breit, Phys. Rev. **35**, (1930), 1447.

[72] E. Rosenthal, G. Breit, Phys. Rev. **41**, (1932), 459.

[73] M. Crawford, A. Schawlow, Phys. Rev. **76**, (1949), 1310.

[74] H. J. Rosenberg, H. H. Stroke, Phys. Rev. **A 5**, (1972), 1992.

[75] S. A. Blundell, C. W. P. Palmer, J. Phys. **B 21**, (1988), 3809.

[76] V. M. Shabaev, J. Phys. **B 27**, (1994), 5825.

[77] M. Finkbeiner, B. Fricke, T. Kühl, Phys. Lett **A 176**, (1993), 113.

[78] S. M. Schneider, J. Schaffner, G. Soff, W. Greiner, J. Phys. **B 26**, (1993), L581.

[79] L. N. Labzowsky, W. R. Johnson, S. M. Schneider, G. Soff, Phys. Rev. **A 51**, (1995), 4597.

[80] M. Tomaselli, S. M. Schneider, E. Kankeleit, T. Kühl, Phys. Rev. **C 51**, (1995), 2989.

[81] S. M. Schneider, G. Soff, W. Greiner, Phys. Rev. **A 50**, (1994), 118.

[82] A. Lindner, *Drehimpulse in der Quantenmechanik*, (Teubner, Stuttgart, 1984)

[83] H. Persson, S. M. Schneider, G. Soff, W. Greiner, I. Lindgren, Phys. Rev. Lett. **76**, (1996), 1433.

[84] N. M. Kroll and F. Pollock, Phys. Rev. **85**, (1952), 876.

[85] R. Karplus, A. Klein, Phys. Rev. **85**, (1952), 927.

[86] D. E. Zwanziger, Phys. Rev. **121**, (1961), 1128.

[87] S. J. Brodsky, G. W. Erickson, Phys. Rev. **148**, (1966), 26.

[88] J. R. Sapirstein, Phys. Rev. Lett. **51**, (1983), 985.

STATE-SPECIFIC MULTI-REFERENCE COUPLED CLUSTER FORMULATIONS : TWO PARADIGMS

UTTAM SINHA MAHAPATRA, BARNALI DATTA,
BARUN BANDYOPADHYAY and DEBASHIS MUKHERJEE
Department of Physical Chemistry,
Indian Association for the Cultivation of Science,
Calcutta 700032, INDIA

ABSTRACT

The traditional multi-reference coupled cluster (MRCC) methods are based on effective hamiltonian formalism and often suffer from the problem of intruders. A state-specific MRCC approach, focusing on only one state, offers the attractive possibility of avoiding intruders while at the same time incorporating the nondynamical correlation in a size-extensive manner. In this paper we discuss two alternative paradigms which allow us to achieve this goal. The first, to be called the decontracted description, we deliberately retain certain linearly dependent cluster amplitudes and allow the combining coefficients of the reference determinants to be updated to their values for the exact function. The presence of the linearly dependent cluster amplitudes requires imposition of suitable sufficiency conditions, which are invoked in a manner which naturally ensures size-extensivity. In the second approach, to be called the contracted description, we would generate a cluster expansion with respect to the entire reference function consisting of a combination of reference determinants and retain only the linearly independent cluster-amplitudes in the cluster expansion. For an efficient implementation of the formalism, we shall introduce the notion of extended normal ordering and an analogue of Wick's theorem which uses the entire reference function as the multi-determinantal analogue of the vacuum. This necessarily imposes the restriction that the combining coefficients appearing in the reference function have to remain frozen in the equations for the cluster amplitudes. Relaxation of the coefficients can be acheived only after the cluster-amplitudes with the current coefficients are solved.

ADVANCES IN QUANTUM CHEMISTRY, VOLUME 30

0065-3276/98 $25.00

I. INTRODUCTION

The search for viable size-extensive methods which are computationally tractable and stable over a wide range of nuclear geometries, including quasi-degeneracies leading to avoided crossings, remains a frontier area of quantum chemical research. It seems to be generally agreed upon that a cluster expansion starting from a multi-determinantal reference function is the most promising strategy, but its rigorous realization remains a formidable challenge. There are several theoretical aspects which should be carefully resolved before a satisfactory solution can emerge. Firstly, the cluster expansion must be well-defined in the sense that there should be well-defined equations determining the cluster amplitudes. A straightforward cluster expansion from a multi-determinant reference function entails redundancy and linear dependence of the cluster amplitudes. Thus either suitable sufficiency conditions or some supplementary conditions motivated by physical appeal have to be invoked to resolve the redundancy, or, the redundancy should have to be systematically eliminated without sacrificing generality. Secondly, the cluster expansion should confer enough flexibility to the wave function such that the combining coefficients of the reference determinants take on the values they should have in the full function. There are two paradigms which strive to attain these two requirements in two different ways. In one paradigm, which we shall call a *decontracted description*, the combining coefficients are sought to be obtained by diagonalizing a matrix in the space of reference determinants and the redundancy of the cluster-amplitudes is exploited to ensure size-extensivity. In the other, the cluster expansion is attempted from the entire combination of determinants rather than the individual determinants themselves. The cluster operators are defined as excitations from the reference determinants. The prior fixing of the coefficients confers a *contracted description* of the reference function which has then to be relaxed to generate the proper mixing coefficients in the full final function. The redundancy of the cluster amplitudes is eliminated by extracting only the linearly independent amplitudes. It is interesting to note that although these two paradigms are not new, attempts for their systematic and rigorous implementation to generate size-extensive formalisms were rather late in coming. This happened primarily because of the theoretical difficulties mentioned. Multi-reference approaches focusing on several wave-functions in an effective hamiltonian framework were the first rigorous size-extensive formulations to appear, since size-extensivity is more easily monitored in this formulation, though their practical implementation was still problematic due to the intruder state problem. It is only in trying to remove the difficulty of the effective hamiltonian based multi-reference approaches that the newer formalisms started emerging. Since these focus on one state, yet start from several reference determinants, we may collectively call them state-specific multi-reference cluster expansion methods. Our intention is to highlight two such recent developments in this paper of which one belongs to the first paradigm and the other to the second. To motivate towards our formulations and to set our formulations in the perspective of the traditional multi-reference approaches, we would first present a resumé of the general aspects of cluster expansion representation.

II. CLUSTER EXPANSION STRATEGY : A BRIEF SURVEY

Cluster expansion representation of a wave-function built from a single determinant reference function [1] has been eminently successful in treating electron correlation effects with high accuracy for closed shell atoms and molecules. The cluster expansion approach provides size-extensive energies and is thus the method of choice for large systems. The two principal modes of cluster expansion developments in Quantum Chemistry have been the use of single reference many-body perturbation theory (SR-MBPT) [2] and the non-perturbative single reference Coupled Cluster (SRCC) theory [3,4]. While the former is computationally economical for the first few orders of the perturbation expansion [5], for treating higher order correlation effects or, in particular, for selectively summing a set of important higher order terms, the SRCC theory has the advantage of combining computational economy and accuracy [6,7]. The literature in this field is very extensive and we have mentioned above only some representative references.

For quasi-degenerate situations, where it is necessary to develop cluster expansion from multi-determinant reference functions, the theoretical possibilities are numerous. Using the approach of effective hamiltonians [8], there have been formulations both in perturbation theory and the Coupled Cluster theory. The perturbative many-body formalisms [9] were all based on the concept of a model space spanned by the determinants comprising the reference functions and as such they may all be classified as multi-reference many-body perturbative methods (MR-MBPT). Although the analogous multi-reference Coupled Cluster (MRCC) formulations also made use of effective hamiltonian approach, they were rather more varied in their scope of applications. They may be broadly classified under valence-universal [10-13] and valence-specific [14-15] categories. The former methods use a single wave-operator for not only the parent model space of interest but also for all the lower valence model spaces of various degrees of ionizations. They are thus useful for obtaining not only the state energies *per se* but also energy differences of spectroscopic interest. The valence-specific methods, on the other hand, generate the state energies of the functions generated from the model space of interest and are thus more suited to study potential surfaces.

The above multi-reference developments were all based on complete model spaces – a restriction mainly dictated by the desire to maintain size-extensivity of the energies obtained from the effective hamiltonian. Practical implementations of all these methods are seriously beset with the problem of intruders [16,17]. Although they can sometimes be avoided by clever choices of the single particle potentials defining the orbitals [18,19], it seems appropriate to look for methods which bypass intruders by a more flexible choice of model spaces. Size-extensive formulations have been developed recently with incomplete model spaces, by abandoning the intermediate normalization convention of the wave-operator and by imposing appropriate decoupling conditions for the transformed hamiltonian induced via the wave-operator [20-26]. Both valence-universal [20-24] and valence - specific [25,26] MRCC formulations using incomplete model spaces have now appeared.

It seems that intruders can be avoided almost entirely for states computed at fixed nuclear geometries by a judicious choice of the incomplete model space [27,28]. The situation is far less satisfactory, however, for generating potential surfaces. Here different intruder states plague the computations at different regions of the potential surface, and there is neither a unique nor a natural choice of an incomplete model space which avoids intruders at all geometries. It thus becomes necessary to switch over to different model spaces for describing the different regions of the surface [29].

The problem of intruders stems essentially from the effective hamiltonian approach for generating energies which requires that, for all $N \times N$ model space, all the N energies obtained on diagonalization are eigenvalues of the hamiltonian. This is a serious limitation, since due to the presence of intruders, not all of the N roots of the effective hamiltonians will have the model space functions as their dominant components. A much more natural approach would have been to abandon this restriction and to generate only those functions which are unencumbered by intruders. Such a possibilty was first explored in a perturbative development by Kirtman [30] who partitioned the model space into a main and an intermediate buffer subspace : the buffer subspace interacts strongly with intruders while the primary subspace does not (we use the modern terminology for these subspaces). The functions of the main subspace was allowed to mix with the virtual functions via the wave-operator as in the traditional effective hamiltonian method, while those of the intermediate subspace remained uncorrelated, thus avoiding intruders. The method was not fully size-extensive, but it showed a way to handle intruders. Malrieu *et al* [31] proposed the first systematic and general approach using the above partitioning concept which they termed as the intermediate hamiltonian approach. They also provided the general working equations for the pseudo wave-operator that provides eigenvalues for functions dominated by the primary space and some extraneous roots for the rest, dominated by the intermediate subspace. Since the extraneous roots are completely arbitrary, the intermediate hamiltonian approach is very flexible. This was exploited by the use of a suitable constant "shift" in the unperturbed energies of the intermediate components but allowing them to be perturbed in the same way as the main components. This is a definite improvement over the approach of Kirtman [30] who left the intermediate components uncorrelated. The early formulations as well as some later variants [31,32] were of configuration space-oriented non-many-body variety, and did not offer size-extensive energies.

Coupled Cluster based size-extensive intermediate hamiltonian formalisms were developed by our group [33-35] by way of transcribing a size-extensive CC formulation in an incomplete model space in the framework of intermediate hamiltonians. In this method, there are cluster operators correlating the main model space. There are no cluster operators for the intermediate space. This formulation thus is conceptually closer to the perturbative version of Kirtman [30], but unlike the latter, is fully size-extensive. Koch [36] derived a many-body intermediate hamiltonian method using different shift operators for the different intermediate functions. Koch's choice of the main model space was not fully flexible and he only surmised but could not prove size-extensivity of his method. Mukhopadhyay *et al* [33] developed a rigorously size-extensive CC formalism for intermediate hamiltonians using different shift operators for different functions via certain projectors. This method may be looked upon as the size-extensive CC implementation of the general strategy of intermediate hamiltonians [31-32]. Although not initially recognized, another size-extensive realization of the intermediate hamiltonians was achieved via a different route [34]. In an attempt to bypass intruders in the effective hamiltonian formulations of MRCC theory, the MRCC equations were reformulated as a pseudo eigenvalue problem in the union space comprised of the $N \times N$ model space of the MRCC theory and the space spanned by the virtuals reached by the action of the cluster operators on the model space functions. The N roots dominated by the model space functions turn out to be the same as those obtained from the conventional formulation of the MRCC theory, but the rest of the roots, though extraneous, are also reasonably good approximations to the higher roots of the hamiltonian. The union space may be looked upon as the composition of the model space, serving as the main subspace, and the virtuals, serving as the intermediate space. This was used for generating main as well as satellite peaks of ionization and Auger spectra. A more recent formulation by Datta *et al* [34] makes the connection of this approach with the intermediate hamiltonians particularly transparent. All these developments fall in the category of MRCC approaches to intermediate hamiltonians.

Essentially the same strategy, viz. of converting a set of CC equations to an equivalent pseudo eigenvalue problem, has been extensively followed by Malrieu *et al* [37] to generate intermediate hamiltonians for the ground state and for excited states spanned by single and double excitations from the ground state. The main difference of these methods with the MRCC approaches is the focus on generating only specific roots, but retaining the cluster-expansion structure of the wave-function. They are thus state-specific. In their more recent formulations, however, they do not start from a wave-operator description followed by a transcription to the pseudo eigenvalue form, and thus the presence of a cluster expansion representation of the wave-operator remains implicit [38]. As mentioned before, if a cluster expansion is attempted from a multi-determinantal function, there are redundant cluster amplitudes. By positing equations for them from without, suitable working schemes may be developed. Malrieu *et al* take precisely such an approach, guessing reasonable sufficiency equations from perturbative arguments [39]. We have also proposed an alternative formalism [40] where the corresponding sufficiency conditions follow naturally from an *Ansatz* for the wave-operator which decouples the dynamical and the non-dynamical correlation effects. We shall undertake in this paper the development of an explicitly wave-operator based strategy and, motivated by physical reasonings, shall propose suitable supplementary *non-perturbative* conditions. This method conforms to the first paradigm, viz. the decontracted description, and will be described in Sec. 3. Unlike the methods of ref [40], the present formulation does not invoke the artifact of separating the dynamical and non-dynamical correlations.

Silverstone and Sinanoglu [41] were the first to propose a cluster expansion starting from a combination of functions. This was the first attempt to formulate a multi-reference state-specific theory belonging to the second paradigm. To eliminate the redundancy of the cluster amplitudes in a straightforward cluster expansion, they proposed the anonymous parentage approximation to equate certain cluster amplitudes. This approximation is physically an untenable one and the method fell into disuse for quite a long time. Even now there are not many methods using this approach. Meanwhile, perturbative methods starting from a combination of functions came into the fore as practical tools to treat quasi-degeneracy [42-44]. The perturbtive developments require a suitable unperturbed hamiltonian about which the perturbation is performed. Several such choices have been considered in [42-44]. It seems that the methods are straightforward only at the second and third order. Higher order generalizations are unwieldy. Moreover, monitoring size-extensivity in an order-by-order expansion seems to be very difficult.

We have recently developed a rigorous cluster expansion formulation, making use of certain generalizations of the notions of normal ordering and Wick-like reduction theorem with respect to a multi-determinantal reference state [45-46]. In Sec. 4, we shall describe a state-specific formulation where we shall make use of a wave-operator written as an exponential of a cluster operator under this new normal ordering. The cluster operator will contain only the linearly independent cluster amplitudes defined with respect to the entire multi-reference combination as the starting function. This formulation obviously corresponds to the contracted description and belongs to the second paradigm. By the very nature of the formulation, the current combining coefficients cannot be updated during the determination of the cluster-amplitudes. They can be relaxed only after the current cluster-amplitudes are determined.

III. A STATE-SPECIFIC COUPLED CLUSTER THEORY WITH A DECONTRACTED REFERENCE FUNCTION

In this section, we shall describe the formulation of a state-specific Coupled Cluster formalism which uses a decontracted description of the reference function in the sense that the combining cefficients of the reference determinants comprising the reference function are iteratively updated along with the cluster amplitudes themselves. Both in the present formalism and the contracted description – to be covered in Sec.4 – we shall start with a set of determinants which ensure a proper dissociation of a molecular state into appropriate fragments. Several choices of such reference functions are possible, depending on the degree of sophistication demanded at the zeroth order description. The most common choices are (a) a complete active space (CAS) configuration interaction function (CAS-CI) or the more elaborate CAS-based multi-configuration self-consistent field (CAS-SCF) function, (b) a CI or an MCSCF function on a quasi-complete active space (QCAS) - a very useful active space respecting size-extensivity of its energy, studied by Lindgren [47] and others [21,22,26], (c) a strongly orthogonal generalized valence-bond (SO-GVB) function. Following the usual convention, we shall call the doubly filled orbitals as inactive "core" orbitals, and the partially filled occupied ones as active "valence" orbitals. The virtual orbitals not contained in the reference function would be termed as inactive "particle" orbitals. We shall designate the core by the Greek indices α, β, $\cdots$ etc, the valence orbitals by English letters u, v, w, ... etc, and the virtuals by p, q, r, ... etc. Arbitrary orbitals will be denoted by $a, b, \cdots$ etc. In all the three choices, the energy obtained from the functions is size-extensive, and the functions support dissociation of the associated molecular state into fragments describable by orbitals obtained by a localizing transformation of the inactive "core" orbitals and the active "valence" orbitals onto the fragment components. Consequently, in all the three choices, a cluster expansion decomposition of the combining coefficients of the reference determinants always leads to *connected cluster-amplitudes* - a very crucial requirement for us to show size-extensivity and size-consistency of both the decontracted and the contracted formalisms. There are some essential simplifications in our formalism if the active space is complete (i.e. the choice (a)) , and we shall develop our formulation with this choice. More restricted choices (i.e.of the types (b) and (c))of the active space will be covered in a forthcoming publication.

A CAS-CI or a CAS-SCF type of reference function, can be generically represented as a combination of reference determinants ϕ_μ 's

$$\psi_0 = \sum_\mu c_\mu \phi_\mu \tag{1}$$

As emphasized above the coefficients can be represented in terms of a cluster expansion involving connected cluster amplitudes, using any ϕ_μ acting as the pivotal function :

$$c_\nu / c_\mu = \langle \phi_\nu | \exp(\sigma_\mu) | \phi_\mu \rangle \tag{2}$$

where σ_μ excites electrons from the active orbitals of ϕ_μ to other active orbitals characterizing various ϕ_ν's ($\nu \neq \mu$). Introducing the resolution of identity $\sum_\nu |\phi_\nu\rangle\langle\phi_\nu|$ in the model space, and using eq.(1), we can write ψ_0 in terms of the pivotal function ϕ_μ as :

$$\begin{aligned} |\psi_0\rangle &= \sum_\nu \phi_\nu c_\nu = \sum_\nu |\phi_\nu\rangle\langle\phi_\nu| exp(\sigma_\mu)|\phi_\mu\rangle c_\mu \\ &= \exp(\sigma_\mu)|\phi_\mu\rangle c_\mu \end{aligned} \tag{3}$$

with the cluster-amplitudes σ_μ manifestly connected.

In the decontracted description of the state specific coupled cluster theory, we shall posit the following *Ansatz* for the exact function :

$$\psi = \sum_\mu \exp(T^\mu) \phi_\mu c_\mu \tag{4}$$

where each T^μ induces all possible excitations to virtual determinants orthogonal to the reference determinants. T^μ involves destruction operators deleting electrons from ϕ_μ and creation operators generating virtual orbital occupancy. There is no need of "spectator" orbitals. Such a cluster expansion representation is strongly reminiscent of the Silverstone-Sinanoglu approach [41] in the state specific context and has been used earlier by Malrieu *et al* [39]. Since each T^μ excites to all the virtual functions from ϕ_μ, we have redundancy of the cluster amplitudes. We shall resolve the redundancy, as explained in Sec.II, by positing suitable supplementary conditions which would be both physically appealing and would guarantee extensivity of the cluster amplitudes.

Since each ϕ_μ has different sets of active orbitals, any specific core-to-particle excitation would lead to a *different* virtual determinant from each ϕ_μ. It then follows that the cluster operators of the form $\langle pq \cdots |t^\mu| \alpha\beta \cdots \rangle a_p^\dagger a_q^\dagger \cdots a_\beta a_\alpha$, inducing core to particle excitations are all *linearly independent.* This is however not so for excitations involving active orbitals. For example, if two determinants ϕ_μ and ϕ_ν differ by a set of n active orbitals, excitations from these sets ϕ_μ and ϕ_ν to a common set of particle orbitals would generate the same virtual determinant. Thus, we would encounter redundancy for the cluster operators involving active orbitals.

Let us assume that ψ satisfies the Schrődinger equation with the eigenvalue E :

$$H\psi \equiv H \sum_\mu \exp(T^\mu)\phi_\mu c_\mu = E\psi \tag{5}$$

The combining coefficients for a CAS-CI or a CAS-SCF based ψ can be obtained by projecting eq.(5) onto the reference determinants:

$$\sum_\nu \langle \phi_\mu | H \exp(T^\nu) | \phi_\nu \rangle c_\nu = E c_\mu \tag{6}$$

While evaluating the matrix elements of $H \exp(T^\nu)$ between ϕ_μ and ϕ_ν, it becomes convenient to rewrite H in normal order with respect to ϕ_ν as vacuum. This simplification of computation of matrix-elements was noted earlier by Hose and Kaldor [17] and has since been exploited by many workers [14,15,25,26]. Since T^ν has only excitations out of ϕ_ν, it has only hole-particle creation operators defined with respect to ϕ_ν and consequently $\exp(T^\nu)$ is in normal order with respect to ϕ_ν. Using Wick's theorem, we then find

$$H \exp(T^\nu) = \{\overline{H \exp(T^\nu)} \exp(T^\nu)\}_\nu \tag{7}$$

$$= \exp(T^\nu)\{\overline{H \exp(T^\nu)}\}_\nu \tag{8}$$

The entity $\overline{Hexp(T^\nu)}$ denotes all terms obtained by joining the operators in H with those of T^ν. The notation $\{\}_\nu$ signifies that the operator inside the curly bracket has been written in normal order with respect to ϕ_ν as the vacuum. The equality (8) follows from equation (7) because T^ν has no destruction operators in the hole-particle form and hence can be taken out of the normal order term of equation (7) from the left. Since T^ν always excites ϕ_ν to the virtual manifold, it follows that, for a CAS-CI or a CAS-SCF function,

$$\langle \phi_\mu | H \exp(T^\nu) | \phi_\nu \rangle = \langle \phi_\mu | \{\overline{H \exp(T^\nu)}\}_\nu | \phi_\nu \rangle \tag{9}$$

From now on, we shall denote the above matrix-element by the symbol $\tilde{H}_{\mu\nu}$. Clearly, $\tilde{H}_{\mu\nu}$ is a connected term if the operator T^{ν} is connected, which we shall prove below. Using equation (9), the equation determining the eigenvalue E is given by

$$\sum_{\nu} \tilde{H}_{\mu\nu} c_{\nu} = E c_{\mu} \tag{10}$$

To generate the equations determining the cluster-ampliudes, we rewrite eq.(5) in normal order, taking each ϕ_{μ} as the vacuum, for evaluating $H \exp(T^{\mu})$. Using eqs.(7-8) we then find

$$\sum_{\mu} \exp(T^{\mu}) \{\overline{H \exp(T^{\mu})}\}_{\mu} |\phi_{\mu}\rangle c_{\mu} = E \sum_{\mu} \exp(T^{\mu}) |\phi_{\mu}\rangle c_{\mu} \tag{11}$$

Inserting the resolution of identity

$$I = Q + \sum_{\mu} |\phi_{\mu}\rangle\langle\phi_{\mu}| \tag{12}$$

where Q is the projector onto the virtual space and the sum $\sum_{\mu} |\phi_{\mu}\rangle\langle\phi_{\mu}|$ as the projector onto the space of reference determinants, and using eq.(9) we get

$$\sum_{\mu} \exp(T^{\mu}) Q \{\overline{H \exp(T_{\mu})}\}_{\mu} |\phi_{\mu}\rangle c_{\mu} + \sum_{\mu,\nu} \exp(T^{\mu}) |\phi_{\nu}\rangle \tilde{H}_{\nu\mu} c_{\mu} = E \sum_{\mu} \exp(T^{\mu}) |\phi_{\mu}\rangle c_{\mu} \tag{13}$$

Due to the presence of linearly dependent cluster amplitudes, eq.(13) by itself generates an insufficient number of equations for determining all the cluster amplitudes. This aspect was recognized earlier by several workers. The most relevant among them are the works of Meller *et al* [39] and Mahapatra *et al* [40] who used different sufficiency conditions for generating the cluster amplitudes. Although explicated only for the leading terms, the conditions used by Meller *et al* [39] amounted to equating the Q-projection of the right- and left- hand sides of eq.(11) for each μ. These then produced as many equations as the number of cluster amplitudes. Mahapatra *et al* [40] adopted an alternative strategy of separating the dynamical from the nondynamical correlation effects, and used a single vacuum for generating the cluster equations. While it is difficult to discern from [39] what form the working equations will take for the complete expansion, [40] spells out in general, the structure of their working equations. In either case, however, a general proof of the connectivity of the cluster operators as well as of the energy E is somewhat belabored. This leads one to surmise that more convenient sufficiency conditions exist.

We shall present in this paper precisely such a set of working equations for which the proof of the extensivity of the theory is particularly transparent. In this sense, the present formulation is more natural. In order to arrive at these equations, we interchange the dummy indices μ and ν in the second term on the right side of eq.(13) and equate the Q-projection of the two sides of the rearranged equation, for each μ. We then have

$$Q[exp(T^{\mu})Q\{\overline{H\exp(T_{\mu})}\}_{\mu}|\phi_{\mu}\rangle c_{\mu}+\sum_{\nu}\exp(T^{\nu})|\phi_{\mu}\rangle\tilde{H}_{\mu\nu}c_{\nu}] = EQ\ \exp(T^{\mu})|\phi_{\mu}\rangle c_{\mu} \tag{14}$$

Representing the Q-space functions by $\langle\chi_l|\exp(-T^{\mu})$, where $\{\langle\chi_l|\}$ are the set of virtual space determinants, we find that

$$\langle\chi_l|\ \{\overline{H\exp(T_{\mu})}\}_{\mu}|\phi_{\mu}\rangle c_{\mu}+\sum_{\nu}\langle\chi_l|\exp(-T^{\mu})\exp(T^{\nu})|\phi_{\mu}\rangle\tilde{H}_{\mu\nu}c_{\nu}=0\ \ \forall\, l,\mu \tag{15}$$

The set of equations (15) are our stipulated working equations for determining the cluster amplitudes T^{μ} .

We now prove the connectedness of the cluster amplitudes from eq.(15). We note that the matrix elements $\langle\chi_l|\{\overline{H\exp(T_{\mu})}\}_{\mu}|\phi_{\mu}\rangle$ and $\tilde{H}_{\mu\nu}$ are connected entities joining H and $\exp(T^{\mu})$ by construction and are extensive quantities if T^{μ} itself is connected. To proceed further, we multiply eq.(15) throughout by c_{μ}^{-1} and use the explicit cluster representation for c_{ν} from eq.(2):

$$\langle\chi_l|\ \{\overline{H\exp(T_{\mu})}\}_{\mu}|\phi_{\mu}\rangle+\sum_{\nu}\langle\chi_l|\exp(-T^{\mu})\exp(T^{\nu})|\phi_{\mu}\rangle\tilde{H}_{\mu\nu}\langle\phi_{\nu}|\exp(\sigma_{\mu})|\phi_{\mu}\rangle=0 \tag{16}$$

Now $\tilde{H}_{\mu\nu}$ depends explicitly on all the active orbitals which distinguish ϕ_{ν} and ϕ_{μ}. This is so because the operator $\{\overline{H\exp(T^{\nu})}\}_{\nu}$ in $\tilde{H}_{\mu\nu}$ causes a transition from ϕ_{ν} to ϕ_{μ}. Since $\langle\phi_{\nu}|\exp(\sigma_{\mu})|\phi_{\mu}\rangle$ also depends explicitly on all the active orbitals distinguishing ϕ_{ν} and ϕ_{μ}, it follows that each σ_{μ} contributing to $\langle\phi_{\nu}|\exp(\sigma_{\mu})|\phi_{\mu}\rangle$ has at least one active orbital distinguishing ϕ_{μ} and ϕ_{ν}. Since each σ_{μ} is a connected operator because ψ_0 is a CAS-type function, it implies also that $\tilde{H}_{\mu\nu}\ \langle\phi_{\nu}|\exp(\sigma_{\mu})|\phi_{\mu}\rangle$ is then a connected quantity if $\tilde{H}_{\mu\nu}$ is connected. If we can show now that the entire second term in eq.(16) is connected, our task will be accomplished. To show this we shall rewrite the product $\exp(-T^{\mu})\exp(T^{\nu})$ using the Baker-Campbell-Hausdorf formula :

$$\begin{aligned}\exp(-T^{\mu})\exp(T^{\nu}) = \exp[(T^{\nu}-T^{\mu})+\frac{1}{2}[T^{\nu},T^{\mu}]+\frac{1}{12}[[T^{\nu},T^{\mu}],T^{\mu}]\\ -\frac{1}{12}[[T^{\nu},T^{\mu}],T^{\nu}]+\cdots\end{aligned} \tag{17}$$

Apart from the difference $(T^\nu - T^\mu)$, all other entities in the exponential on the left-hand side appear as commutators. Since the commutators $[T^\nu, T^\mu]$ are nonvanishing only when they generate anticommutators involving creation and destruction operators with the same orbital labels , it follows that they are non- vanishing when one operator in the pair in (T^ν, T^μ) has some creation operators of active orbitals and the other has the corresponding destruction operators. In this case, the active orbitals belong to the set which distinguish the determinants ϕ_μ and ϕ_ν; for in one operator in the pair (T^ν, T^μ) these belong to the occupied set (appearing in the destruction operator) and in the other, to the unoccupied set (appearing in the creation operator). As an illustration, let us consider the case where ϕ_μ and ϕ_ν have two active orbitals (u,v) and (u,x) respectively. For components of T^μ and T^ν of the types $\langle pq|t^\mu|u\alpha\rangle$ $a_p^\dagger a_q^\dagger a_\alpha a_u$ and $\langle rs|t^\nu|u\beta\rangle$ $a_r^\dagger a_s^\dagger a_\beta a_u$, a_u always appear as destruction operator since this is occupied in both ϕ_μ and ϕ_ν and thus these terms commute. On the other hand, the components of T^μ and T^ν of the types $\langle px|t^\mu|\alpha u\rangle a_p^\dagger a_x^\dagger a_u a_\alpha$ and $\langle pq|t^\nu|\alpha x\rangle a_p^\dagger a_q^\dagger a_x a_\alpha$ do not commute since the orbital x is occupied in ϕ_ν but unoccupied in ϕ_μ. There is thus $a_x^\dagger$ in the T^μ component but a_x in T^ν. Hence all the terms from the expansion of the exponential on the left-hand side of eq.(17) involving commutators will have at least some active orbital levels distinguishing ϕ_ν and ϕ_μ, and thus will have common orbitals with $\tilde{H}_{\mu\nu}\langle\phi_\nu|\exp(\sigma_\mu)|\phi_\mu\rangle$ appearing in eq.(16)- the latter being explicitly labelled by *all* the active orbitals distinguishing ϕ_ν and ϕ_μ .

But T^ν and T^μ in general may involve active orbitals common to both ϕ_ν and ϕ_μ; moreover they may not even involve any active orbitals at all. Thus the individual terms from the expansion involving the difference $(T^\nu - T^\mu)$ multiplying $\tilde{H}_{\mu\nu}\langle\phi_\nu|\exp(\sigma_\mu)|\phi_\mu\rangle$ may not involve common orbital labels and hence, in principle be disconnected. We shall, however, prove now that although the individual terms such as $T^\nu\tilde{H}_{\mu\nu}\langle\phi_\nu|\exp(\sigma_\mu)|\phi_\mu\rangle$ may be disconnected, the terms containing the difference $(T^\nu - T^\mu)$, such as powers of $(T^\nu - T^\mu)$ with $\tilde{H}_{\mu\nu}\langle\phi_\nu|\exp(\sigma_\mu)|\phi_\mu\rangle$ are conneceted. The proof follows from the fact that T^μ and T^ν should have the same functional form, differing in their explicit dependence only on the orbitals distinguishing ϕ_ν and ϕ_μ i.e. on the active orbitals distinguishing ϕ_ν and ϕ_μ. Hence on taking the difference, the expression for $(T^\nu - T^\mu)$ have only those terms surviving which are different for T^ν and T^μ, i.e. those involve active orbitals distinguishing ϕ_ν and ϕ_μ. Thus $(T^\nu - T^\mu)$ depends on some active orbitals distinguishing ϕ_ν and ϕ_μ and hence products of powers of $(T^\nu - T^\mu)$ with $\tilde{H}_{\mu\nu}\langle\phi_\nu|\exp(\sigma_\mu)|\phi_\mu\rangle$ have common orbital labels and hence are connected. It thus follows from eq.(16) that there exists one set of solutions for T^μ's where all the cluster amplitudes are connected. Hence $\tilde{H}_{\mu\nu}$'s are also connected. Since the reference determinants form a complete active space, it also follows that the energy obtained as the eigenvalue of eq.(10) is also size extensive. Since the complete active space spanned by the reference determinants is invariant under localizing transformations separately among the holes and active orbitals, the extensivity of the energy also implies correct separation into two fragments generated from the active orbitals, and hence leading to size-consistency.

We should emphasize here that in practice we will solve eq.(15) for the cluster amplitudes. The transcription of eq.(15) to eq.(16) is only for the purpose of proving the connectedness of the cluster-amplitudes. We shall never need to use σ_μ's anywhere for the actual application of our formalism, and all the ϕ_μ's are treated on the same footing.

Let us note that in solving the equations for the cluster amplitudes, eq.(15), we need the knowledge of the coefficients c_μ's. They are thus coupled. We propose to determine both the sets in a nested iterative loop. In the outer loop, we propose to solve a current set of eigenvalue equation, eq.(10), to generate the current set of coefficients c_μ. With this set, we propose to initiate the inner iteration of solving for the cluster amplitudes. We go to the loop for outer iteration again and continue till convergence. The initial iteration may start with a set of c_μ's obtained by diagonalizing the matrix $\langle\phi_\mu|H|\phi_\nu\rangle$. From the very mode of formulation of the theory, it is clear that the combining coefficients are iteratively updated to their values they should have in exact state ψ, and hence our formulation belongs to the first paradigm for the MRCC formulation, viz. the decontracted variety. The development has the additional advantage of flexibility in the sense that we are not obliged to change the combining coefficients if we do not deem it to be necessary. Thus, this formulation can also be used in the context of the second paradigm, i.e. can be used as in a contracted description.

We conclude this section with some comments pertaining to the related formulations of Meller *et al* [39] and Mahapatra *et al* [40] which were mentioned earlier. The method of Meller *et al* uses the equality of Q-projections of eq.(13) for each μ as sufficiency conditions. This seems like the most straightforward sufficiency conditions to be invoked, but the proof of the connectivity of T^μ's becomes rather difficult to achieve. Meller *et al* [39] show the size-consistency of their formulation under certain approximations. Mahapatra *et al* [40] posit another type of sufficiency conditions, but unlike [39], they use a cluster expansion of the wave-operator with respect to a single vacuum. The vacuum function is taken to be a dominant reference space determinant. They use sufficiency conditions following from the assumption that the total energy E is a sum of dynamical (E_d) and nondynamical (E_{nd}) correlation energies. The dynamical energy stems from the virtual functions which are reached by cluster operators acting on the vacuum function. The cluster operators acting on the other reference determinants generate contributions to the nondynamical correlation energies. The sufficiency conditions in [40] are thus different from those in [39]. Moreover since the wave-operator in [40] is defined with respect to a single vacuum, as opposed to what is done in [39], there is the need to keep spectator active orbitals in [40]. This necessitates positing additional sufficiency conditions for each cluster operator exciting to the same χ_l from a given ϕ_μ, but containing different number of "spectator" orbitals. In contrast, the present development follows from a rearranged form of eq.(13). The resulting working equations can be shown to lead to connected cluster amplitudes T^μ in a rather straightforward manner for the general case. The entire exercise hinges on the connectedness of $\exp(-T^\mu)\exp(T^\nu)\tilde{H}_{\mu\nu}\langle\phi_\nu|\exp(\sigma_\mu)|\phi_\mu\rangle$. Since we also use multiple vacuua to represent the wave-operator, there are no spectator orbitals. Recently Adamowicz *et al* [48,49] suggested a state-specific formulation where a special role is played by one reference determinant which is more dominant than the rest. The entire cluster expansion is performed with respect to this determinant. The nondynamical correlation is brought into the formulation by including in the cluster operator three- and four-body terms with at least one active orbital different from those present in the dominant determinant. Since all the cluster operators are defined with respect to this determinant, this formulation is not really an MRCC approach. Presence of three- and four-body operators makes the organization of the equations rather complex. Our present formulation does not require the presence of a dominant determinant in the reference function and in this sense is both more general and more amenable to computer coding.

IV. A STATE-SPECIFIC COUPLED CLUSTER THEORY WITH A CONTRACTED REFERENCE FUNCTION

We want to delineate in this section the development of a state-specific Coupled Cluster theory which uses the entire combination ψ_0 of the reference determinants as a reference function for effecting the cluster expansion leading to the exact function ψ. Several formulations exist in the literature [42-44] which start from such a multi-determinant ψ_0 for generating the Rayleigh-Schrodinger perturbation theories at low orders (viz. second and third order energy). As we have discussed in Sec.II, there are no discernable systematics as to how one can obtain explicit cluster expansion of the wave-function at higher orders, rendering the development of a proper Coupled Cluster type formulation somewhat unwieldy.

In the formulation we are going to present here, all the quantities of interest, viz. the hamiltonian, the cluster amplitudes, the one- and many- particle density matrices are all expressed with respect to the entire reference function ψ_0. Just as in the traditional many-body formulations, it becomes convenient to introduce the notion of a new "normal ordering" with respect to ψ_0 itself. The expectation value of any normal ordered product in this new notion will have vanishing value with respect to ψ_0. It will transpire that we can also rewrite any ordinary product of creation/annihilation operators as a sum of normal products involving all possible "contractions". This will then be the analogue of the traditional Wick's theorem. Unlike in the traditional theorem, however, the contractions will not just involve one creation and one annihilation operator; rather they will generally have an arbitrary number n of creation and an equal number of annihilation opeartors. It turns out that the contractions involving n creation and n annihilation operators are the analogues of the n-body cumulants introduced earlier by Kubo [50]. We can also generalize these concepts to derive the analogue of the generalized Wick's theorem where we can rewrite products of normal products of operators as a sum of normal ordered products involving all possible contractions, but excluding contractions among the groups of operators which are already in normal order. Preliminary versions of the formulation have already appeared [45,46]. We shall present here a more expanded formulation.

Just as for the formulation described in Sec.III, we shall assume that ψ_0 is a CAS-SCF or CAS-CI function, although the generalization to encompass QCAS-CI or QCAS-SCF or an SO-GVB type of function is fairly straightforward in the present formulation. Since in the present formulation all the density-matrices entering the theory are calculated with respect to the reference function ψ_0, the most convenient starting point is a CAS-SCF ψ_0 rather than a CAS-CI. For a CAS-SCF function, the energy is invariant with respect to transformations among the core and among the valence orbitals, and this along with the generalized Brillouin's theorem can be exploited to our advantage by choosing the orbitals to be natural, e.g. those diagonalizing the one-particle density matrix. This will lead to considerable simplifications in our developments to follow.

It is obvious that ψ_0 cannot serve as a vacuum in the strict sense of the traditional hole-particle formalism, since the valence orbitals in ψ_0 are partially occupied. A straightforward cluster expansion in the occupation number representation from ψ_0 would thus entail two problems:(a) there is no natural choice of vacuum to effect a cluster expansion, and (b) the occupation number representation of cluster operators would refer to orbital excitations with respect to the entire ψ_0, thus necessitating the considerations of virtual functions which are by themselves combination of functions. If we want to formulate a many-body theory using ψ_0 as the reference function, we need constructs where these cause no problems.

As mentioned before, we have recently shown [45,46] that it is possible to define a convenient normal ordering and the analogue of the Wick's theorem for a product of creation/annihilation operators. Let us denote the creation/annihilation operators generically by letters q_i. Let us also rewrite ψ_0 as a cluster expansion in terms of any reference determinant ϕ_μ taken as the reference function as in eq.(3). For our purpose, it is useful here to choose the $\exp(\sigma_\mu)$ as unitary. Also, in what follows, we shall suppress the subscript μ from σ_μ and the operators generated by the unitary operator $\exp(\sigma_\mu)$, since it will turn out that the entities which eventually emerge as building blocks are independent of any specific ϕ_μ chosen as the pivotal function.

We want to show that, for a product of creation/annihilation operators, the following relation follows as an algebraic identity:

$$\begin{aligned} q_iq_jq_kq_l\cdots &= \{q_iq_jq_kq_l\cdots\}_0 + \{\overline{q_iq_j}q_kq_l\cdots\}_0 + \cdots \\ &+ \{\overline{q_iq_j}\;\overline{q_kq_l}\cdots\}_0 + \cdots + \{\overline{q_iq_jq_kq_l}\cdots\}_0 + \cdots \end{aligned} \tag{18}$$

where $\{\ \}_0$ denotes the new normal ordering with respect to ψ_0. The overbar inside $\{\ \}_0$ implies presence of contractions with respect to ψ_0. Thus a term like $\{\overline{q_iq_j}q_kq_l\cdots\}_0$ is really a symbol for $\{\overline{q_iq_j}\}_0\ \{q_kq_l\cdots\}_0$, where $\overline{q_iq_j}_0$ is a pair-contraction given by

$$\{\overline{q_iq_j}\}_0 = \langle\psi_0|q_iq_j|\psi_0\rangle \tag{19}$$

The pair-contractions are the exact analogues of contractions appearing in the traditional Wick's theorem. Due to the difficulty of type-setting expressions with overlapping bars indicating contractions among nonadjacent operators, we shall have to use additional notations for contractions among nonadjacent operators. We have not shown them in eq.(18) but they are also present. They will be denoted by putting some superscripts like centred circles or filled circles and so on. This will be illustrated below with examples.

The new quantities entering in eq.(18) are the 2n-tuple contractions like $\overline{q_iq_j\cdots q_kq_l}$ having n creation and n annihilation operators. As we shall show later (viz. in eq.(36)), they are the n-body cumulants [50]. For four operators, for example, the two-body cumulant turns out to be

$$\begin{aligned} \overline{q_iq_jq_kq_l} &= \langle\psi_0|q_iq_jq_kq_l|\psi_0\rangle - \langle\psi_0|q_iq_j|\psi_0\rangle\langle\psi_0|q_kq_l|\psi_0\rangle \\ &+ \langle\psi_0|q_iq_k|\psi_0\rangle\langle\psi_0|q_jq_l|\psi_0\rangle - \langle\psi_0|q_iq_l|\psi_0\rangle\langle\psi_0|q_jq_k|\psi_0\rangle \end{aligned} \tag{20}$$

$$\equiv \langle\psi_0|q_iq_jq_kq_l|\psi_0\rangle - \{\overline{q_iq_j}\;\overline{q_kq_l}\}_0 - \{q_i^{\odot}q_j^{\bullet}q_k^{\odot}q_l^{\bullet}\}_0 - \{q_i^{\bullet}q_j^{\odot}q_k^{\odot}q_l^{\bullet}\}_0 \tag{21}$$

In the third term in eq.(21), the contraction between the operators q_i and q_k is denoted by centred circles and for q_j and q_l, by filled circles. The meaning of the terms with nonadjacent contracted operators is similar to the one in traditional Wick's theorem :

$$\{q_i^{\odot}q_j^{\bullet}q_k^{\odot}q_l^{\bullet}\}_0 = -\{\overline{q_iq_k}\;\overline{q_jq_l}\}_0 \tag{22}$$

When nonadjacent operators are contracted we will use consistently the superscript type of notation, but when they are brought adjacently, the usual convention of overbar will be used. The meaning of the left side of eq.(22) is provided by the right side where the parity factor needed to bring the pair operators adjacent to each other explicitly appears. A similar meaning holds true for the fourth term of eq.(21).

For product six operators and higher, a similar relation follows: each 2n-tuple contraction is an n-body cumulant which is equal to the expectation value of the products minus the products of all possible lower order cumulants where the operators involved in the lower-order contractions may be nonadjacent. They have to be interpreted exactly as for the pair-contraction. As an example, we have

$$\{q_i^{\bullet} q_j^{\bullet} q_k^{\odot} q_l^{\bullet} q_m^{\odot} q_n^{\bullet}\}_0 = -\{\overline{q_i q_j q_l q_n}\ \overline{q_k q_m}\ \}_0 \tag{23}$$

Similar relations apply also to products in normal order containing contractions:

$$\{q_i^{\bullet} q_j^{\bullet} q_k q_l q_m^{\bullet} q_n^{\bullet}\}_0 = \{\overline{q_i q_j q_m q_n} q_k q_l\}_0 \tag{24}$$

To prove eq.(18), we first note that, using eq.(3) for ψ_0 involving a unitary cluster expansion with respect to a specific ϕ_μ, we can rewrite expectation values like $\langle\psi_0|q_i q_j \cdots|\psi_0\rangle$ in terms of the expectation values involving ϕ_μ and the transformed operators Q_i's:

$$\begin{aligned}\langle\psi_0|q_i q_j \cdots|\psi_0\rangle &= \langle\phi_\mu|\exp(-\sigma) q_i q_j \cdots \exp(\sigma)|\phi_\mu\rangle \\ &\equiv \langle\phi_\mu|Q_i Q_j \cdots|\phi_\mu\rangle\end{aligned} \tag{25}$$

where

$$Q_i = \exp(-\sigma) q_i \exp(\sigma) \tag{26}$$

Since σ is a connected operator for a CAS-SCF ψ_0, Q_i is also connected due to the validity of the Lie-formula:

$$Q_i = \exp(-\sigma) q_i \exp(\sigma) = q_i + [q_i, \sigma] + \frac{1}{2}[[q_i, \sigma], \sigma] + \cdots \tag{27}$$

We can rewrite Q_i's in normal order with respect to ϕ_μ by working out the multicommutator expansion in eq.(27). Since ψ_0 is related to ϕ_μ by a unitary transformation, it seems reasonable to define the new normal product as a reverse unitary transform of a normal product with respect to ϕ_μ.

To discern the algebraic structure of the new normal ordered product, we write the ordinary product as unitary transforms of Q-products in ordinary normal order. Since Q's are sums of products of odd number of creation/annihilation operators (as follows from eq.(27)), we shall use different symbols to indicate connections between Q's where only some operators of them are joined and *complete connections*, whereby the connected Q's generate a number. We shall use overbars and centred- or filled- circles to indicate complete contractions as in eq.(21), and carets or open circles to denote joining by only some operators in Q's. Carets will be used when the connected operators are adjacent, open circles will be used otherwise. Thus we have :

$$\begin{aligned} q_i q_j q_k q_l \cdots &= \exp(\sigma) Q_i Q_j Q_k Q_l \cdots \exp(-\sigma) \\ &\equiv \exp(\sigma)[\{Q_i Q_j Q_k Q_l\} + \{\widehat{Q_i Q_j}\widehat{Q_k Q_l}\cdots\} + \{Q_i^{\circ} Q_j^{\circ\circ} Q_k^{\circ} Q_l^{\circ\circ}\cdots\} \\ &+ \{\overline{Q_i Q_j} Q_k Q_l \cdots\} + \{\overline{Q_i Q_j}\widehat{Q_k Q_l}\cdots\} \\ &+ \{\overline{Q_i Q_j}\;\overline{Q_k Q_l}\cdots\} + \{\overline{Q_i Q_j Q_k Q_l}\cdots\} + \cdots]\exp(\sigma) \end{aligned} \tag{28}$$

In eq.(28), we have first rewritten q_i's in terms of Q_i, and then reordered the product $Q_i Q_j Q_k Q_l \cdots$ as a sum of normal products, using ϕ_μ as the vacuum. The traditional Wick's theorem applied to the products of Q_i's will lead to pair contractions in the traditional sense between groups of creation-annihilation operators in one Q_i with one or more Q_j's. If these contractions completely exhaust all the operators present in the composites Q_i, $Q_j \cdots$ etc. involved in the contraction, we denote them by bars and centred or filled circles. Joining by some operators will lead to terms with carets and open circles. Thus, the second and the third term in the braces involve incomplete contractions. The second term has connections between operators of Q_i and Q_j and between Q_k and Q_l. The third term involves connections between Q_i and Q_k and between Q_j and Q_l. The fourth term involves complete connections between all the operators of Q_i and Q_j. The fifth term involves contraction of all the operators of Q_i with those of Q_j and of some between Q_k and Q_l. The sixth term involves complete contractions between operators of Q_i and Q_j and of Q_k and Q_l. The seventh term $\{\overline{Q_i Q_j Q_k Q_l}\cdots\}$ indicates that all the operators of Q_i, Q_j, Q_k and Q_l are contracted among themselves which cannot be factored out to pairs such as $\overline{Q_i Q_j}\;\overline{Q_k Q_l}$ etc.

It also follows from the property of ordinary contractions that

$$\{Q_i^{\odot} Q_j Q_k^{\odot} Q_l \cdots\} = -\{\overline{Q_i Q_k} Q_j Q_l \cdots\} \tag{29}$$

$$\{Q_i^{\odot} Q_j^{\circ} Q_k^{\odot} Q_l^{\circ} \cdots\} = -\{\overline{Q_i Q_k}\widehat{Q_j Q_l}\cdots\} \tag{30}$$

The parity factor in eqs.(29) and (30) are the same as for q's, since each Q is a sum of odd number of creation / annihilation operators.

We now note that each term of the form $\exp(\sigma)\{\ \}\exp(-\sigma)$, with $\{\ \}$ a normal ordered product in the ordinary sense, will have vanishing expectation value with respect to ψ_0 if there are uncontracted operators in $\{\ \}$:

$$\langle\psi_0|\exp(\sigma)\{\ \}\exp(-\sigma)|\psi_0\rangle \equiv \langle\phi_\mu|\{\ \}|\phi_\mu\rangle = 0 \tag{31}$$

This, coupled with the natural requirement that we want our Wick-like theorem to be defined in terms of the new normal ordered products with contractions (i.e. numbers), indicates that the most natural definition of the new normal ordered products is the following:

$$\begin{aligned}\{q_iq_jq_kq_l\cdots\}_0 = \exp(\sigma)[&\{Q_iQ_jQ_kQ_l\cdots\} \\ &+ \{\widehat{Q_iQ_j}Q_kQ_l\cdots\} + \{Q_i^\circ Q_jQ_k^\circ Q_l\cdots\} \\ &+ \textit{several } Q'\textit{s joined by carets or open circles} \\ &\quad \textit{in all possible manner}]\exp(-\sigma)\end{aligned} \tag{32}$$

We note that the new normal ordered product has been defined as a unitary transform of a sum of products in the traditional normal order which has not generated any completely connected entities of the types $\overline{Q_iQ_j}$ etc. Similarly, a normal ordered product with contractions can be defined as

$$\begin{aligned}\{\overline{q_iq_j}q_kq_l\cdots\}_0 = \exp(\sigma)[&\{\overline{Q_iQ_j}Q_kQ_l\cdots\} \\ &+ \{\overline{Q_iQ_j}\widehat{Q_kQ_l}\cdots\} + \cdots\ \textit{all terms having} \\ &\overline{Q_iQ_j}\ \textit{and the rest joined by carets or by} \\ &\textit{open circles in all possible manner}]\exp(-\sigma)\end{aligned} \tag{33}$$

Similar definitions can be invoked for terms with general 2n-tuple contractions and their products. Using the definitions (32), (33) etc., we can regroup eq.(28) in such a manner that eq.(18) follows. It should also be clear from the definitions that the contractions satisfy the following relations :

$$\{\overline{q_iq_jq_kq_l}\}_0 = \exp(\sigma)\{\overline{Q_iQ_jQ_kQ_l}\}\exp(-\sigma) \tag{34}$$

$$\equiv \{\overline{Q_iQ_jQ_kQ_l}\} \tag{35}$$

Eq.(35) follows from eq.(34) since $\{\overline{Q_iQ_jQ_kQ_l}\}$ is a number and we may bring $\exp(\sigma)$ on the left. Using ordinary Wick's theorem for a product of $Q_iQ_jQ_kQ_l$, we find that

$$\begin{aligned}\langle\psi_0|q_iq_jq_kq_l|\psi_0\rangle &\equiv \langle\phi_\mu|Q_iQ_jQ_kQ_l|\phi_\mu\rangle \\ &= \{\overline{Q_iQ_jQ_kQ_l}\} + \{\overline{Q_iQ_j}\ \overline{Q_kQ_l}\} \\ &+ \{Q_i^\odot Q_j^\bullet Q_k^\odot Q_l^\bullet\} + \{Q_i^\odot Q_j^\bullet Q_k^\bullet Q_l^\odot\}\end{aligned} \tag{36}$$

wherefrom it follows that $\{q_iq_jq_kq_l\}_0 \equiv \{\overline{Q_iQ_jQ_kQ_l}\}$ is a cumulant *à la* Kubo [50]. Thus we verify the relation eq.(21) introduced earlier. Similar considerations apply to more than four operartors. Using the ordinary Wick's theorem for products of operators we can also prove that

$$\begin{aligned}\{q_i q_j \cdots\}_0 \{q_k q_l \cdots\}_0 = & \{q_i q_j \cdots q_k q_l \cdots\}_0 + \{q_i^{\odot} q_j \cdots q_k^{\odot} q_l \cdots\}_0 + \cdots \\ & \{q_i^{\odot} q_j^{\odot} \cdots q_k^{\odot} q_l^{\odot} \cdots\}_0 + \cdots \\ & \textit{all other contractions involving groups} \\ & \textit{of operators from } (q_i q_j \cdots) \textit{and } (q_k q_l \cdots), \textit{ excluding} \\ & \textit{contractions only among the numbers of each group}\end{aligned} \tag{37}$$

Because of the validity of eqs. (29) and (30), the completely contracted q's in any $\{\ \}_0$ can be brought adjacent to one another with a proper parity factor. We shall make use of this generalized version of our new Wick's theorem in our development of the Coupled Cluster theory in terms of ψ_0. The hamiltonian for the electronic system may be written as

$$H = \sum_{ab} h_b^a a_a^{\dagger} a_b + \frac{1}{2} \sum_{abcd} v_{cd}^{ab} a_a^{\dagger} a_b^{\dagger} a_d a_c \tag{38}$$

Let us rewrite H in the new normal order with respect to ψ_0 as the reference function:

$$\begin{aligned} H = & \sum_{ab} h_b^a \{a_a^{\dagger} a_b\}_0 + \sum_{ab} h_b^a \{\overline{a_a^{\dagger} a_b}\}_0 + \frac{1}{2} \sum_{abcd} v_{cd}^{ab} [\{a_a^{\dagger} a_b^{\dagger} a_d a_c\}_0 \\ & + \{a_a^{\bullet\dagger} a_b^{\dagger} a_d a_c^{\bullet}\}_0 + \{a_a^{\bullet\dagger} a_b^{\dagger} a_d^{\bullet} a_c\}_0 + \{a_a^{\dagger} a_b^{\bullet\dagger} a_d a_c^{\bullet}\}_0 + \{a_a^{\dagger} a_b^{\bullet\dagger} a_d^{\bullet} a_c\}_0 \\ & + \{a_a^{\bullet\dagger} a_b^{\odot\dagger} a_d^{\odot} a_c^{\bullet}\}_0 + \{a_a^{\bullet\dagger} a_b^{\odot\dagger} a_d^{\bullet} a_c^{\odot}\}_0 + \{\overline{a_a^{\dagger} a_b^{\dagger} a_d a_c}\}_0] \end{aligned} \tag{39}$$

The completely contracted terms (viz. the last three) in eq.(39) provide the expectation value $\langle\psi_0|H|\psi_0\rangle$. Denoting the pair- and four- particle contractions as ρ and R, we have

$$\langle\psi_0|H|\psi_0\rangle = \sum_{ab} h_b^a \rho_b^a + \frac{1}{2} \sum_{abcd} v_{cd}^{ab} [\rho_c^a \rho_d^b - \rho_d^a \rho_c^b + R_{cd}^{ab}] \tag{40}$$

Let us group the operators with a pair of uncontracted operators to define a new one-body operator f:

$$f_b^a = h_b^a + \sum_{\alpha} (v_{b\alpha}^{a\alpha} - v_{\alpha b}^{a\alpha}) + \sum_{uv} (v_{bv}^{au} - v_{vb}^{au}) \{\overline{a_u^{\dagger} a_v}\}_0 \tag{41}$$

The operator f has an obvious physical interpretation: it is a generalized fock-like operator involving the average potential generated by the one-particle density matrix of ψ_0. Since we have chosen our orbitals to be natural, the density matrix ρ_b^a is diagonal. For the core orbitals, the diagonal elements are unity, while for the valence orbitals they are the occupancy of these orbitals:

$$\rho^u_v \equiv \rho^u_v \delta^u_v = \langle \psi_0 | a^\dagger_u a_v | \psi_0 \rangle \equiv \{\overline{a^\dagger_u a_v}\}_0 \tag{42}$$

The choice of natural orbitals also leads to simplifications in the expression (40) for $\langle \psi_0 | H | \psi_0 \rangle$ as well. Let us also note that in the completely contracted terms generated from $\langle \psi_0 | H | \psi_0 \rangle$, we have both products of pair contractions (the last but one term in eq.(39)) and the four body cumulant (the last term).

In a CAS-SCF or a CAS-CI function ψ_0, all the n-body density matrices with hole labels factorize into antisymmetric products of one-body densities. As a result, the two and higher-body cumulants $R^{ab\cdots}_{cd\cdots}$ are zero when all the labels $(a, b, c, d \cdots)$ in $R^{ab\cdots}_{cd\cdots}$ are holes. The only non-vanishing four- or higher-body cumulants are those with valence labels only. The mixed n-body densities with some holes and some valences are zero unless the number and indices of hole labels in the destruction operators match with those in the creation operators. In case they match, these density matrices factorize into antisymmetric products of one-body density matrices with hole labels and the various cumulants with valence labels.

For the particle densities, we have the non-vanishing entries of the form:

$$\sigma^p_q = \langle \psi_0 | \{a_p a^\dagger_q\}_0 | \psi_0 \rangle = \delta^p_q \tag{43}$$

The higher order particle density matrices are all factorizable into antisymmetric products of one body σ's. Mixed densities with particle as well as valence have a similar factorizability involving one-body σ's and R's with valence labels.

If we generically introduce the matrix-elements σ^a_b as

$$\sigma^a_b = \langle \psi_0 | \{\overline{a_a a^\dagger_b}\}_0 | \psi_0 \rangle \tag{44}$$

then we note that for (a, b) as holes, then $\rho^a_b = \delta^a_b$, and $\sigma^a_b = 0$, and for particles $\sigma^a_b = \delta^a_b$ and $\rho^a_b = 0$. For valence orbitals, both ρ^u_v and σ^u_v are generally non-zero. We now posit on our wave-operator Ω the following normal ordered exponential structure [45,46]:

$$\Omega = \{\exp(T)\}_0 \tag{45}$$

where T consists of the various n-body excitations out of ψ_0:

$$T = \sum_n T_n \tag{46}$$

$$T_n = \frac{1}{n!} \sum_{cd\cdots}^{ab\cdots} t_n{}^{ab\cdots}_{cd\cdots} \{a^\dagger_a a^\dagger_b \cdots a_d a_c\}_0 \tag{47}$$

where $c, d, \cdots$ run over holes and valences and $a, b, \cdots$ run over particles and valences. Let us note that, owing to the new normal ordering in the exponential in eq.(45), there are no contractions among the T's in Ω. This leads not only to compact and simpler expressions for the equations determining T but also to *finite power series expansions* in T of these equations. We should mention that the utility of normal ordered exponentials in the MRCC context was first emphasized and exploited by Lindgren using the traditional vacuum [11]. Inspired by the success of this *Ansatz*, we have generalized this to encompass the multi-determinantal ψ_0 also.

As emphasized earlier, the various excited states generated by the action of operator products such as $\{a_a^\dagger a_b^\dagger a_d a_c\}_0$ etc. are not all linearly independent. When there are no valence labels, such as $\{a_p^\dagger a_q^\dagger a_\beta a_\alpha\}_0$, the excited functions $\{a_p^\dagger a_q^\dagger a_\beta a_\alpha\}_0 \psi_0$ are linearly independent. If there are some valence-orbitals, however, they are generally linearly dependent. We then have to include in T only those excitations leading to linearly independent excitations.

To extract the linearly independent excitations, we shall have to use the so-called singular value decomposition of the valence density matrices generated by the creation/annihilation operators with valence-labels which are present in the particular excitation operator in T. To illustrate this aspect, let us take an example. For any excitation operator containing the destruction of a pair of active orbitals from ψ_0, the overlap matrix of all such excited functions factorize, due to our new Wick's theorem, into antisymmetric products of one-body densities with non-valence labels and a two-particle density matrix:

$$\begin{aligned} &\langle\psi_0|\{a_w^\dagger a_x^\dagger \cdots\}_0\{\cdots a_v a_u\}_0|\psi_0\rangle = \\ &products\ of\ one-body\ density\ with\ non-valence\ labels \times \\ &\langle\psi_0|\{a_w^\dagger a_x^\dagger\}_0\{a_v a_u\}_0|\psi_0\rangle \end{aligned} \tag{48}$$

where the last term generates a sum of products of pair contractions and a term R_{uv}^{wx} by the new Wick's theorem. The $\cdots$ in eq.(48) signify the presence of other non-valence orbitals. The possible linear dependence of the functions $\{\cdots a_v a_u\}_0|\psi_0\rangle$ for the various u, v's will be reflected in the *singular nature* of the matrix $\triangle_{uv}^{ax} = \langle\psi_0|\{a_w^\dagger a_x^\dagger\}_0\{a_v a_u\}_0|\psi_0\rangle$. If the matrix $\triangle$ is singular, we can extract the linearly independent components by bringing $\triangle$ to the diagonal form by a unitary matrix X, retain only the non-zero block, and transform back by the inverse transformation by $X^\dagger$. Thus if $\triangle_d$ is the diagonal from of $\triangle$ induced by X, and $\overline{\triangle_d}$ is its non-zero block, then pre- and post-multiplying $\triangle$ by the identity matrix $I = XX^\dagger$, we find that

$$\triangle = XX^\dagger \triangle XX^\dagger \equiv X \triangle_d X^\dagger = \overline{X}\, \overline{\triangle_d}\, \overline{X^\dagger} \tag{49}$$

where $\overline{X}$ denotes the eigenvectors of $\triangle$ corresponding to the nonzero block $\overline{\triangle_d}$. The product $\Theta = \overline{X}\, \overline{X^\dagger}$ then projects onto the nonzero block of the overlap matrix Δ, and thus extracts the linearly independent components of the functions of the type $\{\cdots a_v a_u\}_0|\psi_0\rangle$. Denoting the non-zero entries of $\overline{\Delta_d}$ by the capital indices U, $V \cdots$ etc., we can then extract the linearly independent components ψ_{UV} as a particular combination of the following form:

$$\psi_{UV} = \sum_{wx} X^{\dagger}{}^{wx}_{UV} \{a_x a_w\}_0 |\psi_0\rangle \tag{50}$$

$$\equiv \sum_{wx} X^{UV}_{wx} \{a_x a_w\}_0 |\psi_0\rangle \tag{51}$$

In an exactly similar manner, we can extract the linearly independent components of the sets $\{\psi^{UV}\}$ and $\{\psi^{U}_{V}\}$.

$$\psi^{UV} = \sum_{wx} Y^{wx}_{UV} \{a^\dagger_w a^\dagger_x\}_0 |\psi_0\rangle \tag{52}$$

$$\psi^{U}_{V} = \sum_{wx} Z^{Vw}_{Ux} \{a^\dagger_w a_x\}_0 |\psi_0\rangle \tag{53}$$

where Y and Z are the unitary matrices bringing the overlap matrices of the sets $\{\{a^\dagger_w a^\dagger_x\}_0|\psi_0\rangle\}$ and $\{\{a^\dagger_w a_x\}_0|\psi_0\rangle\}$ to the diagonal form. Using the same strategy, we may extract the linearly independent components of T. As an illustration, let us present the linearly independent components of T_2. Some typical terms of T_2 are shown below:

$$T_2 \Rightarrow \frac{1}{2}\sum_{pq\alpha\beta} t^{pq}_{\alpha\beta}\{a^\dagger_p a^\dagger_q a_\beta a_\alpha\}_0 + \frac{1}{2}\sum^{UV}_{wxpq} t^{pq}_{UV} X^{UV}_{wx}\{a^\dagger_p a^\dagger_q a_x a_w\}_0$$
$$+\frac{1}{2}\sum^{UV}_{wx\alpha\beta} t^{UV}_{\alpha\beta} Y^{wx}_{UV}\{a^\dagger_w a^\dagger_x a_\beta a_\alpha\}_0 + \frac{1}{2}\sum^{UV}_{wx\alpha p} t^{pU}_{\alpha V} Z^{Vw}_{Ux}\{a^\dagger_p a^\dagger_w a_x a_\alpha\}_0 \tag{54}$$

Similar strategy may be used to extract the linearly independent components of an n-body operator T_n. Whenever there are creation or annihilation operators involving valence labels, we have to extract the linearly independent components from the excitation operators.

Let us now start from the Schrődinger equation for ψ, using the representation, in eq.(45), for Ω, we have

$$H\{\exp(T)\}_0|\psi_0\rangle = E\{\exp(T)\}_0|\psi_0\rangle \tag{55}$$

To generate the equations for all the linearly independent cluster-amplitudes, we have to project eq.(55) onto the linearly independent excited states. Thus, for example, the excited states to be used for determining the cluster-amplitudes $t^{pq}_{\alpha\beta}$, t^{pq}_{UV}, $t^{UV}_{\alpha\beta}$ and $t^{pU}_{\alpha V}$ should be of the following forms:

$$\langle\psi^{pq}_{\alpha\beta}| \equiv \langle\psi_0|\{a^\dagger_\alpha a^\dagger_\beta a_q a_p\}_0 \tag{56}$$

$$\langle\psi^{pq}_{UV}| \equiv \sum_{wx}\langle\psi_0|\{a^\dagger_w a^\dagger_x a_q a_p\}_0 X^{wx}_{UV} \tag{57}$$

$$\langle\psi^{UV}_{\alpha\beta}| \equiv \sum_{wx}\langle\psi_0|\{a^\dagger_\alpha a^\dagger_\beta a_x a_w\}_0 Y^{UV}_{wx} \tag{58}$$

$$\langle\psi^{pU}_{\alpha V}| \equiv \sum_{wx}\langle\psi_0|\{a^\dagger_\alpha a^\dagger_x a_w a_p\}_0 Z^{Ux}_{Vw} \tag{59}$$

Denoting a typical member of the excited state by the generic notation $\langle\psi^r| \equiv \langle\psi_0|Y_r$, where r is a collective index such as $(pq,\alpha\beta)$, (pq,UV) etc. we may write the typical equations determining the cluster amplitudes as

$$\langle\psi_0|Y_r H\{\exp(T)\}_0|\psi_0\rangle = E\langle\psi_0|Y_r\{\exp(T)\}_0|\psi_0\rangle \ \forall r \tag{60}$$

Since each term of the operators Y_r, H and $\{\exp(T)\}_0$ is in normal order, using the generalization of our extended Wick's theorem– eq.(36) we may write each of the expectation values on the left and right side of eq.(60) as a sum over completely contracted quantities. This should exclude contractions within each group of Y_r, H and $\{\exp(T)\}_0$.

The completely contracted terms on the left side of eq.(60) can be classified into two categories. In one, H is contracted completely with various powers of T coming from $\{\exp(T)\}_0$, and the rest of the terms of $\{\exp(T)\}_0$ – not contracted to H– contracted completely with Y_r. In the other, H is contracted completely with both Y_r and $\{\exp(T)\}_0$. Thus, the left side of eq.(60) may be written as

$$\langle\psi_0|Y_r H\{\exp(T)\}_0|\psi_0\rangle = \langle\psi_0|\overline{Y_r\{\exp(T)\}_0}|\psi_0\rangle\langle\psi_0|\overline{H\{\exp(T)\}_0}|\psi_0\rangle$$

$$+ \langle\psi_0|\overline{Y_r H\{\exp(T)\}_0}|\psi_0\rangle \tag{61}$$

where we have used the generalized extended Wick's theorem to generate the two types of completely contracted terms of the left side of eq.(60), and brought the contracted operators adjacent to each other as permitted by our theorem. Since each of Y_r, H and T contain an even number of creation/annihilation operators, there are no phase factors stemming from the permutations needed to bring the operators in the adjacent positions. The right side of eq.(60) similarly leads to

$$E\langle\psi_0|Y_r\{\exp(T)\}_0|\psi_0\rangle = E\langle\psi_0|\overline{Y_r\{\exp(T)\}_0}|\psi_0\rangle \tag{62}$$

Also, projecting eq.(55) onto $\langle\psi_0|$, we have

$$\langle\psi_0|H\{\exp(T)\}_0|\psi_0\rangle \equiv \langle\psi_0|\overline{H\{\exp(T)\}_0}|\psi_0\rangle \quad (63)$$
$$= E\langle\psi_0|\{\exp(T)\}_0|\psi_0\rangle = E \quad (64)$$

using eq.(63), we may equate the first term on the right side of eq.(61) with the expression (62) and hence get

$$\langle\psi_0|\overline{Y_r H\{\exp(T)\}_0}|\psi_0\rangle = 0, \forall\, r \quad (65)$$

Eq.(65) are the set of working equations for determining the cluster amplitudes. Since each term of $\langle\psi_0|\overline{Y_r H\{\exp(T)\}_0}|\psi_0\rangle$ consists of a connected cumulant joining Y_r, H and $\{\exp(T)\}_0$, the cluster amplitudes obtained from eq.(65) will be connected. Since E from eqs.(64) and (63) is generated also from the connected cumulants of $\langle\psi_0|\overline{H\{\exp(T)\}_0}|\psi_)\rangle$, it is also connected. This proves the size-extensivity of the formalism. Also, since ψ_0 separates correctly to the fragments – being a CAS–type function, it also follows that the formalism is size-consistent.

We have not yet discussed the possiblity of relaxing of the combining coefficients c_μ of ψ_0. Since the working equations for T contain the cumulants defined with respect to ψ_0, we have no direct and simple way to update c_μ's. The straightforward way to do it is of course to introduce valence to valence excitations in T , effectively changing the coefficients in the process. But this seems to be impractical for a general n-valence ψ_0, since we need an n-valence, valence to valence excitation operator, viz. an n-body term in T to effect this relaxation. Partial and indirect relaxation of the coefficients is, however, always possible by including a set of valence- to valence excitation of rather low particle rank in T which excites from a pivotal ϕ_μ to other ϕ_ν's. If these valence excitations are included, then a dominant portion of the relaxation of the coefficients can be incorporated. In an actual application, we must truncate the cluster operators after a certain particle rank. A good truncation scheme could be an approximation where T is approximated as $T \simeq T_1 + T_2$, which is the multi-reference analogue of the CCSD approximation. In this scheme, we have to consider the following types of cluster-amplitudes:

$A.t^p_\alpha : hole \longrightarrow particle$

$B.t^U_\alpha : hole \longrightarrow valence$

$C.t^p_U : valence \longrightarrow particle$

$D.t^{pq}_{\alpha\beta} : hole,\ hole \longrightarrow particle,\ particle$

$E.t^{pU}_{\alpha\beta} : hole,\ hole \longrightarrow particle,\ valence$

$F.t^{pq}_{\alpha U} : hole,\ valence \longrightarrow particle,\ particle$

$$
\begin{aligned}
&G.t_{\alpha U}^{pV} : hole,\ valence \longrightarrow particle,\ valence \\
&H.t_{UV}^{pW} : valence,\ valence \longrightarrow particle,\ valence \\
&I.t_{\alpha W}^{UV} : hole,\ valence \longrightarrow valence,\ valence \\
&J.t_{UV}^{pq} : valence,\ valence \longrightarrow particle,\ particle \\
&K.t_{\alpha\beta}^{UV} : hole,\ hole \longrightarrow valence,\ valence
\end{aligned}
\tag{66}
$$

In case we are interested to relax the coefficients c_μ's within the CCSD truncation scheme, we should include in T the following two valence-excitation amplitudes:

$$
\begin{aligned}
&L.t_{U}^{W} : valence \longrightarrow valence \\
&M.t_{UV}^{WX} : valence,\ valence \longrightarrow valence,\ valence
\end{aligned}
\tag{67}
$$

where U, V etc. refers to one specific ϕ_μ taken as the pivotal function and W, X refer to labels not appearing in ϕ_μ. Let us note carefully that we have always used capital valence indices in all the $T-$ amplitudes, since we should work only with the linearly independent cluster-amplitudes. Let us also note that, exactly analogous to what one obtains for the SCF-based single reference CC theory, we also expect the T_1 amplitudes to be small because of the validity of the generalized Brillouin condition.

V. CONCLUDING REMARKS

We have presented in this paper two different methodologies in the multi-reference approach which bypass the intruder problem by focusing on only one specific state. The two state-specific MRCC formalisms belong naturally to two different paradigms. In one, the decontracted description, the combining coefficients c_μ of the reference determinants ϕ_μ comprising the reference function ψ_0 are iteratively relaxed – eventually leading to the fully relaxed values of the coefficients in the exact ψ. In the other, the contracted description, we start from a fixed combination of ϕ_μ's forming ψ_0, and effect a cluster expansion with respect to entire ψ_0. This approach necessitates the notion of a new normal ordering and a corresponding extended Wick's theorem which allows us to define a normal ordered exponential in the new normal ordering – leading to explicitly connected equations for the cluster-amplitudes and energy. The constructs entering the formalism are cumulants analogous to those introduced earlier by Kubo [50] and correspond to certain connected density matrices defined in such a manner that they are manifestly connected.

In the first paradigm, we deliberately retain the linearly dependent (redundant) cluster-amplitudes in the wave-operator, but invoke suitable sufficiency conditions which not only provides the extra equations needed for their determination but also ensures that the cluster amplitudes are manifestly connected. In the second paradigm, we use a singular-value decomposition technique to extract only the linearly independent cluster-amplitudes in as unbiased a manner as is possible – without any pivotal ϕ_μ implied in the generation of the linearly independent amplitudes.

ACKNOWLEDGEMENT

The authors thank the CSIR (New Delhi) and the UGC (New Delhi) for providing financial support. D.M. thanks the S. N. Bose National Centre for conferring on him an honorary Professorship.

REFERENCES

1. F. Coester, *Nucl. Phys.* 1, 421 (1958)
   F. Coester and H. Kümmel, *Nucl. Phys.* 17, 477 (1960)
   O. Sinanoglu, *Adv. Chem. Phys.* 6, 315 (1964)

2. R. J. Bartlett and W. D. Silver, *Int. J. Quantum Chem* S9, 183 (1975)
   J. A. Pople, J. S. Binkley and R. Seeger, *Int. J. Quantum Chem* S10, 1 (1976)

3. J. Cizek, *J. Chem. Phys.* 45, 4256 (1966);
   J. Cizek, *Adv. Chem. Phys.* 14, 35 (1969)
   J. Paldus, J. Cizek abd I. Shavitt, *Phys. Rev. A* 5, 50 (1972)

4. R. J. Bartlett and G. D. Purvis, *Int. J. Quantum Chem.* 14, 561 (1978)
   J. A. Pople, R. Krishnan, H. B. Schlegel and J. S. Binkley, *Int. J. Quantum Chem.* S14, 545 (1978)

5. See e.g. R. J. Bartlett, *Ann. Rev. Phys. Chem.* 32, 359 (1981)
   S. A. Kucharski and R. J. Bartlett, *Adv. Quantum Chem.* 18, 281 (1986)

6. R. J. Bartlett, *J. Phys. Chem.* 93, 1697 (1989)

7. R. J. Bartlett in Modern Electronic Structure Theory, Part-I (Ed : D. A. Yarkony, World Scientific, Singapore) (1995)

8. See e.g. Ph. Durand and J. P. Malrieu, *Adv. Chem. Phys.* 67, 321 (1987)

9. B. Brandow, *Rev. Mod. Phys.* 39, 771 (1967)
P. G. H. Sanders, *Adv. Chem. Phys.* 14, 365 (1969)
I. Lindgren, *J. Phys. B* 7, 2441 (1974)
V. Kvasnicka, *Czech. J. Phys. B* 24, 605 (1974)

10. D. Mukherjee, R. K. Moitra and A. Mukhopadhyay, *Mol. Phys.* 33, 955 (1977)
A. Haque and D. Mukherjee, *J. Chem. Phys.* 80, 5058 (1984)

11. I. Lindgren, *Int. J. Quantum Chem.* S12, 33 (1978)

12. R. Offerman, W. Ey and H. Kümmel, *Nucl. Phys.* A273, 349 (1976)
R. Offerman, *Nucl. Phys.* A273, 368 (1976)
W. Ey, *Nucl. Phys.* A296, 189 (1978)

13. W. Kutzelnigg, *J. Chem. Phys.* 77, 3081 (1982)

14. B. Jeziorski and H. J. Monkhorst, *Phys. Rev. A* 24, 1668 (1981)
B. Jeziorski and J. Paldus, *J. Chem. Phys.* 88, 5673 (1988)

15. L. Meissner, K. Jankowski and J. Wasilewski, *Int. J. Quantum Chem.* 34, 535 (1988)
A. Balkova, S. A. Kucharski, L. Meissner and R. J. Bartlett, *Theo. Chim. Acta* 80, 335 (1991)

16. T. A. Schucan and H. A. Weidenmüller, *Ann. Phys.* 73, 108 (1972); *ibid* 76, 483 (1973)

17. G. Hose and U. Kaldor, *J. Phys. B* 12, 3827 (1979)
G. Hose and U. Kaldor *J. Phys. Chem.* 86, 2133 (1982)

18. H. Sun, N. G. Shepard, K. F. Freed and M. F. Harman, *Chem. Phys. Lett.* 77, 555 (1981)
H. Sun and K. F. Freed, *Chem. Phys. Lett.* 81, 279 (1981)

19. R. Sawatzki and L. S. Cederbaum, *Chem. Phys. Lett.* 126, 430 (1986)

20. D. Mukherjee, *Chem. Phys. Lett.* 125, 207 (1986)
D. Mukherjee, *Int. J. Quantum. Chem* S20, 409 (1986)
D. Sinha, S. K. Mukhopadhyay and D. Mukherjee, *Chem. Phys. Lett.* 129, 369 (1986)

21. I. Lindgren and D. Mukherjee *Phys. Rep.* 151, 93 (1987)

22. W. Kutzelnigg, D. Mukherjee and S. Koch, *J. Chem. Phys.* 87, 5902 (1987)
D. Mukherjee, W. Kutzelnigg and S. Koch, *J. Chem. Phys.* 87, 5911 (1987)

23. S. Pal, M. Rittby, R. J. Bartlett, D. Sinha and D. Mukherjee, *J. Chem. Phys.* 88, 9357 (1988)
S. Pal, M. Rittby, R. J. Bartlett, D. Sinha and D. Mukherjee, *Chem. Phys. Lett.* 137, 272 (1987)

24. I. Lindgren *J. Phys. B* 24, 1143 (1991)

25. D. Mukhopadhyay and D. Mukherjee, *Chem. Phys. Lett.*, 163, 171 (1989)
D. Mukhopadhyay and D. Mukherjee, *Chem. Phys. Lett.*, 177, 441 (1991)
D. Mukhopadhyay and D. Mukherjee, in Applied Many-Body Methods in Spectrscopy and Electronic Structure (Ed: D. Mukherjee, Plenum Press, New York), (1992)

26. L. Meissner, S. Kucharski and R. J. Bartlett, *J. Chem. Phys.* 91, 6187 (1989)
L. Meissner and R. J. Bartlett, *J. Chem. Phys* 92, 561 (1990)

27. U. Kaldor, in Lecture Notes in Chemistry, Vol. 52 (ED. U. Kaldor, Springer-Verlag, Berlin) (1989)
S. Ben-Shlomo and U. Kaldor, *J. Chem. Phys.* 89, 956 (1989)
U. Kaldor, in Applied Many-body Methods in Spectroscopy and Electronic Structure (Ed. D. Mukherjee, Plenum Press, N.Y.) (1992)

28. D. Mukhopadhyay, R. Chaudhuri and D. Mukherjee, *Chem. Phys. Lett.* 172, 515 (1990)
D. Sinha, S. Mukhopadhyay, R. Chaudhuri and D. Mukherjee, *Chem. Phys. Let.* 154, 544 (1989)

29. U. Kaldor, *Chem. Phys.* 140, 1 (1990)

30. B. Kirtman, *J. Chem. Phys.* 75, 798 (1981)

31. J. P. Malrieu, Ph. Durand and J. P. Daudey, *J. Phys. A Math. Gen.* 18, 809 (1985)
J. P. Malrieu, J. L. Heully and A. Zaitsevskii, *Theo. Chim. Acta.* 90, 167 (1995)
J. L. Heully and J. P. Daudey, *J. Chem. Phys.* 105, 6887 (1996)

32. J. L. Heully and J. P. Daudey, *J. Chem. Phys.* 88, 1046 (1988)
A. V. Zaitsevskii and J. L. Heully, *J. Phys. B* 25, 603 (1992)

33. D. Mukhopadhyay, B. Datta and D. Mukherjee *Chem. Phys. Lett.* 197, 236 (1992)

34. B. Datta, R. Chaudhury and D. Mukherjee, *J. Mol. Structure (Theochem)* 361, 21 (1996)

35. D. Sinha, S. Mukhopadhyay, R. Chaudhuri and D. Mukherjee, *Chem. Phys. Lett.* 154, 544 (1989)

36. See also the second entry of ref. [20]

37. S. Koch, *Theo. Chim. Acta.* 81, 169 (1991)

38. I. Nebot-Gil, J. Sanchez-Marin, J. P. Malrieu, J. L. Heully and D. Maynau, *J. Chem. Phys.* 103, 2576 (1995)
J. P. Malrieu, J. P. Daudey and R. Caballol, *J. Chem. Phys.* 101, 8908 (1994)

39. J. Meller, J. P. Malrieu and R. Caballol, *J. Chem. Phys.* 104, 4068 (1996)

40. U. Sinha Mahapatra, B. Datta and D. Mukherjee, in Modern Ideas in Coupled-Cluster Methods. (Ed: R. J. Bartlett, World Scientific, Singapore, 1997)

41. H. J. Silverstone and O. Sinanoglu, *J. Chem. Phys.* 44, 1899, 3608 (1966)

42. K. Wolinski, H. L. Sellers and P. Pulay, *Chem. Phys. Lett.* 140, 225 (1987)

43. K. Andersson, P. Malmquist and B. O. Roos, *J. Chem. Phys.* 97, 1218 (1992)

44. R. B. Murphy and R. P. Messmer, *J. Chem. Phys.* 97, 4170 (1992)
K. Hirao, *Chem. Phys. Lett.* 190, 374 (1992)
K. Hirao, *Int. J. Quantum Chem.* S26, 517 (1992)
K. G. Dyall, *J. Chem. Phys.* 102, 4904 (1995)

45. D. Mukherjee in Recent Progress in Many Body Theories, Vol. 4 (Ed : E. Schachinger, H. Mitter and H. Sormann, Plenum Press, N.Y.) (1995)

46. D. Mukherjee, submitted to *Chem. Phys. Lett.*

47. K. B. Ghose, P. Piecuch, L. Adamowicz, *J. Chem. Phys.* 103, 24 (1995)

48. I. Lindgren, *Phys. Scripta.* 32, 291 (1985)
I. Lindgren, *Phys. Scripta.* 32, 611 (1985)

49. P. Piecuch, N. Oliphant and L. Adamowicz, *J. Chem. Phys.* 99, 1875 (1993)

50. R. Kubo, *J. Phys. Soc. (Japan)* . 17, 1100 (1962)

Note added in proof:

We may remark here that the new Wick's theorem proved by us is really the multi-determinantal analogue of the traditional Wick's theorem in the following sense : we first define the multideterminantal analogues of contractions; then we prove that an ordinary product can be expressed as a sum of the new normal ordered products with all possible contractions. Once we realize that such a reordering formula exists [45,46], we may envisage an alternative route where we define only the contractions and recursively **define** normal ordered products and normal ordered products with contractions from the ordinary product. In this latter way of looking at the problem, there is no "Wick's theorem" as such but only a recursive defining relation. Such a notion has recently been elaborated upon by W. Kutzelnigg and D. Mukherjee, *J. Chem. Phys.* (in press).

CORRELATIONS IN THE VACUUM

Walter Greiner
Institut für Theoretische Physik
J.W. Goethe-Universität
D-60054 Frankfurt am Main
Germany

Abstract

It is shortly reminded on the structure in the electron-positron vacuum for overcritical external fields. Emphasis is given to the structure of the baryon vacuum in the vicinity of a nucleus, particular if it is compressed : thousands of bound states of negative energy emerging out of the negative energy continum represent the highly correlated vacuum in this case. The implication of this correlation structure for the production of the antimatter (i.e. antinuclei) and similarly highly strange hypernuclei is shortly mentioned.

*This article is dedicated to Professor Ingvar Lindgren on occasion of his 65th birthday. Ingvar received the Alexander von Humboldt price in 1993 and, subsequently spent frequent visits at our Institut für Theoretische Physik. His contributions to Quantum Electrodynamics were most inspiring, and out of the collaboration with him and his associate and former student Dr. Hans Persson as well as with my former students Professor Gerhard Soff and Dr. Stefan Schneider emerged several topical research papers. We can say that not only scientific insight, but also friendship developed.

ADVANCES IN QUANTUM CHEMISTRY, VOLUME 30

0065-3276/98 $25.00

1. Introductions and reflections on the status of the supercritical QED

One example of strongly correlated structure in a vacuum state is the diving of an electron bound state in to the negative energy continuum after over critical Coulomb fields [1]. It has been extensively studied theoritically and experimentally. The experimentalists reported the observation of narrow e^+e^- line structure in coincidence measurements[2, 3]. However, recently, due to reports from an Argonne group, claiming not to find those earlier reported structures[4], some of the GSI experimentalists have withdrawn an essential part of their fifteen years work [5]. The present experimental situation is highly unclear and unsatisfactory. This is so, because there are published cases where exactly the same narrow line structures have been found in independently carried out and analysed experiments for the same heavy ion system at the same heavy ion energy window. Though the statistics is small, it is difficult to understand how this can be a statictical fluctuation, as it is now claimed [5]. Neverthless,even if these earlier claimed structures were real, they can not be signatures of new particles, as some of the involved experimentalists have dreamed, and they certainly are no indication for the supercritical decay, which was the prime motivation for these investigation at the beginning. As I stated many times during the last 12 years [6], narrow line structures could most likely be some kind of pair conversion process, eventually even out of a shortliving giant nuclear system [6].

Fig. 1 reminds once more to the overcriticality situation in giant atoms, Z being their central charge. The observation of spontaneous positron decay of the neutral vacuum into a charged vacuum is a key important task, because this phenomenon has close analogons in the Klein Paradox [7], the Hawking radiation from very strong gravitational fields (black holes) and in the quasi-confinement problem of quarks [15]. In constrast to these cases supercritical QED would be the only area where experimental verification of the phenomena has a real chance. Therefore I urgue experimentalists to take up this important problem again.

Let me now discuss the strong correlation in the baryon vacuum around normal and compressed nuclei.

The relativistic mean meson field (RMF) theory formulated by Teller and others [8, 9, 10] and by Walecka [11] is quite successful in both infinite nuclear matter and finite nuclei[12, 13, 14]. In the RMF model, only positive-energy baryonic states are considered to study the properties of ordinary nuclei. This is the so-called "no-sea-approximation". However, an interesting feature of the RMF theory is the existence of bound negative-energy baryonic states. This happens because the interaction with the vector field generated by the baryon-

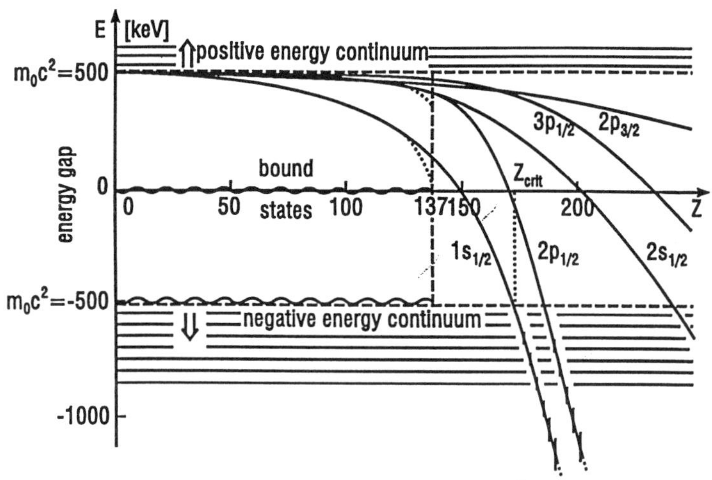

Fig.1. Electron levels in superstrong central fields. The ls state is "diving" into the negative energy continuum at $Z_{crit} \sim 164$, which signals the change in the structure of the vacuum. The neutral vacuum decays into a charged one by spontaneous positron emission.

rich medium is repulsive for baryons and attractive for antibaryons. The vector and scalar potentials are large in magnitude for baryons but nearly cancel each other at normal nuclear density, resulting in a rather shallow nuclear potential depth of about 63 MeV. This is the nuclear shell model. However, the antinucleons feel a much deeper potential of about 630 MeV at the same positive baryon density.

Let the follow up a recent theoretical vision [15] and study the anti-baryon bound states using the axially deformed relativistic mean field (RMF) formalism, involving interacting mesons and nucleons to gain an understanding of the strong vacuum correlations, i.e., the antiproton and antineutron bound states. The relativistic mean field (RMF) theory has the advantage that, with the proper relativistic kinematics and with the mesons and their properties already known or fixed from the properties of a small number of nuclei [14,16,17], the method gives excellent results for the binding energies, root mean square (rms) radii, quadrupole and hexadecapole deformations and other nuclear properties not only of spheri-cal, but also of deformed nuclei. The same parameter sets also describe well the properties of nuclear matter. One of the major attractive features of the RMF approach is that the spin-orbit interaction and the associated nuclear shell structure automatically arise from the meson-nucleon interaction [18]. The inclusion of the p meson takes care of the neutron-proton asymmetry. We can thus expect that RMF calculations provide useful information about the structure of baryonic and hopefully also of anti-baryonic systems.

The relativistic Lagrangian density for a nucleon-meson many-body system [19, 14, 20] reads

$$\begin{aligned}
\mathcal{L} &= \bar{\psi}(i\gamma^\mu\partial_\mu - M_B)\psi \\
&+ \frac{1}{2}\partial^\mu\sigma\partial_\mu\sigma - \frac{1}{2}m_\sigma^2\sigma^2 + \frac{1}{3}g_2\sigma^3 + \frac{1}{4}g_3\sigma^4 - g_s\bar{\psi}\psi\sigma \\
&- \frac{1}{4}\Omega^{\mu\nu}\Omega_{\mu\nu} + \frac{1}{2}m_\omega^2 V^\mu V_\mu + \frac{1}{4}c_3(V_\mu V^\mu)^2 - g_\omega\bar{\psi}\gamma^\mu\psi V_\mu \\
&- \frac{1}{4}\vec{B}^{\mu\nu}.\vec{B}_{\mu\nu} + \frac{1}{2}m_\rho^2\vec{R}^\mu\cdot\vec{R}_\mu - g_\rho\bar{\psi}\gamma^\mu\vec{\tau}\psi\cdot\vec{R}^\mu \\
&- \frac{1}{4}F^{\mu\nu}F_{\mu\nu} - e\bar{\psi}\gamma^\mu\frac{(1-\tau_3)}{2}\psi A_\mu
\end{aligned} \tag{1}$$

The field for the σ-meson is denoted by σ, that of the ω-meson by V_μ and of the isovector ρ-meson by $\vec{R}_\mu$. A^μ denotes the electromagnetic field. ψ is the Dirac field for the nucleons. The third component of isospin is denoted by τ_3. Here g_s, g_w, g_ρ and $e^2/4\pi = 1/137$ are the coupling constants for σ, ω, ρ mesons and photon respectively. M_B is the mass of the nucleon and m_σ, m_ω and m_ρ are the masses of the σ, ω and ρ-mesons respectively. $\Omega^{\mu\nu}$, $\vec{B}^{\mu\nu}$ and $F^{\mu\nu}$ are the field tensors for the V^μ, $\vec{\rho}^\mu$ and the photon fields respectively [14]. I have no time to discuss here the fixing of the parameters and the pro's and con's of the various sets existing so far. A discussion of the various sets in use and their predictive power concerning discovered presuperheavy elements can be found in ref [21]. The fields presented here are obtained with the so called TM1 parameter set [22].

The spin-orbit force of the Dirac Hamiltonian is obtained by using the Foldy-Wouthuysen transformation as $H_{s.o.} = \alpha(r)\vec{S}.\vec{L}$, where

$$\alpha(r) = \frac{1}{2M_B^2 r}\frac{d}{dr}(U_S \mp U_V). \tag{2}$$

The negative sign is for the baryonic and the positive one for the antibaryonic sector. Obviously the spin-orbit strength is very small in the negative-energy states as compared to that of the positive ones, because for the negative-energy states the potentials nearly cancel in equation (2). (Note that for higher densities the states in the inverted potential well may actually have energies above zero; for simplicity we still call them "negative-energy" states.) In these calculations, the spin-orbit strength of the positive-energy sector is about 100 times greater than the spin-orbit strength of the negative-energy sector, leading to a pronounced shell structure in the negative-energy bound states.

The positive and negative-energy potentials are plotted in Fig. 2a. The upper potential corresponds to the positive-energy sector of the neutrons in

^{208}Pb, and the lower one is the deep negative-energy potential at normal nuclear density. It is clear that the gap between upper and lower continuum states is $2M_B^*$, which is reduced substantially inside the nucleus as compared with the vacuum in empty space. In figure 2a, one can also see the shallow nucleon potential with respect to M_B and the deep anti-nucleon potential with respect to $-M_B$. This reduced gap and the strong attractive interaction for anti-baryons was initially argued by Duerr and Teller [8, 9, 10]. They emphasized the strong attraction in the nucleon-antinucleon system as a possible mechanism for $N\bar{N}$ bound and resonant states.

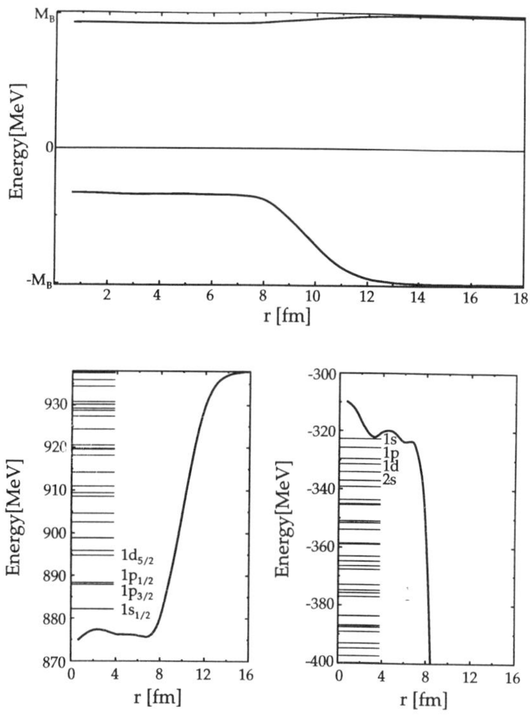

Figure 2: Figure 2a The potentials of the positive and negative-energy states are shown for a ^{208}Pb nucleus. The shallow (upper) one is for positive-energy and the deeper one (bottom) is for negative-energy states. Figure 2b is the diagram of the single particle energies of neutrons and Figure 2c shows the single particle energies of some of the lowest bound negative-energy states. Note that for the negative-energy sector the spin-orbit coupling is so small that the splitting is not visible.

In the case of lead, 126 energy levels are occupied by baryons (neutrons) in the positive potential well as plotted in Fig. 2b, whereas there are about 20000 neutrons bound in negative-energy states. Some of the lowest negative-energy bound states of the single particle energy levels are shown in Fig. 2c to illustrate the pronounced shell structure. Obviously gaps of up to 10 MeV occur.

In nuclear matter all baryonic states obeying the condition

$$E_B(\vec{p}) = -U_V + \sqrt{(\vec{p}^2 + M_B^{*2})} \leq -M_B \tag{3}$$

(i.e. $E^-(-\vec{p}) \geq M_B$) will be unbound. From equation (3) with the equality sign, we have

$$-U_V + \sqrt{p_{\mathrm{max}}^2 + M_B^{*2}} \;=\; -M_B \tag{4}$$

which gives

$$p_{\mathrm{max}} = \sqrt{(U_V + U_S)(U_V - U_S - 2M_B)}. \tag{5}$$

The critical baryon density ρ_c corresponds to $p_{\mathrm{max}} = 0$. This leads to the equations $U_V + U_S = 0$ or $U_V - U_S - 2M_B = 0$ or both. The first condition $U_V + U_S = 0$ is fulfilled earlier than the second one. When the first condition is reached the nucleus becomes unbound, i.e., unstable with respect to emission of nucleons. So it is impossible to compress the nucleus more than the critical density in a self-consistent manner; such densities should occur only as short-lived intermediate stages in a heavy-ion collision. We performed a constraint calculation with the monopole moment [23, 24, 25] , which produced self-consistent solutions up to $\rho_v \approx 3\rho_0$ for the case of ^{208}Pb. A chart for the critical densities of nuclear matter with different parameter sets is given in ref. [26]. We found that the critical densities are much larger for nuclear matter compared to finite nuclei in all available parameter sets. The TM1 parameter set was chosen for our calculations, because it gives a larger critical density of about $3\rho_0$.

In the present context, however, the condition $U_V - U_S - 2M_B = 0$ is desired for baryon anti-baryon pair creation [27] above the critical density. To achieve the higher density we gave up the self-consistent calculations and followed a crude method: we compressed the ^{208}Pb nucleus to 2, 4, 6, and 8.5 times normal nuclear matter density by scaling the vector density with a scaling factor c and the radius with the factor $c^{-1/3}$ to keep baryon number conserved. The scalar potential was scaled with the corresponding nuclear matter factor as follows: we calculated the scalar potential $U_S(\rho_v)$ for nuclear matter as a function of density (see Fig. 3, dotted curve), scaled the vector density ρ_v by a factor c to $c\rho_v$. Now from Fig. 3 we took the value of U_S which corresponds to the density $c\rho_v$. For example, at $\rho_v = 8\rho_0$ (i.e. $\rho_v = 1.16$ fm^{-3}) the value of U_S was taken to be -807 MeV (as $U_S(\rho_v = 8\rho_0) \sim -807$ MeV). (Note that the radius of the nucleus will be reduced by a factor $c^{-1/3}$ while the density increase by a factor c). From the scaled densities and radii, we estimate the potential and the corresponding single-particle energies.

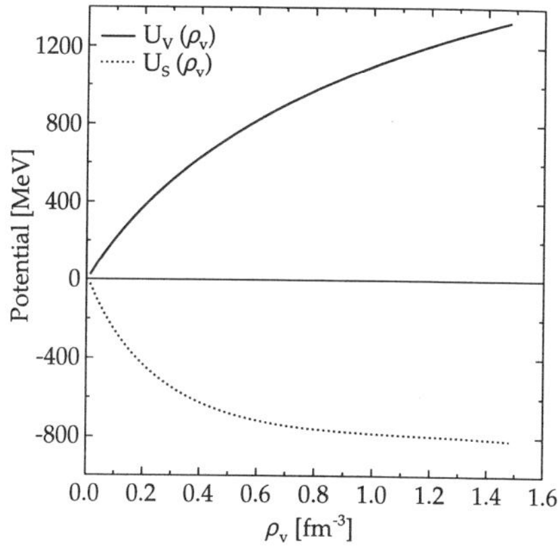

Figure 3: The scalar potential U_S and vector potential U_V for nuclear matter as functions of density ρ_v. The solid line is the vector potential and the dotted line is the scalar potential.

Figure 4 shows the potential for different densities. From the classical relation (2.2b), it follows that the vector potential U_V changes with ρ_v, the baryon density. The scalar potential increases with ρ_v, but more slowly than the vector potential (see fig. 3). In fact, the scalar potential tends to saturate at high baryon densities, since otherwise $M_B^* = M_B + g_s\sigma(r)$ will be negative. Therefore, the energy gap $2M_B^*$, between all baryon and anti-baryon states diminishes (see figure 4). The boundaries of the gap bend upwards with increasing baryon density.

At high enough density, where $U_V - U_S \geq 2M_B$, the upper edge of the Dirac sea rises above the lower edge of the upper continuum. Under such conditions the finite system becomes overcritical (unstable) with respect to spontaneous baryon-antibaryon pair creation at the boundary, where the meson fields vary rapidly. This is analogous to the overcriticality in superstrong electric fields, where spontaneous positron emission sets in and the neutral vacuum decays into a charged one [1]. As a result, a strongly bound anti-baryon appears in the dense nuclear medium accompanied by a low-energy baryon in vacuum. The appearance of anti-baryons in dense baryonic matter lowers the baryon density and therefore provides a negative feedback in the pair creation process. Note that this is a consequence of the spatial separation of the pair, i.e. of the spontaneous emission process for baryons. The instability is terminated when

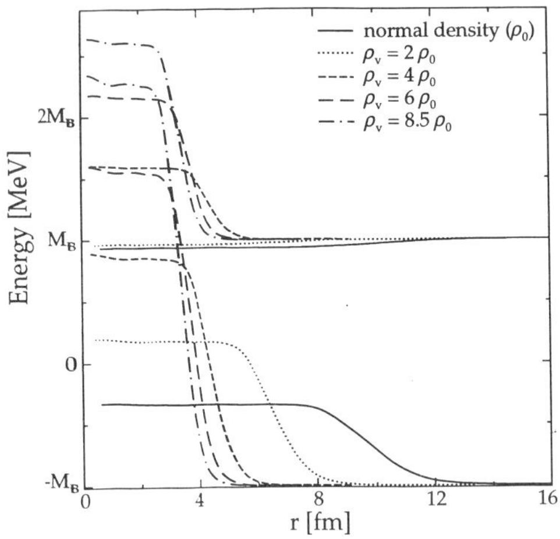

Figure 4: The potentials of the positive and negative-energy sector at different densities. For densities higher than about $4\rho_0$ (see dashed curves) the potentials become overcritical.

ρ_v decreases to ρ_c, the critical density. Many single-particle states become unstable at the critical density in the extended potential well. If the spontaneous decays were very fast compared to the compression and expansion time scales in relativistic heavy-ion collisions, the critical density would be the limiting baryon density, which could not be exceeded in a collision process.

Figure 5 shows the single-particle energies for negative-energy states at $8.5\rho_0$. The energy gap between baryon and antibaryon states becomes smaller. The baryon potential even bends upwards with increasing baryon densities. Even though there is no gap any more between the antibaryon and baryon states, they are still separated by a potential barrier. For overcritical densities the upper edge of the Dirac sea rises above $+M_B$. Obviously, as already explained above, there exists a critical baryon density above which antiparticles can tunnel through the barrier and spontaneous baryon and antibaryon pair creation sets in. In table 1. the number of bound antiparticles and the energy of the lowest level in the negative-energy sector (anti-baryon potential) are given at different densities. Both increase with density.

The bound states in the antibaryon potentials "dive up" from the negative-energy continuum. These "dived up" bound states and the negative-energy continuum states represent the vacuum. Owing to the many "dived up" bound states we thus are dealing with a *highly correlated vacuum*. The picture of hole theory may be helpful here: those "dived up" bound states, like the ones in the negative-energy continuum, are occupied with baryons. A hole in those

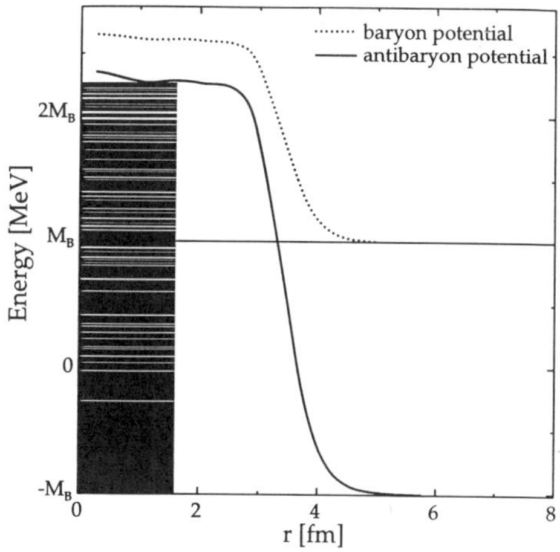

Figure 5: The single particle energies for negative-energy states of ^{208}Pb at density ρ_v=8.5 ρ_0.

Density of baryons	energy of 1s state in antibaryon potential (MeV)	number of bound particles
normal	-322.8	20404
2 ρ_0	160.1	22888
4 ρ_0	784.3	26046
6 ρ_0	1445.9	34644
8.5 ρ_0	2108.1	41240

Table 1: The energy of the most tightly bound 1s state in the antibaryon potential and the number of particles in the potential well emerging out of the lower continuum at various densities.

states represents an antibaryon. Holes in the bound states of the antibaryon potential, i.e., in the "dived up" states represent bound antibaryons. Such holes can be produced (1) due to the time dependence of the potentials in a heavy-ion collision (HIC), i.e., due to heavy-ion dynamics or, (2) due to the thermal excitation which might occur. Several bound hole states may represent an antibaryon cluster, e.g., $\bar{d}$, $\bar{He}$, $\bar{O}$ etc. One should also note that when the nucleus achieves such a high density in a heavy ion collision, the temperature of the system is generally also high, i.e., of the order of $100 - 300$ MeV. Therefore holes (antibaryons) are most likely produced thermally. To estimate the probability of formation of holes (i.e. the anti-particle cluster) in the deep negative-energy potential well at different temperatures, we calculated the values of the chemical potential μ from [28]

$$\rho_v = \frac{\gamma}{(2\pi)^3}\int \frac{d^3k}{1+e^{(\epsilon^*-\mu)/T}} - \frac{\gamma}{V}\sum_i \frac{1}{1+e^{(\epsilon_i+\mu-U_V)/T}} - \frac{\gamma}{(2\pi)^3}\int \frac{d^3k}{1+e^{(\epsilon^*+\mu)/T}} \tag{6}$$

Here i runs over all bound occupied states, ϵ_i is the single-particle energy of the i-th state, γ is the spin-isospin degeneracy factor, $\epsilon^* = \sqrt{{M_B^*}^2 + \vec{k}^2}$ (M_B^* is the effective mass and $\vec{k}$ is the momentum), T is the temperature and V is the volume of the compressed nucleus. Equation (6) will be more transparent if one examines Fig. 5. For example, at high enough density the baryons are in the positive continuum, i.e. above the dotted line in figure 5. These baryons are sitting momentarily in unbound states, still "inertially confined". They are simply pushed out of their shell model orbits due to the compression during the nucleus-nucleus collision. Thus to get the baryon number density one has to use the first term on the right hand side in the above equation (6). The second term is the number density of antibaryons which have single-particle energies ϵ_i and the antibaryonic states lie between $-M_B$ and the solid line in the figure 5. Here all the single-particle energies are discrete and the number density of antibaryons in this region is obtained from the second term of equation (6). The third term in equation (6) takes into account the number density of antibaryons which are in the negative continuum of the highly dense system. These antibaryons lie below $-M_B$ in figure 5.

Knowing the values of μ, we estimated the number of holes formed in the states of the highly correlated vacuum ("negative-energy" states) [28].

In Fig. 6 we have plotted the probability distribution of particles at temperatures T= 0, 20, 100, 200 and 300 MeV. We counted the number of holes at various temperatures with the constant density $\rho_v = 8.5\rho_0$. The number of holes are 2018 and 5548 at T=200 and 300 MeV, respectively. With increasing temperature the number of holes in the negative-energy sector increases, so that a large number of antiparticles, especially with energies greater than M_B are formed, which could in principle escape from the system, if formed near the surface where annihilation may be less probable.

Those antibaryons (antibaryon clusters), being deeply bound, can be annihilated through collisions with surrounding baryons. There is, however, also the fascinating possibility that, in the dynamically expanding environment (decompression), the bound antimatter clusters are driven into the lower (negative-energy) continuum and, thus escape. This is a *production of antimatter clusters directly out of the highly correlated vacuum.* It is to be contrasted to antimatter cluster production through fusion of multiples of antibaryons in phase space, a process which is expected to be smaller by many orders of magnitude as compared to the former one.

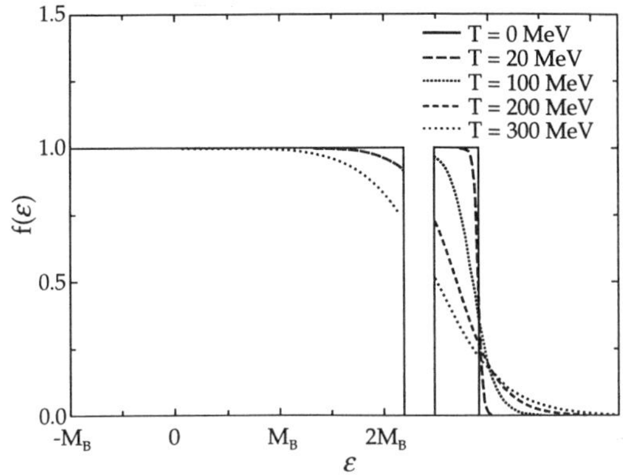

Figure 6: Occupation probabilities of particles $f(\epsilon)$ at different temperatures i.e. at T=0, 20, 100, 200 and 300 MeV.

Conclusions

A large number of negative-energy bound states (i.e $\sim$ 20000) for anti-neutrons and about the same number of states for anti-protons are predicted in a normal ^{208}Pb nucleus. Such negative-energy bound states suggested by relativistic mean field calculations simply do not exist in the conventional nonrelativistic approach. The spin-orbit interaction strength of the negative-energy bound states is about 100 times weaker than the positive-energy states. The number of bound negative-energy states increases with increasing density.

Here the ^{208}Pb nucleus has been compressed up to $\sim 3\rho_0$ utilising a monopole constraint self-consistent calculation. The critical density, i.e., the density were the external nucleus is no more bound, depends very much on the force used. The TM1 parameter set used in [29] gives a relatively high value of critical density. The quartic term of the σ meson coupling is a positive value in this parameter set, whereas it is negative in most of the other successful parameter sets. This positive sign of the quartic coupling constant prevents unphysical behaviour at high densities. Since it was impossible to go beyond the critical density in a self-consistent manner, we switched to a scaling procedure to achieve higher densities. The number of particles in bound "negative-energy" states, i.e., in bound states emerging out of the negative-energy continuum, are found to be 22888, 26046, 34644 and 41240 at $2\rho_0$, $4\rho_0$, $6\rho_0$ and $8.5\rho_0$, respectively. We are thus dealing with a highly correlated vacuum structure, whose consequences are not yet explored fully. However, apart from the above predictions, the following important conclusions have been drawn from the present calculations.

1. In high energy heavy ion collisions it might be possible to produce antimatter clusters like $\bar{d}$, $^4\bar{He}$, $^{12}\bar{C}$ etc. out of the highly correlated vacuum in contrast to their conventional production by fusing antibaryons, step by step, in phase space.
2. There is also the possibility of obtaining overcritical baryonic potentials. In this case the possibility of spontaneous baryon-antibaryon pair creation occurs. This process is, however perhaps less relevant in heavy encounters because of the short time scale for the overcritical situations (short time scale of the overcritical compression).
3. Many observable signals proposed for the quark-gluon plasma, in particular enhanced antibaryon yields may be interpreted in the purely hadronic scenario, if the reduction of hadron masses at high density is taken into account[30].
4. Dynamical calculations which take into account the vacuum rearrangement effects in time-dependent meson fields are required to quantify these predictions.
5. It is very important to check experimentally the depth of the lower potential well emerging out of the negative-energy continuum. Up to now there exists practically no information about it. The use of $e\ e'p$, $p\ 2p$, $\bar{p}$ annihilation on nuclei comes to mind. They will be reported in a separate article [31].
These novel effects due to the highly correlated vacuum are more pronounced at higher baryon densities/temperatures which may be reached during heavy ion collision. In particular the antimatter cluster production out of the highly correlated vacuum state in violent nucleus-nucleus dynamics during and after the encounter deserves prime attention. Work in this direction is continuing and will appear in a forthcoming publication. We finally mention that those strong correlations in the vacuum also exist for hyperons [15] and, similarly to the mechanism for antimatter cluster production suggested here, may be used to produce, e.g., anti-Δ hypernuclei. This may open the possibility to extend the periodic system in to new sectors of antimatter and strangeness.

Acknowledgements

We thank Dr. J. Reinhardt, Dr. C. Greiner, Dr. J. Schaffner, Dr. P.K. Panda, Dr. S.K. Patra, Prof. J. Maruhn and Prof. H. Stöcker for valuable discussions.

References

[1] W. Greiner, B. Müller and J. Rafelski, "Quantum Electrodynamics of Strong Fields", Springer-Verlag (1985).

[2] T.E. Cowan, H. Backe, K. Bethge, H. Bokemeyer, H. Folger, J.S.Greenberg, K. Sakaguchi, D. Schwalm, J. Schweppe, K.E. Stiebing and P. Vincent, Phys. Rev. Lett. **56** 444 (1986); Phys. Lett. **B245** 153 (1990)

[3] I. Koenig, E. Berdermann, F. Bosch, W. Koenig, P. Kienle, C. Kozhuharov, A. Schröter and H. Tsertos, Z. Phys. **A346** 153 (1993), and references therein.

[4] I. Ahmad et al., Phys. Rev. Lett. **75** 2658 (1995); J.S. Greenberg and Guangsheng Xu, "Correlated e^+e^- Emission in Heavy-Ion Collisions", Yale University Preprint (1996).

[5] R. Ganz, R. Bär, A. Balanda, J. Baumann, W. Berg, K. Bethge, A. Billmeier, H. Bokemeyer, H. Fehlhaber, H. Folger, J. Foryciar, O. Fröhlich, O. Hartung, M. Rhein, M. Samek, P. Salabura, W. Schön, D. Schwalm, K.E. Stiebing and P. Thee, GSI-Preprint 96-40, August 1996, in print in Physics Letters.

[6] W. Greiner, "Exclusions of quasibound e^+ e^- states at MeV energies with in the framework of QED", Plenary talk held at the DPG spring meeting, Berlin, March 1988.

[7] O. Klein Z. Phys. **53** (1929) 157.

[8] H.P. Dürr and E. Teller, Phys. Rev. **101** (1956) 494.

[9] M.H. Johnson and E. Teller, Phys. Rev. **98** (1955) 783.

[10] H.P. Dürr, Phys. Rev. **103** (1956) 469.

[11] J.D. Walecka, Ann. Phys. (N.Y.) **83** (1974) 491.

[12] P.-G. Reinhard, Rep. Prog. Phys. **52** (1989) 439.

[13] B.D. Serot, Rep. Prog. Phys. **55** (1992) 1855.

[14] M. Rufa, P.-G. Reinhard, J.A. Maruhn, W. Greiner and M.R. Strayer, Phys. Rev. **C38** (1988) 390; P.-G. Reinhard, Z. Phys. **A329** (1988) 257; Y.K. Gambhir, P. Ring and A. Thimet, Ann. Phys. (N.Y.) **198** (1990) 132; M.M. Sharma, M.A. Nagarajan and P. Ring, Phys. Lett. **B312** (1993) 377.

[15] W. Greiner, Int. J. Mod. Phys. **E5** (1996) 1.

[16] P. G. Reinhard, M. Rufa, J. Maruhn, W. Greiner and J. Friedrich, Z.Phys. **A323** (1986) 13.

[17] S.K. Patra and C.R. Praharaj, Phys. Rev. **C44** (1991) 2552.

[18] C.J. Horowitz and B.D. Serot, Nucl. Phys. **A368** (1981) 503.

[19] B.D. Serot and J.D. Walecka, Adv. Nucl. Phys. **16** (1986) 1.

[20] J. Boguta and A.R. Bodmer, Nucl. Phys. **A292** (1977) 413.

[21] K. Rutz, M. Bender, T. Bürvenich, T. Schilling, P.-G. Reinhard, J. Maruhn, W. Greiner, "Superheavy nuclei in selfconsistent nuclear calculations", Preprint (1996), Institut für Theoretische Physik, JWG Universitat, Frankfurt a M, to be published.

[22] P.K. Panda, S.K. Patra, J. Reinhardt, J. Maruhn, H. Stöcker and W. Greiner "Strong Correlation of Vacuum in the Mean Field Theory", Preprint, Institut für Theoretische Physik, JWG Universitat, Frankfurt a M, August 1996, to be published.

[23] D. Hirata, H. Toki, I. Tanihata and P. Ring, Phys. Lett. **B314** (1993) 168.

[24] W. Koepf and P. Ring, Phys. Lett. **B212** (1988) 397.

[25] H. Flocard, P. Quentin and D. Vautherin, Phys. Lett. **B46** (1973) 304.

[26] I.N. Mishustin, L.M. Satarov, J. Schaffner, H. Stöcker and W. Greiner, J. Phys. **G19** (1993) 1303.

[27] N. Auerbach, Phys. Rev. **C50** (1994) 1606.

[28] W. Greiner, L. Neise and H. Stöcker, "Thermodynamics and Statistical Mechanics", Springer-Verlag (1994).

[29] Y. Suhagara and H. Toki, Nucl. Phys. **A579** (1994) 557.

[30] U. Heinz, P.R. Subramanian, H. Stöcker and W.Greiner, J. Phys. **G12** (1986) 1237.

[31] J. Reinhardt and W. Greiner, to be published.

High-Power Lasers and Their Applications

S. Svanberg

Department of Physics, Lund Institute of Technology
and
Lund Laser Centre, Lund University
P.O. Box 118, S-221 00 Lund, Sweden

Abstract

Through the development of tuneable high-power and short-pulse laser systems atomic and molecular physics has recently been enriched with new tools for laser/atom interaction experiments at high optical field intensities and at short wavelengths. Experiments requiring ultra-intense laser radiation have recently become much more practical through the development of chirped pulse amplification. By using titanium sapphire as the gain medium, very compact and high-repetition rate systems reaching terawatt power levels can be constructed. Within the Lund High Power Laser Facility we operate such a system (100 fs, 2.0 TW, 10 Hz, tuneable 760-840 nm), which is being employed in basic and applied studies requiring high optical powers. On the more basic side, detailed studies of high-harmonics generation in jets of inert gas are being performed and spectrally well adapted harmonics radiation is also being applied in atomic physics experiments. Detailed studies of ionisation of atoms and fragmentation of molecules are also being pursued. Studies of X-ray laser generation are being made. By focusing high-power radiation in water or onto a high-nuclear-charge metal, white light generation and continuum hard X-rays are generated, respectively, being of considerable importance for applications. Thus, white laser light has been used for time-resolved studies of photon propagation in body tissues, leaves and paper sheets, while hard X-rays have been employed in emerging medical X-ray applications.

ADVANCES IN QUANTUM CHEMISTRY, VOLUME 30

0065-3276/98 $25.00

1. Introduction

Chirped pulse amplification is a new technique to achieve ultra-high optical powers in compact laser systems. The technique becomes particularly powerful when used in conjunction with titanium-doped sapphire (Ti:S), which is a broadband gain material, allowing the generation of short pulses (~20 - 100 fs). 10 Hz systems can be designed operating at the terawatt power level. This technology makes high-power experiments readily accessible, with the convenience typical of normal laboratory 10 Hz systems. Different types of high-power experiments are exemplified below:

* Atomic physics at high optical field strengths (10^{10} V/cm $\Leftrightarrow$ 10^{18} W/cm^2)
* High harmonic generation in jets of inert gases, yielding coherent soft X-rays
* X-ray laser pumping, especially optical-field-ionisation pumping
* Molecular fragmentation and ionisation
* Generation of broadband X-rays from a solid-density plasma with medical applications
* White-light generation with applications in chemistry, biology and medicine

Coherent short-wavelength radiation can also be applied as a source for exciting atoms and ions in the XUV region. For this purpose a longer-pulse tuneable system, which allows more narrow-band radiation generation compatible with typical Doppler widths, is more advantageous to use in high-harmonic conversion. Such a system was constructed in Lund and was applied in demonstration measurements in the XUV spectral region. The field of high-power lasers and their applications will be illustrated by work performed at the Lund Institute of Technology.

High-power lasers are described in Sect. 2. Selected topics in high-power laser/matter interaction are then treated: high-harmonics generation, X-ray laser research and molecular fragmentation in Sections 3, 4 and 5, respectively. White-light generation and applications are covered in Sect. 6. Then the generation and application of laser-produced broad-band X-rays are discussed in Sect. 7. Finally, some conclusions and an outlook for the future are presented.

2. High-power lasers

2.1 Terawatt chirped pulse amplification system

Chirped pulse amplification (CPA) [1,2] is a new technique to achieve extremely high optical intensities in compact laser systems. The basic idea, which has its counterpart in radar technology, is illustrated in Fig. 1 [3]. An initial, short pulse

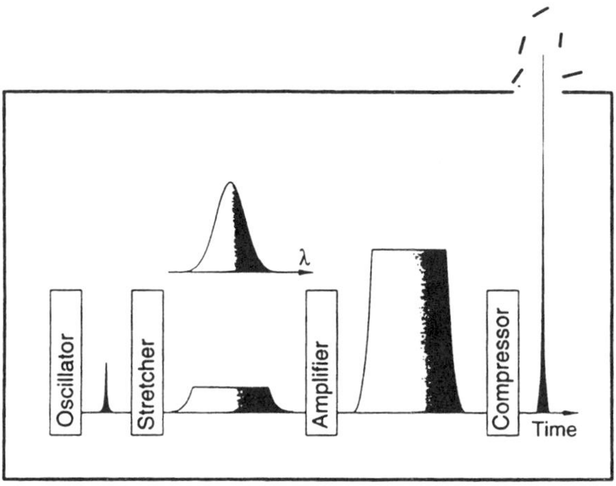

Fig. 1. Basic principle for chirped pulse amplification showing the stretching, amplification and compression stages (From Ref. [3]).

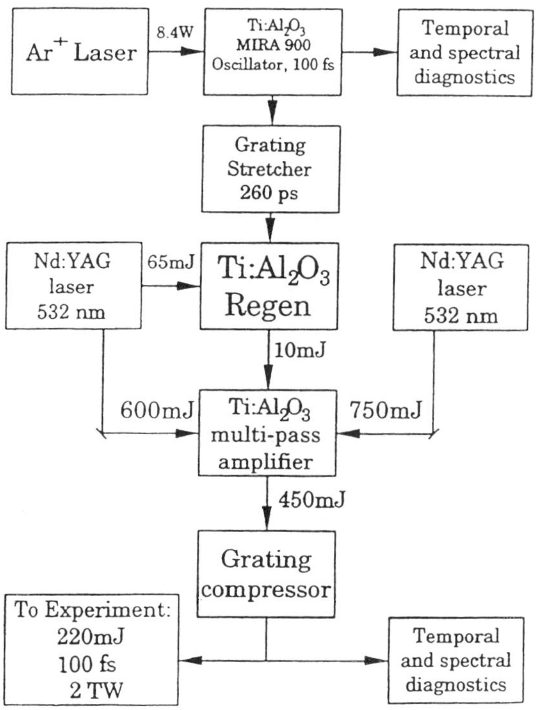

Fig. 2. Lay-out of the Lund High-Power Laser Facility terawatt system, which operates with chirped-pulse amplification in titanium-doped sapphire (Adopted from Ref. [3]).

is temporally stretched before amplification in order to reduce the peak power in the amplifier stage for a given pulse energy, avoiding the need for large beam diameters. Pulse compression is then performed in a grating arrangement to achieve ultra-high powers. The CPA technique has enabled the construction of compact terawatt lasers, sometimes called T^3 (table-top terawatt) systems. The technique was first used for Nd:Glass amplifier systems, that, however, are restricted to low repetition rates. Through the introduction of titanium-doped sapphire (Ti:S) as a gain medium [4-6], particularly convenient systems can be constructed. Ti:S can readily be pumped by frequency-doubled (532 nm) Nd:YAG radiation. The material has a wide gain profile and can thus support amplification of ultrashort pulses (down to tens of fs), which results in lower pulse-energy demands for reaching terawatt power levels. The thermal properties allow a 10 Hz repetition rate. Important quality factors describing a terawatt laser are, apart from the pulse power, the pulse duration, the focusability and the contrast between the main pulse and possible prepulses or a background level (pedestal).

The main laser at the Lund High-Power Laser Facility is a Ti:S CPA terawatt system, which was developed by a commercial company (Continuum Inc.), with system configuration and performance according to our specifications. Different aspects of the underlying technology are treated in Ref. [7]. A general lay-out of the system is shown in Fig. 2.

Nearly transform-limited pulses of 100 fs duration are generated from an Ar^+-laser-pumped Kerr-lens mode locked Ti:S oscillator (Coherent Mira 900). Pulses at a repetition rate of 76 MHz and at an average power of about 1 W (10 nJ/pulse, 100 kW peak power) are obtained. The 100 fs pulsewidth is determined by remaining, non-compensated group-velocity dispersion in the oscillator cavity and can be further reduced.

The 100 fs output pulses from the oscillator are temporally stretched by a factor of about 2500 in a grating and lens arrangement that is double-passed, as illustrated in Fig. 3a. Since the pulse is short it contains different colours, that are diffracted differently by the grating. It can be seen in the figure that the low-frequency components travel a shorter distance and will come out first after reflection back through the system. The emerging pulse is thus colour-coded in time in a well-defined way, allowing, in principle, an exact play back later in the compressor stage. However, this has as a prerequisite that a uniform amplification of all frequency components occurs. Otherwise pulse stretching and the formation of an unwanted intensity pedestal occur.

The stretched oscillator pulse is injected into a regenerative Ti:S amplifier (polarisation switching using an intra-cavity Pockels cell and thin-film polarisers). This unit is basically a Ti:S laser which is pumped by 65 mJ of green light from a frequency doubled Nd:YAG laser. After 12 double passes and 10^6 times amplification, the pulse is ejected, again by polarisation switching. A further Pockels cell is used to suppress any cavity leakage of other pulses than the main

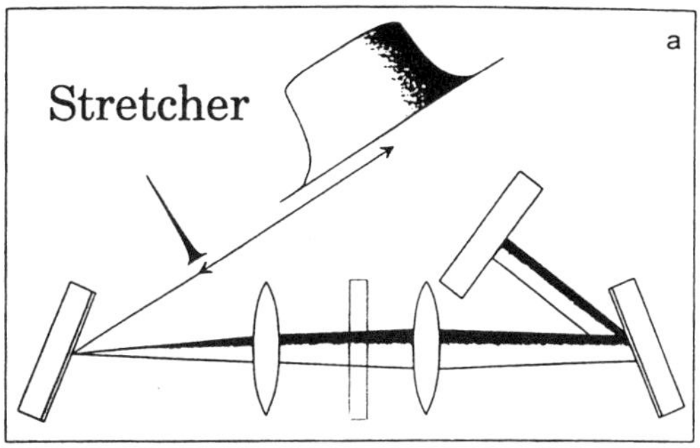

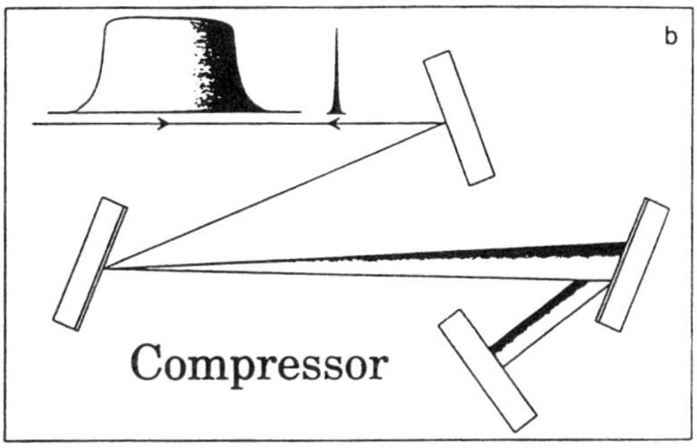

Fig. 3. Operational principles for grating stretcher (a) and compressor (b) (From Ref. [3]).

one. Final power boosting is performed in a four-pass (butterfly) Ti:S amplifier crystal to reach a level of up to 450 mJ. Two frequency-doubled high-energy Nd:YAG lasers (Continuum NY82-10) pump the amplifiers at a total energy level of 1.3 J at 10 Hz. Relay imaging of the pump laser-rod surface using evacuated beam-transport telescopes is used to achieve a uniform pumping. A beam diameter of 8 mm is used through the final Ti:S crystal. After beam expansion to 50 mm diameter, in order to reduce the power density, the pulse is compressed using two parallel, 11 x 11 cm gold coated holographic gratings with 1800 grooves/mm and a first-order diffraction efficiency of about 90 %. The action of the grating compressor is illustrated in Fig. 3b. After double passing the grating arrangement the low-frequency leading-edge light is delayed and the high-frequency trailing edge is catching up to emerge simultaneously from the compressor. 100 fs pulses of powers up to 2 TW can be obtained in the 760-840 nm region. This corresponds to a pulse energy of about 200 mJ. The maximum output is achieved at 792 nm, which is chosen as the operating wavelength when tuning is not important. Tuning is achieved by changing the oscillator centre frequency by a Lyot filter. Geometrical adjustments on the stretcher and compressor are also required.

In order to avoid phase-distortions and associated beam degradation due to the non-linear refractive index of air, the compressed beam can be propagated in beam transport tubes, which occasionally are filled with Ar, which has a lower non-linear index than air. The terawatt laser output can be injected into different beam lines, presently four, for experiments on, e.g. high-harmonic generation, X-ray laser experiments, broadband X-ray generation or applications following white-light generation.

The system is presently being upgraded with an additional amplifier stage, pumped by a large 10 Hz Nd:YAG laser with an output pulse energy at 532 nm of 5 J. A vacuum compressor will be connected to new high-power experimental stations. Design goals for the present upgrade are 80 fs 8 TW pulses maintained at 10 Hz.

In order to characterise the system, the oscillator output is analysed spectrally with a small spectrometer equipped with a diode-array detector for real-time spectral display, and temporally with a scanning autocorrelator. The final output beam is analysed with a single-shot, second-order auto-correlator, and will be supplemented with a third-order unit, which allows the leading and trailing pulse edges to be distinguished. The beam can be focused down to about three times the diffraction-limited spot, i.e. about 60 μm diameter for an f=100 cm lens, or to 3 μm using an f=5 cm off-axis parabola.

2.2 Distributed feedback, direct amplification system

For short-wavelength spectroscopy of free atoms and ions, which can become feasible through high-harmonic conversion, the extremely short pulse length and the associated line broadening of the terawatt system are not very suitable. Then a fully tuneable and more narrow-band laser with a pulse length of few tens of ps is more desirable. Such a system has been constructed at the Lund High-Power Laser Facility [8,9]. As an oscillator for the system a distributed feed-back dye laser (DFDL), pumped by the frequency-doubled output of an 80 ps Nd:YAG laser is used. The dye laser output pulses are amplified in two dye stages and are finally amplified in a 4-pass titanium-sapphire stage, pumped by a large-frame nanosecond Nd:YAG laser. The system lay out is shown in Fig. 4. The resulting pulses of 50 mJ energy and 80 ps duration are well suited for the generation of low-order spectroscopic harmonics. Since the CPA technology is not used, power densities close to the damage threshold for optical surfaces are reached, but on the other hand tuning is easy since the whole system just follows the oscillator. A pulsed wavemeter is needed to monitor the tuning by angle adjustment on the pumping mirrors or by temperature changes in the dye.

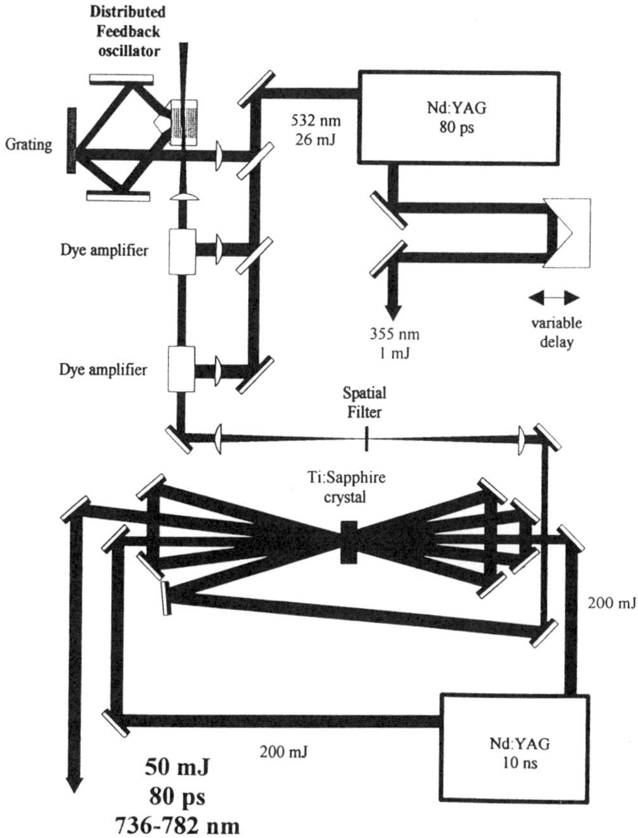

Fig. 4. Lay-out of a spectroscopic high-peak-power, fully tunable combined dye/Ti:S laser system, suitable for the generation of narrow-band, tunable harmonics (From Refs. [8,9]).

3. High Harmonic Generation and Applications

3.1 The generation and characterisation of high over-tones

When a high-power laser beam is focused into a gas jet of inert gas atoms, odd high harmonics are generated. High-harmonic generation is now relatively well understood, and overviews are given in e.g. [10,11]. The first step is a tunnelling of an external electron through the barrier due to the Coulomb field combined with the electric field of the laser light. The electron is accelerated in the optical field and is then forced back towards the nucleus when the field switches direction. Passing in the vicinity of the nucleus it recombines with the ion and harmonic radiation is emitted [12]. The highest order harmonics are obtained for the light inert gases such as He and Ne, but the most efficient generation of lower harmonics is in the heavy gases such as Kr and Xe. Experimental recordings of high-harmonics are shown in Fig. 5 [13]. The shortest wavelength obtained through high-harmonic conversion is of the order of 7 nm. Thus, the 109th harmonic for a titanium sapphire laser at 806 nm [14], the 143rd harmonic for a Nd:Glass system at 1053 nm [15] and the 37th harmonic of a KrF laser at 248 nm [16] have been achieved. Much effort has been put into the understanding of the high-frequency cut-off for the harmonics. Considering the single-atom response only, the cut-off is given by [12] $E_{cut\text{-}off} = I_p + 3.2U_p$, with the ponderomotive potential U_p given by $U_p = e^2E^2/4m\omega^2$ (e and m are the electronic charge and mass, respectively, E is the light electrical field strength and ω is the angular frequency of the light field).

Experimental results for neon are given in Fig. 6 [13]. It is noted that the relation given above is obeyed (I_p for Ne is 22 eV), but an experimental slope factor of 2.4 is obtained instead of the theoretical value 3.2. This is because of phase mismatch in the collective generation by an ensemble of atoms. Also this effect is now well understood.

The cut-off relation given above would suggest that shorter-wavelength radiation could be achieved using materials with maximum I_p. This would suggest ions (e.g. the neon-like ion Na^+ has an ionisation potential of 47 eV). Experimentally this is, however, not achieved [17,18]. Ions work less well mainly because of the strong influence of the free electrons on beam focusability as well as propagation phase mismatch.

In order to develop high harmonics into an attractive general purpose source of spectroscopic radiation it is necessary to optimise the generation efficiency with regard to focus geometry, gas density etc., and many studies along these lines as well as for the temporal and spatial characterisation have been performed [19-24].

The harmonics are particularly attractive in terms of the radiation peak power. Thus, it should open the way for non-linear laser spectroscopy in the short-wavelength (10 nm) regime. A competing technology requiring much larger installations is Self Amplified Spontaneous Emission used for Free-Electron-Laser action (SASE-FEL) [25,26].

Harmonics are tuneable when the primary laser radiation is obtained from a tuneable laser, such as a titanium-doped sapphire (Ti:S) unit. However, the need for wavelength-dependent adjustments in the CPA chain can make the tuning procedure tedious. The laser system described in Sect. 2.2 is easier to tune. Since CPA techniques are not used the entire system tunes easily together with the DFDL oscillator. The peak power is still enough for this system to generate low

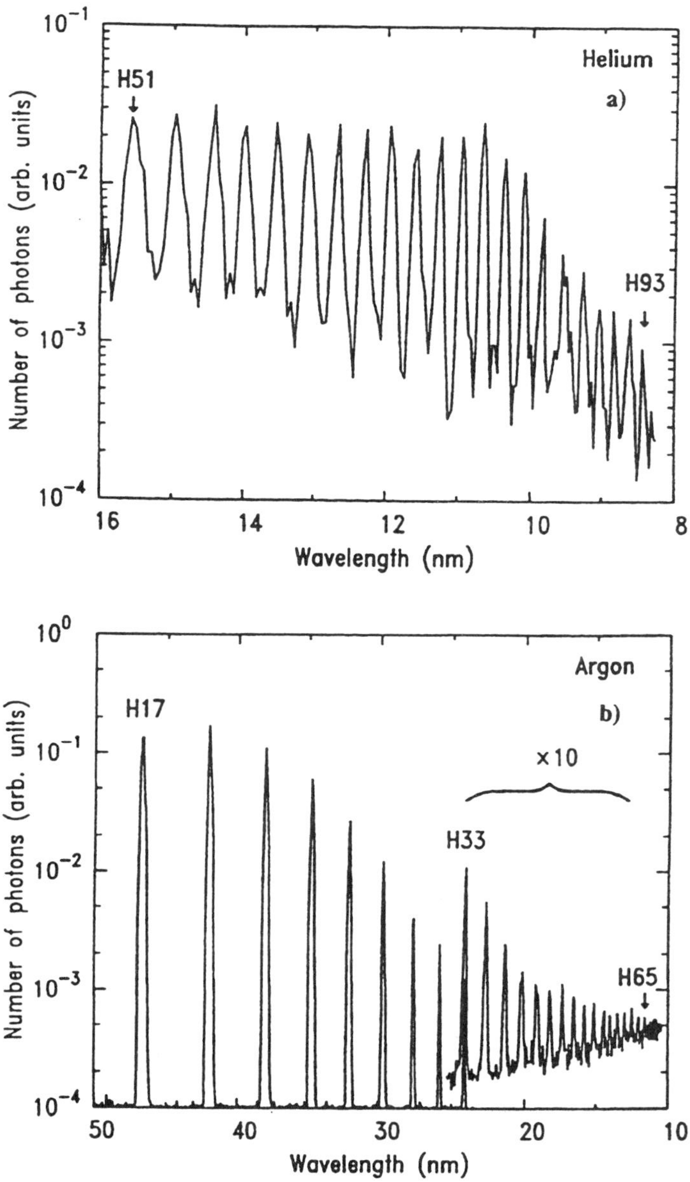

Fig. 5. Harmonics from He and Ar pulsed jets when the focused laser intensity was approximately 10^{15} W/cm^2 (From Ref. [13]).

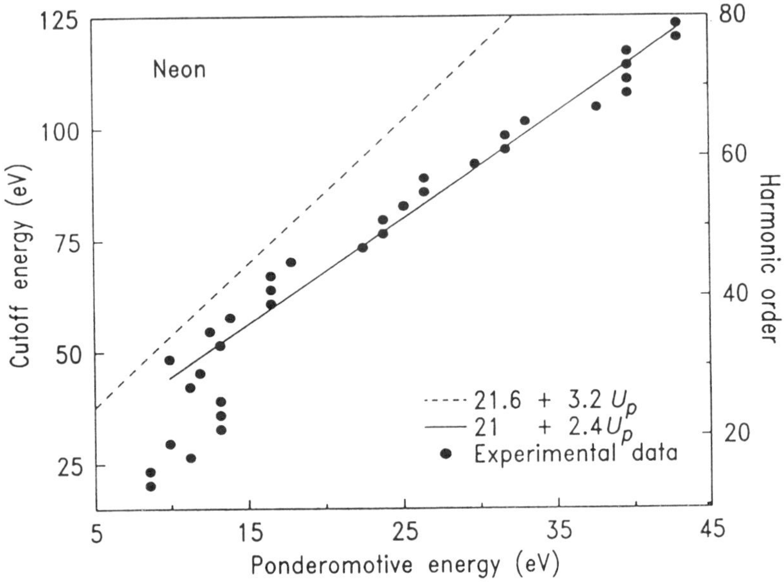

Fig. 6. Intensity dependence of different harmonic orders from neon (From Ref. [13]).

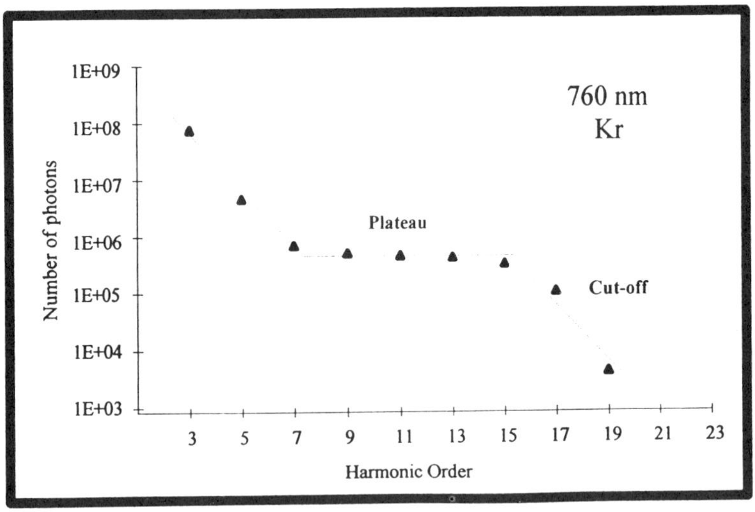

Fig. 7. Harmonics generated in Kr from a DFDL/Ti:S system (From Ref. [28]).

harmonics (7 - 21th) in a jet of Kr or Xe [27]. In this way a very interesting fully tuneable laser source for the XUV region has been achieved. An example of harmonic generation with this source is given in Fig. 7 [28], where the plateau and the cut-off regions are indicated.

Still an alternative way for tuning is to mix fix-frequency terawatt radiation (ω_f) with tuneable radiation ω_t from an optical parametric oscillator/amplifier [29], which is pumped by a fraction of the fixed-frequency output. This is done in a pulsed rare-gas jet, where new frequency components $n\omega_f \pm m\omega_t$ are generated. Here, n normally is a large number and m a small number. Promising results have been obtained [30,31].

An interesting aspect of high-order harmonics is the possibility of generating a train of attosecond pulses [32,33], recognising the conceptual similarity between the energetically equidistant high-order harmonics and the cavity modes in a CW laser, which can be locked together for femtosecond pulse generation. Pulses of a duration of about 10^{-16} s (100 as) should be achievable opening up a new field of ultra-fast atomic spectroscopy.

3.2 Application of high-order harmonics

Experimentally determined radiative properties of free atoms and ions are of great interest for testing atomic theories and more practically for astrophysics, plasma and laser physics. With the availability of the Hubble Space Telescope the VUV spectral region, out of reach for terrestrial observation, has become accessible and there is a great need for accurate data in the short-wavelength region. The most straight-forward way to measure excited-state lifetimes is to perform a selective laser excitation and then observe the exponential decay of the light intensity from the excited state. Such measurements are described, e.g. by Berzinsh and Svanberg ([34], this volume). The VUV region can be assessed by stimulated Raman scattering or by four-wave mixing. For still shorter wavelengths (the XUV region) high harmonics can be used as recently demonstrated using the laser system described in Sect. 2.2. Experimentally recorded harmonic intensitites are shown in Fig. 7. The available energy for the excitation is low, and direct fluorescence light observation is not possible. Instead, a more efficient detection method using a pump/probe scheme was used to follow the decay. Such a technique using optical delays of the probe beam also eliminates problems with limited time response for the detection system. A demonstration experiment to measure the radiative lifetimes of the 2p and 3p $^1P_1$ states of helium was performed [27,10]. The experimental set-up used is shown in Fig. 8 [27]. To excite the 2p state (with a transition energy of 21 eV, i. e. 58.4 nm), the DFDL laser was tuned to 760 nm and the 13th harmonic was selected with a normal-incidence spherical grating. For the 3p case, requiring 54 nm (23 eV), the DFDL was tuned to 752 nm, the laser radiation was frequency-doubled in a KDP crystal and the 7th harmonic of this

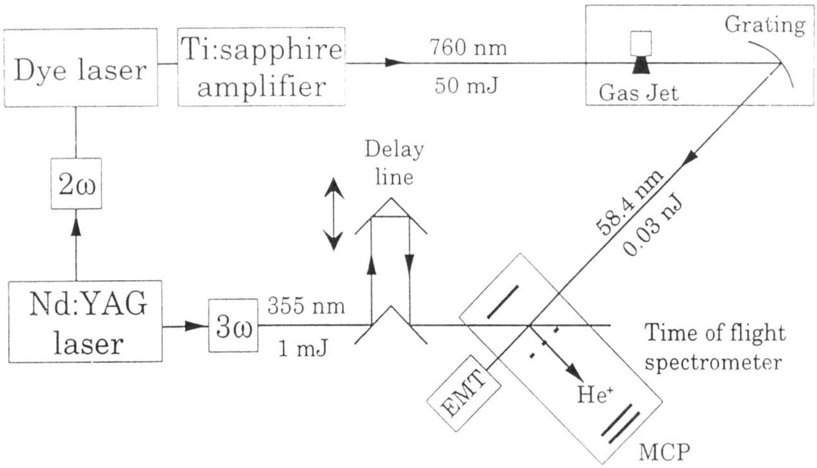

Fig. 8. Set-up for delayed photo-ionization measurements on helium atoms, excited by high-harmonic radiation (From Ref. [27]).

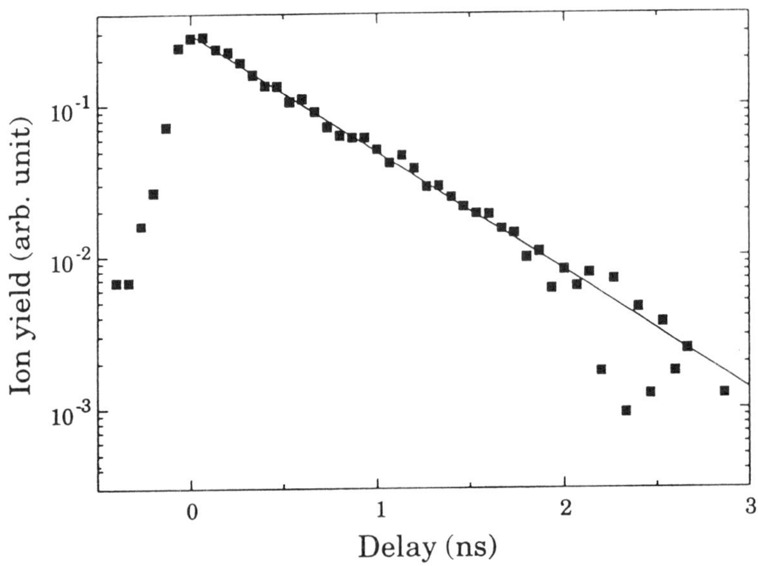

Fig. 9. Experimental decay curve for the 2p $^1P_1$ state of He (From Ref. [27]).

UV radiation was selected. The linewidth of the harmonics is 0.01 nm, which is still a few tens of times the Fourier transform limit. However, a reasonable match to the 0.0003 nm Doppler width of the helium atoms is obtained. Helium atoms were excited and the number of exited atoms was probed by photoionisation using a temporally delayed pulse, chosen to be the second or the third harmonic of the Nd:YAG pump laser. The resulting ions were detected by a microchannel plate detector after passage through a time-of-flight spectrometer that was tuned to the helium mass. An experimental decay curve for the 2p state is given in Fig. 9 [27]. We performed detailed studies with the helium atomic density varying over 3 orders of magnitude and could show that there was no influence of resonance trapping or collisions on the measured lifetime. Thus the XUV pulse intensities achieved with our system are adequate for performing general-purpose time-resolved laser spectroscopy in the XUV region.

The high spectral resolution of the system could also be utilized for measuring autoionisation widths of the Rydberg-valence state complex of the NO molecule in the 11-12 eV region [35].

4. X-ray Laser Pumping

Soft X-ray lasing was first demonstrated in neon-like selenium, yielding emission at about 21 nm [36]. Since then an intense research activity has been pursued involving mainly collisional and recombination schemes. Initially, very large lasers with low repetition rates had to be used for achieving X-ray lasing. The shortest wavelength achieved so far is 3.5 nm for collisional pumping of nickel-like gold [37]. Of particular interest in terms of the development of practical sources is research aiming at the utilization of new high-repetition-rate and moderate pulse-energy systems as described in Sect. 2.1 [38-41]. Special attention has been given to optical field-ionization schemes using gaseous media. Detailed spectroscopy of laser-produced plasmas is needed for locating gain line candidates [42]. The experimental set-up is then very similar to the one used in the studies of high-harmonic generation. Search for gain can be performed by observing the emission of the laser-produced gas plasma in the forward and in a perpendicular direction as shown in Fig. 10. There indications of gain by stronger forward emission for certain lines are also shown for the case of ionized nitrogen [40,43]. Particularly spectacular is the demonstration of strong amplification in Xe^{8+} at about 42 nm [38]. Recent progress in the field of X-ray lasers is comprehensively covered in the proceedings of the latest International Conferences on X-ray Lasers [44-46].

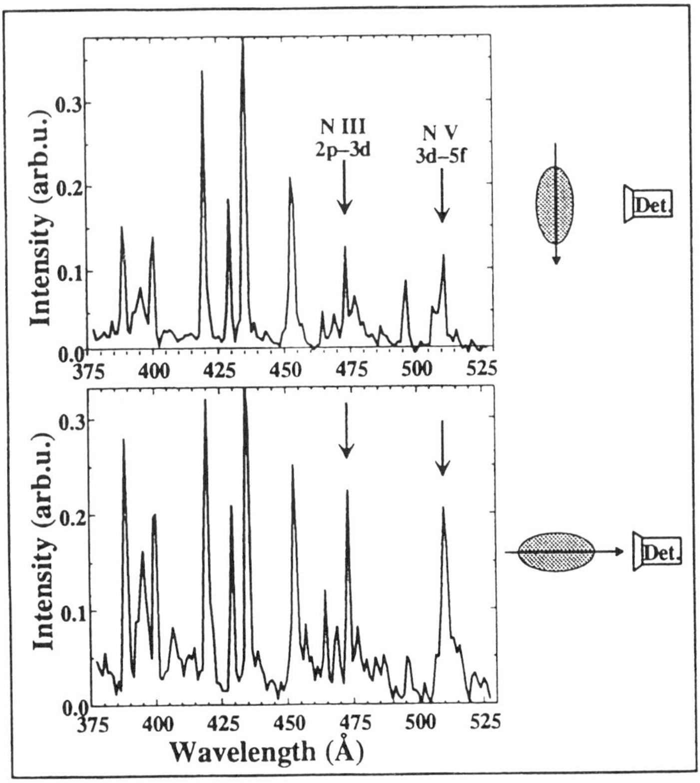

Fig.10. Longitudinally and transversely recorded spectra from a N_2 jet, subjected to focused terawatt laser radiation (From Refs [40,43]).

5. Molecular Fragmentation and Ionization

When high-power laser radiation is focused at molecules, fragmentation and ionization occurs. The fragments can best be studied using time-of-flight spectrometers. Some experiments of this kind have been performed at the Lund High-Power Laser Facility. Of special interest are experiments on the buckminster fullerene clusters C_{60} [47]. Free clusters are generated by thermal evaporation of commercially available powder. Mass spectra following fragmentation with 100 fs pulses at a power density of 10^{13}-10^{14} W/cm^2 are shown in Fig. 11. By studying the power dependence of the fragment yield it is possible to infer the dynamics of the process. The production of $C_{60}{}^+$ follows a I^{13} dependence, i.e. 13 photons are

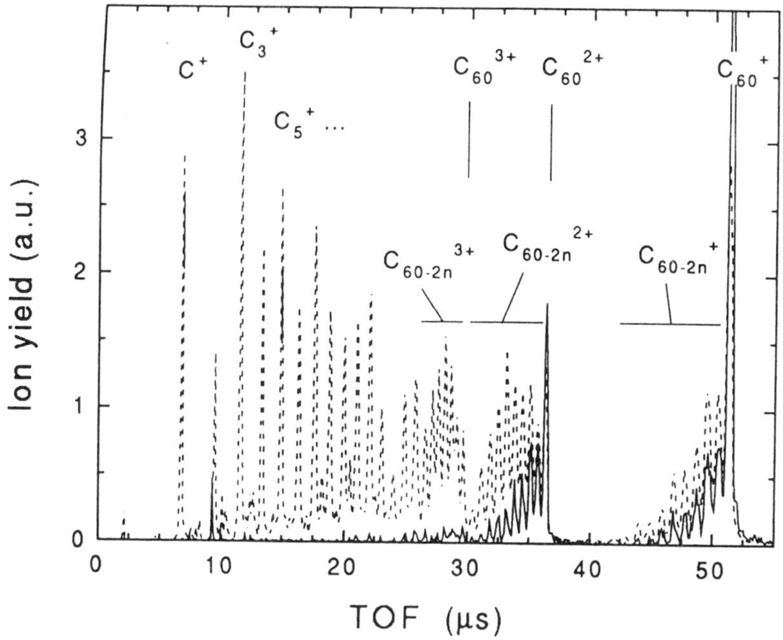

Fig. 11. Fragmentation and ionization of C_{60} measured with a time-of-flight spectrometer (From Ref. [47]).

absorbed in the multi-photon ionization process. This corresponds to a total energy of 13 x 1.57 eV = 20.4 eV corresponding to the wellknown 20 eV plasmon resonance in C_{60}. Correspondingly, to achieve C_{60}^{2+} formation an I^{24} dependence was found corresponding to a two-plasmon resonance. The molecule can fragment or multiply ionize only if double excitation of this resonance occurs.

Much insight in the dissociation patterns of molecules can be obtained by using the so called covariance mapping technique [48]. Terawatt laser excitation allowed such experiments for CO_2 to be performed in a non-explored power density regime [49].

6. White-light Generation and Applications

By focusing high-power radiation in water, self-phase modulation leads to intense generation of short-pulse white light, that can be used for many types of temporally resolved experiments. By using a fast streak camera (~1ps) in conjunction with a spectrometer, simultaneous recording of the temporal response for a distribution of wavelengths, dispersed along the entrance slit of the streak

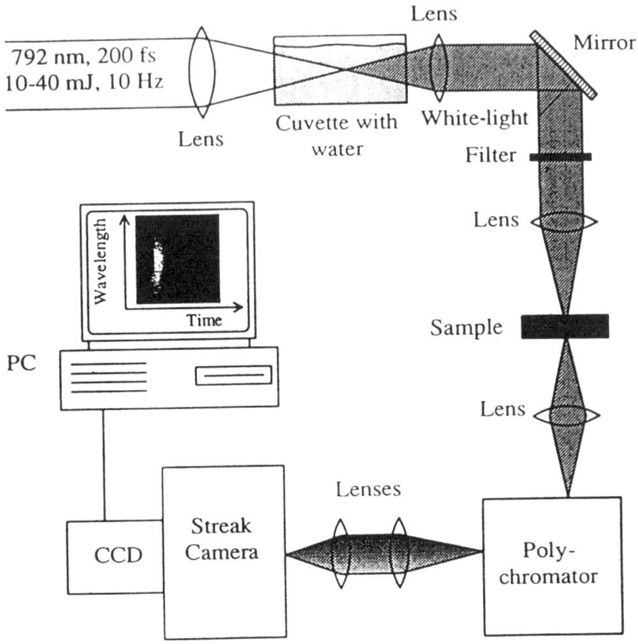

Fig. 12. Set-up for picosecond transillumination experiments with white light. (From Ref. [50]).

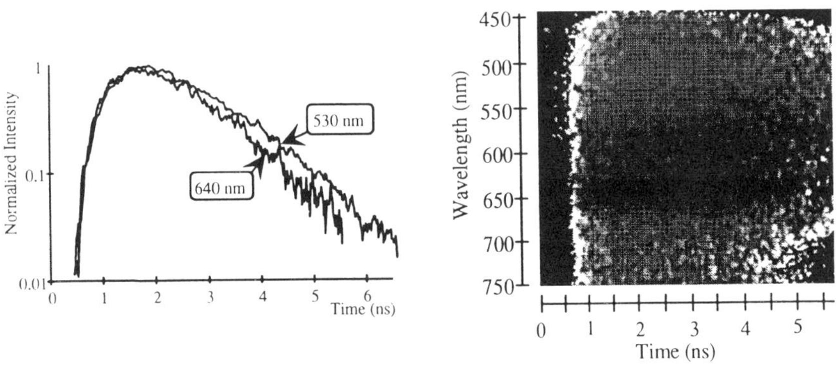

Fig. 13. Recordings of photon migration through a 3 cm thick scattering tissue phantom. Left: Temporal dispersion curves for two wavelengths. Right: Ratio of two streak-camera images; one recorded through a 5 mm diameter cavity filled with a dye solution and one 10 mm beside it. (From Ref. [55]).

camera can be achieved. A 2D CCD detector is used for read out of the temporal response versus wavelength. Such a system, illustrated in Fig. 12, has been used for photon migration studies in tissue [50]. The motion of photons in strongly scattering media is a field of intense current interest.

Tissue transillumination experiments allow multi-spectral determination of tissue optical constants and light fluxes, with application to optical mammography [51], photodynamic tumour therapy [52,53] and brain oxygenation measurements [54]. These are all fields of great medical interest. Data for transillumination of a 3 cm thick slab of a tissue phantom with a 5 mm diameter simulated tumour loaded with Rhodamine 700 (a dye with peak absorption at 640 nm) are shown in Fig. 13 [55]. To the left the temporal history of photons at 640 nm are shown to differ substantially from that for photons at 530 nm. (Note, that a lot of photons emerge even 5 ns after the pulse entry into the sample, which corresponds to a pathlength of about 1 m!) The influence of the agent absorption can be clearly seen in the right part of the figure, where the full CCD readouts for recordings through the hole and 10 mm beside it have been divided by each other. The potential use of optical contrast agents is illustrated. By gating on the first photons exiting from the sample, multiple scattering can be suppressed and enhanced contrast in transillumination imaging can be achieved [56-58]. Transillumination experiments have also been performed on green leaves [59] and on sheets of paper [60] allowing assessment of the scattering and absorptive properties of such materials. The white-light generation scheme discussed here is very useful for exploratory multi-colour measurements of light propagation in strongly scattering media and can form a starting point for the construction of more simple and dedicated set-ups employing monochromatic sources at an optimum wavelength.

7. Broadband X-ray Generation and Applications

By focusing sub-picosecond terawatt pulses on solid targets, intense X-ray radiation can be obtained. In contrast to the situation when nanosecond pulses are employed, there is no time for the formation of an expanding plasma that would shield off the main part of the impinging pulse. Thus, femtosecond laser pulses may lead to the formation of a solid-density plasma. The plasma radiates at all wavelenghts from microwaves to hard X-rays. It has been shown that photon energies approaching MeV can be obtained [61]. The measurement of the spectrum poses difficult problems since the radiation is produced in ultra-short (10^{-12} s), extremely intense bursts followed by intermissions that are typically 10^{11} times longer. Thus solid-state detectors, developed for low intensity, conventional X-ray and γ-ray spectroscopy are hampered by heavy pile-up problems. By using energy-dispersive detectors in Compton-scattering geometry and a specially developed X-ray spectrometer it was still possible to characterise

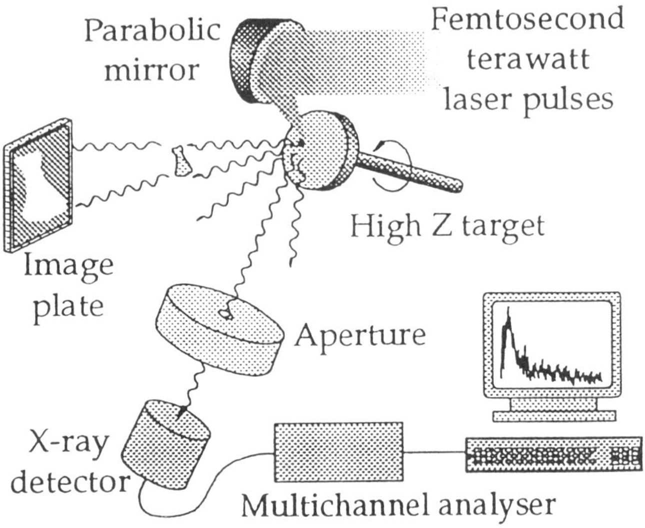

Fig. 14. Experimental set-up for the generation of ultra-short broadband hard X-ray radiation (From Ref. [67]).

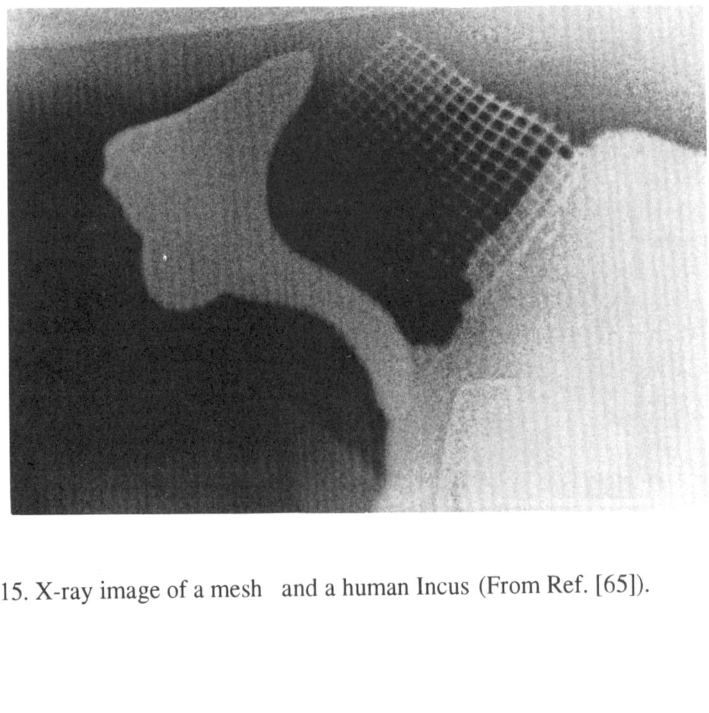

Fig. 15. X-ray image of a mesh and a human Incus (From Ref. [65]).

the spectrum [62,63], which consists of a broad continuum with superimposed characteristic lines. The new X-ray source has special properties and can have medical applications. We have demonstrated imaging on standard radiological image plates using this source [64]. The experimental set up is shown in Fig. 14. The extremely small source size allows magnification radiography (x80 magnification demonstrated). Single-pulse image recording was also found to be feasible. Such a recording is shown in Fig. 15, which is an image of a fine metal mesh (periodicity 0.25 mm) and a human ear-bone (Incus) [65].

Differential absorption radiography using monochromatic X-rays above and below the K-absorption edge of a contrast agent such as iodine (33 keV) has been demonstrated using synchrotron radiation for angiographic applications [66]. We have attempted such measurements using the characteristic lines from Gd and W targets bridging the K-edge (at 50 keV) of gadolinium, which is a contrast agent used in magnetic resonance imaging. The differential effect can be seen in Fig. 16, displaying two rats, to the stomachs of which cerium and gadolinium, respectively, had been administered [67].

The extremely short duration of the X-ray burst can be utilised for gated viewing through tissue in order to eliminate Compton-scattered X-rays dimming the image. While multiple scattering is massive in the optical regime calling for the development of scattering reduced methods for optical mammography as discussed above, elimination of the less prominent X-ray scattering, particularly for thick tissue imaging, can be advantageous from a dose-reduction point of view as put forward in Ref. [68]. We have performed extensive Monte Carlo simulations on X-ray scattering in tissue and also performed experiments using an X-ray streak camera for one-dimensional imaging through thick tissue phantoms [69,70]. In Fig. 17, the shadow of a small lead object is recorded time-resolved through water layers of increasing thickness. It can be clearly seen, that a sharp shadow is obtained using the ballistic X-rays, even when imaging through 15 cm of water, while the time-integrated images in the lower part of the figure show a strong loss of contrast [70]. The future of this technique will be strongly coupled to the potential development of efficient gateable imaging X-ray detectors.

High-speed X-ray imaging can also be applied for other scientific and technical tasks, i.e. high-speed X-ray crystallography and technical imaging of fast events in optically opaque materials. Clearly, the new source of short-pulse, ultra-intense X-ray radiation also has many potential applications in basic atomic physics. Different kinds of pump/probe experiments on inner shell processes can be envisaged.

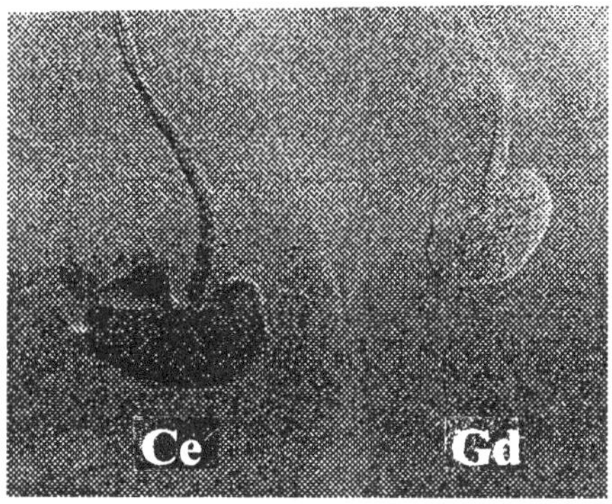

Fig. 16. Division of two images of rat stomachs, one containing a gadolinium and the other one a cerium solution. One of the images was recorded with tantalum as the target material while for the other gadolinium was used (From Ref. [67]).

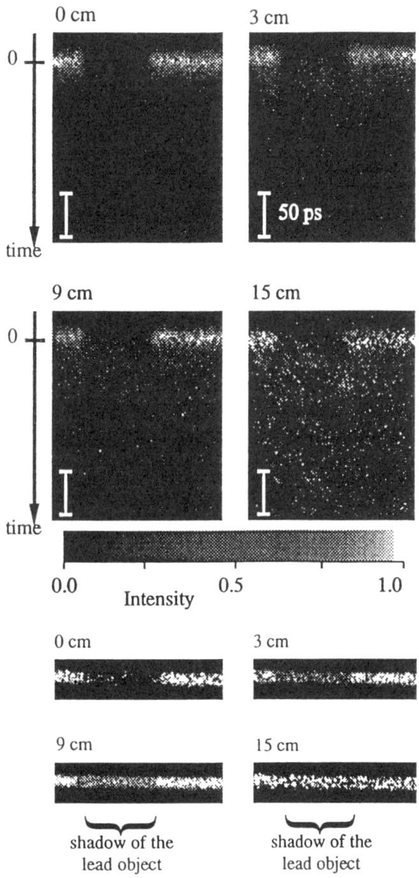

Fig. 17. Gated viewing recordings of a small lead object using short-pulse X-ray bursts passing through water layers of different thicknesses. In the lower part of the figure time-integrated representations are shown clearly demonstrating the loss of contrast for thicker samples (From Ref. [70]).

8. Conclusion

As illustrated in the present paper, the easy access to ultra-high intensity laser light from table-top systems enables a whole host of experiments that otherwise would be very hard to perform. Basic research in atomic, molecular and solid-state physics employing the new sources is rapidly expanding [71,72]. Powers on the tens of TW level are becoming available and even PW systems are in construction. High peak powers are favoured in short-pulse systems (~10 fs) but for retaining a reasonable line-width for spectroscopy experiments a longer pulse length is required. Both in the time and in the frequency domains a fast development of techniques can be anticipated. A strong developmental trend in spectroscopy is the thrust to achieve coherent control in dynamic studies, where the phase of the atomic wave functions as well as the laser field is important.

As also illustrated in the present paper applications of high-power lasers are also emerging, e.g. in the medical field. Practical applications will strongly depend on the development of compact, high-repetition rate systems. With diode-laser pumping of solid-state laser materials now coming of age, very realistic high-power laser systems seem to be in sight.

Acknowledgments

On the occasion of the 65th birthday of Professor Ingvar Lindgren the author would like to express his sincere gratitude to his teacher for a very valuable combination of scientific guidance and fostering for independent research. This paper is dedicated to Ingvar.

The author further acknowledges a very fruitful collaboration with a large number of colleagues and graduate students through the years. Special thanks are due to A. L'Huillier and C.-G. Wahlström. The support from the Knut and Alice Wallenberg Foundation, the Swedish Natural Science Research Council, the Swedish Engineering Sciences Research Council, the Swedish Medical Research Council, the Swedish Board for Technical and Industrial Development and the Swedish Cancer Society is much appreciated.

References

[1] D. Strickland and G. Mourou, Opt. Commun. **56**, 219 (1985).

[2] P. Maine, D. Strickland, P. Bado, M. Pessot, and G. Mourou, IEEE J. Quantum Electron. **24**, 398 (1988).

[3] S. Svanberg, J. Larsson, A. Persson and C.-G. Wahlström, Physica Scripta **49**, 187 (1994).

[4] J.D. Kmetec, J.J. Macklin, and J.F Young, Opt. Lett. **16**, 1001 (1991).

[5] A. Sullivan, H. Hamster, H.C. Kapteyn, S. Gordon, W. White, H. Nathel, J.R. Blair, and R.W. Falcone, Opt. Lett. **16**, 1406 (1991).

[6] C. Le Blanc, G. Grillon, J.P. Chambaret, A. Migus and A. Antonetti, Opt. Lett. **18**, 140 (1993).

[7] P.H. Chiu, A. Magana, K. Yamakawa, and J.D. Kmetec, Proc. SPIE, Los Angeles, paper 1861-05 (1993).

[8] J. Larsson, to appear.

[9] R. Zerne, Time-Resolved Studies of Atoms and Ions in the Short-Wavelength Region, PhD Thesis LRAP-195 (Lund Institute of Technology, Lund 1996).

[10] A. L´Huillier, T. Auguste, Ph. Balcou, B. Carré, P. Monot, P. Salières, C. Altucci, M. Gaarde, J. Larsson, E. Mevel, T. Starczewski, S. Svanberg, C.-G. Wahlström, R. Zerne, K.S. Budil, T. Ditmire, and M.D. Perry, Int. J. Nonlinear Opt. **4** (1995) 647.

[11] C.-G. Wahlström, in X-ray Lasers 1994, D.C. Eder and D.L Matthews (eds) (AIP, Woodbury 1994) p. 312.

[12] J.L. Krause, K.J. Schafer and K.C. Kulander, Phys. Rev. Lett. **68**, 3535 (1992).

[13] C.-G. Wahlström, J. Larsson, A. Persson, T. Starczewski, S. Svanberg, P. Salieres, P. Balcou and A. L'Huillier, Phys. Rev. **A48**, 4709 (1993).

[14] J.J. Macklin, J.M. Kmetec and C.L. Gordon III, Phys. Rev. Lett. **70**, 766 (1993).

[15] M.D. Perry and G. Mourou, Science **264**, 917 (1994); A. L´Huillier and Ph. Balcou, Phys. Rev. Lett. **70**, 774 (1993).

[16] S.G. Preston, A. Sanpera, M. Zepf, W.J. Blyth, C.G. Smith, J.S. Wark, M.H. Key, K. Burnett, M. Nakai, D. Neely, A.A. Offenberger, Phys. Rev. **A53**, 31 (1996).

[17] Y. Akiyama, K. Midorikawa, Y. Matsunawa, Y. Nagata, M. Obarqa, H. Tashiro and K. Toyoda, Phys. Rev. Lett. **69**, 2176 (1992).

[18] C.-G. Wahlström, S. Borgström, J. Larsson and S.-G. Pettersson, Phys. Rev. **A51**, 585 (1995).

[19] C. Altucci, T. Starczewski, C.-G. Wahlström, E. Mevel, B. Carré and A. L´Huillier, J. Opt. Soc. Am. **B13**, 148 (1996).

[20] J.M. Schins, P. Breger, P. Agostini, R.C. Constantinescu, H.G. Muller, A. Bouhal, G. Grillon, A. Antonetti, and A. Mysyrowicz, J. Opt. Soc. Am. **B13**, 197 (1996).

[21] T. Starczewski, J. Larsson, C.-G. Wahlström, M.H.R. Hutchinson, J.E. Muffett, R.A. Smith, and J.W.G. Tisch J. Phys. **B27**, 3291 (1994).

[22] T.E. Glover, R.W. Schoenlein, A.H. Chin, and C.V. Shank, Phys. Rev. Lett. **76**, 2468 (1996).

[23] P. Salières, T. Ditmire, K.S. Budil, M.D. Perry, and A. L'Huillier J. Phys. **B26**, L217 (1994).

[24] P. Salières, A. L'Huillier, and M. Lewenstein, Phys. Rev. Lett. **75**, 3776 (1995).

[25] R. Bonifacio, C. Pellegrini and I.M. Narducci, Opt. Commun. **50**, 373 (1984).

[26] A VUV Free Electron Laser at the TESLA Test Facility at DESY - Conceptual Design Report (DESY Print, Hamburg 1995).

[27] J. Larsson, E. Mevel, R. Zerne, A. L´Huillier, C.-G. Wahlström, and S. Svanberg, J. Phys. **B28**, L53 (1995).

[28] E. Mevel, J. Larsson, M.B. Gaarde, R. Zerne, A. L´Huillier, C.-G. Wahlström and S. Svanberg, Fifth European Conference on Atomic and Molecular Physics (ECAMP), Edinburgh 1995.

[29] R. Danelius, A. Piskarskas, A. Persson and S. Svanberg, Lith. J. Phys. **33**, 328 (1993).

[30] H. Eichmann, S. Meyer, K. Riepl, C. Momma and B. Wellegehausen, Phys. Rev. **A50**, R2834 (1994).

[31] M.B. Gaarde, P. Antoine, A. Persson, B. Carré, A. L´Huillier and C.-G. Wahlström, J. Phys. **B29**, L163 (1996).

[32] P.B. Corkum, N.H. Burnett, and M.Y. Ivanov, Opt. Lett. **19**, 1870 (1994).

[33] M.Y. Ivanov, P.B. Corkum, T. Zuo, and A. Bandrauk, Phys. Rev. Lett. **74**, 2933 (1995).

[34] U. Berzinsh and S. Svanberg, Adv. Quant. Chem., this volume

[35] P. Erman, A. Karawajczyk, E. Rachlew-Källne, E. Mevel, R. Zerne, A. L´Huillier and C.-G. Wahlström, Chem Phys. Lett. **239**, 6 (1995).

[36] D.L. Matthews, P.L. Hagelstein, M.D. Rosen, M.J. Eckart, N.M. Ceglio, A.U. Hazi, H. Medecki, B.J. MacGowan, J.E. Trebes, B.L. Whitten, E.M. Campbell, C.W. Hatcher, A.M. Hawryluk, R.L. Kauffman, L.D. Pleasance, G. Rambach, J.H. Scofield, G. Stone, and T.A. Weaver, Phys. Rev. Lett. **54**, 110 (1985).

[37] B.J. MacGowan, L.B. Da Silva, D.J. Fields, C.J. Keane, J.A. Koch, R.A. London, D.L. Matthews, S. Maxon, S. Mrowka, A.L. Osterheld, J.H. Scofield, G. Smimkaveg, J.E. Trebes, R.S. Walling, Phys. Fluids **B4**, 2326 (1992).

[38] B.E. Lemoff, G.Y. Yin, C.L. Gordon III, C.P.J. Barty and S.E. Harris, Phys. Rev. Lett. **75**, 1574 (1995).

[39] Y. Nagata, K. Midorikawa , M. Obara, H. Tashiro, and K. Toyoda, Phys. Rev. Lett. **71**, 3774 (1993).

[40] S. Borgström, E. Fill, J. Larsson, T. Starczewski, J. Steingruber, S. Svanberg and C.-G. Wahlström, Inst. Phys. Conf. Ser. No 140 (IOP Publishing, Bristol 1995), p. 141.

[41] B.N. Chichkov, A. Egbert, H. Eichmann, C. Momma, S. Nolte and B. Wellegehausen, Phys. Rev. **A52**, 1629 (1995).

[42] E. Fill, S. Borgström, J. Larsson, T. Starczewski, C.-G. Wahlström and S. Svanberg, Phys. Rev. E **51**, 6061 (1995).

[43] S. Borgström, E. Fill, T. Starczewski, J. Steingruber, S. Svanberg and C.-G. Wahlström, Laser and Particle Beams **13**, 459 (1995).

[44] E.E. Fill (ed.), *X-Ray Lasers 1992* (IOP Publishing, Bristol 1992).

[45] D.E. Eder and D.L. Matthews (eds), *X-Ray Lasers 1994* (AIP Publishing, Woodbury 1994)

[46] S. Svanberg and C.-G. Wahlström (eds), *X-Ray Lasers 1996*, IOP Conference Series No. 151, (IOP Publishing, Bristol 1996)

[47] S. Hunsche, T. Starczewski, A. L'Huillier, A. Persson, C.-G. Wahlström, B. van Linden van den Heuvell and S. Svanberg, Phys. Rev. Lett. **77**, 1966 (1996).

[48] L.J. Fransinski, K. Codling and P.A. Hatherly, Science **246**, 973 (1989).

[49] L.J. Fransinski, P.A. Hatherly, K. Codling, M. Larsson, A. Persson and C.-G. Wahlström, J. Phys. B**27**, L109 (1994).

[50] S. Andersson-Engels, R. Berg, A. Persson, and S. Svanberg, Opt. Lett. **18**, 1697 (1993).

[51] G. Müller, B. Chance, R. Alfano, S. Arridge, J. Beuthan, E. Gratton, M. Kaschke, B. Masters, S. Svanberg, and P. van der Zee (Eds), *Medical Optical Tomography: Functional Imaging and Monitoring,* SPIE Institute Series Vol. 11 (SPIE, Bellingham 1993).

[52] S.L. Marcus, p.1 in: "*Future Directions and Applications on Photodynamic Therapy*" Bellingham: SPIE, (Edited by C. Gomer). The International Society for Optical Engineering, 1990.

[53] K. Svanberg, T. Andersson, D. Killander, I. Wang, U. Stenram, S. Andersson-Engels, R. Berg, J. Johansson, and S. Svanberg, British J. of Dermatology **130**, 743 (1994).

[54] B. Chance, J.S. Leigh, H. Miyake, D.S. Smith, S. Nioka, R. Greenfeld, M. Finander, K. Kaufmann, W. Levy, M. Young, P. Cohen, H. Yoshioka, and R. Boretsky, Proc. Natl. Acad. Sci. USA **85**, 4971 (1988).

[55] R. Berg, S. Andersson-Engels, C. af Klinteberg and S. Svanberg, in *OSA Proc. in Optical Imaging and Photon Migration*, Vol. 21 (OSA, Wash. D.C., 1993) p. 126.

[56] S. Andersson-Engels, R. Berg, S. Svanberg, and O. Jarlman, Opt. Letters **15**, 1179 (1990).

[57] S. Andersson-Engels, R. Berg, and S. Svanberg, J. Photochem. Photobiol. **16**, 155 (1992).

[58] R. Berg, S. Andersson-Engels, O. Jarlman and S. Svanberg, Appl. Opt. **35**, 3432 (1996).

[59] J. Johansson, Berg, R., A. Pifferi, S. Svanberg, and L.O. Björn, to appear.

[60] J. Carlsson, P. Hellentin, L. Malmqvist, A. Persson, W. Persson and C.-G. Wahlström, Opt. Lett. **34**, 1528 (1995)

[61] J.D. Kmetec, C.L. Gordon III, J.J. Macklin, B.E. Lemoff, G.S. Brown and S.E. Harris, Phys. Rev. Lett. **68**, 1527 (1992).

[62] C. Tillman, S.Å. Johansson, B. Erlandsson, M. Grätz, B. Hembal, A. Almén, S. Mattson and S. Svanberg, Nucl. Instr. Meth., in press.

[63] G. Hölzer, E. Förster, M. Grätz, C. Tillman and S. Svanberg, J. X-ray Sci. Tech., in press.

[64] K. Herrlin, G. Svahn, C. Olsson, H. Petterson, C. Tillman, A. Persson, C.-G. Wahlström and S. Svanberg, Radiology **189**, 65 (1993).

[65] C. Tillman, A. Persson, C.-G. Wahlström, S. Svanberg and K. Herrlin, Appl. Phys. **61**, 333 (1995).

[66] W.R. Dix, Prog. Biophys. Molec. Biol. **63**, 159 (1995).

[67] C. Tillman, I. Mercer, S. Svanberg and K. Herrlin, J. Opt. Soc. Am. **B13**, 209 (1996).

[68] C.L. Gordon, III, G.Y. Yin, B.E. Lemoff, P.B. Bell and C.P.J. Barty, Opt. Lett. **20**, 1056 (1995).

[69] M. Grätz, A. Pifferi, C. Tillman, C.-G. Wahlström and S. Svanberg, in Ref. [46], p. 539.

[70] M. Grätz, A. Pifferi, C.-G. Wahlström and S. Svanberg, IEEE J. Sel. Top. Quant. Electr. (Special issue on Lasers in Medicine and Biology), in press.

[71] M. Gavrila, Ed., *Atoms In Intense Laser Fields*, (Academic Press, San Diego (1992).

[72] G. Mainfray and C. Manus, Rep. Prog. Phys. **54**, 1333 (1991).

A Periodic Table in Three Dimensions: A Sightseeing Tour in the Nanometer World

Arne Rosén
Department of Physics
Göteborg University and Chalmers University of Technology, S-412 96 Göteborg, Sweden

Abstract

Access to techniques to produce and characterize free clusters built up from two to thousands of atoms has during the last decades generated several exciting discoveries and established cluster science as a research field of its own. This field is highly interdisciplinary, and knowledge from different areas of physics and chemistry has been of considerable importance for this rapid development. One of the objectives is to gain an understanding of the material growth i. e. how many atoms are needed in a cluster to make its physical or chemical properties similar to what is known for the corresponding solid. Studies of various properties for some clusters have, however, revealed large fluctuations and periodicities which can be interpreted either as geometric or electronic shell closings, with the appearance of so-called magic numbers. This kind of periodicity is quite different from what is known from the periodic table in atomic physics but has some similarities with the magic numbers in nuclear physics. In addition to clusters characterized by shell structure, also very unique clusters exist such as the fullerenes, in particular Buckminsterfullerene or C_{60} discovered in 1985, which was a new form of carbon different from the earlier known forms of graphite and diamond. The field of fullerenes has, especially, after the invention of a method for production of macroscopic amounts in 1990, developed in an extremely exciting way with several serendipitous advances including fabrication of crystals, films and new materials with unique properties such as the nanotubes, carbon onions and metcars. Many of these discoveries have opened up new areas of modern mesoscopic physics and materials science. Particularly interesting is how some data available for clusters and fullerenes show periodicities which might be classified in "A Periodic Table in Three Dimensions".

ADVANCES IN QUANTUM CHEMISTRY, VOLUME 30

0065-3276/98 $25.00

I. INTRODUCTION

A general procedure in the characterization of matter has been to find systematics of different properties and to use the collected data to find empirical relations which at a later stage have been theoretically explained. The first layout of the "Periodic Table of Elements" [1] was based on the grouping of elements by mass and on bracketing those elements with similar chemical behavior in the same group. Later measurements of X-rays and use of Moseley's law gave the atomic numbers and the ordering and grouping of the elements known today. Historically the first successful theoretical model of the periodic table was developed by Bohr in 1922 [2] with the "building-up" (or "Aufbau") principle. The understanding was, however, somewhat limited and it was only after the discovery of the exclusion principle by Pauli [3] that concepts such as electronic shells and electronic configurations were fully understood. Further knowledge of different properties for the known elements was obtained from optical and X-ray spectroscopy and from studies of the chemical character in the formation of different compounds. Introduction of quantum mechanics made it possible to even solve the Schrödinger equation for an atomic system in a self-consistent way as first performed by Hartree in 1928 [4].

A good survey of atomic physics and the status of the field for the interpretation of line spectra for atoms in those early days was given in the book by Condon and Shortley [5] first published in 1935. This book, which has been reprinted many times, has been one of the standard textbooks for graduate students and researchers in atomic spectroscopy. The book also gives an overview of the central field approximation and how this method could be used in the solution of the Schrödinger and Dirac equations. Solution of these equations gives properties such as the ground state configuration, ionization energies, X-rays, optical transitions and different expectation values. The possibilities for detailed calculations were however limited in the early days due to the absence of powerful computers. This has, however, changed during the last three decades and the access, especially to work stations, have made it possible to do extensive calculations for atomic systems in general [8–19]. Moreover, it is feasible with these kinds of calculation to extend the knowledge of elements in the periodic table and also make predictions of properties for superheavy elements with an atomic number of up to 172 [17]. The existence of superheavy elements was discussed within nuclear physics in the late sixties and seventies [20] and the recent discoveries by the group at GSI Darmstadt [21,22] of elements with an atomic number up to 112 represent a true success. Separation of superheavy elements requires detailed knowledge of the electronic structure and bonding properties which, in the case of Ha with an atomic number of 105, has been obtained from relativistic molecular calculations of the halides

[23]. The goal to extend the periodic table including the superheavy elements is one example of a general trend of science to explore unknown areas which might give possibilities of testing existing models and formulating new ones.

A similar type of research took place also in nuclear physics during the thirties with a systematic characterization of different properties for a number of atomic nuclei [24]. As an example can be mentioned the studies of the neutron cross sections as a function of the number of neutrons or protons in the nuclei, which showed systematic variations with very small values at certain numbers corresponding to nuclei with 20, 50, 82 and 126 neutrons. This discovered periodicity was rather different compared with the periodicity of atomic properties as the first ionization potential and electron affinity for alkali and noble gas atoms. Speaking at a meeting of the Chemical Society on April 19, 1934, the centenary of the birth of Mendeléev, Rutherford concluded, "*It may be that a* Mendeléev *of the future may address the Fellows of this Society on the 'Natural Order of Atomic Nuclei' and history may repeat itself* " [25]. Measurements of for example nuclear spins for a number of isotopes also showed a similar type of periodicity as found in neutron cross sections. This kind of periodicity could not at that time be understood from the commonly used liquid drop model [26] but based on the single particle model formulated by Mayer, Haxel, Jensen and Suess in 1949 [27].

Further studies of level structures of nuclei also showed a special behaviour of rotational levels, which indicated that nuclei might be deformed. Actually, the existence of deformed nuclei was realized already in 1935 by Schüler and Schmidt [28] who, from optical hyperfine structure investigations of europium, determined the spectroscopic quadrupole moments for ^{151}Eu and ^{153}Eu to be 1.5 b and 3.2 b, respectively. An increase in the neutron number from 88 to 90 neutrons for these isotopes, which both have the same nuclear spin $I = \frac{5}{2}$, was thus associated with an increase of the quadrupole moment by a factor of two. The moments were evaluated from the measured hyperfine interaction constants using expectation values of r^{-3} determined from spin orbit interaction and relativistic correction factors evaluated from hydrogenic wavefunctions as summarized in the classical book by Casimir [29]. The interpretation of the collected experimental data for deformed nuclei was readily understood with the introduction of the unified nuclear model by Bohr and Mottelson [30–32]. The earlier mentioned shell model for spherical nuclei was also later extended by S.G. Nilsson to deformed nuclei by solution of the Schrödinger equation for a spheroidal harmonic oscillator potential [33]. This model showed that the neutrons of ^{151}Eu and ^{153}Eu isotopes occupied different single particle orbitals with the same nuclear spin at different quadrupole deformations.

Systematic and extensive measurements of nuclear spins, hyperfine interaction constants such as the dipole, the quadrupole have for a long time been a very active research area in atomic physics. The main goal of this research, which in the early days was obtained from atomic spectroscopy, was to test the existing nuclear models. Introduction of radio frequency methods with the Atomic Beam Magnetic Resonance, ABMR, methods by Rabi and coworkers [34], as summarized in the book by Ramsey [35], made this type of measurements more accurate. Use of this technique made it possible to determine octupole and hexadecapole interaction constants and to carry out systematic measurements of nuclear spins, dipole and quadrupole interaction constants for radioactive nuclei [36]. The ABMR- technique was introduced in Sweden by Lindgren in the late 1950s in his thesis work [37,38] and has since that time been used at Uppsala, Göteborg and at CERN Geneva using different set-ups [39,40].

The focus of the research in those early days was to test the nucelar models such as the above mentioned shell and Nilsson models. Complementary information about nuclear properties could be obtained from studies of excited atomic states as obtained in atomic spectroscopy but also from use of double resonance techniques [41], level crossing spectroscopy [42] and, during the last twenty years from, laser spectroscopy [43,44]. Exploration of this area of atomic physics was started by Sune Svanberg when Ingvar Lindgren, Karl-Erik Ädelroth and I moved from Uppsala to Göteborg in 1966. The progress for evaluation of accurate moments from measured hyperfine structure constants in atomic ground and excited states has required access to accurate treatment of the atomic part of the hyperfine interaction as discussed by many authors [6,7,29,19]. I had the opportunity to work in this area for a number of years, as summarized in Appendix 1.

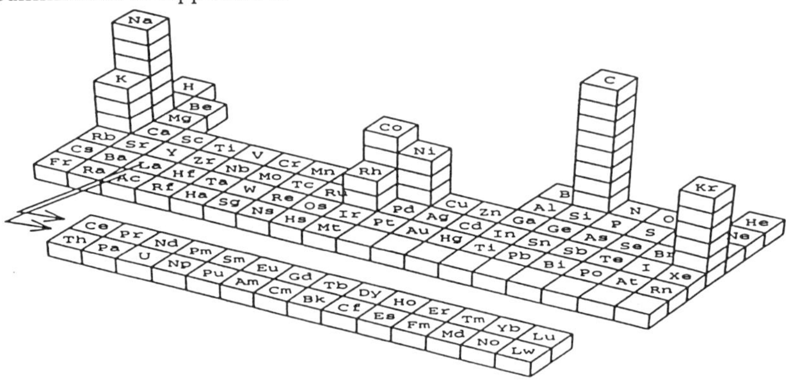

FIG. 1. A schematic overview of how clusters built up from one element can be placed in a periodic table in three dimensions. In the case that the height in the third dimension will contain 6.023×10^{23} atoms we will have species corresponding to one mole of the element, approaching what is known for the solid.

After working within atomic and nuclear physics for many years I have gradually moved in the direction of molecular physics, and in recent years into another area of science known as cluster science. A cluster can be characterized as an assembly of atoms or molecules which are bonded together in larger units. Systematic studies of clusters can be seen as the exploration of the "*The Periodic Table in a Third Dimension*", as shown in Fig. 1. This figure gives an overview of the standard periodic table, where the first plane represents our knowledge of atoms and, the second plane what is known for diatomic molecules [45]. Successive adding of atoms in a third dimension will give still larger clusters with the solids at the far end. Today we have a good knowledge of the species we start with, i. e. the atoms, diatomic molecules and the end products the solids [46], while the knowledge of the species between the clusters is growing. One could perhaps assume that the values of different properties would change smoothly when the number of atoms increases in a cluster from a few up to very large clusters comprising a large number of atoms and ferminons. We will, however, see that this is in general not the case due to the existence of different types of periodicity. The whole field of clusters represents such exciting areas of modern science which might have an impact on modern technology, materials science and mesoscopic physics. This paper gives a short personal overview of this area of clusters with a few snapshots where new types of periodicity appear compared with the periodicity known for atoms in the Periodic Table. Extensive overviews of different aspects of clusters can be found in a number of articles, books and proceedings from conferences etc [47–59].

II. SOME GENERAL PROPERTIES OF CLUSTERS

Clusters are characterized by their size-dependent large surface/volume ratio which may be estimated using different models. If we assume that a cluster containing N atoms is built up of spheres each with a radius r = 1Å and closely packed together to a bigger sphere with a radius R, one obtains for the volume of the cluster $V = \frac{4}{3}\,\pi R^3$ and for the volume of each atom $v = \frac{4}{3}\,\pi \mathrm{r}^3$ the relation, $V = \frac{4}{3}\pi(R)^3 = Nv = N\frac{4}{3}\,\pi(r)^3$

If the number of atoms on the surface of the cluster is N_s with the total surface area of $\mathrm{N}_s\pi r^2$, one obtains the following ratio between the number of atoms on the surface N_s and the total number of atoms N in the cluster $F = \frac{N_s}{N} = 4(\frac{r}{R})$

An overview of different values of these quantities as a function of cluster size is given in Table 1. We notice the large surface to volume ratio for the small clusters which, will influence their properties.

TABLE I. An overview of the variation of the sizes, volume and especially the ratio of surface atoms N_s to the total number of atoms N in a cluster.

N	R(Å)	N_s/N	V(Å^3)
125	5	0.8	5.2x10^2
10^3	10	0.4	4.2x10^3
10^4	21.5	0.2	4.2x10^4
10^6	100	0.04	4.2x10^6

Examples of the dependence of the energies on the s and d bands of Cu_N^- and Au_N^- clusters are given in Fig. 2 [60,61]. These curves are characterized by a general slope, which for large clusters approaches the value known for the solid but with in many cases strong fluctuations for the smaller clusters. A somewhat different behaviour is obtained for the size dependence of the ionization potentials of Hg_n clusters, [62] for which a transition exists between van der Waals bonding for small clusters with an approach to metallic bonding in the solid for clusters larger than $\sim$ 70 atoms.

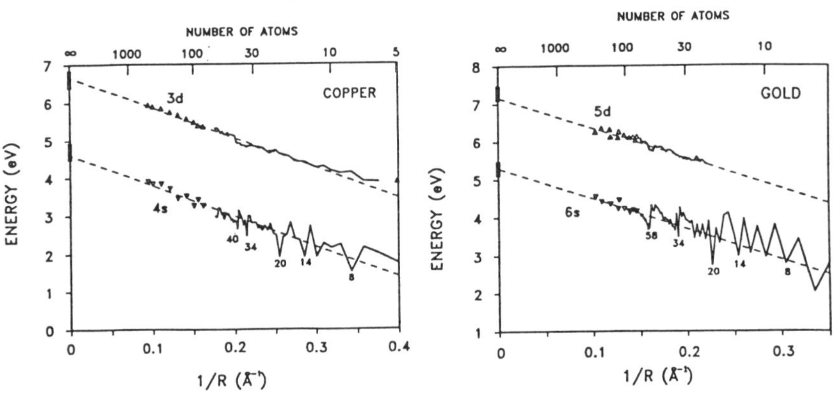

FIG. 2. The dependence of the energies on the s and d bands of Cu_N^- (N =1-411) and Au_n^- (n =1-233) clusters [60,61] as a function of 1/R which is proportional to $n^{-\frac{1}{3}}$. The scaling relation will in the limit of big clusters approach the corresponding values known for the solid of Cu and Au.

It is in connection with studies of this kind of size dependency of different quantities for clusters common to use the following scaling relation

$$X(n) = X(\infty) + An^{-\beta}. \quad (1)$$

where X(∞) represents the value for the solid phase. Eq. (1) is visualized in Fig. 3, where we notice the rather large fluctuations for the smaller clusters in particular.

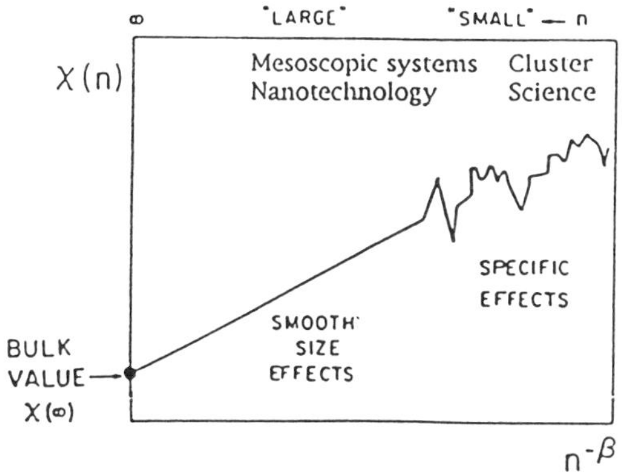

FIG. 3. The cluster size dependency of a cluster property X(n) on the number, n, of the number of constituents in the cluster. The data are plotted vs $n^{-\beta}$ where $0<\beta<1$ [51]. Size specific effects appear for the small clusters while "large" clusters exhibit a smooth size dependency for many properties when $X(n)$ converges what is known for $n \to \infty$ to the bulk value. The transition from small to large will also have connection to the field of mesoscopic physics, where the coherence of electron motion will be of importance.

A more detailed analysis of these fluctuations shows, however, that there often exist some periodicities in this behaviour which in some cases can be understood as shell closings, geometric [63] or electronic [64], with the appearance of so-called "magic numbers". Studies of clusters composed of a few hundred or up to thousands of atoms [65,66] also showed that the periodicity could be extended with the existence of a new type of shells known as supershells representing a transition from pure quantum phenomena towards the limit of large quantum numbers, where a correspondence should exist between classical and quantum motion [71,72]. We will also see how there will be connections to powder technology [50,57,58] and to the mesoscopic world of nanotechnology with a two-dimensional electron gas [73–75]. In addition to these clusters characterized by shell structure also very unique species of carbon have been discovered such as the fullerenes [67], nanotubes [68], carbon onions [69], and metcars [70].

The field of clusters and fullerenes represents areas of modern science where the properties are determined by the reduced coordination. This will modify the functional properties when clusters are used in disperse forms or as units in cluster assembled materials. Examples of applications can be; catalysts, sensor materials, units in nanophase/nanocrystalline materials with improved mechanical, electrical, magnetic or optical properties, of cluster based materials for sun protection, solar energy conversion, as an alternative to quantum dots produced with traditional techniques, fabrication of mesoscopic systems etc. The hope is *to tune the properties with cluster size*, making cluster based materials with characteristics more advanced than those of conventional materials. Production of these types of cluster and exploration of their properties of free as well as deposited clusters are a challenging task of basic and applied science which will be covered in the following sections of this article.

III. EXPERIMENTAL TECHNIQUES

A. Production Methods

As early as in the Middle Ages people discovered that stained glass with beautiful colours could be fabricated by special treatment of the metal-containing glass. Rayleigh discovered that the colours of such stained glass were due to the scattering of light by the small metal particles, which were embedded in the glass. The optical properties of these type of small particles were analyzed in terms of electrodynamic theory by Mie and Debye [76,77] using Maxwells equations. This area of metal clusters and metal particles has recently been summarized by Kreibig and Wollmer with special emphasis on optical properties [78]. The discovered unique optical properties of small particles made it interesting to develop versatile techniques, which could be used for fabrication of small or ultrafine particles from various elements. As early as in 1976 Granqvist and Buhrman [79] developed the inert gas evaporation technique which is frequently used today for commercial fabrication of Ultrafine Particles, UFP.

More exact production methods have been developed during the last decade for clusters in beams but with the drawback that these methods only produce microscopic amounts of clusters. These developments go back to the classical work on molecular and atomic beams by Ramsey [35] and recent overviews have been given by Scoles [80], de Heer [54] and Haberland [55]. The first modern production of metal and carbon clusters was accomplished by Furstenau and Hillenkamp [81] using the laser microprobe mass spectrometer, LAMMA, technique, i. e. clusters were sputtered out in vacuum and detected with a time of flight spectrometer. This technique also has been extended using sputtering with inert gas ion beams [82]. The other sources commonly used today can be classified as gas aggregation sources and laser vaporization techniques with supersonic jets [84,85].

In the gas-aggregation source a metal is vaporized and introduced in a flow of cold inert gas in which the vapor becomes highly supersaturated. Clusters are mainly produced by successive single-atom addition in the build-up of larger species. This type of source has been used to produce continuous cluster beams of alkali elements. By using two separate ovens in the source, each containing separate materials, clusters with two elements can be produced as C_{60} covered with alkali metals [83]. The limitation of this type of source is that only metals with a low melting point can be studied.

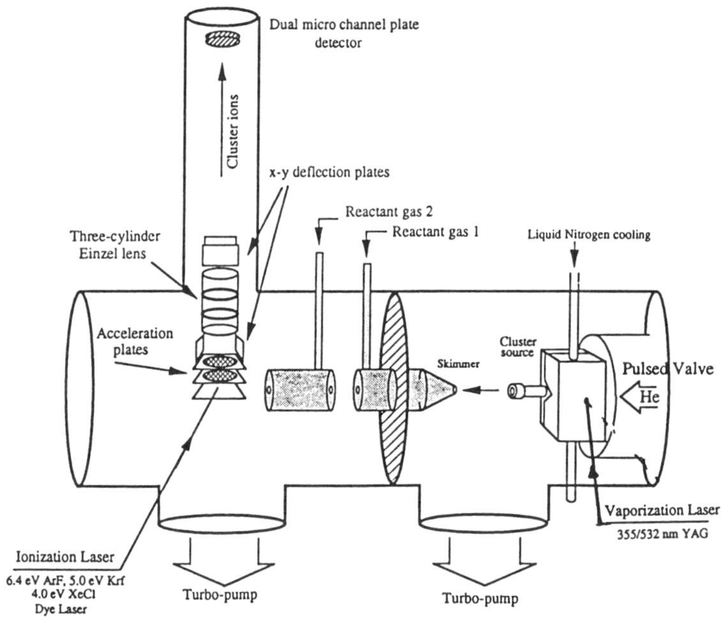

FIG. 4. Overview of the set-up used in Göteborg for production of clusters with the laser vaporization technique using the arrangement shown in Fig. 5. The clusters produced in section one is transferred through the skimmer to section two, where they are ionized and detected with the time-of-flight spectrometer. This section also contains two reaction cells for studies of the chemical reactivity or sticking probability of the clusters.

In the sources based on supersonic jets, metal is vaporized by resistive heating in an oven or by laser vaporization and mixed with or "seeded" in an inert gas, which, with a pressure of several atmospheres, is introduced to the region with atoms and ions. The mixture of metal vapor and gas is ejected into vacuum through a small hole or nozzle producing a supersonic beam. The expansion into the vacuum is adiabatic which causes a cooling of the mixture of buffer gas and supersaturated metal vapor and condenses in the form of clusters in a beam. A schematic overview of such a cluster beam apparatus in Göteborg is given in Fig. 4. This set-up is built up in the form of a two chamber system with a laser vaporization source and a time-of-flight mass spectrometer [88,89]. Both chambers are pumped by turbomolecular pumps.

The cluster source is a normal pulsed laser (10 Hz) vaporization source, as shown in Fig. 5. The buffer gas of He is injected through a 1 mm diameter channel into the source by a pulsed nossle. For vaporization, the light from the third harmonic (λ = 355 nm) of a Nd:YAG laser is focused with a lens onto the sample in the form of a rotating disc. The cluster source is attached to a heat reservoir constructed as a hollow copper body through which water or liquid nitrogen can be circulated to control and stabilize the temperature of the cluster source. After expanding from the source, the cluster beam travels 12 cm in the first chamber to the skimmer in the form of a 1 mm opening. The clusters are detected with photoionization and mass spectrometry in the second chamber using the light from an excimer laser, ArF (λ = 193 nm) , or a dye laser for ionization and a time-of-flight spectrometer for detection of the produced ions.

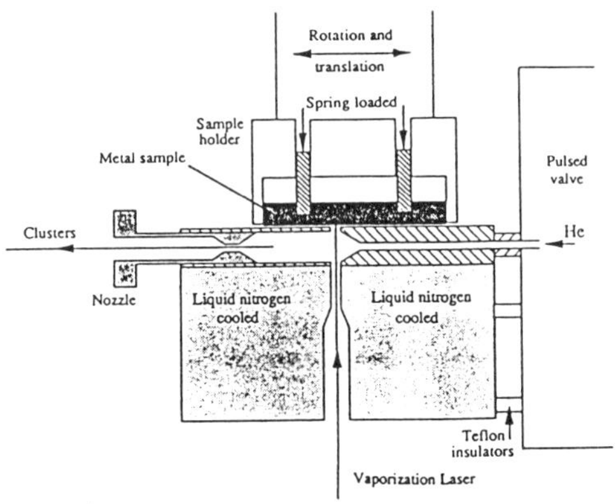

FIG. 5. Schematic overview of the set-up with the laser vaporization source and the nozzle for supersonic expansion used in the production of clusters at Göteborg. The figure shows the laser vaporization source with the target material, the laser beam for evaporation, the small volume where a plasma of atoms and ions exist and the region where the clusters are formed in the expansion through the nozzle.

B. Studies of Properties

As discussed above clusters are characterized by their size-dependent, large surface/volume ratio, and this ratio will also for rather big clusters, in a first approximation, determine the properties, although each cluster is unique with a certain geometrical and electronic structure. Today a number of properties have been studied for clusters of different elements using techniques developed in atomic and molceular physics a long time ago. The following examples of properties can be mentioned; **Abundance spectra, Static dipole polarizabilities, Photoionization and ionization potentials, Photoelectron and electron affinities, Optical properties, Stability and fragmentation, Magnetic moments and Reactivity.**

In the very early experiments on clusters as reviewed by Kaldor, Cox and Zakin [90] and Riley [91] a reaction tube was introduced after the production chamber with either pure helium or a reactant/helium mixture. The reaction products as well as the pure clusters were detected with the TOF spectrometer as described above. This type of arrangement with clusters and reaction products flowing together in the buffer gas will give a system in thermal equilibrium which means that heat released in the chemisorption process will be cooled away by the buffer gas.

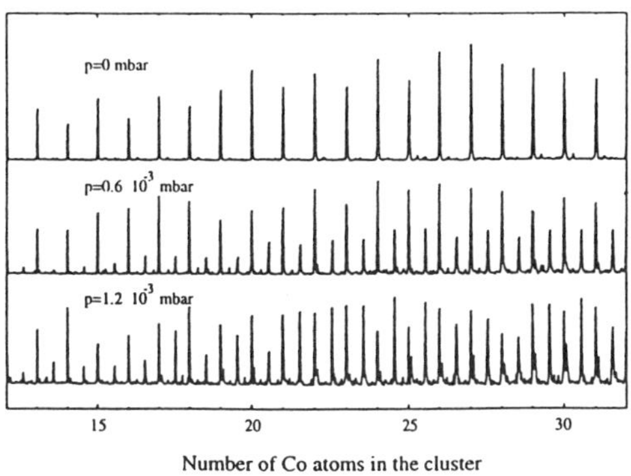

FIG. 6. Example of the intensity distribution in a mass spectrum for clusters of Co produced with the laser vaporization technique and ionized with the light from an ArF laser with a photon energy of 6.4 eV. The top spectrum has been recorded without oxygen in the reaction cell. The lower spectra have been recorded for increasing oxygen pressure in the reaction cell.

In our studies of reactivity of clusters performed at Göteborg quite a different arrangement is chosen with two reaction cells placed in a separate chamber.

A constant pressure of reactive gas is maintained by a continuous flow of gas through the cells which have aperture diameters of 1 mm and lengths of 106 and 50 mm, respectively. During a reaction experiment, the cell pressure is varied in the range 10^{-4} – 10^{-2} mbar, corresponding to an average number of collisions ranging from less than one to a few. Use of a higher pressure would decrease the total number of clusters coming through the cell due to scattering. To quantify the scattering, reference measurements are always made with an unreactive gas in parallel with the reactivity measurements. An example of a mass spectrum recorded for clusters of Co in the range from 10 – 40 atoms is shown in Fig. 6. Increasing the pressure of oxygen changes the spectra; new peaks appear between those corresponding to the pure clusters.

IV. EXAMPLES OF SOME CLUSTERS

It is not feasible in this article to cover all that is known today experimentally as well as theoretically, about the properties of clusters. I will rather in this section give some selected experimental data for clusters of different characters to illustrate different types of bonding. Extensive overviews can be found in [47,55].

A. Clusters of noble gas atoms give geometrical shell structure

The noble gas atoms are characterized by closed electronic shells and, in a first approximation, the atoms are spherical and have only a small tendency to form stable molecules with other elements. Noble gas atoms could therefore be considered as hard spheres, which under certain thermodynamic conditions can form larger units. In 1981 Echt, Sattler and Recknagel [63] performed experiments with a supersonic expansion of xenon gas at 175 K and a pressure of 300 mbar through a capillary into vacuum and recorded the mass spectrum of Xe clusters shown in Fig. 7.

The abundance vary in rather peculiar ways with a slow decrease with increasing size but is not a monotonic function of size. The most prominent intensity drops occur after clusters with sizes of 13, 19, 25, 55, 71, 87, and 147 atoms. Echt *et al.* [92] were also able in a subsequent experiment to show that similar changes in intensity occurred for clusters with 309, 561 and 923 atoms. They found that some of these sizes corresponded to the elements in a series given as

$$N = 1 + \sum_{p=1}^{n}(10p^2 + 2). \tag{2}$$

where for n = 1, 2, 3, 4, 5, 6 one will have sizes corresponding to 13, 55 ,147, 309, 561. This series describes the packing of spheres in a family of closed icosahedral arrangements, named after Mackay [93]. Farges *et al.* [94] were also able to use electron diffraction experiments on clusters in a beam to show that the clusters coming from the nozzle agreed with what could be expected for icosahedral packing, Fig. 8.

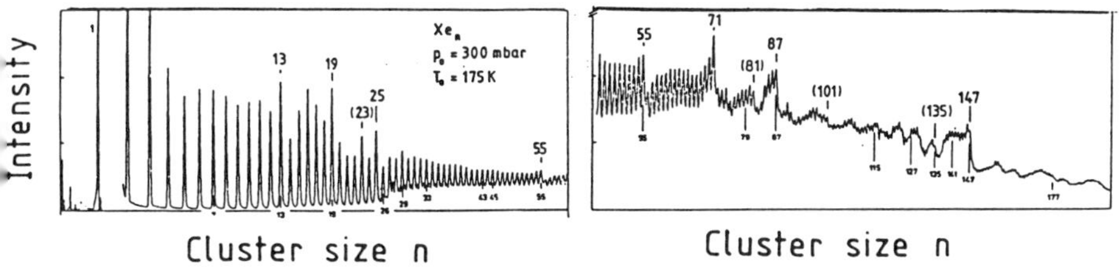

FIG. 7. Mass spectrum of Xe clusters formed in the supersonic expansion of Xe-gas through a nozzle at a temperature of 175 K and a pressure of 300 mbar. The neutral clusters produced in this experiment are ionized by electron impact and analyzed in a time-of-flight spectrometer [63].

FIG. 8. Examples of the smallest Mackay icosahedra, from sizes of 13, 55, 147, 309 and 561. The larger clusters show the formation of crystal surfaces [93].

The noble gases krypton, argon and neon have essentially a similar behaviour as observed for xenon. The growth of the lightest noble gas atom helium with the isotopes ^{3}He and ^{4}He is different and behaves more like a fermion gas, as has been extensively discussed by Bjørnholm [97].

B. Clusters of alkali atoms give shell closings

A new area of cluster science started in 1984 when Knight and his group at Berkeley [64] produced and detected clusters of alkali metals with up to 100 atoms. An example of their measured mass spectrum is shown in Fig. 9

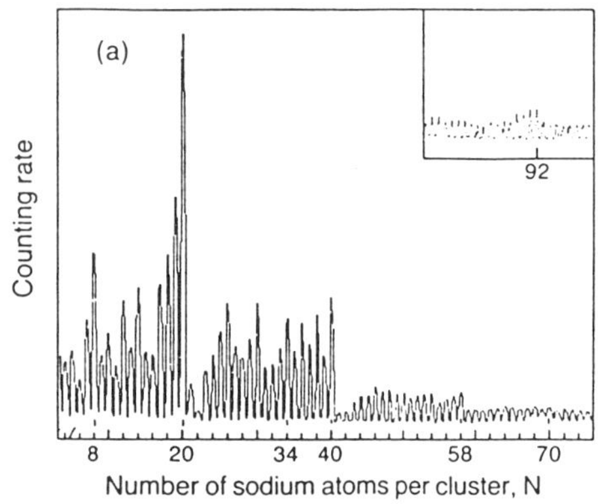

FIG. 9. Sodium cluster abundance spectrum detected by Knight [64]

These clusters are hot, and loosely bound atoms therefore evaporate from the clusters resulting in these especially stable clusters having a high intensity in the beam. Clusters composed of 8, 20, 40 and 58 atoms and with the same number of valence electrons are especially stable. These numbers are similar to the magic numbers known from nuclear physics and it is therefore attractive in a first approximation to use this approach as will be discussed in the next section.

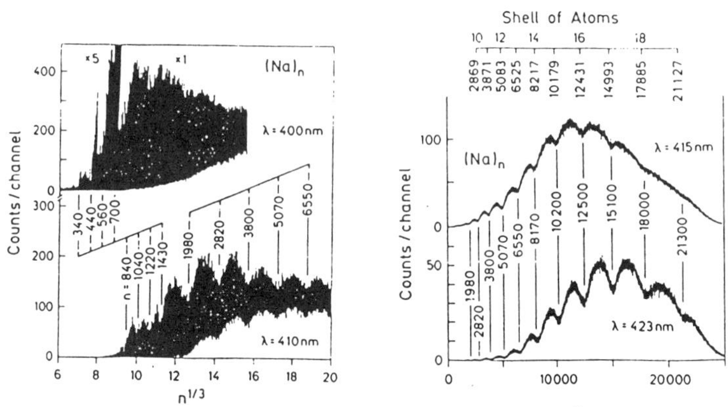

FIG. 10. Clusters of Na showing shells of electrons for the smaller ones and shells of atoms for the bigger ones detected by the group of Martin [98]. The clusters are photoionized using light with different wavelength to show the periodicity.

These types of study were extended to clusters containing up to 500 Na atoms by the group of Bjørnholm at Copenhagen [65] and in a joint co-operation project between the groups of Bjørnholm, Martin at Stuttgart and

Brechignac at Orsay, when they detected even larger clusters [66]. The group of Martin produced still larger clusters containing up to 22000 Na atoms [98]. The mass spectrum in Fig. 10 shows the existence of a pattern described by shells of electrons for clusters less than 2000 atoms in size while the pattern changes into a pattern described by shells of atoms above 2000 atoms [98]. Just like normal bulk materials can exist in the solid or liquid phase one would expect that the same can happen with clusters. Martin and coworkers found that when the clusters were heated with light from an Ar ion laser, the geometrical packing with shells of atoms changed into a geometry governed by shells of electrons [98].

C. Clusters of noble metals also give periodicities

The mono-valent metals have also been found to show a certain type of periodicity dominated by an odd-even alternation. One example of the odd-even alternation is abundance spectra, where the even clusters are more abundant than the neighbouring odd ones. Ionization potentials have also been measured for clusters of Cu [99] to be higher for the even than for the odd sizes, Fig. 11. In addition to the odd-even alternation in the ionization energies, as shown in the figure there is an additional periodicity with high values for sizes 20, 34 and 40, which are the very numbers found for clusters of alkali elements. Values obtained are strongly dependent on the temperature of the cluster and a prerequisite for resolving this type of odd-even alternation is that the clusters are sufficiently cold.

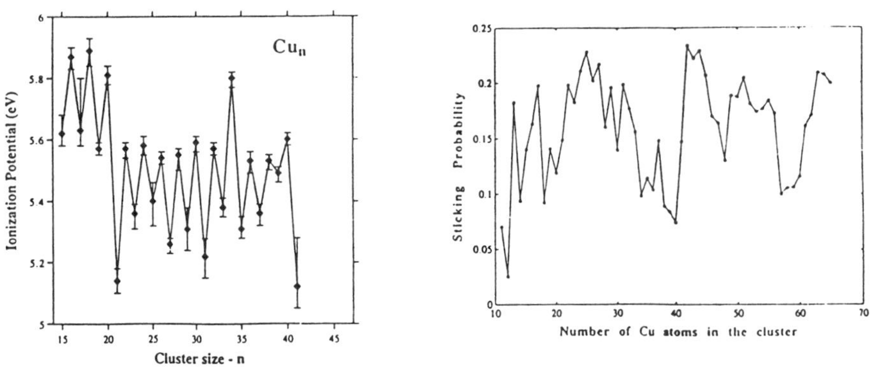

FIG. 11. Measured ionization potentials of Cu clusters to the left [99] and the reactivity towards oxygen to the right [89].

Studies of the reactivity versus cluster size N [89], Fig. 11, also show a number of repeated maxima and minima with a sticking probability in the range from 0.05 to 0.25. The periodicity has a similar behavior as obtained for the ionization potentials. Particularly small numbers of the sticking probability exist for clusters close to sizes of 20, 40 and 58, while the somewhat bigger clusters are more reactive. The reactivity is also correlated with the measured ionization energies [99]. This shows that the reactivity towards oxygen is mainly controlled by Cu 4s electrons and that the oxidation is a charge transfer reaction.

D. Clusters of carbon

At the same time as Smalley and students at Rice University, Houston Texas, developed the laser vaporization method for production of clusters [84], a similar set-up was built at Exxon's Research Laboratory, New Jersey, USA, by the group of Kaldor and Cox [102,103]. They studied in particular transition metal clusters but also produced clusters of carbon containing up to more than hundreds of atoms as shown in the mass spectrum in Fig. 12.

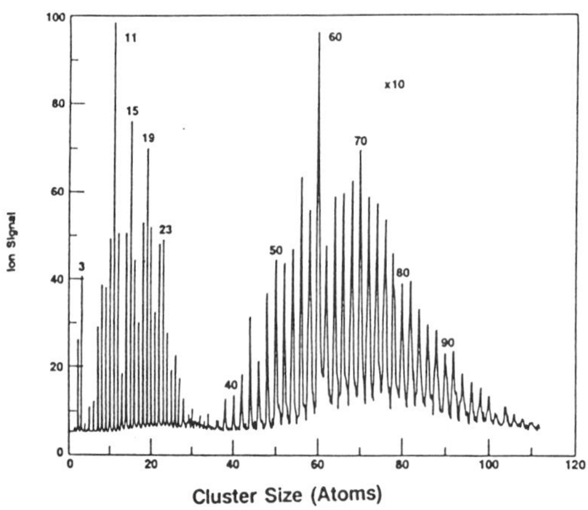

FIG. 12. Time-of-flight mass spectrum of carbon clusters by Rohlfing, Cox and Kaldor produced in a supersonic beam by laser vaporization of graphite [103].

The mass spectrum in Fig. 12 consists of two distributions with two maxima at sizes in the range from 10-20 atoms separated with a minimum at 30 with a new maximum at 50-80 atoms. The first distribution consists of clusters containing even as well as odd number of atoms, while the second distribution contains only clusters with an even number of atoms. Rohlfing, Cox and

Kaldor [103] had the idea that the clusters in the first group had the form of long carbon chains, as predicted by theoretical calculations. It is in retrospect interesting to notice that the peaks for C_{60} and C_{70} appear to be somewhat more abundant compared with the close lying ones. This was however not recognized at the time.

At Sussex, England, Kroto and coworkers had since the early 1980s [104] synthesized carbon chains, cyanopolyynes, of the form HC_nN. These long carbon chain molecules were characterized by their microwave spectra and also detected by radioastronomy in space, which was a real breakthrough. Kroto wanted to produce even longer chains and realized after contact with Curl at Rice University that the new laser vaporization set-up of Smalley would be the most suitable one. Laser vaporization experiments on graphite at Rice University resulted in the discovery of these long carbon chains, but still more exciting was a quite different mass spectrum compared with the one recorded by Rohlfing, Cox and Kaldor a few years before, [103], Fig. 12. The mass spectrum which was recorded at somewhat different pressures in the source gave a picture which was totally dominated by peaks corresponding to C_{60} and C_{70}. The new carbon molecule C_{60} was proposed to have a closed cage structure like a soccerball or truncated icosahedral with 60 vertices, where the carbon atoms were located, 12 pentagonal faces, 20 hexagonal faces and 90 bonds [67]. The structure can be obtained by cutting each corner on the regular icosaheder. Such a spherical structure which was estimated to have a diameter of 7Å was expected to have many unique properties. Fig. 13 gives a schematic overview of C_{60}, obtained from a stick model and from electronic structure calculations.

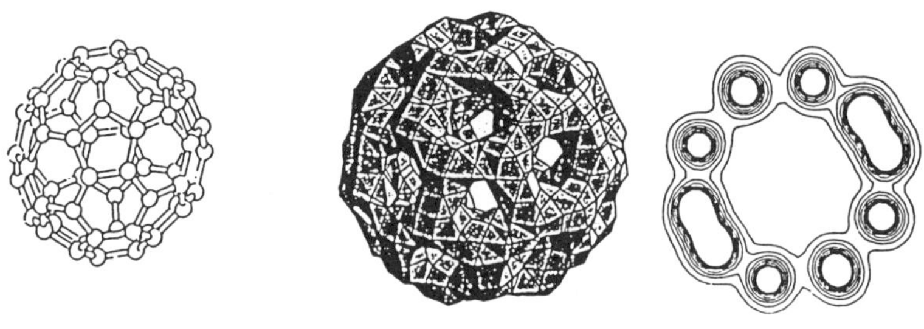

FIG. 13. A schematic overview of C_{60} represented by a stick model, 3D and 2D contour plots of the electron density. In the 3D plot in the middle the single contour has been chosen to show how the electrons are distributed in the bonds. The 2D contour plot shows the electron density in a plane that includes the center of the molecule. We clearly see that there is a void, which means that C_{60} constitutes a spherical shell.

C_{60} should for example have the electronic and spectroscopic features which were known for a π -electron system originating from the 2p electrons of carbon on the cage and at the same time have very special properties due to its spherical symmetry. This type of spherical structure had already been patented by Buckminster Fuller [105] in the construction of geodesic domes at the world exhibition at Montreal in 1967. The new C_{60} molecule with this caged structure was given the name of *Buckminsterfullerene* in recognition of Buckminster Fuller. The clusters around C_{60} were also believed to have a cage structure and were named fullerenes.

The quantities produced with the laser vaporization method were however not even sufficient for doing experiments to verify the proposed structure. This required the development of a new production method compared with the laser vaporization technique which only produced small amounts. Independently, Krätschmer from Heidelberg Germany and Huffman from Arizona USA had as early as in 1982, [106–108] i.e. three years before the discovery of C_{60} in 1985, produced C_{60} without knowing it. They used an electric arc in an helium atmosphere of 150 torr and produced a special kind of soot with a unique type of optical absorption known as "*the camel hump smoke*" in the UV region. The objective of their research was to explain a very strong absorption in the UV region of interstellar dust [109] at about 220 nm or 5.6 eV. They were, however, not able to understand the origin of their recorded spectrum. Only in 1987 did some indication of a solution emerge with the publication by Larsson, Volosov and A. Rosén (the author) [110] of several predictions of a spectrum of C_{60} which fitted rather nicely to the observed triple-humped structure or "*the camel hump smoke*" in the UV-visible spectrum [111]. After a number of trials [112], they found in 1990 that the "the special carbon soot" could be dissolved in benzene, which provided the possibility to separate C_{60} from the carbon particles [113]. The contribution from the background of carbon particles in their earlier recording was eliminated with the appearance of very strong characteristic absorption peaks at 216, 264 and 339 nm to be compared with the spectrum from "the special carbon soot" with the "*the camel hump smoke*". Krätschmer, Lamb, Fostiropolous and Huffman [113] fabricated crystals of C_{60} and C_{70} and determined the crystal structure. Suddenly a new kind of carbon material had been found in addition to the commonly known diamond and graphite.

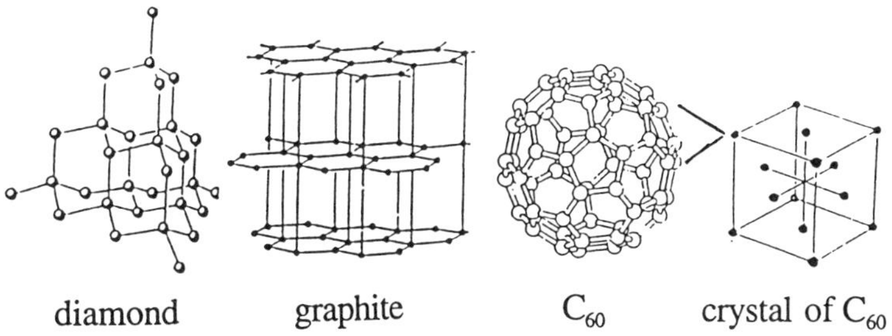

FIG. 14. Overview of different forms of carbon, diamond, graphite, C_{60} and crystals of C_{60}

The unique structure of C_{60} and access to an effective method for the production of macroscopic quantities opened up a new era of carbon research compared with what was known up to that date [114,115]. Access to more material made it possible to use NMR [116,117] to verify the structure proposed by Kroto *et al.* in 1985. It was also possible to use a number of spectroscopic methods to determine binding energies for the free C_{60} molecule, films and crystals. Very exciting was the discovery that films of C_{60} became conducting when doped with alkali atoms [118] as well as superconducting [119] at 33 K, which was very high for an organic superconductor. It also became possible to study the earlier discovered endohedral C_{60} species [120] more accurately using different techniques as excellently reviewed by Bethune [121]. The field of fullerenes is a truly interdisciplinary field which has been covered in a number of proceedings, review articles and books [122–133]. I will in this article be able to present only a few examples to give some idea of the highly interdisciplinary nature.

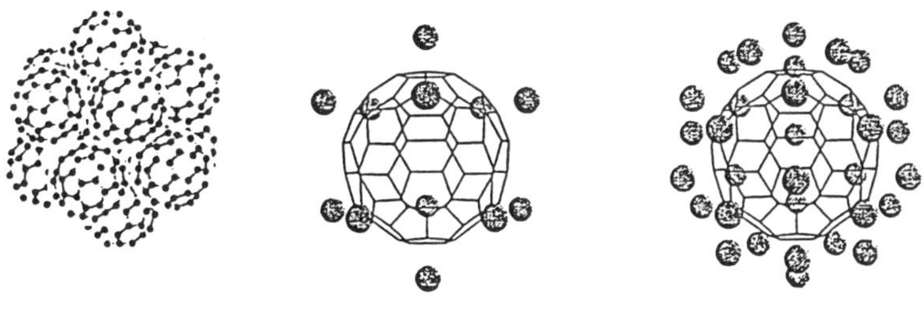

FIG. 15. Clusters of C_{60} [59,136] and two specific models of metal covered C_{60} with 12 and 32 metal atoms outside the pentagons and hexagons.

A commonly discussed topic has been the geometrical structure of carbon species including the doped ones. Do the species have the form such as; chains, rings, multirings or three dimensional cage structures like the C_{60} molecule. Some answers to these problems have been provided by the groups of Jarrold [134] and Bowers [135] by injection of the ionized carbon species in a drifttube with He and studying their mobility. T. P. Martin at Stuttgart [136] has also been able to produce clusters of C_{60} using a gas evaporation source. By coevaporating C_{60} and different metals as Li, Ca, Sr and Ba, Martin and coworkers [137,138] were also able to produce species covered with metal as $Li_{12}C_{60}$, $Ca_{32}C_{60}$, Fig. 15.

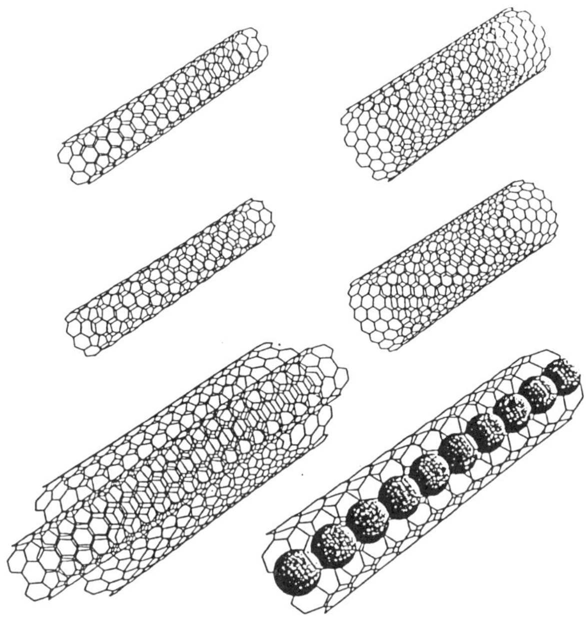

FIG. 16. Examples of "armchair" single nanotubes at the top, "zigzag" single nanotubes in the middle [139]. The popular terms "armchair" and "zigzag" refer to the way the carbon hexagons have been rolled into the cylindrical walls and the difference can be recognized from the patterns at the open ends. The carbon hexagons can also form spiral patterns along the tube leading to chiral nanotubes, which can be a metal or insulator depending on the chirality [131]. To the left at the bottom is shown a multi-walled nanotube and to the right a single nanotube filled with lead atoms [139]. Many of these tubes have been produced and characterized today as summarized by Dresselhauss and Eklund [131] and by Ebbesen [140]

Another exciting discovery was the nanotubes first found in 1991 by Iijima at NEC in Japan [68]. These tubes can be formed if a graphite sheet is wrapped around to form single wall, multiwall structures. The tubes can be open but also closed and even filled with different elements. Still another discovery was the carbon onions or "Russian Dolls" discoverd by Ugarte in 1992 [69]. Many of these species have been discussed in detail by Dresselhaus and Eklund [131] and in a recent thesis by one of my students, Daniel Östling [139]. A schematic overview of different nanotubes taken from the thesis of Östling is given in Fig. 16. Other carbon structures with enhanced stability are the metallo-carbohedrenes, Met-Cars, found by the group of Castleman [70]. This structure can, for example, be formed by twelve carbon and eight metal atoms, $C_{12}M_8$ where, the metal atom is a transition metal at the beginning of the 3d, 4d and 5d series.

V. THEORETICAL APPROACH

Access to clusters of principally all elements in the periodic table as outlined above has to some extent fulfilled the wish expressed by J Friedel in 1976 [47] "*Access to free clusters is a dream for a theoretician*", although there are still a number of unknown values of different properties for these clusters. Furthermore, although one today has access to fast computers and a number of *ab intio* as well as semi-empirical methods for solution of the Schrödinger and Dirac equations [141–144], calculations for clusters of more than ten to twenty atoms are still unfeasible in many cases. The general goal for many of the calculations has been to determine the geometry for the ground state for series of clusters as exemplified for niobium in Fig. 17 [145–147].

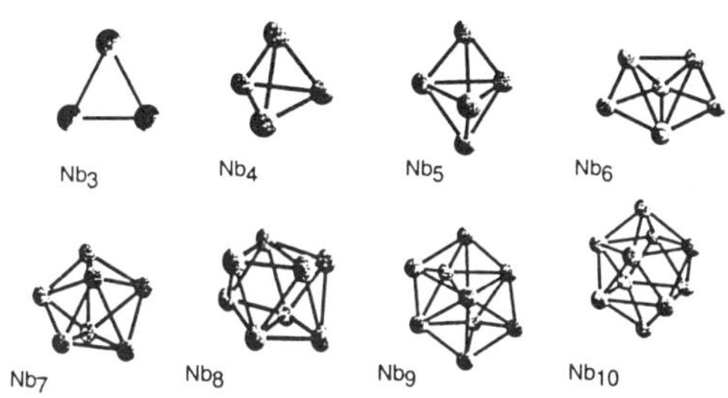

FIG. 17. Low-energy structures for niobium clusters, from results of LCAO calculations [146,147].

These structures have no similarities with the bcc structure known for the solid phase. The general hope is that extension of these types of calculation to bigger clusters would converge to geometries and properties known for the solid phase [46]. Covering the whole region from the atom to the solid phase with a description of for example a phase transformation, as observed for the earlier mentioned clusters of Hg is, however, not straightforward.

The geometrical structure for clusters in general will depend on the balance between the different terms in the hamiltonian, i.e. electron-nuclei, electron-electron and nuclei-nuclei interaction, including the valence as well as the core electrons in the constituent atoms. The full hamiltonian $\hat{H}$ for the cluster is normally separated into a hamiltonian $\hat{H}_N$ for the nuclei and another one $\hat{H}_{el}$ for the electrons with fixed positions for the nuclei according to the Born Oppenheimer approximation [148].

A very common approach for evaluation of the electronic structure for small and large clusters has been to use the density functional theory, DFT, with the electronic density $\rho(\mathbf{r})$ as the key parameter. The modern form of DFT is based on the two theorems by Hohenberg, Kohn and Sham from 1964-1965 [149,150]. Extensive treatment of DFT can be found in the literature [151]. Using the DFT approach the many electron Schrödinger equations are reduced to a set of one-electron Kohn Sham, KS, equations,

$$\hat{h}_{el}(\mathbf{r})\psi_i(\mathbf{r}) = \epsilon_i\psi_i(\mathbf{r}) \tag{3}$$

where $\psi_i(\mathbf{r})$ represents the wavefunction and $\hat{h}_{el}(\mathbf{r})$ is given as:

$$\hat{h}_{el}(\mathbf{r}) = -\frac{1}{2}\nabla^2 + \sum_{\nu=1}^{N} \frac{-Z_\nu}{|\mathbf{r} - \mathbf{R}_\nu|} + \int \frac{\rho(\mathbf{r}')\, d\mathbf{r}'}{|\mathbf{r} - \mathbf{r}'|} + V_{xc}(\rho(\mathbf{r})) \,. \tag{4}$$

where $\rho(\mathbf{r})$ is the total electronic charge density. The first term represents the kinetic energy part, the second term the nuclei-electron interaction and the third term the electron-electron interaction. The last term takes into account the exchange and correlation effects for the electrons.

One way to solve the Kohn-Sham equations Eq. (3) is to expand the molecular orbital wavefunctions in a set of symmetry adapted functions $\chi_j(\mathbf{r})$ which is expanded as a linear combination of atomic orbitals $u_{nl}Y_{lm}$ with coefficients $W^{j\nu}_{nlm}$ i.e.

$$\psi_i(\mathbf{r}) = \sum_j \chi_j(\mathbf{r})\, C_{ji} \,. \tag{5}$$

$$\chi_j(\mathbf{r}) = \sum_{\nu,n,l,m} W^{j\nu}_{nlm}\, u^\nu_{nl}(r_\nu)\, Y_{lm}(\hat{r}_\nu) \,. \tag{6}$$

The variational coefficients, C_{ij}, are obtained from the matrix secular equation

$$(\mathbf{H} - \epsilon\mathbf{S})\,\mathbf{C} = \mathbf{0}, \tag{7}$$

where **H** and **S** are the Hamiltonian and overlap matrices, respectively. For the construction of the symmetry orbitals we use numerical atomic basis functions which are obtained by solving the corresponding KS equations for the free atoms and ions.

A combined method for solving both the geometric and electronic problem simultaneously is the Car-Parrinello method, which is a DFT dynamics method [152]. This method uses a plane wave expansion for the density, and the inner ions are replaced by pseudo-potentials.

A. Model potentials

i) The Jellium Model

Within the jellium model for metal clusters [95,53] as described in the introduction, the positive background potential is in a first approximation normally chosen as a spherical shape of the following form

$$\rho^+(r) = \rho_0\,\Theta(r - R) \tag{8}$$

where $\rho_0 = 3/4\pi r_s^3$ and the radius is $R = r_s N^{1/3}$. N is the number of valence electrons in the neutral cluster and r_s is the Wigner-Seitz radius. Using this approximation the geometrical structure of the cluster is neglected and one obtains for the nuclear ion potential in Eq.(4);

$$V_{ion} = \begin{cases} -\frac{N}{2R}[3 - (\frac{r}{R})^2] & r \le R \\ -N/r & r \ge R. \end{cases}$$

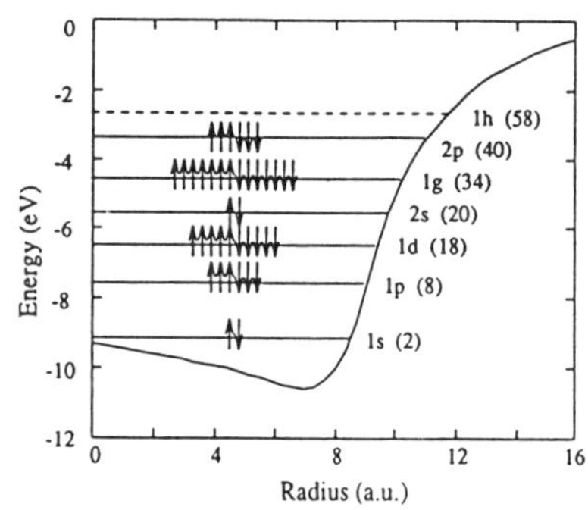

FIG. 18. Self-consistent electron potential, calculated for a spherical cluster of Cu with forty atoms. The calculated single-electron energy levels are also shown. The one-electron energy levels are denoted by the radial quantum number and angular momentum. The energy levels are occupied by electrons according to the Pauli principle.

Ekardt [95] and Beck [96] solved the Schrödinger equation self-consistently for this type of spherical potential for sodium clusters of various size. An example of calculations for a Cu cluster of 40 atoms [147] is shown in Fig. 18. We see how the closed shells in this description of the cluster corresponds to the magic numbers found by Knight *et al.* [64] in Fig. 9.

ii) Three dimensional spherical potentials

It is interesting that the ordering of the energy levels obtained within the jellium model is almost the same as obtained with the non-selfconsistent harmonic oscillator, radial square well potential and Wood Saxon potential used by Goeppert Mayer, Haxel, Jensen and Suess in the formulation of the shell model for nuclei [27]. They had however to add the contribution of spin-orbit coupling by perturbation theory to describe nuclear properties. In the first analysis of the observed periodicity in the mass spectra of Na clusters Knight *et al.* [64] used these non-selfconsistent potentials and found an almost complete agreement for the majority of the observed peaks in the mass spectrum except for a few ones such as the size of 34. Clemenger and coworkers [154,52] found, however, that the agreement for clusters could be improved using an ellipsoidal form as employed within the unified model in nuclear physics by Bohr and Mottelson [32] and S. G. Nilsson [33] in the solution of the Schrödinger equation for nuclei with quadrupole deformation. It was also possible to obtain information about the deformation by measurements of the splitting of the Mie plasma resonance. This resonance splits into two peaks when the cluster is deformed into a prolate or oblate shape. Results from such studies [155,156] are given in Fig. 19.

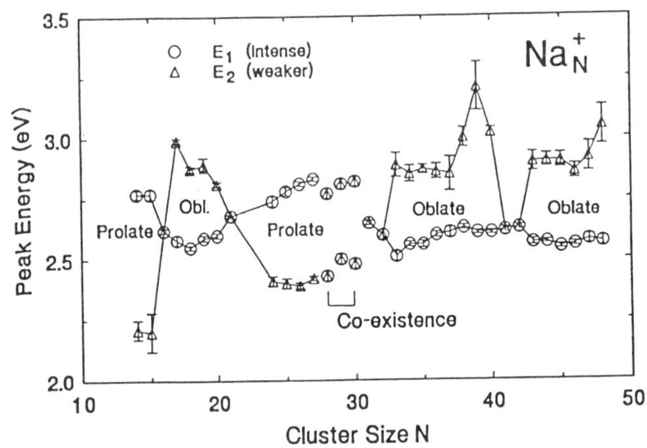

FIG. 19. Overview of deformations of sodium clusters determined from studies of the optical absorption of the plasmon resonance [155,156].

Nishioka, Hansen and Mottelson [71] extended also the calculations to very large clusters using a Woods-Saxon potential with the goal to analyze what would happen in the classical limit. The total binding energy E(n) in this approximation is given by the sum of eigenvalues [153]; $E(n) = \sum_j e_i$. This should be compared with the liquid drop expression for the binding energy which is given by, $\hat{E}(n) = -an + bn^{\frac{2}{3}}$. The first term represents the binding energy per particle inside the drop and the second term represents a correction term for the surface energy, accounting for the reduced binding of the particles on the surface of the drop. The difference between these two expressions for the total binding energy will give the contribution due to the shell structure i.e., $E_{shell}(n) = \hat{E}(n) - E(n)$. A plot of this difference as a function of the shell radius $n^{\frac{1}{3}}$ taken from the work of Nishioka, Hansen and Mottelson [71] is given in Fig. 20.

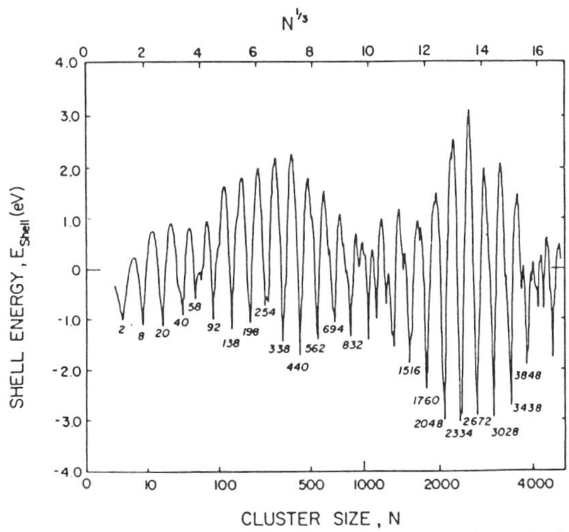

FIG. 20. The periodically varying contribution to the calculated binding energy of spherical sodium cluster. The *periodic variations* are due to the quantized motion of the electrons in the background potential [71,153].

The shells are regularly spaced with a pattern corresponding to the shell closings or magic numbers. There is in addition a superimposed beat pattern of a similar shape as that one obtained when two waves with a somewhat different wavelength interfere. The existence of this beat pattern has been interpreted as the result of quantization of triangular and square electron orbits inside the jellium droplet [72,153]. This kind of periodicity is very similar to what has recently been found from studies of a two dimensional electron gas 2DEG [157].

VI. OUTLOOK AND CONCLUSIONS

Extension of the periodic table in the third dimension has opened up a new area of science from the microscopic to the macroscopic world. The species in this world have been found to have very unique properties with the appearance of clusters which can be described as built up of shells of atoms or as shells of electrons. In addition, very unique species have been discovered such as the fullerenes, endohedral fullerenes, nanotubes, carbon onions, metcars, metal covered species etc. Access to all these new types of species has opened up a totally new direction of modern science where basic knowledge from atomic and molecular physics has been crucial for the present state-of-the-science in this area. Particularly exciting from the basic science point of view has been the possibilities to test models for species of a large number of fermions in the form of many thousands of atoms. With these kinds of species it has been possible to establish a connection with mesoscopic physics but also to form a link between the microscopic world and the world described using classical mechanics. The wish expressed by J. Friedel, *"Access to free clusters is a dream for a theoretician"* has in some respect been fulfilled in areas he never envisaged while there is still a great deal of information missing for example the geometrical structure for most of the clusters with the exception of the diatomics.

Clusters containing a large number of atoms also have connection with materials which are used today in commercial applications such as in powder technology and in catalysts. The Consecutive Executive Officer, CEO, of a company may therefore be pleased with the present status of the field of cluster science. He knows, for example, how important the material properties of small particles are for production of hard materials. He might therefore have a hope to use nano particles with *tuned properties* to fabricate materials with improved materials compared with the existing ones. As other examples can be mentioned the production of single and multiwalled nanotubes of controlled sizes [121,158–160] for which the electronic behaviour depends on the chirality of the tube [161,162]. Recently [163] nanotubes have also been found to have high values of Young modulus, comparable with what is known for carbon fibres. The CEO knows that carbon fibres is a commercial product today and there is therefore some hope for new improved materials in this area.

The discovery that the properties of the clusters vary in particular for the smaller ones is also an incentive to look for clusters of certain sizes for which the value of a certain property will fit in with some application. Studies of various properties of clusters have also given examples of how cluster science has connection to the traditionally known world of nanotechnology or the mesoscopic physics, where quantum interference will lead to deviations from predictions of classical transport theory. These types of effect will appear when an electron travels in different paths between two points or moving in some trajectory, acquiring a phase difference along the path. The final effects will depend on whether the phase coherence length is lost or maintained. A number of experiments in this field have been made for systems described as two-dimensional electron gas, 2DEG. Recently Brack and coworkers [157] have used the jellium model for large clusters to describe an experiment on the Aharaonov-Bohm effect for a 2DEG of GaAs. We have recently in our group [164] adopted a numerically stable technique, introduced by Salomonsson and Öster [165] in atomic calculations, to obtain the complete spectrum of occupied and unoccupied eigenstates of systems with cylindrical symmetry as nanotubes and nanowires. The calculations are based on the *ab initio* density functional formalism. Applying this technique to such systems with extreme spatial variations of the charge density, the electronic structure arising from the interacting $2s$ and $2p$ valence electrons of carbon are replaced by the point charges of the individual C^{4+} ions in the graphitic walls by two-dimensional charged "sheets" ("2D jellium background") of cylindrical symmetry, with uniform surface-charge density σ. For a graphitic honeycomb lattice with a C–C bond length of 2.68 a.u. we obtain $\sigma = +0.428$ e/a.u.^{2}. The same formalism can also be applied to nanowires. These calculations for nanotubes give electronic density of states with a quantization which is similar to what is obtained in mesoscopic systems. Actually, clusters have connections with many areas of science, as visualized in Fig. 21.

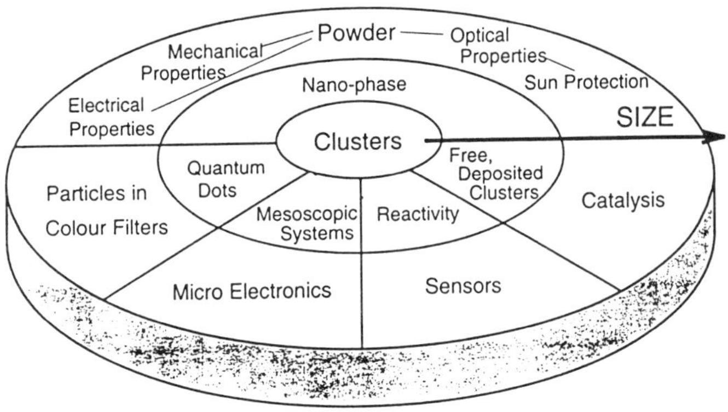

FIG. 21. A schematic overview of how clusters are related to many areas of modern material science, micro electronics, mesoscopic physics when the sizes of the clusters increase.

Many of the discoveries which have been found during the last few years and are discussed today were predicted by Richard Feynman as early as in 1959

[166]. For people working in atomic physics Richard Feynman is probably most well-known for *The Feynman Lectures in Physics* and for the *Theory of QED.* Feynman presented the following visions at a lecture for the American Physical Society at Caltec in 1959 where he talked about the creation of new materials with new properties by manipulation of matter on a small scale. He presented the famous vision; **There's plenty of room at the bottom;** *"I would like to describe a field, in which little has been done, but in which an enormous amount can be done in principle. The field is not quite the same as others in that it will not tell us of fundamental physics (in the sense of "What are the strange particles ?"), but it is more like solid state physics in the sense that it might tell us much about the strange phenomena that occur in complex situations. Furthermore, a point that is most important is that it would have an enormous number of technical applications. What I want to talk about is the problem of manipulating and controlling things on a small scale. As soon as I mention this, people tell me about miniaturization, and how far it has progressed today. They tell me about motors that are the size of the nail on your small finger. And there's a device on the market, they tell me, by which you can write the Lord's Prayer on the head of a pin. But that's nothing; that's the most primitive, halting step in the direction I intend to discuss. It is a staggeringly small world that is below."* Feynman continues by discussing the visions of; **How to write the Encyclopedia Britannica on the head of a pin. Getting even smaller. Rearranging atoms** and **Atoms in a small world**

The visions given by Feynman have been fulfilled in many respects today and even more has happened with the development of science to a highly interdisciplinary field. Particularly fascinating is how research in one area will be of great use in quite different areas of science. In 1985, studies of clusters of carbon initiated the discovery of the whole new class of carbon species known as fullerenes, for which Curl, Kroto and Smalley were awarded the Nobel Prize in Chemistry in 1996. In 1990, Krätschmer and Huffman published the method for production of a special type of soot, containing especially C_{60} in macroscopic amounts, which could be separated from the soot and used for fabrication of new materials. The early research on fullerenes was never focused upon or financed by grants in materials science but from grants for basic research in astrophysics with the goal to understand the interstellar absorption [109] and the existence of small molecules in space [123]. As an example of a very recent discovery can be mentioned the fabrication of an abacus of C_{60} molecules [167] or the earlier found quantum coralls [168]. These are just a few examples of how the visions of Feynman have been fulfilled but it probably only represents the beginning to build new functional units in nanotechnology as in the earlier *Stonehenge Area* when people used big stones as building blocks. That is, the beauty of modern science which many scientists would wish that politicians and industrial leaders would realize. One should however remember that there is a need for focused research in basic science, which will certainly spring a number of surprises but it may take time to develop for applications. No doubt modern forms of atomic physics will be very important in this area.

VII. APPENDIX 1: HISTORICAL OVERVIEW

When I started in 1964 as a graduate student in the group supervised by Associate Professor Ingvar Lindgren at the Department of Physics, Uppsala University, the first computer IBM 1620 had been installed a few years before. A new computer CD 3600 was installed later that year. Associate professor Mats Olsmats and the graduate students, Karl Erik Ädelroth, Leif Sanner, Björn Wannberg and Leif Jönsson, also included in the group were busy using the computer for analysis of their experimental data and design of new experimental set-ups. A new ABMR- apparatus was built and some of the students were involved in the building of electronics for this ABMR-apparatus. The old AMBR-apparatus [37] was also in operation for studies of isotopes of manganese and chromium. This apparatus, which was designed by Lindgren doing calculations of trajectories by hand, had for a number of years been used for measurements of nuclear spins and hyperfine structure constants for bismuth, thallium, polonium, tellurium and gold isotopes. Before that period the analysis of the experimental data was done manually using first, second or third order perturbation theory.

After about six month with the group I had the opportunity to participate in the Seventh "Brookhaven" conference which was held at Uppsala end of June in 1964. Participation in this conference was meant meeting a number of scientists and listening to their presentation of up-to-date research results. Brian Judd gave lectures on "Hyperfine Interaction" and use of tensor operators for analysis of hyperfine structure and Raymond Sörensen lectured on "Theory Nuclear Structure" with a discussion of nuclear models. The main goal of the hyperfine structure research was to evaluate accurate nuclear moments for test of nuclear models.

I participated as a graduate student in the experimental activities of the group but also given the task to learn about electronic structure calculations. The standard procedure for evaluation of nuclear moments from measured hyperfine structure constants was to obtain a value of r^{-3} from the spin-orbit coupling constant, using relativistic correction factors. The formulas as well as the relativistic correction factors were given in the book by Kopfermann [6]. These factors were first evaluated by Breit in 1930 [169] using relativistic hydrogenic wavefunctions. It was questionable how accurate these correction factors were for atoms and ions with a number of electrons. Another crucial problem was to estimate the core polarization contribution and, in particular the screening in the evaluation of quadrupole moments. Sternheimer had as early as in 1950 [170], done some pioneering work to evaluate a correction factor denoted mostly by R and known as the Sternheimer correction factor. Many scientists believed in all these relativistic correction factors, screening constants and evalution procedures for calculation of accurate values of nuclear moments, while others were rather suspicious of the procedure.

Ingvar started in this period to write a computer program for self-consistent calculations for atoms. Herman and Skillman [10] had just published their pioneering calculations and compilation of wavefunctions and eigenvalues for atoms with an atomic number of up to 103. These calculations were based on Slater's approximation from 1951 [171,172] or the Hartree Fock Slater, HFS, the exchange expression for an approximation of the exchange in the Hartree Fock equations. The meaning of the eigenvalues obtained in these calculations and how they could be used in comparison with experimental binding energies was a common issue discussed in those days. Actually a rather good agreement was obtained with experimental binding energies if one used the eigenvalues directly from HFS calculations, which was not the case if one used the eigenvalues from HF calculations as for example done by Hartree in 1928. To solve this problem, Ingvar introduced a new method which was based on the HFS expressions with a few parameters and known as Optimized Hartree Fock Slater, OHFS, [173,174]. Using wave functions calculated with this method, the total energy was calculated using the HF expression but with HFS wavefunctions. It became then possible to evaluate binding energies from differences in total energies, in this way calculating relaxation effects and estimating the accuracy of Koopmans' theorem [175]. I had the opportunity to take part in the extension of this method to relativistic calculations which gave the Relativistic Optimized Hartree Fock Slater, ROHFS, method [176,13]. Access to these computer programs made it also possible to evaluate binding energies and chemical shifts for comparison with the data from Electron Spectroscopy Group of Kai Siegbahn and Carl Nordling at the Department of Physics, Uppsala. These wave functions could also be used for evaluation of expectation valuee of r^{-3} to be used in the analysis of hyperfine structure data.

In 1966 Ingvar was appointed Professor in Physics at Göteborg University and Chalmers University of Technology. Karl-Erik Ådelroth and I joined him in Göteborg with the new ABMR-apparatus and left the old one at Uppsala. This new apparatus was given the nickname, Hippopotamus or Hippo by some of the students. We started in Göteborg a very interesting project with measurements of nuclear spins for rare earth isotopes which we obtained not only from the cyclotron at Uppsala but also from the cyclotron at Dubna Russia. Two new graduate students, Curt Ekström and Håkan Nyqvist, joined the group in the measurements of nuclear spins and hyperfine structure constants. Interpretation of the nuclear spins was done using the Nilsson model [33]. At that time Sune Svanberg also started using resonance methods for studying hyperfine interaction in excited states.

There was no good computer facility in Göteborg and we had therefore initially to send input data by mail to the computing centre at Uppsala. Within a few years IBM 360/50 and IBM 360/65 computers were purchased for use in Göteborg which represented a huge improvement. We continued the development of the computer programs in particular to obtain good values of r^{-3} for evaluation of nuclear moments from hyperfine interaction constants. Another interesting theory which appeared at that time was the use of effective operators, as outlined in the very famous paper by Sandars and Beck [177]. Using this theory, relativistic effects in hyperfine structure could be evaluated with LS coupled wavefunctions and an effective operator with effective values of $r_{ij}-3$ compared with the nonrelativistic value of r^{-3}. Use of this formalism with self-consistent wave functions and measurements of nuclear spins and hyperfine structure constants became the theme of my thesis for a doctorate in Philosophy in 1973 [178]. The experience obtained from atomic physics in those early days has been very valuable into my future research into other areas of science, as outlined in this article.

VIII. ACKNOWLEDGMENTS

The author wants to thank Ingvar Lindgren for introducing me to the field of atoms and hyperfine structure and in particular to the interplay between experiments and theory. My present research on clusters have benefitted from contributions of earlier and present graduate students. Special thanks should be given Mats Andersson, Henrik Grönbeck, Lotta Holmgren, Daniel Östling and John Persson for valuable comments and help with figures in this article. Financial support for the cluster research is acknowledged from NUTEK, The National Swedish Board for Industrial and Technical Development, and NFR, The Swedish Natural Science Research Council Materials Science Consortium, "Clusters and Ultrafine Particles".

IX. REFERENCES

[1] J. W. van Spronsen, *The Periodic System of Chemical Elements; A History of the First Hundred Years*, Elsevier, Amsterdam London New York 1969.

[2] N. Bohr, Z. Phys. **9**,1 (1922).

[3] W. Pauli, Z. Phys. **31**, 765 (1926).

[4] D. R. Hartree, Proc. Cambr. Phil. Soc. **24**, 89,111 (1928).

[5] E. U. Condon and G. H. Shortley, *The Theory of Atomic Spectra*, Cambridge University Press, Cambridge 1935.

[6] H. Kopfermann, *Nuclear Moments* Academic Press, New York 1958.

[7] L. Armstrong Jr *Theory of the Hyperfine Structure of Free Atoms* Wiley-Interscience, New York 1971.

[8] D. R. Hartree, *The Calculation of Atomic Structures*, John-Wiley and Sons New York, 1957.

[9] J. C. Slater, *Quantum Theory of Atomic Structure*, McGraw-Hill 1960.

[10] F. Herman S. Skillman, *Atomic Structure Calculations* Prentice-Hall Inc., New Jersey (1963).

[11] C. Froese Fischer, *The Hartree-Fock Method for Atoms*, Wiley New York 1977.

[12] C. Froese Fischer, Comput. Phys. Reports **3**,274(1986). Comput. Phys. Commun. **64**,369 (1991).

[13] I. Lindgren and A. Rosén, Case Stud At. Phys. **4**, 93 (1974).

[14] R. D. Cowan,*The Theory of Atomic Structure and Spectra*, University of Californina Press, Berkeley 1981.

[15] I. P. Grant, B. J. McKenzie, P. H. Norrington, D. F. Mayers and N. C. Pyper, Comput. Phys. Commun. **21** 207 (1980).

[16] J. P. Desclaux, Comput. Phys. Commun. **9**, 31 (1975).

[17] B. Fricke, Struct. Bonding, Berlin, **21**, 89 (1975).

[18] P. Pyykkö, Chem. Rev. **88**, 563 (1988).

[19] I. Lindgren and J. Morrison, *Atomic Many-Body Theory*, Springer-Verlag Berlin 1982.

[20] S. Nilsson, C. Tsang, A. Sobiczweski, Z. Szymanski, S. Wycech, C. Gustafson, I. L. Lamm and P. Möller, Nucl. Phys. A **131**, 1 (1969).

[21] S. Hoffmann *et al.* Z. Phys. A, **350**, 277,281 (1995).

[22] S. Hoffmannn *et al.* Z. Phys. A **354**, 229 (1996).

[23] V. Pershina, W.-D Sepp, B. Fricke and Rosén, J. Chem. Phys. **96**, 8367 (1992).

[24] W. E. Burcham, *Nuclear Physics*, Longmans, London 1963.

[25] Rutherford in Burcham, ref. 24 page 361

[26] N. Bohr, Naturwiss. **24**, 241 (1936).

[27] M. Mayer, *Elementary Theory of Nuclear Shell Structure*, Wiley, New York 1965.

[28] H. Schüler and T. Schmidt, Z. Physik **94**, 457 (1935).

[29] H. B. G. Casimir, *On the Interaction between Atomic Nuclei and Electrons*, Teyler's Tweede Genootschap, Haarlem, 1936; New printing, Freeman and Comp. San Francisco 1963.

[30] A. Bohr, Phys. Rev. **81**, 134 (1951).

[31] A. Bohr and B. R. Mottelson, K. Vidensk. Selsk. Mat. Phys. Medd. **27**, No. 16 (1953).

[32] A. Bohr and B. R. Mottelson: *Nuclear Structure, Vol I and II*, Benjamin New York, 1969, 1975.

[33] S. G. Nilsson, K. Vidensk. Selsk. Mat. Phys. Medd. **29**, 237 (1955).

[34] I. I. Rabi, J. R. Zacharias, S. Millman and P. Kusch, Phys. Rev. **53**, 318 (1938)

[35] N. F. Ramsey, *Molecular Beams*, Oxford University Press Oxford 1956.

[36] I. Lindgren, *Table of Nuclear Spins and Moments*, In Alpha-, Beta-, and Gamma-Ray Spectroscopy, Ed. K. Siegbahn, North Holland Publ. Comp. Amsterdam 1965.

[37] I. Lindgren, Nucl. Instr. Methods **3**, 1 (1958).

[38] I. Lindgren, *Nuclear Spins and Moments of Radioactive Isotopes Investigated by an Atomic Beam Method*, Thesis of Doctor of Philosophy, Uppsala University, May 21 1959.

[39] C. Ekström, M. Olsmats and B. Wannberg, Nucl. Instr. Methods.**103**,13 (1972).

[40] H. T. Duong, C. Ekström *et al.*, Nucl. Instr. Methods., **A325**, 465 (1993).

[41] A. Kastler and J. Brossel, Compt. Rend. **229**, 1213 (1949).

[42] W. Hanle, Z. Phys. **30**, 93 (1924); Erg. ex. Naturwiss., **4**, 214 (1925).

[43] S. Svanberg, *Atomic and Molecular Spectroscopy*, Springer Verlag Berlin 1991.

[44] W. Demtröder, *Laser Spectroscopy*, Springer Verlag Berlin 1981, 1996.

[45] K. P. Huber and G. Herzberg, *Molecular Spectra and Molecular Structure, Constans of Diatomic Molecules*, van Nostrand Reinhold Comp New York 1979.

[46] N. W. Ashcroft and N. D. Mermin, *Solid State Physics* , Saunders College, CBS Publ. Asia Ltd. Philadelphia 1976.

[47] ISSPIC 1, J. Phys. **38**, (1977); ISSPIC 2, Surf. Sci. **106**, (1981); ISSPIC 3, Surf. Sci. **156**, (1981); ISSPIC 4, Z. Phys. D **12**, (1989); ISSPIC 5, Z. Phys. D **19**, (1991); ISSPIC 6, Z. Phys. D **26**, (1993); ISSPIC 7, Surf. Rev. and Lett. **3**, (1996); ISSPIC 8, Z. Phys. D , (1997)

[48] G. Benedek, T. P. Martin and G. Pacchioni, Eds. *Elemental and molecular clusters*, Proc. of the 13th International School, Erice Italy, 1987. Springer Series in Materials Science Vol. 6 Berlin, Heidelberg, New York Springer 1988.

[49] M. Brenner, T. Lönnroth and F. B. Malik, Eds. *Clustering Phenomena in Atoms and Nuclei*, Springer Series in Nuclear and Particle Physics, Berlin, Heidelberg, New York Springer 1992.

[50] *Nanophase Materials Synthesis-Properties-Applications*, Eds. G. C. Hadji-

panayis and R. W. Siegel, Series E: Applied Sciences - Vol. 260 Kluwer Academic Publ, Dordrecht 1994.

[51] J. Jortner, Z. Phys. D **24**, 247 (1992).

[52] W.A. de Heer, W. A. Knight, M. Y. Chou, and M. L. Cohen, Eds. H. Ehrenreich and D. Turnbull, *Solid State Physics*, Academic Press , New York Vol. 40, p 93 1987.

[53] M. Brack, Rev. Mod. Phys. **65**, 677 (1993).

[54] W. de Heer, Rev. Mod. Phys. **65**, 611 (1993).

[55] *Clusters of Atoms and Molecules I and II*, Ed. H. Haberland, Springer Series in Chemical Physics 52 and 56 1994

[56] A. W. Castleman, Jr. and K. Bowen, J. Phys. Chem. **100**, 12911 (1996).

[57] R. W. Siegel, Physics Today **Oct 93**, 64 (1993).

[58] R. W. Siegel, Nanostr. Mater. **4**, 121 (1994).

[59] T. P. Martin, Physics Reports. **273**, 199 (1996).

[60] O. Cheshnovsky, k. J. Taylor, J. Conceicao and R. J. Smalley, Phys. Rev. Lett. **64**, 1785 (1990).

[61] K. J. Taylor, C. L. Pettiette-Hall, O. Cheshnovsky and R. J. Smalley, J. Chem. Phys. **96**, 3319 (1992).

[62] K. Rademann, B. Kaiser, U. Even, F. Hensel, Phys. Rev. Lett. **59**, 2319 (1987).

[63] O. Echt, K. Sattler and E. Recknagel, Phys. Rev. Lett. **47**, 1121 (1981).

[64] W. Knight, K. Clemenger, W. de Heer, W. Saunders, M. Chou and M. Cohen, Phys. Rev. Lett. **52**, 2141 (1984).

[65] J. Pedersen, S. Björnholm, J. Borggren, K. Hansen, T. P. Martin, and H. D. Rasmusen, Nature, **353**, 733 (1991).

[66] T. P. Martin, S. Björnholm, J. Borggren, C. Brechignac, Ph. Cahuzac, K. Hansen and J. Pedersen, Chem. Phys. Lett. **186**, 53 (1991).

[67] H.W. Kroto, J.R. Heath, S.C. O'Brien, R.F. Curl and R.E. Smalley, Nature **318**, 162 (1985).

[68] S. Ijima, Nature **354**, 56 (1991).

[69] D. Ugarte Nature **359**, 707 (1992).

[70] B. C. Guo, K. P. Kerns and A. W. Castleman, Jr., Science **255**, 1411 (1992).

[71] H. Nishioka, K. Hansen and B. R. Mottelson, Phys. Rev. B **42**, 9377 (1990)

[72] R. Balian and C. Bloch, Ann. Phys. **69**, 76 (1971).

[73] C. W. J. Beenakker and H. van Houten, *Quantum Transport in Semiconductor Nanostructures*, in Soild State Physics, Vol. 44, p. 1-228, Eds. H. Ehrenreich and D. Turnbull, Academic Press San Diego 1991.

[74] *Transport Phenomena in Mesoscopic Systems*, Eds. H. Fukuyama and T. Ando, Springer Verlag, 1992.

[75] *Nanostructures and Mesoscopic Systems*, Eds. W. P. Kirk, and M. A. Reed, Academic Press, San Diego, 1992.

[76] G. Mie, Ann. d Physik **25**, 377 (1908)

[77] P. Debye, Ann d Physik **30**,57 (1909)

[78] *Optical Properties of Metal Clusters*, U. Kreibig and M. Wollmer, Springer Series in Materials Science 25, Springer Verlag, 1995.

[79] C. G. Granqvist and R. A. Buhrman J. Appl. Phys. **47**, 2200 (1976).

[80] G. Scoles, *Atomic and Molecular Beam Methods*, Vol 1, Oxford University Press, New York 1988.

[81] N. Furstenau and F. Hillenkamp, Int. J. Mass. Spectr. and Ion Phys. **37**, 135 (1981).

[82] P. Fayet, J. P. Wolf and L. Wöste, Phys. Rev. B, **33**, 6792 (1986).

[83] T.P.Martin, N. Malinowski, U. Zimmerman, U Naher and H. Schaber, J. Chem. Phys. **99**, 4210 (1993). U. Zimmermann, N. Malinowski, U. Naher, S. Frank and T. P, Martin, Phys. Rev. Lett. **72**, 3542 (1994).

[84] R. E. Smalley, Laser Chem. **2**, 167 (1983). T. G. Dietz, M. A. Duncan, D. E. Powers and R. E. Smalley, J. Chem. Phys. **74**, 6511 (1981). J. B. Hopkins, p. R. R. Landgridge-Smith, M. D. Morse and R. E. Smalley, J. Chem. Phys. **78**, 1267 (1984).

[85] V. E. Bondybey and J. H. English, J. Chem. Phys. **76**, 2165 (1982).

[86] H. G. Limberger and T. P. Martin, J. Phys. Chem. **90**, 2979 (1988).

[87] U. Zimmermann, N. Malinowski, U. Naher, S. Frank and T. P. Martin Z. Phys. D **31**, 85 (1994).

[88] M. Andersson, *Metal Cluster Reactivity: Adsorption of small molecules in bimolecular collisions*, Thesis Chalmers University of Technology, Göteborg, 1995.

[89] M. Andersson, J. L. Persson and A. Rosén, J. Phys. Chem. **100**, 1222 (1996).

[90] A. Kaldor, D.M. Cox, M.R. Zakin, Adv. in Chem. Phys. **70**, 211 (1988).

[91] S. Riley in Ref. 55

[92] O. Echt, O. Kandler, T. Leisner, W. Miehle, and E. Recknagel, Faraday Symposium No. 25 on Large Gas Phase Clusters, University of Warwick, December 12-14,1989

[93] A. L. Mackay, Acta. Crystallogr., **15**, 916 (1962).

[94] J. Farges, M. F. De Feaudy, B. Raoult and G. Torchet, J. Chem. Phys., **84**, 3491 (1986).

[95] W. Ekardt, Phys. Rev. B **29**, 1558 (1984).

[96] D. E. Beck, Solid. State. Commun. **49**, 381 (1984).

[97] S. Bjørnholm, Contempary Physics **31**, 309 (1990).

[98] T. P. Martin, T. Bergmann, H. Göhlich and T. Lange, Chem Phys. Lett. **172**, 209 (1990).

[99] J. L. Persson, M. Andersson,L. Holmgren, T. Åklint and A. Rosén, Chem. Phys. Lett. in press (1997).

[100] S. J. Riley, E. K. Parks, G. C. Nieman, L. G. Pobo and S. Wexler, J. Chem. Phys. **80**, 1360 (1984)

[101] S. C. Richtsmeier, E. K. Parks, K. Liu, L. G. Pobo and S. J. Riley, J. Chem. Phys. **82**, 3659 (1985).

[102] E. A. Rohlfing, D. M. Cox and A. Kaldor, Chem. Phys. Lett. **99**, 161 (1983); J. Phys. Chem. **88**, 4497 (1983).

[103] E.A. Rohlfing, D. M. Cox, A. Kaldor, J. Chem. Phys. **81**, 3322 (1984).

[104] H. Kroto, Science **242**, 1139 (1988).

[105] R. W. Marks, *The Dymaxion World of Buckminster Fuller*, Reinhold, New York, 1960.

[106] D. R. Huffman, Physics Today, **Nov 91**, 22 (1991).

[107] W. Krätschmer, Z. Phys. **D19**, 405 (1991).

[108] D. R. Huffman and W. Krätschmer, Mater. Res. Soc. Symp. **206**, 601 (1991).

[109] D. R. Huffman, Adv. Phys. **26**, 129 (1977)

[110] S. Larsson, A. Volosov and A. Rosén Chem. Phys. Lett. **137**, 501 (1987).

[111] W. Krätschmer, K. Fostiropoulos, D.R. Huffman, *Dusty Objects in the Universe*, Eds. E. Bussoletti and A. A. Vittine, Kluwer Academic Publ. Dordrect, Boston London, 1988.

[112] W. Krätschmer, K. Fostiropoulos, D.R. Huffman, Chem. Phys. Lett. **170**, 167 (1990).

[113] W. Krätschmer, L.D. Lamb, K. Fostiropoulos, D.R. Huffman, Nature **347**, 354 (1990).

[114] H. Palmer and M. Shelef, *Chemistry and Physics of Carbon*, Walker, J. R. , Jr., Ed. Marcel Dekker: New York, 1967, Vol. 4 pp 85-135.

[115] W. Weltner and R. J. Van Zee, Chem. Rev. **89**, 1713 (1990).

[116] R. Taylor, J. P. Hare, A. K. Abdul-Sada and H.W. Kroto, J. Chem. Soc. Chem. Commun. **20**, 1423 (1990).

[117] R.D. Johnson, G. Meijer and D. S. Bethune, J. Am. Chem. Soc. **112**, 8983 (1990).

[118] R. C. Haddon, *et al.*, Nature **350**, 320 (1991).

[119] A. F. Hebard, *et al.*, Nature **350**,600 (1991).

[120] J. R. Heath, S. C. O. Brien, Q. Zhang, Y. Liu, R. F. Curl, H. W. Kroto and R.E. Smalley, J. Am. Chem. Soc. **107**, 7779 (1985).

[121] D. S. Bethune, R.D. Johnson, J.R. Salem, M.S. de Vries and C. S. Yannoni, Nature **366**, 123 (1993).

[122] H. Kroto, A. W. Allaf and S. P. Balm, *C_{60}: Buckminsterfullerene*, Chem. Rev. **91**,1213 (1991).

[123] H. Kroto, *C_{60}: Buckminsterfullerene, The Celestical Sphere that Fell to Earth*, Angew. Chem. **31**,111 (1992).

[124] *Special Issue on Fullerenes*, Ed. H. Kroto, Pergamon Press, New York 1992, Carbon, **30**, 1139 (1992).

[125] *Clusters and Fullerenes*, Proc. Adriatico. Res. Conf. Eds. V. Kumar, T. P. Martin and E. Tosatti, World Scientific, Singapore 1992, Int. J. Mod. Phys. B

6, (1992).
[126] *Fullerenes*, Ed. H. Kroto, MRS Bulletin, Nov **XIX**, No 11, (1994).
[127] A. Hirsch, *The Chemistry of the Fullerenes*, Georg Thieme Verlag 1994.
[128] *The Fullerenes*, Eds. H.W. Kroto, J.E. Fischer and D.E. Cox, Pergamon Press, Exeter, Great Britain 1993.
[129] *Physics and Chemistry of the Fullerenes*, Ed. K. Prassides, NATO ASI series, Kluwer Acad. Pub., Dordrecht, The Netherlands 1994.
[130] Eds. H. Ehrenreich and F. Spaepen, *Solid State Physics (Fullerenes)*, **48**, Academic Press, London 1994.
[131] M.S. Dresselhaus, G. Dresselhaus, and P.C. Eklund, *Science of Fullerenes and Carbon Nanotubes* Academic Press, San Diego, 1996.
[132] *The Chemical Physics of Fullerenes 10 (and 5) Years Later*, Ed. W. Andreoni, NATO ASI series, Kluwer Acad. Pub., Dordrecht, The Netherlands 1996.
[133] *Perfect Symmetry*, J. Baggott, Oxford University Press, Oxford 1996.
[134] K.B. Shelimov, D.E. Clemmer and M.F. Jarrold, J. Phys. Chem., **98**, 12819 (1994).
[135] G. von Helden, N. G. Gotts, M.T.Bowers, Nature **363**, 60 (1993).
[136] T.P. Martin, U. Näher, H. Schaber and U. Zimmermann, *Phys. Rev. Lett.* **70**, 3079 (1993).
[137] T.P. Martin, N. Malinowski, U. Zimmermann, U. Näher and H. Schaber, *J. Chem. Phys.* **99**, 4210 (1993).
[138] U. Zimmermann, N. Malinowski, A. Burkhardt and T.P. Martin, *Carbon* **33**, 995 (1995).
[139] D. Östling, *Electronic Strucuture and Optical Properties of C_{60} Nanotubes and Carbon Onions*, PhD Thesis, Göteborg University 1996.
[140] T. W. Ebbesen, Physics Today **June**, (1996)
[141] J. C Slater,*Quantum Theory of Molecules and Solids*, Vol 1, McGraw-Hill, New York, 1963.
[142] H. F. Schaefer III, *The Electronic Structure of Atoms and Molecules; A Survey of Rigorous Quantum Mechanics Results*, Addison-Wesley Publ. Comp. 1972.
[143] *Density Functional Methods in Chemistry*, Eds. J. K. Labanowski, J. C. Andzelm, Springer-Verlag, New York, Berlin, 1991.
[144] V. Bonačić-Koutecký, P. Fantucci, J. Koutecký, Chem Rev. **91**, 1035 (1991).
[145] L. Goodwin and D. R. Salahub, Phys. Rev. **A47**, 774 (1993).
[146] H. Grönbeck and A. Rosén, Phys. Rev. **B54**, 1549 (1996).
[147] H. Grönbeck, *On the structure of and bonding in metal clusters* PhD Thesis, Göteborg University 1996.
[148] M. Born and R. Oppenheimer, Ann. Physik **87**, 457 (1929).
[149] P. Hohenberg, W. Kohn, Phys. Rev. **136**, 864 (1964).
[150] W. Kohn, L.J. Sham, Phys. Rev. **140**, A1133 (1965).
[151] There are various reviews which demonstrate the strength (and weakness) of

the DFT, for example, W. Kohn, A.D. Becke, R.G. Parr, J. Phys. Chem. **100**, 12974 (1996).

[152] R. Car and M. Parrinello, Phys. Rev. Lett. **55**, 2471 (1985).

[153] S. Bjørnholm, in ref. 55

[154] K. Clemenger, Phys. Rev. B **32**, 1359 (1985).

[155] J. Pedersen *et al.*, Z. Phys. D **26**, 281 (1993).

[156] J. Borggreen *et al.*, Phys. Rev. B **48**, 17507 (1993).

[157] S. M. Reimann, M. Brack, A. G. Magner and M.V.N. Murphy, Surf.Rev.Lett. **3**, 19 (1996)

[158] P. M. Ajayan *et al.*, Chem. Phys. Lett. **215**, 509 (1993)

[159] P. M. Ajayan and S. Iijima, Nature **361**, 333 (1993)

[160] P. M. Ajayan *et al.*, Nature **362**, 522 (1993)

[161] S. Seraphin *et al.*, Nature **362**, 503 (1993)

[162] A. Thess *et al.*, Science **273**,483 (1996)

[163] T. W. Ebbesen, Nature **382**, 5 (1996)

[164] D. Östling, D. Tománek and A. Rosén, Phys. Rev. **B55** , (1997).

[165] S. Salomonson and P. Öster, Phys. Rev. **A40** , 5559 (1989).

[166] R. Feynman, Science **254**, 1330 (1991).

[167] M.T.Cuberes, *et al.*, Appl. Phys. Lett. **69**, 3016 (1996)

[168] M. F. Crommie, C. P. Lutz, and D. M. Eigler, Science **262**, 218 (1993)

[169] G. Breit, Phys. Rev. **35**, 1447 (1930).

[170] R. Sternheimer, Phys. Rev. **80**, 102 (1950); Phys. Rev. **84**, 244 (1951);Phys. Rev. **86**,316 (1952).

[171] J. C. Slater, Phys. Rev. **81**, 385 (1951).

[172] J. C. Slater, Phys. Rev. **82**, 538 (1951).

[173] I . Lindgren, Phys. Lett. **19**, 382 (1965).

[174] I . Lindgren, Arkiv Fysik **31**, 59 (1966).

[175] T. Koopmans, Physica, **1**, 104 (1933).

[176] A. Rosén and I. Lindgren, Phys. Rev., **176**, 114 (1968).

[177] P. G. H. Sandars and J. Beck, Proc. Roy. Soc., **289**, 97 (1965).

[178] A. Rosén , *A Study on Atomic and Nuclear Structure by means of Experimental and Theoretical Investigations of Atomic Hyperfine Interactions*, Abstract of Gothenburg Dissertations in Science 35, Acta Universitatis Gothoburgensis 1973.

RELATIVISTIC AND DYNAMIC CONTRIBUTIONS IN ION-ATOM AND ION-SOLID COLLISIONS[§]

B. Fricke, J. Anton, K. Schulze, W.-D. Sepp
Fachbereich Physik, Universität Kassel, D-34109 Kassel, Germany

and

P. Kürpick
J.R. Macdonald Laboratory, Department of Physics, Kansas State University, Manhattan, Kansas 66506-2604, USA

Abstract:

We solve the time-dependent relativistic Dirac-equation for complicated ion-atom systems in the Dirac-Fock-Slater approximation including the many-electron effects by means of an inclusive probability description. Various gas-gas and gas-solid target collision systems are discussed. We find the dominant effect in the observed excitation probability of inner shells to be a relativistic dynamic coupling of various levels which in a non-relativistic description is zero. The effect of excitation and transfer during the passage within the solid allows us to understand the gas-solid target systems. An ab-initio calculation of K MO X-rays is presented for the system Cl^{16+} on Ar.

1 Introduction

From a very general point of view every ion-atom collision system has to be treated as a correlated many-body time-dependent quantum system. To solve this from an ab initio point of view is still impossible. So, one has to rely on various approximations. Nowadays the best method which can be applied to realistic collision systems (which we discuss here) is on the level of the non-selfconsistent time-dependent Hartree-Fock-Slater or, in the relativistic case, the Dirac-Fock-Slater method. Up-to-now no correlation beyond this approximation can be taken into account in the case of 3 or more electrons. (This is in accordance with the definition of correlation given by Löwdin [1] in 1956) In addition no QED contributions, i.e. no correction to the 1/r Coulomb interaction between the electrons, ever have been taken into account, although in very heavy collision systems this effect may become important. This will be discussed in section 5. A short survey of the theory used is followed by our results on impact parameter dependent electron transfer and excitation calculations of ion-atom and ion-solid collisions as well as first results of an ab initio calculation of MO X-rays in such complicated many particle scattering systems.

[§] dedicated to Prof. Lindgren on the occasion of his 65. birthday

ADVANCES IN QUANTUM CHEMISTRY, VOLUME 30

0065-3276/98 $25.00

2 Description of Time-Dependent Ion-Atom Collisions

2.1 The general problem

Any ion-atom collision can be described by the simple equation

$$A^{n+} + B \rightarrow A^{*p+} + B^{*m+} + q\,e^{-} + \text{photons}\,.$$

The experimental variables in such a system are the energy of the colliding ion and the initial ionization. If the energy is large enough to describe the motion of the nucleus by a classical trajectory, the impact parameter will become a third variable. During and after such a collision X-rays and electrons (Auger- and δ-electrons) can be emitted. If the emission occurs in the interactive region of the collision the X-rays and Auger-electrons are called MO X-rays and MO Auger-electrons.

2.2 Description by a correlation diagram

If the collision energy is large enough to describe the nuclear motion by a classical trajectory and the collision energy small enough to allow a molecular quantum mechanical description of the electrons, then correlation diagrams are a good physical way to describe the collision. Such correlation diagrams describe the total energy of the quantum mechanical system as a function of internuclear distance as well as of the quantum numbers for the full electronic system.

In case of the very heavy ion-atom collisions with many electrons, however, we use the simpler (independent particle) form of the correlation diagram where the single-particle levels are given as function of the internuclear distance. Fig. 1 shows such a schematic correlation diagram as an example.

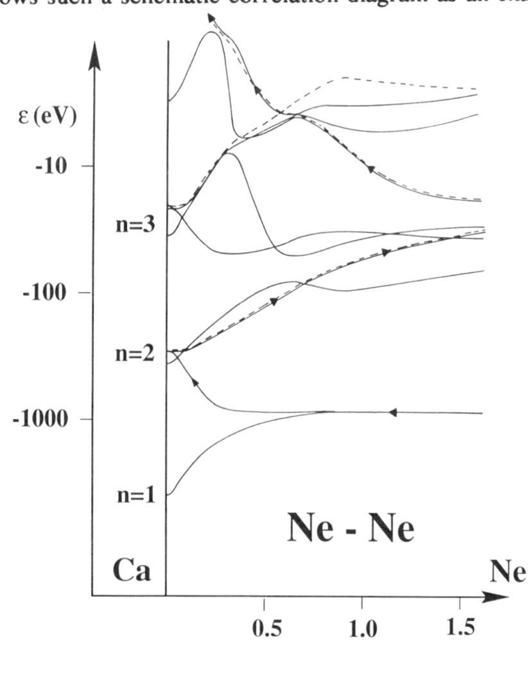

Fig. 1 Schematic correlation diagram of the system Ne - Ne. The arrows present the one-particle processes to understand excitation and transfer as well as ionization in such a picture.

This diagram also shows the two main processes which occur in such a system. The first is the inner electron transfer and excitation from the K-shell of one of the partners via the rotational dynamic coupling (described below) to the L- shell of the other collision partner. The other possibility is the transfer of a L-electron to a higher shell via a molecular level which is connected to levels in the vicinity of the continuum and thus ionizes.

2.3 Theoretical description

To describe such a complicated many-electron ion-atom collision properly the following has to be taken into account

- all electrons in the collision with good energy eigenvalues and a good potential function to describe the motion of the nucleus
- a relativistic description, because for high Z systems any non-relativistic description is wrong from the very beginning
- the time-dependence of the collision process.

Thus, one has to solve the time-dependent Dirac-equation

$$H \Psi = i h \, \partial/\partial t \, \Psi \quad . \tag{1}$$

where H is the many-particle Hamiltonian. For the many-particle wavefunction Ψ we choose an anti-symmetric Slater-Determinant built from single-particle time-dependent wavefunctions $\psi_i(t)$. These wavefunctions are calculated as solutions of the Time-Dependent Dirac-Fock-Slater equations (TDDFS) obtained by introducing the Slater-Determinant Ψ in equ. 1

$$h^{TDDFS} \psi_i(t) = i h \, \partial/\partial t \, \psi_i(t) \qquad i = 1,N \tag{2}$$

where N is the number of electrons involved. The $\psi_i(t)$ are expanded as linear combinations of solutions ϕ_j of the static molecular DFS equations for a fixed internuclear distance R:

$$\psi_i(r,t) = \Sigma \; a_{ij}(t) \; \phi_j(r;R(t)) \; . \tag{3}$$

If this ansatz is introduced in equ. 2 one gets the coupled channel equations for the amplitudes a_{ij}

$$d/dt \; a_{im}(t) = \Sigma \; - a_{ij}(t) \langle \phi_m \,|\partial/\partial t\,| \phi_j \rangle \exp(-i/h \textstyle\int (\epsilon_i - \epsilon_j) dt) \qquad i = 1,N \; . \tag{4}$$

The time-dependent matrixelements $\langle \phi_m \,|\partial/\partial t\,| \phi_j \rangle$ can be rewritten in the form of rotational- and radial-matrix elements which couple the amplitudes a_{ij} of the levels in the correlation diagram with the same projection of the angular momentum of the levels in case of the radial matrix elements and with +1 or -1 in case of the rotational matrix elements. For details on this procedure see Ref. 2.

The set of amplitudes $\{a_{ij}\}$ allows to reconstruct the many particle wavefunction Ψ introduced in equ. 1 for every kind of many-particle electron configuration as long as Ψ is described as a time-dependent Slater-Determinant. In nearly all experiments the initial and final states are measured only partially. Therefore, one has to sum (average) over all unmeasured states. This method of Inclusive Probability is well described in Ref. 3-5.

3 Results on Ion - Atom collisions with gas targets

3.1 Triple differential X-ray cross-sections

The most complicated example ever calculated by us was the system S^{16+} on Ar which is a gas-type experiment published by Schulz et al. [6]. Fig. 2 gives the results of the experiment and our calculation [7,8] for the triple differential cross section to observe at least two X-rays after the collision from either Ar (P_{AA}) or S (P_{SS}) or one X-ray of both types (P_{SA}). In this case we adapted even the correlation diagram according to the occupation of the levels during the collision.

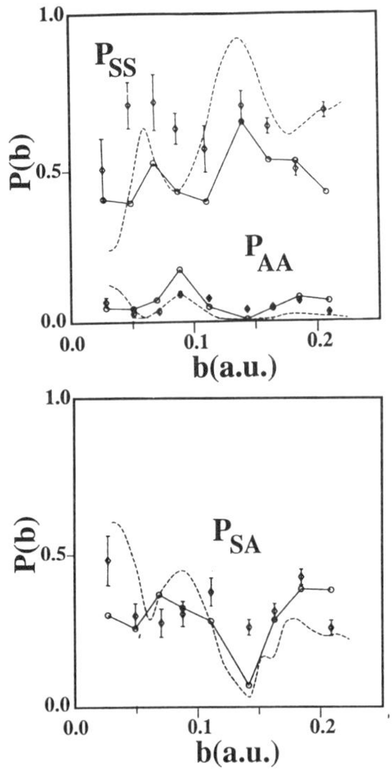

Fig. 2 Impact parameter dependent probability P to observe two simultaneous X-rays of the collision system 16 MeV S^{16+} on Ar. The twoindices S (Sulphur) and A (Argon) denote the origin of the X-rays. The experimental values are from Schulz et al. [6]; the full curves denote our TDDFS results [7] and the broken curves our simpler SCF results [8].

3.2 Results on MO X-rays

The second possibility to get direct information from such a collision system is the observation of MO X-rays. The result of our calculation [9] is given in Fig. 3 where the spectrum of the collision system 20 MeV Cl^{16+} on Ar is compared with the experiment [10]. In this calculation the radiation field was coupled to the system by first order perturbation theory but the wavefunctions were taken from the TDDFS calculations described above. Ref. 9 also gives a detailed description of the theoretical model used.

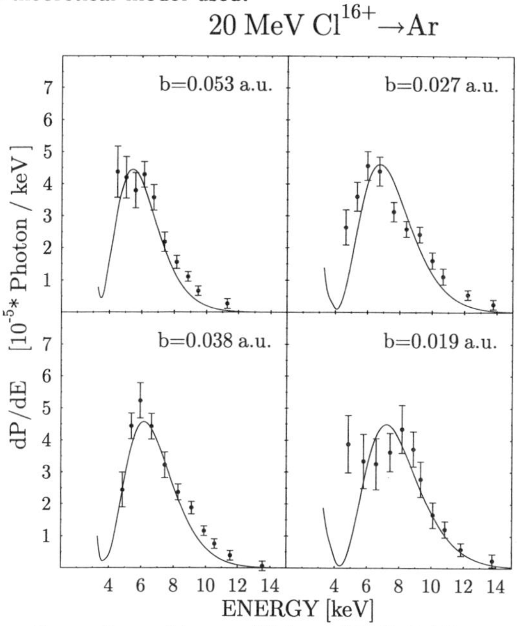

Fig. 3 MO X-ray spectrum for the system 20 MeV Cl^{16+} on Ar for four different impact parameters. The full curve is our calculation [9] scaled with a factor of 1.5. The experiment is from Schuch et al. [10].

4 Results on Ion - Atom collision with Solid State targets

4.1 Difference between gas target and solid state target experiments

In a gas target experiment an ion collides only once with a target atom. This is also the assumption of the theoretical procedure described above. In a solid state target the incoming ion interacts during the passage with every atom in the solid nearer than the interaction distance (approximately the diameter of its electron cloud). The largest collision distance such a ion can have in the solid is half the distance between two nearest neighbour atoms. So every ion experiences a large number of collisions with atoms in the solid at relatively large impact parameters. In order to simulate this we perform a number of large impact paramater coupled channel calculations with the initial electron distribution of the incoming ion. Thus, we learn how the electron cloud of the moving ion in the solid excites or transfers electrons. The time involved in these collisions is so short that this ion has almost no chance to re-arrange, therefore after a few layers it is highly excited and charged differently. If the ion then experiences a collision at very small impact parameters (leading to a measurable deflection) it will be measured in the coincidence experiment. Owing to this excitation and transfer of electrons in the ion, which moves through the solid, the relevant many-particle excitation and tranfer in the small impact parameter collision (which is measured) is quite different to a ion-gas experiment because the ion no longer possesses its original charge nor its original ground state.

4.2 Some examples

One such example with 50 MeV Cu on Ni is given in Fig. 4 where the effect of thick as well as thin targets can be seen. The experiment is by Annett et al. [11] and the calculation is published in ref. 12. The prediction of the $2p_\pi$ -$2p_\sigma$ theory of Taulbjerg et al. [13] is included as well.

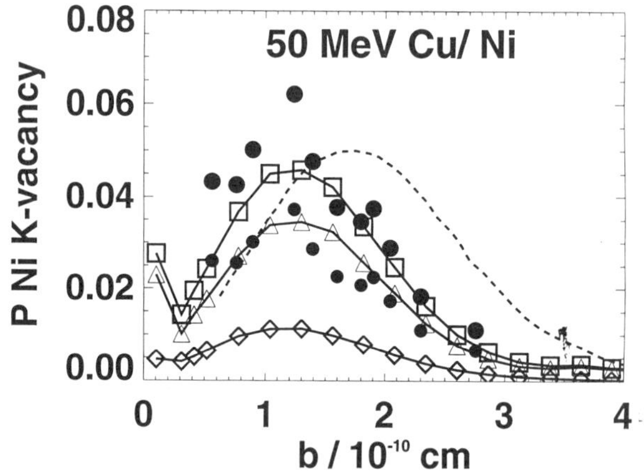

Fig. 4 Ni K-vacancy probability in a 50 MeV Cu on Ni collision as function of impact parameter. The two upper solid curves give the results of a thick and a thin target calculation. The thick and thin dots are the experimental points [11]. The lowest curve is the contribution from inner-shell holes which were created in the solid. The dashed line is the Taulbjerg, Briggs theory [13].

The P(b)-maximum of this simple 2-level theory [13] and the results of many experiments differ in positions for ion-solid target collisions compared to those of ion-gas experiments. The maximum in the gas-gas experiments were more or less at the same position as the simple theory. In an early paper by Schuch et al. [14] this effect was attributed to the solid itself as an unexplained "solid-target" effect. We proved in several publications [15-17] (and Fig. 4 is an example) that our theory is always at or very near the experimental maximum. In the course of this investigation we found the reason for this discrepancy; it is the difference between the non-relativistic and a relativistic description in connection with a different occupation of the holes in the collision: at very small internuclear distances not only the relativistic equivalent of the $2p_\pi$ - $2p_\sigma$ rotational coupling contributes but also the other two levels of the united atom L-shell (which do not couple in a non-relativistic treatment [15]). Fig. 5 presents the example of the K X-ray emission probability in the Br - Ni collision where we have plotted the various contributions from the coupling of the 3 relativistic levels to the 1s shells of the atomic systems. This is the relativistic analogue to the non-relativistic $2p_\pi$ - $2p_\sigma$ coupling. The relativistic 1(3/2)± level coincides in the non-relativistic limit with the $2p_\pi$ level. The other two levels (3(1/2)± and 4(1/2)±) which transfer holes in the lowest levels do not couple in the non-relativistic limit. Especially the contribution of the 3(1/2)± level moves the maximum of the impact parameter dependent transfer probability drastically to smaller b already for systems with a united Z around 50 and even below.

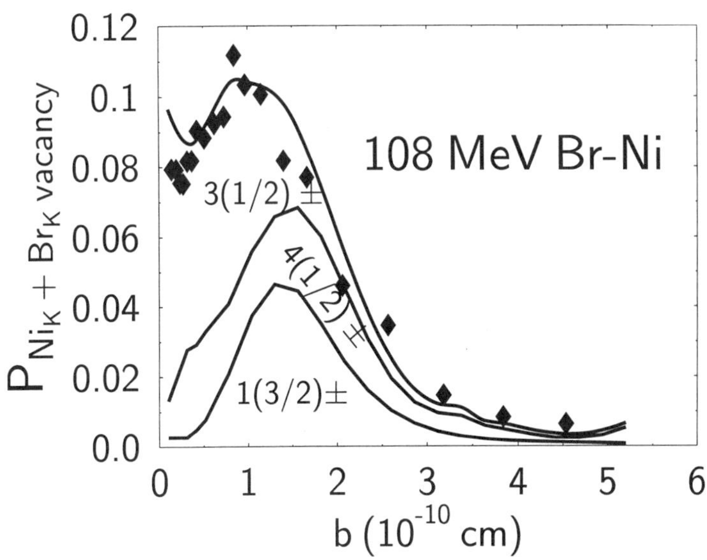

Fig. 5 Partial contributions of the molecular levels to the total K-vacancy production probability for 108 MeV Br on Ni.

5. Future experiments

When investigations on this subject started 25 years ago one of the main questions was how the collision system behaves in the region of a united charge well above 100. For a point-like nucleus the Dirac-equation is known to diverge for the innermost 1s state at Z = 137 - the inverse of the Sommerfeld Constant [18]. Already in 1969 Piper and Greiner [19] investigated that the energy eigenvalue of this level depends very much on the structure of the extended nucleus. Assuming a realistic continuation of the charge distribution we theoretically predicted that the innermost level dives into the negative continuum at about Z = 173 [20]. What happens there? At the beginning of this debate the general opinion was that positrons are generated when the 1s level brings holes into the collision. The Frankfurt group [21] demonstrated that the full dynamic description, and not the static one, leads to the continuum of positrons observed in many experiments. In addition experimentalists claim the existence of relatively narrow positron lines additionally to these dynamic positrons [22]. Theoreticians have worked hard to understand these narrow lines [23]. Nowadays, however, this is very much in question [24]. Nevertheless, there remains the question of what the structure of the innermost levels is like. Static correlation diagrams (from the solution of the static molecular Dirac-equation) provide a first theoretical hint. One of the early correlation diagrams was the system Pb-Pb [25]. Instead of this system we present here the even heavier system U-Pb in Fig. 6. The extreme effect of the strong direct relativistic effect on all $s_{1/2}$ and $p_{1/2}$ levels is quite obvious. At very small internuclear distances the $2p_{1/2}$ level is even more strongly bound than the $2s_{1/2}$ level and thus the whole level structure changes drastically.

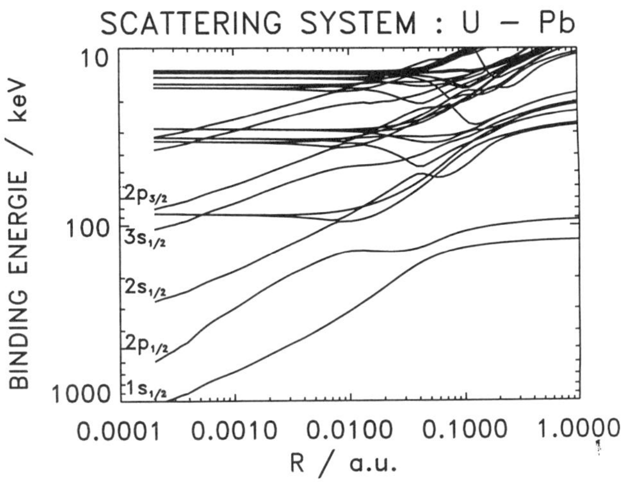

Fig. 6 Correlation diagram of the system U on Pb which takes into account 56 electrons. The point of interest is the behaviour of the innermost levels at small internuclear distances; it is very different compared to correlation diagrams for small Z like the one in Fig.1.

Keeping that in mind it will be absolutely necessary to perform experiments in order to detect the level structure of these systems. The quality of the theoretical description is now so good that

- accurate correlations diagrams can be calculated,
- full coupling treatment, including all rotational and radial couplings of the inner shells, can be performed and

- a many-particle interpretation is possible which allows to take into account the electron excitation and transfer in a collision of a naked projectile in a solid state target.
- In addition MO X-ray calculations and interpreation is now possible

This system U-Pb from Fig. 6 is the collision system which will be used soon for an experiment at GSI where they will try to observe MO X-rays as well as innershell X-rays of the target Pb atoms in the solid. We hope to be able to provide coupled channel calculations soon in order to calculate the X-ray production of the innermost levels of the Pb target as well as MO calculations to give a prediction on the MO X-ray structure for this experiment. First theoretical attempts in this direction had already been made many years ago by Kirsch et al. [26]; they compared Pb-Pb collision systems with small initial ionizations [27] but took into account only the radial coupling.

6 Conclusions

As long as it is not possible to perform experiments on collision systems containing only one electron, many-electron effects are important and must be taken into account when attempting to compare with an experiment. They modify the calculations in two ways: Firstly in the influence on the correlation diagrams and secondly in the theoretical interpretation of the measured cross sections via the Inclusive Probability analysis.

Relativistic effects become strong. They produce not only large spin-orbit splittings and changes of the wavefunctions and their radial and angular behaviour but also a very different radial and rotational coupling behaviour between the relativistic levels. Especially this last effect produces a radical change in the impact-parameter dependent transfer probability for K-shell excitation. This facilitates understanding of ion-solid target collisions as discussed above. Thus, the so called "solid-target effect" (at least for inner shell transfer) is very questionable. A consequence of this improved understanding is the undertaking of experiments which, hopefully, will unveil the secret of the innermost structure of the levels at superheavy collision systems.

Improved measurements and analysis are expected to provide information on QED corrections, level structure or possible other still missing influences.

Acknowledgements

J.A. and K.S. thank the Gesellschaft für Schwerionenforschung (GSI) for financial support. P.K. thanks the Deutsche Forschungsgemeinschaft (DFG) for a postdoctoral grant and partial support by the Division of Chemical Sciences, Office of Basic Energy Sciences, Office of Energy Research, U.S. Department of Energy.

References

1. P. O. Löwdin, Adv. Phys. **5** (1956) 1
2. P. Kürpick, W.-D. Sepp, B. Fricke, Phys. Rev. **A51** (1995) 3693
3. J.F. Reading, Phys. Rev. **A8** (1973) 3262
4. J. Reinhardt, B. Müller, W. Greiner, B. Soff, Phys. Rev. Lett. **43** (1979) 1307
5. P. Kürpick, H.-J. Lüdde, Comp. Phys. Comm. **75** (1993) 127
6. M. Schulz, E. Justiniano, J. Konrad, R. Schuch, A. Salin, J. Phys. **B20** (1987) 2057
7. P. Kürpick, W.-D. Sepp, H.-J. Lüdde, B. Fricke, Nucl. Instr. Meth. **B94** (1994) 183
8. P. Kürpick, B. Thies, W.-D. Sepp, B. Fricke, J. Phys. **B24** (1991) L139
9. K. Schulze, J. Anton, W.-D. Sepp, B. Fricke, P. Kürpick, submitted to J. Phys. B (1996)
10. R. Schuch et al., Phys. Rev. **A37** (1988) 3313
11. C.H. Annett, B. Curnutte, C.L. Cocke, Phys. Rev. **A19** (1979) 1038
12. P. Kürpick, B. Fricke, C.L. Cocke, Phys. Rev. **A53** (1996) R1943
13. K. Taulbjerg, J.S. Briggs, J. Phys. **B8** (1975) 1895 and K. Taulbjerg, J.S. Briggs, J. Vaaben, J. Phys. **B9** (1975) 1351
14. R. Schuch, G. Nolte, H. Schmidt-Böcking, Phys. Rev. **A22** (1980) 1447
15. P. Kürpick, T. Bastug, W.-D. Sepp, B. Fricke, Phys. Rev. **A52** (1995) 2132
16. P. Kürpick et al., Phys. Lett. **A207** (1995) 199
17. P. Kürpick, B. Fricke, W.-D. Sepp, J. Phys. **B28** (1995) L127
18. a review is given in: A. Sommerfeld, Naturwiss. **28**, 417 (1940)
19. W. Piper, W. Greiner, Z. Phys. **218**, 327 (1969)
20. B. Fricke, W. Greiner, J.T. Waber, Theor. chim. Acta **21**, 235 (1971)
21. G. Soff, J. Reinhardt, B. Müller, W. Greiner, Z. Phys. **A294**, 137 (1980)
22. the whole field is discussed in: W. Greiner, B. Müller, J. Rafelski, Quantum Electrodynamics of Strong Fields, Springer Verlag, Berlin (1985)
23. J. Schweppe et al., Phys. Rev. Lett. **51**, 2261 (1983)
24. J. Baumann et al., GSI Scientific Report 1996, p. 106, GSI96-1
25. W.-D. Sepp, B. Fricke, T. Morovic, Phys. Lett. **A81** (1981) 258
26. J. Kirsch, W. Betz, J. Reinhardt, B. Müller, W. Greiner, Z. Phys. **A292** (1979) 227 and J. Kirsch, B. Müller, W. Greiner, Phys. Lett. **A94**, 151 (1983)
27. W.E. Meyerhof, T.K. Saylor, S.M. Lazarus, A. Little, B.B. Triplett, L.F. Chase, R. Anholt, Phys. Lett. **32** (1974) 1279

Atomic Radiative Lifetimes Measured by Pulsed Laser Spectroscopy in the UV/VUV Spectral Region

U. Berzinsh
Institute of Atomic Physics and Spectroscopy, University of Latvia,
19 Rainis Blvd, LV-1586 Riga, Latvia
and
S. Svanberg
Department of Physics, Lund Institute of Technology
and
Lund Laser Center, Lund University
P.O. Box 118, S-221 00 Lund, Sweden

Abstract

Experimental lifetime measurements on free atoms and ions performed by using pulsed laser excitation in the ultraviolet and vacuum-ultraviolet spectral regions are presented. Different methods of laser pulse generation, atomization and light detection are described. A short summary of results obtained is also given. Future possibilities in determination of lifetimes and transition probabilities in the short wavelengths region are discussed.

ADVANCES IN QUANTUM CHEMISTRY, VOLUME 30

0065-3276/98 $25.00

1. Introduction

The structure of atoms and their interaction with the surrounding are reflected in their emission or absorption spectra. Intensities of spectral lines give information not only on atomic properties but also on their concentration in the observed object. The line strengths, oscillator strengths and transition probabilities are radiative constants of atoms related to transitions between different energy states and they govern the intensities of spectral lines. These quantities can be interrelated through simple expressions by using atomic constants describing multiplicities and energy separations. The radiative constants are discussed in many textbooks on atomic physics, for example [1-3], and reviews on the experimental methods of their determination have been published; see e.g. [4-7].

Transition probabilities in fact provide a particularly severe test of atomic calculations, because they are rather sensitive to the wavefunctions of both levels involved in the transition and to the approximations used, especially when electron correlations and relativistic effects are considered. On a more fundamental level, measurements of transition probabilities are also being used to explore the non-conservation of parity predicted by the unified theories of the weak and electromagnetic interaction; here, highly forbidden transitions must be tested for an admixture of allowed transition probabilities [8,9].

Besides the basic interest in atomic quantities, the values of transition probabilities are used in different fields of research and technology: astrophysics, atmospheric physics, plasma physics, thermonuclear-fusion research and development of laser systems. The evaluation of atomic concentrations can frequently be performed from observed light intensities, when direct concentration measurements are not possible. A large data compilation has been performed at NIST [10-13]. Presently, a great part of the data collection and systematization is performed for astrophysical needs within the international Opacity Project and the Iron Project [14,15]. Since 1990, when the Goddard High-Resolution Spectrograph on the Hubble Space Telescope started to work, the demand for transition probabilities for VUV spectral lines dramatically increased [16,17].

Transition probabilities have mainly been determined from calculations and to a much smaller extend from experiments [18]. Accurate experimental data are needed for checking of theoretical models and methods. Furthermore, in many cases, especially for complex heavy atoms, the theoretical models are under development and calculations with sufficient accuracy cannot be performed yet. For such atoms, experimental data are of major importance in practice. Presently, one of the most accurate methods to determine transition probabilities is the use of radiative lifetimes in combination with branching ratios.

Since the advent of lasers, they have been widely developed and explored in studies of atoms [19-20], including lifetime studies. Even high precision methods for lifetime measurements were demonstrated [21-23]. New applications of lasers were developed in different fields. In particular, the laser spectroscopy methods have been applied for studies of atoms in the UV/VUV (100-300 nm) spectral region (See, e.g. [24,25]). Nonlinear methods [26] have been used to transform tunable visible laser radiation into UV/VUV radiation. Experimental arrangements for time-resolved laser spectroscopy in the VUV spectral region have been developed at the Lund Laser Center [27]. There a large number of lifetime measurements have been performed using short-wavelength laser excitation [28].

2. Experimental methods

Pulsed laser excitation of evaporated substances and subsequent fluorescence decay detection have been used in the studies reviewed here. In our opinion this is the most convenient method for lifetime measurements in the UV/VUV spectral range. The principal scheme of an experimental set-up is shown in Fig.1. It has been slightly modified in various measurements, but the main parts are the same and correspond to the three tasks: atomization, excitation and detection. (The excitation laser pulse has to be of short duration and in the UV/VUV region.) Each of these tasks are discussed more in detail below.

2.1 Excitation

Until recently, the excitation of atoms in collisions with fast beams of electrons or in collisions of a fast beam of ions with a thin foil, was the only efficient way to populate highly excited states corresponding to UV/VUV transitions. Such methods have been employed in lifetime measurements of many atoms and highly charged ions [30-33]. But the non-selective excitation and possible influences of cascade decays are serious disadvantages of such methods. The excitation by laser light is selective. In quasi cw measurements using, e.g., the phase shift method, the measurement result becomes dependent on the laser power [34]. The pulsed laser excitation does not have such disadvantages.

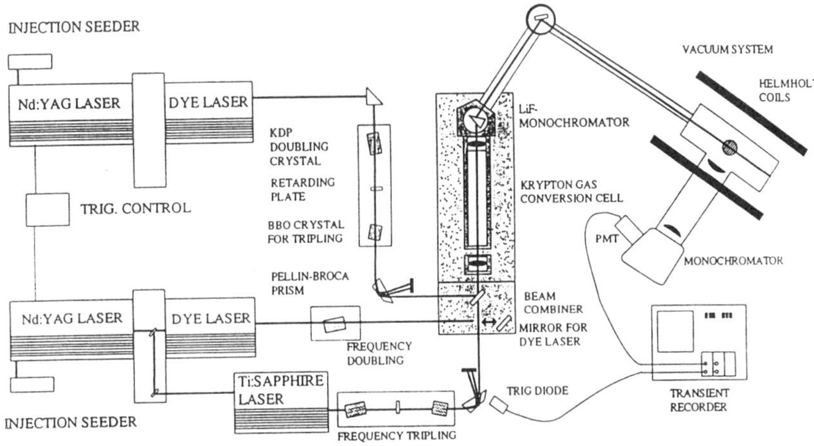

Figure 1. Experimental set-up for time-resolved laser spectroscopy (From Ref. [29]).

2.1.1 Short pulse generation

Normal pulsed lasers, chopped cw laser radiation, or mode-locked dye lasers can be used for the generation of the light pulses. The output powers of the cw and mode locked lasers are rather low and they can be applied only in the visible or, after frequency doubling, in the UV spectral region [35]. They can be used for lifetime measurements with high accuracy [36] Only radiation from pulsed lasers can be transformed down to the VUV spectral region by using non-linear effects.

Excimer or Nd:YAG-pumped dye lasers were used in many studies. The typical pulse duration is about 10 ns, sufficiently short for a large number of measurements (Fig. 2.).

In cases where shorter lifetimes were investigated, the experimental set-up has been modified. The simplest solution is to use a short-pulse nitrogen laser for pumping of the oscillator, while the amplifiers are pumped in the usual way; see Fig.3. The main problem of this scheme is the trigger jitter in the pump lasers, which leads to a non-perfect time overlap of oscillator and amplifier duty cycles, and therefore the outgoing radiation exhibits large fluctuations in intensity. Another way is to apply frequency mixing of the pump laser radiation with the dye laser radiation. The resulting output pulse from a nonlinear crystal is depending on the time overlap of the two input pulses, but there is no jitter. By delaying one pulse with respect to the second one, the pulse lengths can be shortened.

An alternative way is to use a ps laser pumping a distributed feedback dye laser. Such a system with two amplifier stages is shown in Fig. 4. By using a quenched distributed feedback system short pulses can be obtained even with a ns pump laser [39].

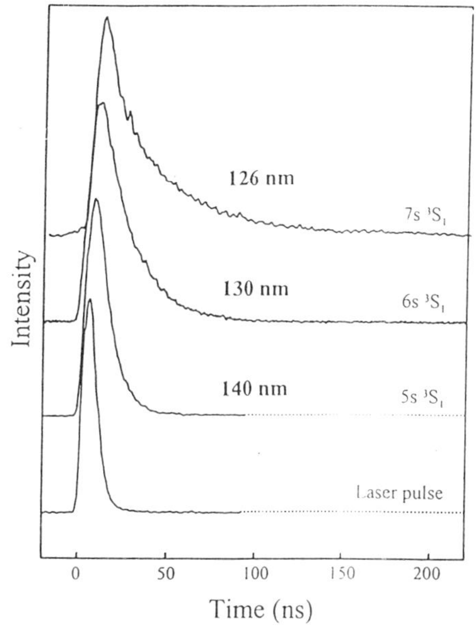

Figure 2. Recordings of fluorescence light transients for sulphur excited states and recording of scattered light from the laser pulse (From Ref. [37]).

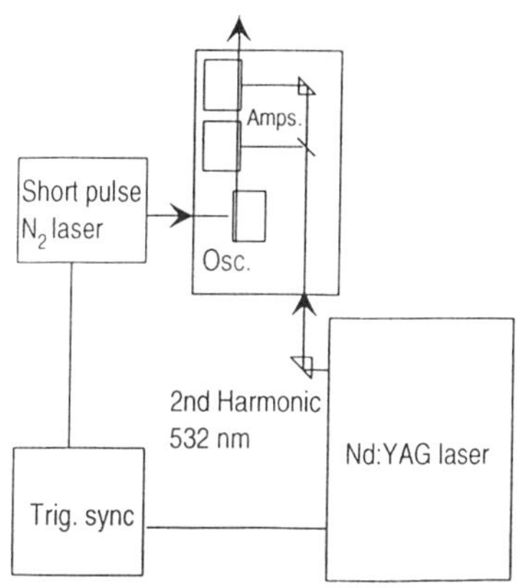

Figure 3. Experimental set-up for short-pulse generation using a nitrogen pump laser (From Ref. [38]).

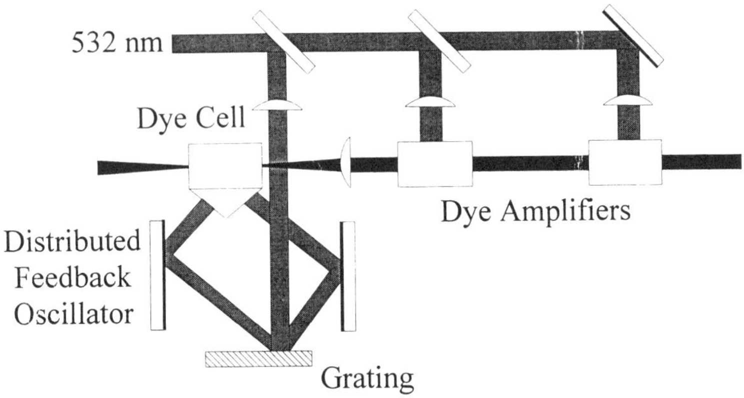

Figure 4. Experimental set-up for short- pulse generation by a distributed feedback dye laser (From Ref. [25]).

2.1.2. Generation of tunable UV/VUV laser radiation.

In early laser spectroscopy experiments the ions were shifted into resonance with lower energy photons through Doppler tuning, by changing the velocity of ions in a fast beam [23,40]. Presently, tunable laser radiation in the UV/VUV spectral region can be obtained from visible radiation by using the following nonlinear phenomena: harmonic generation and frequency mixing in crystals, stimulated anti-Stokes Raman scattering in gases and four-wave mixing in gases. Such generation down to the LiF absorption cut-off is illustrated in Fig. 5. Even shorter wavelengths can be reached by high-order harmonic generation in gases (See, e. g. [24,25]).

Nonlinear crystals are normally used for frequency doubling and mixing down to 200 nm, where the oxygen bands in air start to absorb. The β-barium borate crystal has a 189 nm cut-off, which is determined by the absorption in the crystal itself. Practically it is used down to about 200 nm, where the radiation is absorbed by oxygen in air. For shorter wavelengths the oxygen has to be removed from the path of the laser radiation.

Coherent radiation can be shifted towards shorter wavelengths applying stimulated anti-Stokes Raman scattering in gases. Particularly, the hydrogen molecule has the largest vibrational splitting corresponding to 0.5 eV. By using the second anti-Stokes component it is possible to shift 200 nm radiation down to 170 nm.

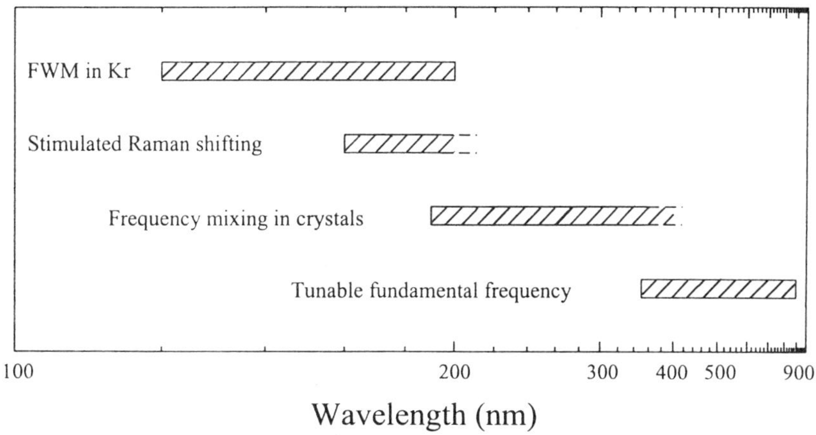

Figure 5. Different methods used for efficient generation of short-wavelength radiation down to the LiF cut-off (From Ref. [41]).

Shorter wavelengths can be reached through sum-difference four-wave mixing in gases. Two lasers are used, one of them in resonance with a two-photon transition in an atom, and the second one tunable. The resulting radiation has a photon energy defined by the difference between the two-photon energy and the tunable photon energy. In principle, using the resonant two-photon transition in krypton at 212.55 nm, any wavelengths longer than 106 nm could be obtained. The shortest obtainable wavelength is defined by the longest wavelength produced by the tunable laser, and it is about 120 nm if a Ti:sapphire or a dye laser is used.

An alternative to previously mentioned methods is high-order harmonic generation in gases. The harmonics are produced by focusing intense laser radiation into a jet of rare gases. Presently, the shortest wavelength obtained is in the order of 7 nm [42,43]. By using tunable dye or Ti:sapphire lasers tunable radiation at XUV and VUV wavelengths can be generated. An alternative way to obtain continuously tunable radiation is to mix high-power fixed frequency radiation with lower-power tunable radiation [44,45]. However, intense and short-pulse lasers have to be used for the generation of the really high harmonics. In spectroscopic applications the line-widths achieved with short-pulse lasers are frequently too large (Heisenberg uncertainty). Then laser systems with longer pulse duration are more useful for spectroscopy.

2.2 Detection

The fluorescence decay following pulsed laser excitation is normally recorded by a transient digitizer. This is a more efficient way to record decay curves in comparison with box-car integration or delayed-coincidence techniques. The fluorescence light is collected perpendicularly to the plane of the laser and atomic beams by quartz or lithium fluoride lenses and is selected by a monochromator or a filter before reaching a photomultiplier tube. Transients are stored in the memory of a fast digital oscilloscope. The time resolution is defined by the rise time (3 ns) of the photomultiplier tube. For faster measurements, a micro-channel plate has been applied. The accuracy of this detection scheme is limited by the linearity of the detectors, which have to measure both high and low intensity signals separated in time by short intervals.

When a very high sensitivity is needed the pump/probe technique using photo-ionization was employed [46]. The atoms were excited by the pump beam and then a delayed probe beam was applied to ionize them from the excited state. The ion signal versus delay time is detected. Advantages of this technique are: a high sensitivity and a time resolution limited only by the duration of the pump and probe pulses, but not by the time resolution of the detector.

2.3. Atomization

Different methods have been used for producing free atoms. For lifetime measurements atoms have to be studied in conditions, where the interaction with the environment is very low. Usually, that is inside a vacuum system. The easiest way of producing free atoms in the vacuum system is by leaking in a gas consisting of the atoms needed [47,48]. But only a few elements are available in the gas phase at room temperature, therefore, the thermal evaporation from an oven inside the vacuum system is often used. Many materials evaporate as molecules and then the dissociation of molecules is a special task. In some cases, the laser light used for excitation serves for dissociation, too [47-49]. When a small amount of substance is available or if it is toxic, sealed-off cells were used in the measurements [50,51]. However, they have several disadvantages: the presence of rest gas in the cells, a large amount of stray light from laser light scattering in the cell windows and the wavelength range of the applied radiation is limited by absorption in the cell windows. When they are made from quartz the limit is around 160 nm.

If the material is difficult to vaporize by thermal heating, bombardment with electrons and extraction from the discharge can be applied in an effusive hollow-cathode arrangement [52,53]. An alternative way is to use focused laser light to produce a plasma containing atoms and ions of the species under investigation. A conventional target consists of a disc, with a surface of the material to be studied [54-56]. The disc is rotated with the purpose to bring a new flat surface in the focus after each laser shot. For some materials the fabrication of such a disk is a rather complicated task and then a powder in a container can be used (Fig. 6.) [57-59]. The container is shaken at a high frequency and the surface of the powder is brought back in the light focus before the next laser pulse. A few μs after the laser hits the target no emission from the plasma itself is observed, and the excitation laser pulse can be applied (Fig. 7.).

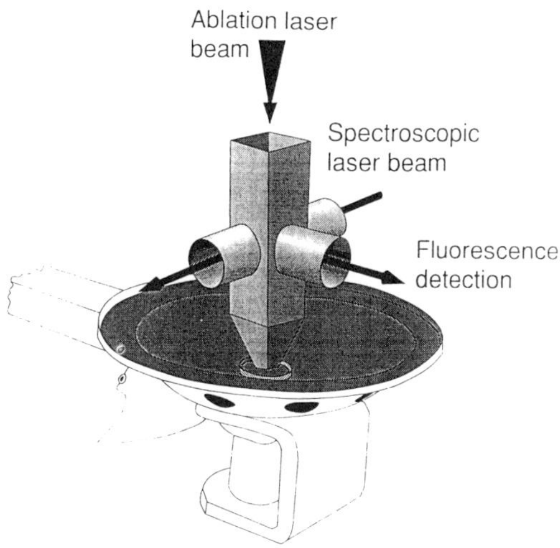

Figure 6. Arrangement for laser-produced plasma generation from a powder (From Ref. [57]).

A sufficient number of atoms has to be produced to be able to detect a fluorescence signal. On the other hand, when the density of atoms is too large, the collisional quenching interferes with the radiative lifetime. Normally, a Stern-Vollmer plot is drawn and the lifetime extrapolated to zero concentration is taken as the radiative lifetime (Fig. 8.). Extrapolation is simple when using a cell containing the atoms.

When a thermal beam or a laser-produced plasma is applied, it is much more difficult to evaluate concentrations. Then the lifetimes are measured at different experimental conditions in order to observe possible influences of collisions on the measured lifetimes; see Fig. 9

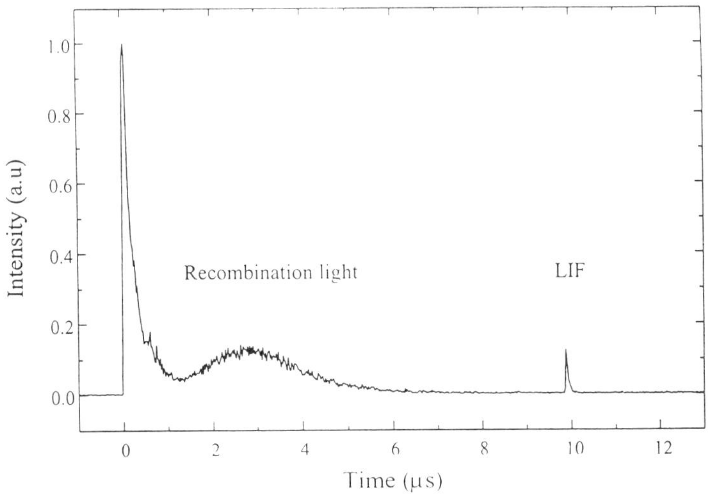

Figure 7. Detected light intensity at 126 nm from a sulphur plasma as a function of the time after the ablation pulse. The signal at about 10 μs is laser-induced fluorescence from the excited state. The signal before 6 μs is due to recombination light from atoms and ions. (From Ref. [37]).

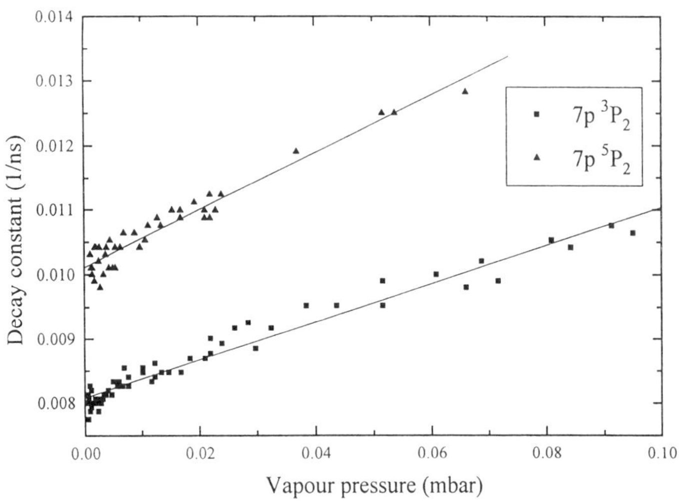

Figure 8. Stern-Vollmer plot for tellurium 7p $^5P_2$,$^3P_2$ state lifetime determination (From Ref. [49]).

a)

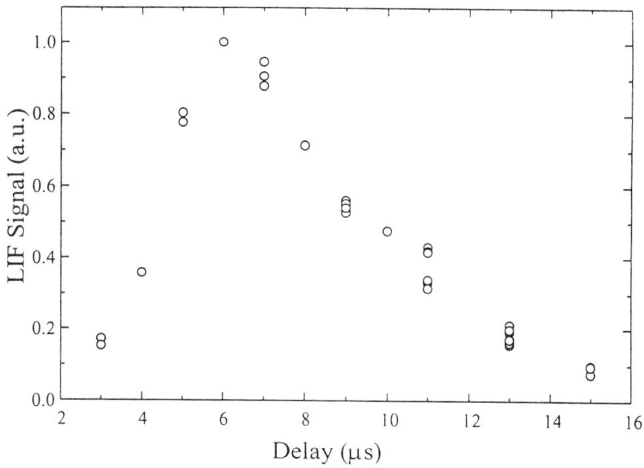

b)

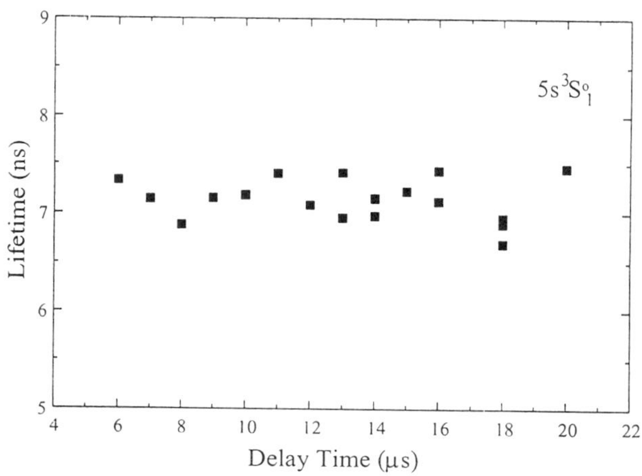

Figure 9. The time dependence of the fluorescence light intensity and decay time for a laser-produced sulphur plasma: a) Laser-induced fluorescence intensity from the 7s $^3S^0_1$ state as a function of the delay time for the VUV laser pulse with respect to the ablation pulse. b) Measured radiative lifetime of the 5s $^3S^0_1$ state as a function of the delay. (From Ref. [37]).

2.4. Sources of possible errors

We already mentioned two possible sources of errors: a non-linearity of the detector, and the collisional quenching of excited states. A non-linearity of the detector is intensity dependent; by recording fluorescence signals at different intensities the influence of possible saturation of the detector can be observed. Radiation trapping is one of the processes changing the number of atoms in excited states. It is a density dependent process and may be taken into account by carrying out measurements at different densities. Another problem is possible blends from other transition lines. This has to be checked very carefully if interference filters are used and molecular lines are overlapping with detection lines. The best way to solve this problem is to use different decay channels for the detection. Influences from quantum beats have been reduced by applying an additional magnetic field over the interaction region. For the higher Rydberg states a possible influence of blackbody radiation has to be accounted for [60].

3. Results

High lying excited states are accessible using UV/VUV radiation. Various types of levels can be excited: Rydberg states for large groups of atoms, resonance states for another group of atoms and ions in the first ionization stage. For many of the states the first lifetimes where measured by the LTH group. These results have served as test data for theoretical calculations. In some cases a comparison between experimental results obtained with different methods was possible and thus serving as a test of the methods. For different atoms and ions the motivation for lifetime measurements were slightly different.

In a large group of papers lifetime measurements of irregularities in long Rydberg sequences were performed for elements Group 2a (Mg I, Sr I) [61,62], Group 3a (Al I, Ga I, In I) [63-65], Group 1b (Cu I, Ag I, Au I,) [66-71] and Yb I [72], Zn I [73] and Bi I[74]. For the Group 2a and 3a elements non-perturbed S sequences and perturbed D sequences were observed. The P sequence of group 1b atoms was found to be strongly perturbed, while S and D sequences were following a cubic rule. Lifetimes of P sequences in ytterbium also show strong influence of a perturbation, which was confirmed by g_j factor measurements. This initiated further Multi-channel Quantum Defect Theory analysis. In zinc, a deviation was observed between measured and calculated data. In bismuth rather good agreement with a cubic rule was observed, with the exception for one state, for which the hyperfine structure was also found to be perturbed.

Measurements of lifetimes in Gd II, Ru II, Pt II, Pd II, As I, P I and S I [53,57,75,76,50,59,58,37] were motivated by a strong interest from astrophysicists. The aim was transition probability values. The measurements were performed on astrophysically interesting states, and data obtained were used for the evaluation of the abundance of elements in astrophysical objects. Experimental and theoretical branching ratio data were used for evaluation of transition probabilities. The possible astrophysical applications as well as the existence of experimental branching ratio data, suggested experiments on selenium and tellurium. For bismuth a revision of transition probability data was performed for more than 20 lines [74].

In some experiments new experimental schemes were applied. In the lifetime measurements of O I, a two-photon excitation in a first step also inducing molecular dissociation, followed by a second excitation step was demonstrated [47]. That allowed to excite atoms to levels otherwise corresponding to direct groundstate transitions with wavelengths <100 nm. In later experiments on N I even higher levels were reached [48]. In an experiment on C I two-photon excitation and lifetime measurements were achieved with detection through cascades via a short-lived state [55], a technique also successfully later applied on S I and P I [58,59]. The application of four-wave mixing was demonstrated in experiments on Mg I [29] and later used for studies on Au I, Zn I and S I [71,73,37]. The application of a new picosecond laser system based on a mode-locked Nd:YAG and a distributed feedback dye laser was demonstrated in paper [71].

4. Future perspectives.

Further progress in experimental techniques will allow lifetime measurements in neutral and singly ionized atoms to be extended to even shorter wavelengths and lifetimes. Any free neutral and singly ionized atom can be produced in a laser-generated plasma. For short-lived states, the pump-probe technique allows to overcome the problems with detector response time. A distributed feedback laser pumped by picosecond pulses from a mode-locked Nd:YAG laser can serve as a light source.

In astrophysics transition probabilities are needed and branching ratios have to be used in conjunction with lifetime data. In the VUV region accurate branching ratio measurements are complicated due to difficulties in calibration of the spectral sensitivity [77]. One way to overcome this problem is to perform transition probability measurements in absorption for lines with a common lower level. This approach, called the Ladenburg method, was already demonstrated in the UV part of the spectrum [78,79]. In view of the strong ongoing effort in space-based short wavelength astronomy there is a strong motivation for future studies of transition probabilities in the UV/VUV part of the spectrum.

Acknowledgments

This work was supported by the Swedish National Science Research Council. One of us (U.B.) would like to thank the Swedish Academy of Sciences for stipends supporting his stays in Sweden.

This paper is dedicated to Professor Ingvar Lindgren on the occasion of his 65th birthday.

References

1. E.U. Condon, G.H. Shortley: *The Theory of Atomic Spectra* (Cambridge Univ. Press, Cambridge 1964).
2. I.I. Sobelman: *Atomic Spectra and Radiative Transitions*, Springer Ser. Chem. Phys., Vol.1 (Springer, Berlin, Heidelberg 1979).
3. R.D. Cowan: *The Theory of Atomic Physics and Spectra* (Univ. California Press, Berkeley 1981).
4. W.L. Wiese, *in Progress in Atomic Spectroscopy, Pt.B*, ed. by W. Hanle, H. Kleinpoppen (Plenum, New York 1979) p.1101.
5. N.P. Penkin, *in Atomic Physics 6*, ed. by R. Damburg (Plenum, New York 1979) p.33.
6. M.C.E. Huber, R.J. Sandeman, Rep. Prog. Phys. **49**, 397 (1986).
7. A. Hibbert, *in Progress in Atomic Spectroscopy, Pt.B*, ed. by W. Hanle, H. Kleinpoppen (Plenum, New York 1979) pp 1.
8. M.A. Bouchiat, L. Portier, *in Laser Spectroscopy III*, ed. by J.L. Hall and J.L. Carlsten (Berlin, Springer 1977) pp 9.
9. P.G.H. Sandars, *in Laser Spectroscopy III*, ed. by J.L. Hall and J.L. Carlsten (Berlin, Springer 1977) pp 21.
10. W.L. Wiese, M.W. Smith, B.M. Glennon: *Atomic Transition Probabilities - Hydrogen through Neon* (U.S. Government Printing Office, Washington D.C. 1966)
11. W.L. Wiese, M.W. Smith, B.M. Miles: *Atomic Transition Probabilities - Sodium through Calcium* (U.S. Government Printing Office, Washington D.C. 1969)
12. G.A. Martin, J.R. Fuhr, W.L. Wiese, J. Phys. Chem. Ref. Data **17**, Supplement 3 (1988).
13. J.R. Fuhr, G.A. Martin, W.L. Wiese, J. Phys. Chem. Ref. Data **17**, Supplement 4 (1988).
14. C. Mendoza, Phys. Scr. **T65**, 198 (1996).
15. K. Butler, Phys. Scr. **T65**, 63, (1996).

16. D.S. Leckrone, S. Johansson, G.M. Wahlgren, C.R. Proffitt, T. Brage, Phys. Scr. **T65**, 110 (1996).
17. D.S. Leckrone, S. Johansson, G.M. Wahlgren, S.J. Adelman, Phys. Scr. **T47**, 149 (1993).
18. W.L. Wiese, Phys.Scr. **T65**, 188, (1996).
19. W. Demtröder: *Laser Spectroscopy*, 2nd edition (Springer, Berlin, Heidelberg 1996).
20. S. Svanberg: *Atomic and Molecular Spectroscopy*, 2nd edition (Springer, Berlin, Heidelberg 1992).
21. W. Demtröder, M. McClintock, R.N. Zare, J. Chem. Phys. **51**, 5495 (1969).
22. K. Sakurai, G. Capelle, H.P. Broida, J. Chem. Phys. **54**, 1220 (1971).
23. H.J. Andrä, A. Gaupp, W. Wittmann, Phys. Rev. Lett. **31**, 501 (1973).
24. R.G. Beausoleil, D.H. McIntyre, C.J. Foot, E.A. Hildum, B.Couillaud, T.W.Hänsch, Phys.Rev. **A35** 4878 (1987); ibid **A39** 4591, (1989); K.S.E. Eikema, W. Ubachs, W. Vassen, W. Hogervorst, Phys. Rev. Lett. **71**, 1690 (1993); ibid **76** 1216 (1996); J. Larsson, S. Svanberg, Appl. Phys. **B59**, 433 (1994).
25. S. Svanberg, A. L´Huillier, C.-G. Wahlström, Nucl. Instr. Meth. (to appear).
26. V.S. Letokhov, V.P. Chebotaev: *Nonlinear Laser Spectroscopy* (Springer, Berlin, Heidelberg 1977).
27. S. Svanberg, J. Larsson, A. Persson, C.-G. Wahlström, Phys. Scr. **49**, 187 (1994).
28. U. Berzinsh, M.B. Gaarde, J. Larsson, A. L'Huillier, H. Lundberg, L. Caiyan, E. Mevel, S. Svanberg, C.-G. Wahlström, R. Zerne, *in Laser Spectroscopy 12*: Ed. by M. Inguscio, M. Allegrini, A. Sasso (World Scientific, Singapore, New Jersey, London, Hong Kong1995) p. 394.
29. J. Larsson, R. Zerne, A. Persson, C.-G. Wahlström, S. Svanberg, Z. Phys. **D 27**, 329 (1993).
30. D.J. Pegg, Methods of Experimental Physics **17**, 529 (1980).
31. I. Martinson, Rep. Prog. Phys. **52**, 157 (1989).
32. E. Träbert, Phys. Scr. **48**, 699 (1993).
33. P. Erman, Nucl. Instr. Methods **110**, 471 (1973).
34. L. Armstrong Jr., S. Feneuille, J. Phys. **B8**, 546 (1975).
35. J. Carlsson, Z. Phys. **D9**, 147 (1988).
36. U. Volz, M. Majerus, H. Liebel, A. Schmitt, H. Schmoranzer, Phys. Rev. Lett. **76**, (1996).
37. U. Berzinsh, Luo Caiyan, R. Zerne, S. Svanberg, E. Biemont, Phys. Rev. A, to appear.
38. G.J. Bengtsson, U. Berzinsh, J. Larsson, S. Svanberg, Z. Physik **D23**, 29 (1992).
39. W. Schade, B. Garbe, V. Helbig, Appl. Opt. **29**, 3950 (1990).

40. H.C. Bryant, B.D. Dieterle, J. Donahue, H. Sharifian, H. Tootoonchi, D.M. Wolfe, P.A.M. Gram, M.A. Yates-Williams, Phys. Rev. Lett. **38**, 228 (1977).
41. R. Zerne, Ph.D. Thesis, Lund Reports on Atomic Physics **195** (1996).
42. J.J. Macklin, J.M. Kmetec, C.L. Gordon, Phys. Rev. Lett. **70**, 766 (1993).
43. A. L'Huillier, Ph.Balcou, Phys. Rev. Lett. **70**, 744 (1993).
44. M.B. Gaarde, P. Antoine, A. Persson, B. Carré, A. L'Huillier, C.-G. Wahlström, J. Phys. **B29**, L163 (1996).
45. H. Eichmann, S. Meyer, K. Riepl, C. Momma, B. Wellegehausen, Phys. Rev. **A50** R2834 (1994)
46. J. Larsson, E. Mevel, R. Zerne, A. L'Huillier, C.-G. Wahlström, and S. Svanberg, J. Phys. B**28**, L53 (1995).
47. S. Kröll, H. Lundberg, A. Persson, S. Svanberg, Phys. Rev. Lett. **55**, 284 (1985).
48. G.J. Bengtsson, K. Hansen, J. Larsson, S. Svanberg, D.D. Wang, Phys Rev. **A45**, 2712 (1992).
49. R. Zerne, U. Berzinsh, S. Svanberg, Z. Physik **D32**, 31 (1994).
50. G.J. Bengtsson, U. Berzinsh, J. Larsson, S. Svanberg, Astron. Astrophys. **263**, 440 (1992).
51. G.J. Bengtsson, U. Berzinsh, J. Larsson, S. Svanberg, R. Zerne, J. de Physique **2**, 773 (1992).
52. H. Bergström, H. Lundberg, A. Persson, W. Schade, Y.Y. Zhao, Phys. Scr. **33**, 513 (1986).
53. H. Bergström, E. Biémont, H. Lundberg, A. Persson, Astron. Astrophys. **192**, 335 (1988)
54. H. Bergström, G.W. Faris, H. Hallstadius, H. Lundberg, A. Persson, C.-G. Wahlström, Z. Phys. **D8**, 17 (1988).
55. H. Bergström, H. Hallstadius, H. Lundberg, A. Persson, Chem. Phys. Lett. **155**, 17 (1989).
56. H. Bergström, G.W. Faris, H. Hallstadius, H. Lundberg, A. Persson, Z. Phys. **D13**, 27 (1988).
57. Se. Johansson, A. Joueizadeh, U. Litzén, J. Larsson, A. Persson, C.-G. Wahlström, S. Svanberg, D.S. Leckrone, G.M. Wahlgren, Astrophys. J. **421**, 809 (1994).
58. R. Zerne, Luo Caiyan, U. Berzinsh, S. Svanberg, Phys. Scr., in press.
59. U. Berzinsh, S. Svanberg, E. Biémont, Astron. Astrophys., in press.
60. T.F. Gallagher and W.E. Cooke, Phys. Rev. Lett. **42**, 835, (1979).
61. G. Jönsson, S. Kröll, A. Persson, S. Svanberg, Phys. Rev. **30**, 2429 (1984).
62. G. Jönsson, C. Levinson, A. Persson, C.-G. Wahlström, Z. Phys. **316A**, 255 (1984).
63. G. Jönsson, S. Kröll, H. Lundberg, S. Svanberg, Z. Physik **A316**, 259 (1984).

64. J. Carlsson, H. Lundberg, W.X. Peng, A. Persson, C.-G. Wahlström, T. Brage, C. Froese Fischer, Z. Phys. **D3**, 345 (1986).
65. G. Jönsson, H. Lundberg, S. Svanberg, Phys. Rev. **A27**, 2930 (1983).
66. J. Carlsson, A. Dönszelmann, H. Lundberg, A. Persson, L. Sturesson, S. Svanberg, Z. Physik **D6**, 125 (1987).
67. R. Zerne, J. Larsson, S. Svanberg, Phys. Rev. **A49**, 128 (1994).
68. J. Bengtsson, J. Larsson, S. Svanberg, Phys. Rev. **A42**, 5457 (1990).
69. Z. Jiang, P. Jönsson, J. Larsson, S. Svanberg, Z. Physik **D17**, 1 (1990).
70. G.J. Bengtsson, P. Jönsson, J. Larsson, S. Svanberg, Z. Physik **D22**, 437 (1991).
71. M.B. Gaarde, R. Zerne, C. Luo, Z. Jiang, J. Larsson, S. Svanberg, Phys. Rev. **A50**, 209 (1994).
72. Z. Jiang, J. Larsson, Z. Physik. **D22**, 1 (1990).
73. R. Zerne, C. Luo, Z. Jiang, J. Larsson, S. Svanberg, Z. Physik. **D32**, 187 (1994).
74. C. Luo, U. Berzinsh, R. Zerne,, S. Svanberg, Phys. Rev. **A52**, 1936 (1995).
75. J. Larsson, R. Zerne, H. Lundberg, Accepted by J. Phys. B.
76. H. Lundberg, Se. Johansson, J. Larsson, D.S. Leckrone, U. Litzén, S. Svanberg, G.M. Wahlgren, R. Zerne, Astrophys. J. **469**:388-392 (1996)
77. J.Z. Kloze, J.M. Bridges, W.R. Ott, NBS Special Publication 250-3, (U.S. Government Printing Office, Washington D.C. 1987).
78. B.L. Cardon, P.L. Smith, W. Whaling, Phys. Rev. **A20**, 2411 (1979).
79. A. Ubelis, U. Berzinsh, Phys. Scr. **43**, 162 (1991).

A theoretical and experimental study of the quartet system of B I*

Yaming Zou[1], *Tomas Brage and Indrek Martinson*

Department of Physics, University of Lund, S-223 62 Lund, Sweden

Abstract

We report on an investigation of the quartet system for neutral boron, B I. Published and unpublished experimental data, largely from beam-foil investigations and absorption spectroscopy are confronted with new theoretical calculations. Good agreement is found between experiment and theory. Some remaining problems are identified and additional studies are suggested. Besides the fundamental atomic-physics relevance, e.g. for understanding the electronic structure for systems with several open subshells, boron now plays a very important role in nuclear astrophysics and thermonuclear fusion devices.

1. Introduction

For atomic spectroscopists using a plasma light source such as the hollow cathode, boron has been a very difficult element to excite, a fact that has severely limited experimental progress. Thus, in 1970 Edlén *et al.* [1] remarked "Our knowledge of the first spectrum of boron, B I, has remained conspiciously incomplete in comparison with that of the other light elements". However, using absorption spectroscopy, these authors were able to locate the $2s^2nf$ series (with n = 4 - 11) of B I which resulted in the first accurate determination of the ionization energy of boron. Subsequently, Brown *et al.* [2], also by means of absorption spectroscopy, observed the $2s^2ns\ ^2S$ and $2s^2nd\ ^2D$ Rydberg series up to n = 10 and n = 39, respectively. Because of the perturbers $2s2p^2\ ^2S$ and ^2D, which are embedded in them, these series show a most interesting structure. Recently, the relevant interactions were thoroughly studied by Carlsson *et al.* [3] who used the multiconfiguration Hartree-Fock (MCHF) approach.

*Dedicated to Professor Ingvar Lindgren

[1]Permanent address: Applied Physics Department, Jiao Tong University, Shanghai, The Peoples Republic of China

ADVANCES IN QUANTUM CHEMISTRY, VOLUME 30

0065-3276/98 $25.00

In B I there is also a quartet system which for many years eluded experimental studies. In Moore's Tables [4] only two quartet terms are given, $2s2p^2\ ^4P$ and $2p^3\ ^4S$. A few years later Burke and Mack [5] found the $2s2p3s\ ^4P$ term, from 6 lines between 2066.40 Å and 2067.22, nicely bracketed by the intense B III resonance doublet at 2065.78 and 2067.24 Å. This called for a comment about "the remarkable coincidence between unrelated B I and B III lines in this proverbially line-poor spectrum" [5]. The lowest quartet term $2s2p^2\ ^4P$ is metastable and its three fine-structure levels decay to the ground term $2s^22p\ ^2P$ with spin-forbidden transitions. The latter have very low rates in neutral atoms and none of the 5 possible intersystem lines has been observed in the case of B I. This is in contrast to the isoelectronic O IV spectrum where similar lines were observed in solar spectra [6]. However, by means of isoelectronic extrapolation Edlén *et al.* [7] have obtained quite accurate values of the excitation energies of the quartets with respect to the ground state, the uncertainty being as low as ± 15 cm^{-1} (0.05 %).

In 1976 Roig and Tondello [8] reported an absorption study of B I by the flash pyrolysis technique. In their source the $2s2p^2\ ^4P$ term was populated which allowed the observation of five previously unknown quartet terms in the photoabsorption spectra from this metastable term. All quartets except $2s2p^2\ ^4P$ lie energetically between the ground state $2s^2\ ^1S$ and the $2s2p\ ^3P$ metastable term of B II, i.e. in the interval 66 928 cm^{-1} - 104 286 cm^{-1} above the B I ground state, cf. the energy level diagram in Fig. 1. However, because of the $\Delta S = 0$ selection rule these quartet levels cannot autoionize into the $2s^2\varepsilon l\ ^2L$ continuum by the fast Coulomb interaction mechanism which has typical rates of 10^{13} - 10^{15} s^{-1}. On the other hand, autoionization by relativistic processes, spin-orbit, spin-other orbit or spin-spin interaction, which are orders of magnitude slower, is still possible.

It is obvious that studies of B I, in particular the quartet system, present challenging problems. In many cases there are three open subshells, the theoretical description of which may call for the use of very sophisticated methods, e.g. many-body techniques [9]. Cases should also be found where the levels have competing de-excitation channels, radiative as well as relativistically induced non-radiative decay.

In astrophysics much interest has in recent years been focussed on boron. Although its cosmic abundance is extremely low, it plays an important role in testing models of Big Bang Nucleosynthesis [10]. Optical spectroscopy is the only method for establishing B abundances in stellar objects, and thus a good knowledge of energy structure, transition probabilities, hyperfine structure and isotope shifts is needed [11].

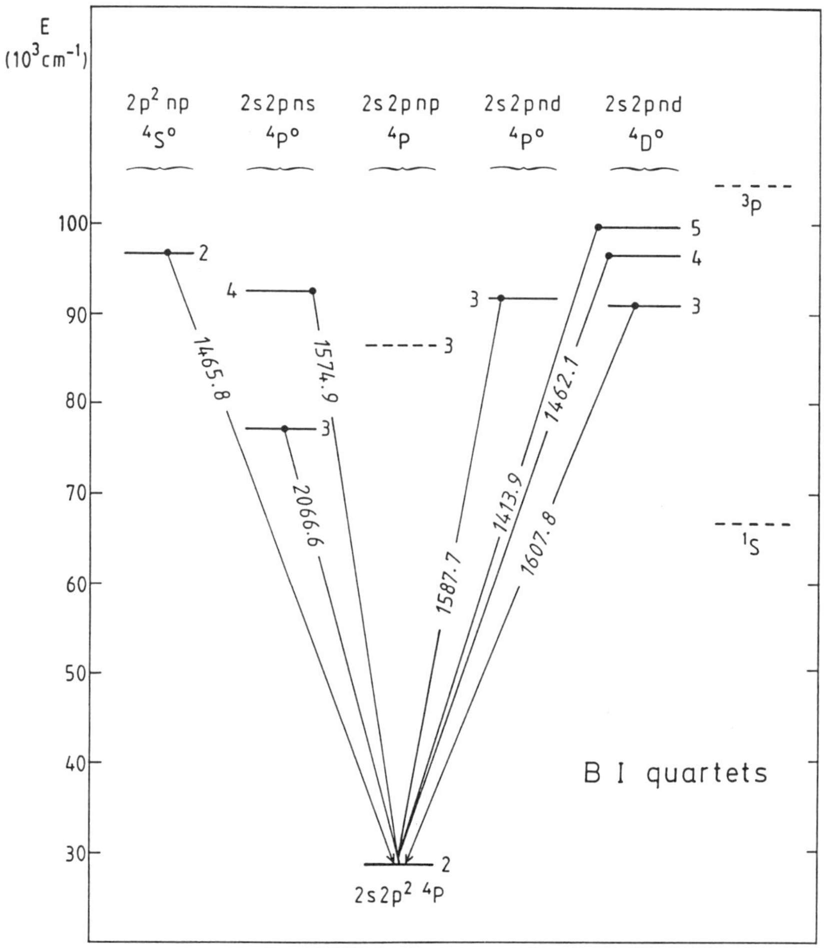

Fig. 1. Energy level diagram of the quartet system of B I. Experimental wavelengths [8] (in Å units, and rounded off) are indicated. No experimental data are available for the 2s2p3p ^4P term. The positions of the $2s^2$ ^1S ground state and the 2s2p ^3P excited term in B II are shown on the right-hand side.

Futhermore, boron also plays a key role in thermonuclear fusion research, belonging to the most abundant impurities in the plasmas of several presently operating tokamaks. Thus, boron is a constituent in carbon-based wall materials which face the plasma, and it can enter the plasma by various plasma-wall interaction mechanisms, sputtering, erosion, etc. [12]. As in the case of astrophysics, fusion plasma diagnostics, by means of optical spectroscopy, utilizes basic experimental and theoretical data on the structure and interaction of boron in various stages of ionization.

2. Survey of beam-foil studies

When accelerated ions (e.g. the beam-foil excitation technique) are used instead of plasma light sources, there are usually no problems in obtaining clean and fairly line-rich spectra of B in various charge states. In one of the earliest comprehensive studies, several new energy levels were found and a substantial number of lifetimes were determined by utilizing the good time-resolution of the beam-foil source [13]. Among the transitions studied was a line at 1607 Å, at that time ascribed to the 2s2p ^{1}P - 2s3s ^{1}S combination in B II [14]. However, a subsequent theoretical calculation, by Weiss [15] using the superposition of configurations (SOC) method, resulted in an f-value more than an order of magnitude lower than the experimental result [13]. Weiss even questioned the assignment of the 1607 Å line, because its energy differed by 2700 cm^{-1} from the calculated value, whereas theory and experiment agreed quite well for other B II lines.

This problem was practically solved when the Roig and Tondello paper [8] appeared, because here it was shown that the 2s2p^2 ^{4}P - 2s2p3d ^{4}D multiplet in B I lies in the region 1607.7-1607.9 Å, see Fig. 1. Also, various beam-foil spectra [13,16,17] contained other quartet transitions, identified in Ref. [8]. In this context also the comprehensive lifetime study by Kernahan *et al.* [18] must be mentioned.

In addition to the beam-foil results already cited [13, 16-18] some more recent experimental data are available [19]. These were obtained using the 380 kV ion accelerator at the Research Institute of Physics in Stockholm (now the Manne Siegbahn Laboratory). The wavelength resolution was here somewhat higher than in previous work. One main difference between Ref. [8] and Refs. [13,16-19] is that in the latter case the transitions were observed in emission and not in absorption spectra. In the beam-foil spectra transitions from the n = 2 and n = 3 quartet terms were found whereas no lines from n = 4 or 5 could be observed. To some extent this may be caused by the fact that the latter transitions were masked by stronger B I - B III lines. Beam-foil work at higher resolution would be needed to identify the higher-lying quartets. The wavelengths obtained in the most recent beam-foil spectra [19] are in very good agreement with the absorption data [8]. However, the latter have superior resolution and all the fine-structures were separated.

3. Theoretical calculations

We approach the problem of computing transition rates for quartet levels of B I with two different methods. For some of the lower levels we use an *ab initio* method, based on the MCHF approach [20]. A larger set of terms is then treated with the SOC code of Cowan [21], which employs the Blume-Watson method [22] for spin-orbit integrals. It was modified for PC by Ralchenko and Kramida [23].

In the MCHF calculation we use a complete active space (CAS) approach [24] to include outer correlation in a systematic way. According to this we generate all possible configuration state functions (CSFs), with the constraint of a closed inner 1s-shell, from an active set of orbitals. The orbital set is increased, stepwise, by one set of given main quantum number n at the time. The resulting wavelengths and transition rates are given in Table I, and we can see that the results have converged.

In the SOC calculation 17 configurations of principal quantum numbers n = 2-5 for each parity were included, in order to take into account the main configuration interactions. The spin-orbit integrals ζ were reduced to 95% of their *ab initio* values, whereas the electrostatic integrals between equivalent electrons F^k, those between non-equivalent electrons F^k and G^k, and the configuration interaction integrals R^k were reduced to 80% to compensate for the interaction from more distant configurations. In this way we calculated the lifetimes of 7 quartet terms and the wavelengths of their transitions to $2s2p^2\,{}^4P$.

Our results are listed in Table II together with available theoretical and experimental data. It is interesting to compare the two approaches used in the present work. While the MCHF calculation includes a larger and more complete treatment of outer correlation, it does not allow for mixing with doublet terms. This is done (somewhat crudely) in the SOC calculations, and the agreement between the two results indicates that spin-dependent mixing has a minor influence on the lifetimes. Previous theoretical work includes calculations of the $2s2p^2\,{}^4P$ - $2p^3\,{}^4S$ transition probability by Weiss [25] and Sibincic [26] who employed the SOC method, obtaining identical results for this multiplet. Also, Brage and Froese Fischer [27] have performed MCHF calculations for transitions from the $2s2p3d\,{}^4P$ and ${}^4D$ terms.

Table I. Wavelengths and transition probabilities in B I.

Transition		n	Wavelength (Å)	Transition probability (10^8 s^{-1})
$2s2p^2\ ^4P$	- $2p^3\ ^4S$	HF	1411.9	24.8
		3	1452.3	20.1
		4	1452.9	19.8
		5	1452.9	19.7
		6	1453.0	19.7
			1465.8[a]	
	- $2s2p3d\ ^4D$	HF	1670.3	2.25
		3	1584.2	2.42
		4	1612.4	2.77
		5	1613.6	2.80
			1607.8[a]	
	- $2s2p3s\ ^4P$	HF	2067.2	1.56
		3	2067.2	1.92
		4	2073.7	2.22
		5	2076.0	2.23
		6	2076.2	2.23
			2066.6[a]	

[a]Roig and Tondello [8]

Table II shows that our SOC lifetime of $2p^3\ ^4S$ is somewhat shorter than the MCHF one and also the previous theoretical predictions [25,26]. The latter agree reasonably well with the most recent experimental result [18], however. A new measurement might still be worth undertaking here. The situation looks better in the case of the 2s2p3d ^4D level, where the present calculations and experiment show fairly good agreement.

Table II. Wavelengths and lifetimes in B I.

Transition			Wavelengths		Lifetime (ns)	
			Exp.[a]	Theory[b]	Experiment	Theory
$2s2p^2\ ^4P$	-	$2p^3\ ^4S$	1465.8	1452	0.61 ± 0.05[d]	0.50[e,f]
	-				0.53 ± 0.03[g]	0.41[b]
						0.51[c]
	-	$2s2p3s\ ^4P$	2066.6	2105		4.5[b,c]
	-	$2s2p4s\ ^4P$	1574.9	1571		27.7[b]
	-	$2s2p3d\ ^4P$	1587.7	1594		4.8[b]
						6.8[h]
	-	$2s2p3d\ ^4D$	1607.8	1611	2.7 ± 0.4[d]	3.0[b]
						3.57[c]
					3.30 ± 0.17[g]	3.89[h]
	-	$2s2p4d\ ^4D$	1469.1	1466		8.6[b]
	-	$2s2p5d\ ^4D$	1413.9	1409		25.5[b]

[a]Roig and Tondello [8]
[b]This work (SOC)
[c]This work (MCHF)
[d]Martinson *et al.* [13]
[e]Weiss [25]
[f]Sibincic [26]
[g]Kernahan *et al.* [18]
[h]Brage and Froese Fischer [27]

4. Concluding remarks

We have found good agreement between experiment and theory for the wavelengths in the B I quartet system. Also, for the two terms for which experimental lifetimes are available, i.e. $2p^3\ ^4S$ and $2s2p3d\ ^4D$, there is satisfactory accord between our calculations and experimental values. This agreement definitely proves that the 1607.8 Å line, for many years ascribed to

B II, actually originates from the B I quartet system. In this context, we may ask about the correct identification of the 2s2p ^{1}P - 2s3s ^{1}S transition in B II. Bashkin *et al.* [17] have tentatively proposed a revised energy value for the ^{1}S level, but other assignments may also be possible. Additional studies of the unpublished beam-foil material [19] might here provide a conclusive answer.

Figure 1 shows that 8 terms of the B I quartet system are experimentally established, by means of absorption spectroscopy [8]. However, in emission spectra, studied by the beam-foil method, transitions between only 4 of these term have been reported. We now plan such measurements at higher resolution which should allow the separation of the quartet lines and those belonging to e.g. the B I doublet system or B III. Note also that only a theoretical value exists for the energy of the 2s2p3p ^{4}P term (dotted line in Fig. 1). Combinations between this even term and the adjacent odd-parity ones, 2s2p3s ^{4}P and 2s2p3d ^{4}P or ^{4}D should lie in the infrared region. Although this region has been studied in the region 1 - 4 μm [28] no quartet transitions seem to have been observed.

The MCHF wavelengths (Table I) converge towards the wrong limit in some cases. This is not due to relativistic effects which have been investigated. It is probably due to left out core-valence correlation. Another important point concerns lifetime measurements. Thus, Miecznik and Brage [29] have reported elaborate calculations of J-dependent lifetimes in the 2s2p3p ^{4}D and 2s2p3d ^{4}F terms for the B-like spectra C II to F V. Here also some experimental data are available [30]. This J-dependence arises from spin-orbit induced mixing of quartet and doublet states, the latter of which can autoionize to the doublet continuum. It would be desirable to search for such effects also for B I, both experimentally and theoretically. It may appear that here such effects are very small - on the other hand they have been observed and theoretically explained in the case of Be I [31] and C I [32].

Acknowledgements

We are grateful to Professor C. Froese Fischer, Dr. S. Mannervik, Dr. H. Cederquist and Dr. U. Litzén for valuable discussions and advice. One of the authors (YZ) acknowledges a grant from the Royal Swedish Academy of Sciences. The work has been supported by the Swedish Natural Science Research Council.

References

1. B. Edlén, A. Ölme, G. Herzberg and J.W.C. Johns, J. Opt. Soc. Am. **60**, 889 (1970).
2. C.M. Brown, S.G. Tilford and M.L. Ginter, J. Opt. Soc. Am. **64**, 877 (1974).
3. J. Carlsson, P. Jönsson, L. Sturesson and C. Froese Fischer, Phys. Rev. A **49**, 3426 (1994).
4. C.E. Moore, Atomic Energy Levels Vol. I, Natl. Bur. Standards Circular No. 467 (1949).
5. E.W. Burke, Jr. and J.E. Mack, J. Opt. Soc. Am. **46**, 100 (1956).
6. W.M. Burton, A. Ridgeley and R. Wilson, Mon. Not. Roy. Astron. Soc. **135**, 207 (1967).
7. B. Edlén, H.P. Palenius, K. Bockasten, R. Hallin and J. Bromander, Solar Phys. **9**, 432 (1969).
8. R. A. Roig and G. Tondello, J. Phys. B: Atom. Molec. Phys. **9**, 2373 (1976).
9. I. Lindgren and J. Morrison, "Atomic Many-Body Theory" (Springer, Berlin, 1982).
10. D.N. Schramm, in "Origin and Evolution of the Elements", edited by N. Prantzos, E. Vangioni-Flam and M. Cassé (Cambridge University Press, 1993) p. 112.
11. S.G. Johansson, U. Litzén, J. Kasten and M. Kock, Astrophys. J. **403**, L25 (1993) and references therein.
12. "Collision Processes of Be and B Atoms and Ions in Fusion Plasmas", edited by R.K. Janev, Physica Scripta **T62**, 1-104 (1996).
13. I. Martinson, W.S. Bickel and A. Ölme, J. Opt. Soc. Am. **60**, 1213 (1970).
14. A. Ölme, Physica Scripta **1**, 256 (1970).
15. A.W. Weiss, Nucl. Instr. Methods **90**, 123 (1970) and private communication (1973).
16. I. Martinson and W.S. Bickel, unpublished (1973).
17. S. Bashkin, L.C. McIntyre, H. v. Buttlar, J.O. Ekberg and I. Martinson, Nucl. Instr. Methods **B9**, 593 (1985).
18. J.A. Kernahan, E.H. Pinnington, A.E. Livingston and D.J.G. Irwin, Physica Scripta **12**, 319 (1975).
19. I. Martinson, S. Mannervik and H. Cederquist, unpublished (1984).
20. C. Froese Fischer, Comput. Phys. Commun. **64**, 369 (1991).
21. R.D. Cowan, "The Theory of Atomic Structure and Atomic Spectra" (University of California Press, Berkeley, 1981).
22. M. Blume and R.E. Watson, Proc. Roy. Soc (London) **A 270**, 565 (1963)
23. Yu. V. Ralchenko and A.E. Kramida, private communication (1996).
24. B.O. Roos, P.R. Taylor, P.E.M. Siegbahn, Chem. Phys. **48**, 157 (1983).

25. A.W. Weiss, Phys. Rev. **188**, 119 (1969).
26. Z. Sibincic, Phys. Rev. **A 5**, 119 (1969).
27. T. Brage and C. Froese Fischer, unpublished.
28. U. Litzén, Physica Scripta **1**, 251 (1970) and private communication (1996)
29. G. Miecznik and T. Brage, J. Phys. B: At. Mol. Opt. Phys. **28**, 4233 (1995).
30. M. Westerlind, P. Bengtsson, M. Kirm and L. Engström, J. Phys. B: At. Mol. Opt. Phys. **27**, 877 (1994) and references therein.
31. D.G. Ellis, I. Martinson and M. Westerlind, Physica Scripta **49**, 561 (1994).
32. R.R. Haar, T.J. Kvale, D.G. Ellis, I. Martinson, L.J. Curtis and D.J. Beideck, Phys. Letters **A 141**, 131 (1989).

Two Electron Dynamics in Photodetachment

D. Hanstorp, G. Haeffler, A. E. Klinkmüller, U. Ljungblad
U. Berzinsh[§], I. Yu. Kiyan[#] and D. J. Pegg[*]

Department of Physics,
Göteborg University and Chalmers University of Technology,
SE-412 96 Göteborg, Sweden
[§]*Department of Spectroscopy, University of Latvia, LV 1586, Riga, Latvia*
[#]*Russian Academy of Sciences, General Physics Institute, 117 942 Moscow, Russia*
[*]*Department of Physics, University of Tennessee, Knoxville, TN 37996 USA*

Abstract

We present the results of experimental studies of photon-negative ion interactions involving the dynamics of two electrons. Resonances associated with doubly excited states of Li^- and He^- have been observed using laser photodetachment spectroscopy. Total and partial photodetachment cross sections have been investigated. In the former case, the residual atoms are detected irrespective of their excitation state, while in the latter case only those atoms in specific states are detected. This was achieved by the use of a state selective detection scheme based on the resonant ionization of the residual atoms. In addition, in the case of Li^- photodetachment, the threshold behavior of the $Li(2^2P)+e^-(ks)$ partial cross section has been used to accurately measure the electron affinity of Li.

ADVANCES IN QUANTUM CHEMISTRY, VOLUME 30

0065-3276/98 $25.00

1. Introduction

We present results of recent studies of the photodetachment of few-electron atomic negative ions. The motivation for these measurements is to better understand the general problem of correlated motion of particles in many-body systems. Correlated motion plays an important role in several fields of physics. In nuclear physics, for example, it manifests itself in halo nuclei, in which the two outermost neutrons contribute to the formation of very diffuse nuclei [1]. In the case of atomic systems, it is known that negative ions, in particular, exhibit an enhanced sensitivity to electron correlation effects due to the suppression of the normally dominant Coulomb force of the core. This sensitivity is further increased in the case of doubly excited states of negative ions where, in general, the outermost electron moves in the field of an excited atom. Such an electron is bound by short range polarization forces, in contrast to the long range Coulomb force characteristic for the binding of atoms and positive ions. The more symmetric the two electron excitation, the more highly correlated the state. A highly excited pair of electrons are no longer independent, rather they begin to move collectively. The autodetaching decay of doubly excited states is manifested as resonances in photodetachment cross sections. Such resonances arise from an interference between two paths to the same final continuum state: direct photodetachment and resonant photodetachment via the intermediate doubly excited state. The results of experiments on negative ions, and particularly those involving resonances associated with excited negative ions, therefore provide sensitive tests of the ability of theory to go beyond the independent electron model.

The prototypical atomic negative ion is H^-. Here, two electrons move in the field of a point core, the proton. In general, if both electrons are excited the outer electron moves in the field of an excited H atom consisting of the nucleus and the inner electron. Since this permanent dipole field differs intrinsically from the long range Coulomb field, the series of resonances will not be strictly Rydberg in nature. Electron correlation becomes more enhanced as the level of excitation increases since the excited H atom becomes more easily polarizable. Resonances representing the decay of such doubly excited states appear in the photodetachment cross section below the excited H atom thresholds. Double excitation of the H^- ion has been studied extensively both experimentally [2] and theoretically [3]. It is clear that these highly correlated states are not well represented by the independent electron model. Instead, new quantum number labels have been introduced that describe the collective motion of the electron pair [4].

We have chosen to study the two electron dynamics in He^- and Li^- ions since they represent the logical extension of the work on the H^- ion. In the case of both He^- and Li^-, the two active electrons move in the field of a finite core. He^- and Li^- ions are therefore simple systems for investigating few-particle interactions since the core, in each case, is finite but small. The spherically symmetric closed core of Li^- might be expected to approximate the H^- ion somewhat better than the open core of He^-. The presence of a core will serve to lift the degeneracy of the excited state thresholds characteristic of the H atom. Doubly excited states will then be bound in the shorter range field of an induced dipole and not a permanent dipole, as is the case for H^-. As a consequence, one expects the photodetachment spectra of these ions to differ from that of the H^- ion, at least at low-to-intermediate levels of excitation. For example, the resonances should be broader since the lifting of the threshold degeneracy results in a larger number of continua available for autodetachment [5, 6]. It is also not clear whether the quantum numbers and propensity rules developed to describe the H^- spectra will remain valid to the same extent in the case of the He^- and Li^- spectra. Of course, as the level of excitation increases the role played by the core will become less significant and the spectra, particularly that of Li^-, should appear more and more like that of H^-. In fact, the Li^- ion can be considered as the "poor man's H^- ion" in the sense that at high levels of excitation their spectra should be similar. The doubly excited states of Li^- are, however, far more accessible. In both Li^- and He^-, high lying states can be studied in high resolution experiments using tunable dye lasers. Corresponding states in the H^- ion, however, have excitation energies about an order of magnitude larger and therefore remain relatively inaccessible.

2. Method

2.1 Total Cross Sections

The total photodetachment cross section describes the probability that an electron is detached from a negative ion following the absorption of a photon, regardless the excitation state of the residual atom or the energy or direction of the emitted electron. A total cross section is the sum of partial cross sections for detachment into each of the energetically allowed continua. This is illustrated in Fig. 1. Here we show the three possible channels accessible to a doubly excited state of Li^- that lies just below the $Li(3^2P)$ detachment threshold. The total cross section may be determined by

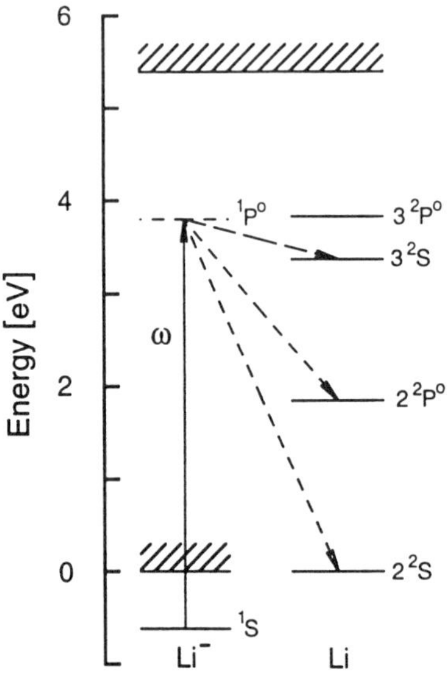

Fig. 1 *Partial energy level diagram for the Li^-/Li systems showing the autodetaching decay of a doubly excited state of Li^- into different continua. Each of the three continuum channel is characterized by the state that the residual Li atom is left in following photodetachment via the doubly excited state. A particular decay channel is isolated by state selectively detecting the excited residual atom.*

monitoring the production of either the detached electrons or the neutral atoms left following photodetachment. In ion beam experiments the residual atoms can be efficiently collected in the forward direction since they are moving with the same velocity as the beam ions. In the present experiments we measure the yield of the residual atoms produced when a beam of negative ions is merged collinearly with a beam of pulsed, tunable laser radiation. Studies of double excitation of negative ions demand both high sensitivity and energy resolution as a consequence of the low production cross sections and the high density of resonances below highly excited thresholds. The choice of a collinear geometry has two major advantages over a crossed beam geometry. First, the interaction volume can be made at least two orders of magnitude larger, resulting in a more efficient production of the doubly excited states. Second, kinematic compression [7] of the longitudinal velocity distribution of the ions significantly improves the energy resolution of the photodetachment spectra.

Typically, doubly excited states autodetach predominantly to just a few continua and the associated resonance structure modulates the corresponding partial cross sections. A problem arises in total cross section measurements as high lying doubly excited states are accessed. Resonance structure in total cross sections tends to "wash out" as many partial cross sections with little or no structure are added to those that exhibit well modulated structure. Under such conditions, it becomes necessary to isolate a particular continuum channel and study resonances in the corresponding partial cross section. Another problem arises if one attempts to study the total cross section for the photodetachment of metastable negative ions such as He^-. In this case a large neutral atom background arises from autodetachment of the unstable ions in their ground state. To avoid these two problems, we have chosen to selectively ionize the residual atoms and detect positive ions in the presence of a much smaller background. By use of this state selective detection scheme, we hence study partial cross sections instead of total cross sections.

2.2 Partial Cross Sections

A partial photodetachment cross section describes the probability that an electron is detached from a negative ion into a particular continuum state. Each decay channel is characterized by the excitation state of the residual atom and the energy and angular momentum of the detached electron. A particular channel can therefore be identified by either measuring the energy of the detached electron, as in photoelectron spectroscopy, or by state selectively detecting the residual atom. We have chosen the latter alternative in the present work. The excitation state of the residual atom can be determined by the use of the well established technique of Resonance Ionization Spectroscopy (RIS). This sensitive and selective technique was developed more than 20 years ago [8] but it has only recently been applied to the study of negative ions [9]. The use of a collinear beam geometry combined with the resonant ionization of the residual atoms presents several distinct advantages over photoelectron spectroscopy in which the detached electrons are energy analyzed, invariably using a crossed beam geometry. First, the positive ions can be collected far more efficiently than photoelectrons since they move in one direction only, the ion beam direction. In contrast, the electrons are ejected over a distribution of angles and therefore only a small fraction can be collected within the finite acceptance angle of a detector. Second, the background associated with the detection of positive ions is generally far smaller than that associated with the detection of photoelectrons. Both of these advantages enhance the sensitivity of state selective measurements on the residual atoms. On the other hand, the major advantage of photoelectron spectroscopy is that all energetically allowed photodetachment channels can be isolated and measured simultaneously. Another attribute is that the angular distribution of the detached electrons can be measured. This additional information provides further insight into the photodetachment process.

In order to extract information on the parameters of a resonance one must fit a parameter function that describes the interaction of a doubly excited state with continua to the experimental data. The most commonly used form is the Beutler-Fano profile [10]

$$\begin{aligned} \sigma(E) &= a + b\frac{(q+\varepsilon)^2}{1+\varepsilon^2} \\ \varepsilon &= 2\frac{(E-E_0)}{\Gamma} \end{aligned} \tag{1}$$

where E is the photon energy, E_0 and Γ are the position and width of the resonance, q is the shape parameter, a is the background and b is the amplitude of the resonance. This form is strictly only valid for the treatment of resonance structure in a total cross section. It can, however, be used to parametrize resonances in a partial cross section if, as in the present case, only the positions and widths of the resonances are required. Starace has extended the Beutler-Fano treatment to cover the case of partial cross sections [11]. The form that Starace has proposed can be written as

$$\begin{aligned} \sigma(\varepsilon) = a + \frac{b}{(1+\varepsilon^2)}[\varepsilon^2 + 2\varepsilon(q\,\mathrm{Re}(\alpha) - \mathrm{Im}(\alpha)) + \\ 1 - 2q\,\mathrm{Im}(\alpha) - 2\,\mathrm{Re}(\alpha) + (q^2+1)|\alpha|^2] \end{aligned} \tag{2}$$

where the parameters have the same significance as in the Beutler-Fano formula. In addition a complex parameter, α, is introduced to describe the branching into a particular channel.

2.3 Experimental Procedure

A detailed description of the interacting beams apparatus used in the present work can be found elsewhere [12]. In short, positive ions are extracted from a plasma-type ion source and accelerated to beam energies that can be varied between 2.5 and 5 keV. Negative ions are produced in the beam by double sequential charge exchange in a cesium vapor. A pair of electrostatic quadrupole deflectors (QD1, QD2) is used to direct the negative ion beam into and out of the path of the laser beams, as shown schematically in Fig. 2. The ion-laser interaction region is defined

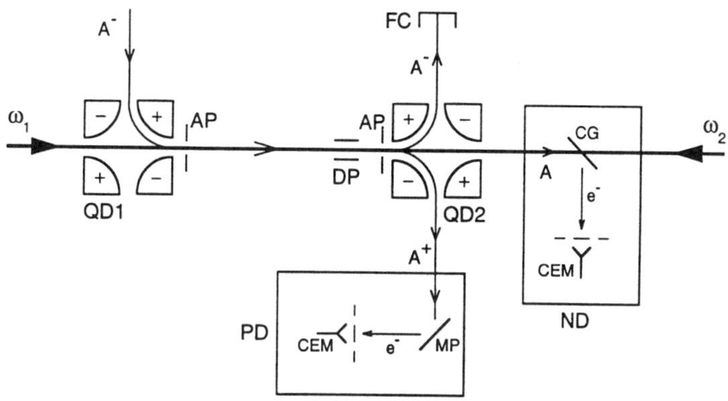

Fig. 2 Portion of the collinear laser-ion beam apparatus. QD1, QD2, electrostatic quadrupole deflectors; CEM, channel electron multiplier; DP, deflection plates; PD, positive ion detector; FC, Faraday cup; ND, neutral particle detector; CG, conducting glass plate; AP, aperture; MP, metal plate; A, any element. The distance between QD1 and QD2 is approximately 0.5 m.

by two apertures (AP) of diameter 3 mm placed 0.5 m apart. The amount of negative ions remaining after passage through the interaction region is monitored for normalization purposes using a Faraday cup (FC) placed adjacent to the second quadrupole deflector. Neutral atoms produced by photodetachment or collisional detachment are unaffected by the electric field in the second quadrupole. They enter the neutral atom detector (ND) situated downstream of the second quadrupole deflector. Here, the impact of the fast beam atoms on a quartz plate (CG) produces secondary electrons that are efficiently detected using a channel electron multiplier

(CEM). The plate is coated with a few atomic layers of platinum in order to prevent charge build-up. The transparency of the plate makes it possible to direct the laser beams both parallel and anti-parallel with respect to the ion beam. By performing both co- and counter-propagating experiments, it is possible to correct for the Doppler shift to all orders. A serious background source with this arrangement, however, is associated with photoelectrons generated on the plate by pulsed UV radiation. This contribution is greatly reduced by modulating the bias on a grid placed between the glass plate and the CEM, as described in detail in ref. [13]. This detector has been shown to work efficiently for laser wavelengths down to 250 nm.

The third detector (PD) in the interaction chamber is used to detect positive ions in the state selective measurements. It is situated adjacent to the second quadrupole, opposite the Faraday cup used to monitor the negative ion current. It operates in the same way as the neutral atom detector just described. In this case, however, the detector is out of the line of sight of the laser beam and so laser-generated photoelectrons pose no problem. The positive ion detector is used to state selectively detect residual atoms in the partial cross section measurements. The electric field of the second quadrupole is used both to field ionize the Rydberg atoms and to bend the resulting positive ions into the detector. The positive ion background is usually small, the major contribution arising from double detachment collisions of negative ions with the residual gas in the interaction chamber. This contribution is minimized by a combination of a good vacuum (approximately 10^{-9} mbar) and a pair of deflection plates (DP) positioned just prior to the entrance of the second quadrupole. The transverse electric field is insufficient to field ionize the atoms prepared in the highly excited Rydberg states but it does serve to sweep out of the beam any positive ions produced upstream in collisional events. This procedure substantially reduces the collisional detachment background since only atoms left in high lying Rydberg states will contribute to the background. Of course, this simultaneously hinders the negative ions from reaching the Faraday cup. The deflection plates are therefore switched off periodically to monitor the negative beam intensity.

The laser system used in the experiments consist of two excimer-pumped dye lasers. The dye lasers can produce narrow bandwidth laser radiation of wavelengths between 340 and 850 nm, which can be extended, with the use of second harmonic generation, down to 205 nm. The bandwidth is nominally about 5 GHz, but this can be reduced to 1 GHz by use of an intra cavity etalon. The energy resolution in the present experiments is limited solely by the linewidth of the laser. The maximum pulse energy is typically a few mJ in the visible wavelength region and a few hundred μJ in the UV, after doubling. The laser intensity is measured after the passage through the interaction chamber for normalization purposes. The wavelength of the laser light is calibrated by observing well known resonance lines in Ne or Ar using optogalvanic spectroscopy in hollow cathode lamps.

3. Results

3.1 Total Cross Section Measurements

The cross section for photodetachment of an electron from the Li^- ion has previously been investigated, both experimentally [14] and theoretically [15]. Earlier laser studies have focused on the region around the first and second detachment thresholds. The latter, also called the first excited state threshold, corresponds to the process in which the residual Li atom is left in the 2^2P state following photodetachment. In contrast to the heavier alkali ions [16], the Li^- photodetachment cross section shows no resonance structure below the first excited state threshold. It is expected, however, that doubly excited states should appear at higher levels of excitation. The atomic wave function then becomes more extended in space, making it easier for an extra electron to share the attractive force from the nucleus, or equivalently, the atom becomes more polarizable. We therefore initiated a search for doubly excited $^1P^o$ states of Li^-, that are bound with respect to the 3^2P state of the Li atom [17]. The results in the energy region 4.2-4.5 eV are shown in Fig. 3. The Li atom signal is proportional to the total cross section which, in this case, is equal to the sum of the partial cross sections for the three channels shown in Fig. 1. The experimental cross section data are not absolute. The measurements have therefore been normalized to a theoretical curve at the $Li(3^2P)$ threshold. The main contribution to the approximately 10% scatter in the experimental data arises from changes in the spatial overlap of the laser and ion beams that occurred as the laser frequency was scanned. The statistical scatter due to counting of the residual atoms was less than 3%. The experimental data exhibits three significant features. A small narrow dip is seen at 4.456 eV, a rather narrow structure is observed at 4.453 eV (labeled b) and finally, a broad resonance is seen to cover essentially the entire energy region between the $Li(3^2S)$ and $Li(3^2P)$ thresholds (labeled a).

The small dip at 4.456 eV is a Wigner cusp that arises from the opening of the $Li(3^2P)+e^-(ks)$ photodetachment channel (hereafter labeled 3^2Pks, for brevity). The other two structures are resonances associated with the decay of doubly excited states. These resonances overlap one another as well as the Wigner cusp, making it

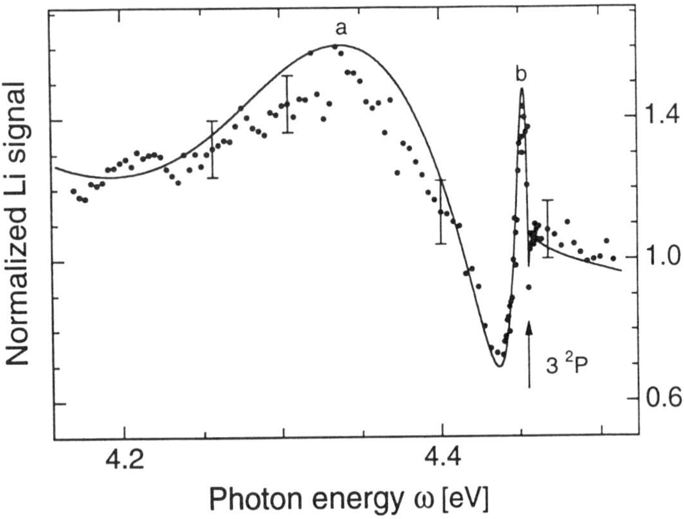

Fig. 3 Yield of Li atoms vs. photon energy in the range 4.2-4.5 eV. The Li signal is proportional to the total cross section for the photodetachment of Li⁻ via the 2^2Skp, 2^2Pks,d and 3^2Skp channels (see Fig. 1). The experimental data (dots) has been normalized to theory (solid line with scale to the right) at the 3^2P threshold.

impossible to extract information on individual positions and widths. The Beutler-Fano formula, which is normally used to extract resonance parameters, cannot be applied in this case since it is only valid for isolated resonances. In order to explain the origin of the observed structures in the Li^- photodetachment cross section, comparisons with two independent calculated cross sections were made. Lindroth [17] used an approach which combines complex rotation [18] with a discrete numerical basis set [19], whereas Pan *et al.* [5] employed the R-matrix method. The two theoretical methods yield essentially identical results (see the solid line in Fig. 3). We observe a rather good agreement between theory and experiment over the whole energy region studied. According to Lindroth´s calculation, the broad structure (a) is identified with an intrashell resonance associated with the symmetrical excitation of the two valence electrons. This resonance was found to be dominated by the 3p3d and 4s3p configurations. The narrow structure (b) at approximately 4.453 eV corresponds to several overlapping intershell resonances in which the valence electrons are asymmetrically excited.

3.2 Partial Cross Section Measurements

The state selective detection scheme based on RIS has been applied to isolate specific autodetaching channels for doubly excited states in both He^- and Li^-. Here we present the results of recent measurements of the positions and widths of doubly excited states which appear as resonances in partial cross sections below certain excited state thresholds. The 1s3s4s ^4S state in He^-, for example, lies just below the He(3^3S) threshold. A Feshbach resonance associated with the autodetaching decay of this state was observed in both the He(1s2s ^3S)+e(ks) and He(1s2p ^3P)+e(kp) partial cross sections(hereafter labeled 2^3Sks and 2^3Pkp) [20]. Similarly, in the case of Li^-, resonances due to doubly excited states below the Li(4^2P) and Li(5^2P) thresholds were observed in the 3^2Skp partial cross section [21]. The state selective detection scheme has also been used to study the threshold behavior of the 2^2Pks partial cross section. This technique enabled Haeffler *et al.* [22] to accurately measure the electron affinity of Li.

3.2.1 Doubly Excited States of He^-

Doubly excited states of He^- of doublet symmetry have been observed in studies of electron impact on He [23]. In contrast, data on quartet states are sparse. Selection rules on photoexcitation from the $^4P^o$ ground state limit excited state production to those of ^4S, ^4P and ^4D symmetry. Recently, the He^- photodetachment cross section below 4 eV has been calculated by Xi and Froese Fischer [24]. They predicted a number of resonances associated with the decay of doubly excited quartet states. Bae and Peterson [25] observed a (1s2p^2 ^4P) shape resonance in the total photodetachment cross section of He^- just above the He(2^3P) threshold. The parameters of this resonance were subsequently measured more accurately by Walter *et al.* [26]. We recently embarked on a program aiming to investigate doubly excited states in He^- below the He(n>2) thresholds. The first measurement involved the 1s3s4s ^4S state that lies below the He(3^3S) threshold. The decay of this state was observed as a Feshbach resonance in both the 2^3Sks and the 2^3Pkp partial cross sections.

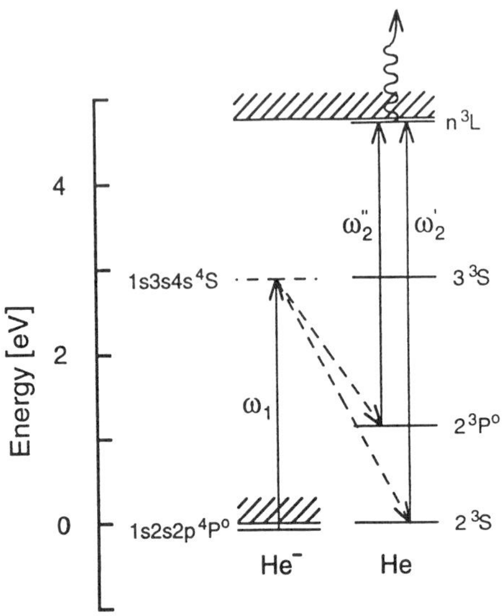

Fig. 4 Partial energy level diagram for the He⁻/He systems. Solid lines represent photoexcitation, the wavy line field ionization and dashed lines autodetachment. The 1s3s4s ^4S state is produced by absorption of photons of frequency ω_1. The 2^3Sks and 2^3Pkp decay channels were isolated by selectively detecting the residual He(2^3S) and He(2^3P) atoms using RIS (laser ω_2' and ω_2'', respectively).

Fig. 4 shows selected energy levels of the He^-/He systems. The 1s3s4s ^4S state of He^- is situated just below the He (3^3S) threshold. This state, which is excited with laser ω_1, rapidly autodetaches via the 2^3Sks and 2^2Pkp channels. Following the decay, the residual He atom will be left in either the 2^3S or 2^3P excited states. Two different laser frequencies ω_2' and ω_2'' were applied separately in the resonance ionization scheme used to monitor the population of the 2^3S and 2^3P states. The frequency ω_2' was chosen to induce a transition between the 2^3S and the 24^3P states of the He atom, when photodetachment into the 2^3Sks channel was studied. The frequency ω_2'' induced a resonance transition between the 2^3P and 26^3D states of He, when photodetachment into the 2^3Pkp channel was studied. The population of both the high lying Rydberg states were efficiently depleted by the electric field of the second quadrupole deflector and He^+ ions thus produced were recorded as a function of frequency of laser ω_1. The output of laser ω_1 was attenuated to avoid saturation of the photodetachment process and the detector.

Fig. 5 shows how the He^+ signal varies as a function of the laser frequency ω_1

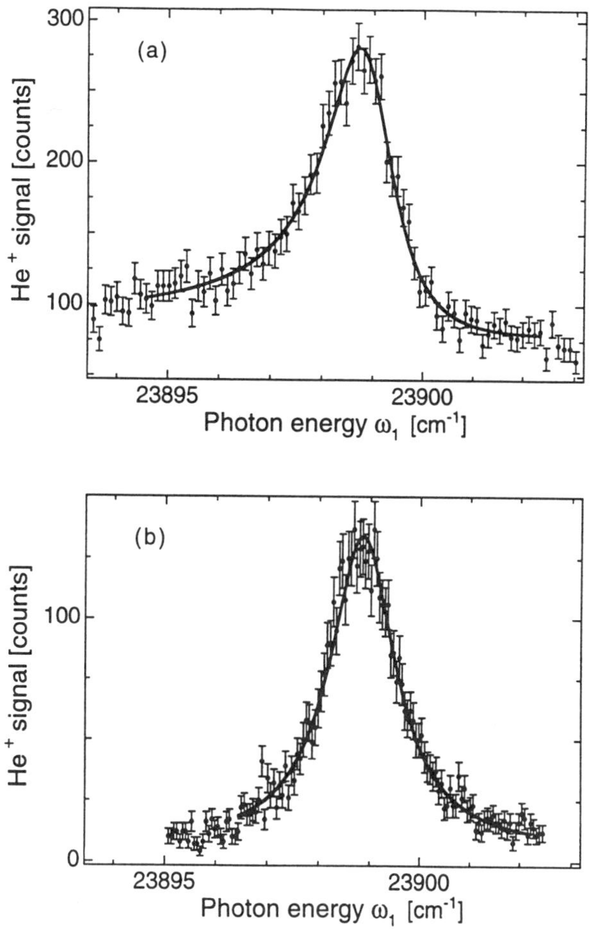

Fig. 5 Yield of He^+ ions vs. photon energy. The He^+ signal is proportional to the partial cross section for the photodetachment of He^- via the $2^3Sks,d$ channel (Fig. 5a) and the 2^3Pkp channel (Fig. 5b). The solid line represents a fit of the Fano function (Eq. 1) to the experimental data (dots).

in the vicinity of the resonance. The signal is proportional to the partial cross sections for photodetachment via the 2^3Sks,d channels (Fig. 5a) and the 2^3Pkp channel (Fig. 5b). Selection rules forbid the autodetachment of the 1s3s4s ^{3}S state via the 2^3Skd channel in the L-S coupling approximation and so in Fig. 5a this channel contributes to the background only (we do not resolve the degenerate s- and d-wave channels). According to recent calculations of Xi and Froese Fischer [27] the cross section for the d-wave channel remains nearly constant at 6 Mb across the resonance, while the s-wave channel has a peak value of 20 Mb on the resonance. The difference in the level of the background in Fig. 5a and Fig. 5b arises essentially from the aforementioned d-wave channel. The background in Fig. 5b, and that remaining in Fig. 5a after the d-wave contribution has been subtracted, arises from several sources. The two most significant background processes that produce He^+ ions, indirectly from He^- ions, are collisional single detachment and photodetachment processes by laser ω_2' (ω_2'') that populate the He(2^3S, 2^3P) states. Once these states are formed by background processes, they will be resonantly ionized by the laser frequencies ω_2' and ω_2'', respectively, and the resulting He^+ ions will be indistinguishable from the signal. The He^+ ions produced directly in double detachment events were essentially eliminated in our experiment by the use of a pair of deflection plates placed just upstream of the second quadrupole deflector (see Fig. 2). The low pressure in the interaction chamber is essential to reduce the background due to collisional detachment processes. In addition, the output of the laser at frequencies ω_2' and ω_2'' was attenuated to reduce the background arising from the photodetachment of He^- at these frequencies.

The resonance energy and width are obtained by fitting the data shown in Fig. 5 to both a Beutler-Fano profile (Eq.1) and to the form proposed by Starace(Eq.2). As expected, the values of E_0 and Γ were the same using both parametrizations. In each case, the measurement was performed with co-propagating and counter-propagating laser (ω_1) and ion beams to eliminate the Doppler effect associated with the fast moving ions. The data shown in Fig. 5, for instance, have been recorded with co-propagating laser (ω_1) and ion beams. The fit to these data yields the blue-shifted resonance position, E_0^b. The measurement was repeated with counter-propagating beams to yield the red-shifted resonance position, E_0^r. The resonance position, corrected for the Doppler effect to all orders, is given by the geometric mean, $E_0 = \sqrt{E_0^b E_0^r}$, of the red- and blue-shifted energies. Table 1 compares the present measurement of the resonance position and width with the result of a recent calculation of Xi and Froese Fischer [24]. One can see that the present results are in good agreement with theory.

Table 1 A comparison of experimental and theoretical values for the positions and widths of resonant states in He⁻ and Li⁻ relative to the ground state of the negative ion.

Resonance	E_0(eV)	Γ(eV)	Reference
$He^-(^4S)$	2.959255(7)	0.00019(3)	Experiment [20]
	2.95907	0.00019	Theory [24]
Li^- (c)	5.1132(4)	0.0074(5)	Experiment [21]
	5.115	0.0073	Theory [6]
Li^- (d)	5.1234(4)	0.0076(11)	Experiment [21]
	5.126	0.0099	Theory [6]
Li^- (j)	5.4485(1)	0.0018(1)	Experiment [21]
Li^- (k)	5.4508(1)	0.0073(3)	Experiment [21]

3.2.2 Doubly Excited States in Li^-

The method of studying partial photodetachment cross sections based on the selective detection of the residual atoms has also been used to investigate resonances associated with the decay of doubly excited states of Li^-. Fig. 6 illustrates a state selective detection scheme designed, in this case, to isolate the 3^2Skp partial cross section in Li^- photodetachment. A laser of frequency ω_1 excites the negative ion into a ^1P doubly excited state that is embedded in several continua. This transient state subsequently autodetaches producing a free electron and a neutral Li atom which, in general, is excited. In the case shown in the figure, the quasi-discrete state is embedded in six continua i.e. there are six possible decay channels that are energetically accessible. These channels are characterized by the residual Li atom being left in 2^2S, 2^2P, 3^2S, 3^2P, 3^2D or 4^2S state following detachment. A second laser of frequency ω_2 is applied to resonantly photoexcite those Li atoms left in the 3^2S state to the 22^3P state. Atoms in such highly excited Rydberg states are efficiently ionized by the electric field of the second quadrupole. The positive ions thus produced can be efficiently detected in the presence of a relatively low background. This procedure isolates the 3^2Skp channel.

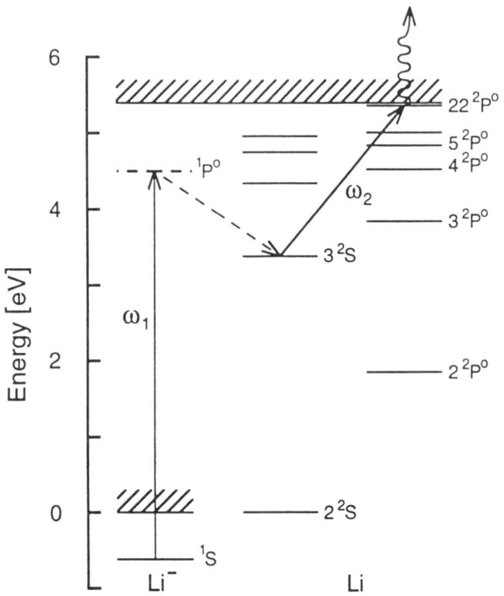

Fig. 6 Partial energy level diagram for the Li/Li⁻ systems. Solid lines represent photoexcitation, the wavy line field ionization and the dashed line autodetachment via the 3^2Skp channel. Doubly excited states of Li⁻ are produced by the absorption of photons of frequency ω_1. The 3^2Skp partial cross section is isolated by selectively detecting the residual atoms left in the 3^2S state using RIS (laser ω_2).

In Fig. 7, we show the cross section for photodetachment of Li^- via the 3^2Skp channel over photon energies of approximately 5.04-5.16 and 5.29-5.46 eV. These ranges cover the regions below, and including, the Li(4^2P) and Li(5^2P) thresholds, respectively. Several Feshbach "window" resonances are observed to lie below these thresholds. In the figure the present measurements are compared with the result of a recent eigenchannel R-matrix calculation by Pan *et al.* [28]. The experimental resolution, which is estimated to be about 25 μeV, is sufficiently high compared to the typical resonance widths that a direct comparison with theory can

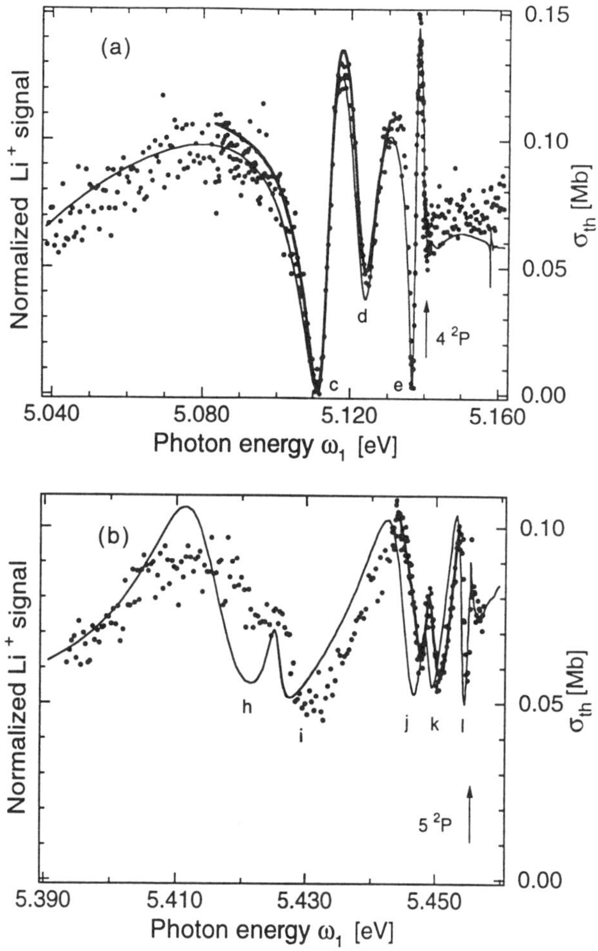

Fig. 7. *Yield of Li^+ ions vs. photon energy in the ranges 5.04-5.16 eV (Fig. 7a) and 5.39-5.46 eV (Fig. 7b). The Li^+ signal is proportional to the partial cross section for photodetachment of Li^- via the 3^2Skp channel (see Fig. 6). The experimental data (dots) has been normalized to theory (solid line with scale at the right). The thick solid lines indicate fits of the double Fano formula (Eq. 3) to the data.*

be made without resorting to deconvolution procedures. At photon energies just below 5.39 eV, the ionization energy of Li, we also observed in-board calibration lines arising from the photoexcitation of ground state Li atoms, by laser ω_1, to high lying Rydberg states that are subsequently field ionized. It was unnecessary in this case to resort to the use of co- and counter-propagating beams to eliminate the Doppler effect since the resonances in the Li^- spectra are much broader than those in the He^- spectra. The calibration procedures described above sufficed to determine the resonance position. The cross section measurements are relative. In Fig. 7a and 7b, they have been normalized to the theoretical results of Pan *et al.* [28] by multiplying the data by factors equal to the ratio of the areas under the experimental and theoretical cross section curves between the limits 5.04-5.16 eV and 5.39-5.46 eV, respectively.

Fig. 7 shows that there is generally a good qualitative agreement between experiment and theory in the sense that all the predicted resonances have been observed and no new resonances are apparent. There is also excellent quantitative agreement in the case of the resonant structure below the Li(4^2P) threshold, between 5.04 and 5.16 eV (Fig. 7a). The situation is slightly different, however, for the resonances below the Li(5^2P) threshold. Here, theory fails to predict the exact energies and strengths of the resonances. For example, there exists a difference in energy between the measured and calculated resonances ranging from 0-2 meV. The size of the discrepancy appears to increase the further the resonance is from the threshold. Experiment and theory overlap, however, close to the threshold. The resonance labeled h is also seen to be considerably weaker than theoretically predicted, while resonance j is not as strong as resonance k, contrary to theory. The more recent calculation by Liu and Starace [29] shows, however, an improved agreement with the experimental data.

The window resonances shown in Fig. 7 partially overlap each other, and it is therefore not possible to fit a single Fano profile to each resonance. We have instead fitted a form representing the sum of two Fano profiles to the data in order to extract the energy and width of each resonance pair:

$$\sigma(E) = a + \sum_{n=1}^{2} b_n \frac{(q_n + \varepsilon_n)^2}{1 + \varepsilon_n^{\,2}}$$

$$\varepsilon_n = 2\frac{(E - E_{0n})}{\Gamma} \qquad (3)$$

Here, the index n represent the parameters for each resonance. Using this method we determined the parameters for resonances c and d in Fig. 7a. The measured positions and widths are shown in Table 1, along with corresponding values calculated by Lindroth [6]. There is a good agreement between the experiment and theory in this case. Lindroth´s resonance parameters are derived directly from a complex rotation calculation. The R-matrix calculation of Pan *et al.* [28] did not explicitly yield the resonance parameters and therefore cannot be used for comparison. Since the Fano formula strictly only applies to total cross sections, the values of the q shape parameters are not entirely meaningful in the context of partial cross sections. This parameter is therefore omitted in the table.

We also attempted, unsuccessfully, to include the resonance labeled e in the fitting procedure. In this case it appears that the resonance is prematurely terminated by the opening of the 4^2Pks channel. The resonance energies in this case depended strongly on the interval of the fit, which was not the case when the c and d resonances were treated as a pair. In a similar manner, the double Fano formula was fit to the two resonances labeled j and k. The parameters obtained from this fit are also included in Table 1. In this case there is no theoretical data available for comparison.

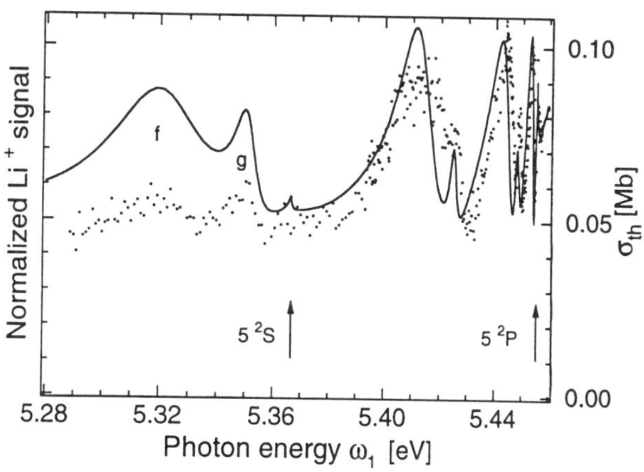

Fig. 8 *Yield of Li^+ ions vs. photon energy in the ranges 5.29-5.46 eV (Fig. 7b). The Li^+ signal is proportional to the partial cross section for photodetachment of Li^- via the 3^2Skp channel (see Fig. 6). The experimental data (dots) has been normalized to theory (solid line with scale at the right).*

In Fig. 8 we show an extended 3^2Skp partial photodetachment cross section which includes measurements below the Li(5^2S) threshold in addition to the data already shown in Fig. 7b. The data in the range 5.29-5.39 eV represents a relatively low statistics survey scan. It is normalized to the calculation of Pan *et al.*[28] using the same factor as for the data between 5.39-5.46 eV. It is clear, even from this relatively low quality data, that the two resonances labeled f and g are observed at approximately the calculated energies but their measured strengths appear to be weaker than predicted. Presumably, the lowest lying resonance, labeled f, is the intrashell resonance representing symmetric excitation of the two valence electrons.

3.2.3 Threshold Investigations of Li^-

Wigner [30] predicted that in the vicinity of a threshold the photodetachment cross section is represented by

$$\sigma(E) \propto (E - E_0)^{l+1/2} \qquad E > E_0 \qquad (4)$$

where $E - E_0$ is the excess energy above threshold and l is the angular momentum of the outgoing electron. This form has been tested experimentally [31] and is now well established. The range of validity of the Wigner law describing threshold behavior, however, varies from system to system. It is primarily determined by the polarizability of the residual atom [32]. Resonances near the threshold also play a role in limiting the energy region over which the threshold law applies. If the outgoing electron is represented by $l = 0$ (s-wave detachment) the cross section rises with an infinite slope at the threshold. This allows one, in principle, to perform very accurate threshold measurements. If, however, $l \geq 1$ the slope of the cross section is zero at threshold, making high precision measurements much more difficult.

An investigation of the threshold behavior of the photodetachment process in a s-wave detachment channel, using tunable laser sources, has the potential for yielding the most accurate values for electron affinities. It was recently demonstrated [22, 33] that, by state selectively detecting the residual atom following photodetachment, it is possible to determine an electron affinity to very high accuracy for, in principle, almost any element. In such a scheme, the residual atoms are resonantly ionized and the resulting positive ions are detected in the presence of a low background. The excellent signal-to-background ratio in this case produces data of high statistical quality, which is especially valuable in the vicinity of a threshold. In addition, the use of a collinear laser-ion beam geometry results in high energy resolution measurements. Again, this is particularly valuable near a threshold where the range of validity of the Wigner law is limited. We demonstrated the potential of this technique [22] in a measurement of the electron affinity of Li. The result is about an order of magnitude more accurate than any previous measurement. In the experiment we selectively detected the residual Li atoms left in the 2^2P state following photodetachment via the 2^2Pks channel. The sharp onset of production of Li^+ ions at this s-wave detachment threshold enabled us to easily fit the Wigner law form to the near-threshold data and extrapolate to the threshold value. The data shown in Fig. 9 were taken with counter-propagating laser and ion beams. The measurement was repeated with co-propagating beams. The threshold energy was determined, corrected for the Doppler effect to all orders, from the geometrical mean of the measured blue- and red-shifted threshold energies. The energy of the threshold of the 2^2Pks channel was found to be 19 888.55(16) cm^{-1}. The well

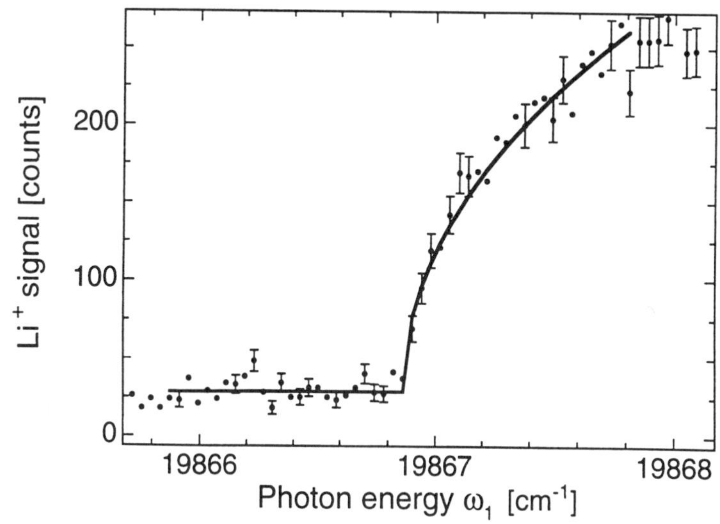

Fig. 9 Yield of Li^+ ions vs. photon energy in the region of the 2^2P threshold. The Li^+ signal is proportional to the partial cross section for photodetachement of Li^- via the 2^2Pks channel (See Fig. 1). The solid line is a fit of the Wigner law (Eq. 4) to the data in the range shown. The error bars on selected data points represent the shot noise.

known separation of the 2^2S-2^2P states of Li [34] was subtracted from the threshold energy to yield an electron affinity of 618.049(20) meV using the conversion factor 8065.5410 cm^{-1}/eV.

4. Conclusions

The structures of singly excited atoms and positive ions are well described within the independent particle approximation. A large base of experimental data is available for such systems. More loosely bound systems such as negative ions and doubly excited systems in general are, however, more sensitive to electron correlation since here the Coulomb field becomes less dominant. In these cases, the independent electron model begins to break down and collective motion becomes apparent. Experimental data on doubly excited states of negative ions, where correlation effects are of paramount importance, are relatively sparse. Until a few years ago, for example, a systematic investigation of doubly excited states had only been made in the case of the simplest two electron negative ion, H^-.

In the present work we have extended the study of double excitation to the He^- and Li^- ions which also have two active valence electrons. We have determined the positions and widths of several states in these systems by analyzing resonances in photodetachment cross sections. The measurements were performed under conditions of high sensitivity and resolution since production cross sections are typically small and resonances are close lying in energy. The sensitivity and resolution were enhanced by combining a collinear laser-ion beam apparatus with the detection of positive ions produced by resonantly ionizing the residual atoms from the photodetachment process. This state selective detection scheme, which is based on the well established technique of Resonance Ionization Spectroscopy, greatly extends the field of negative ion spectroscopy. It is our intention to continue these studies and investigate doubly excited states in He^- and Li^- and their branching into different continua, all the way to the double detachment limits.

5. Acknowledgements

We would like to take the opportunity to acknowledge the support and encouragement that we have received from Professor Ingvar Lindgren throughout this work. As a theorist, his interest in our work most certainly stems from the enhanced role played by electron correlation in negative ion physics. As a former practicing experimentalist, however, we suspect he derives great pleasure from observing how the development of new experimental techniques allows investigations of more and more subtle effects in atomic systems. Thanks Ingvar, we wish you well in your future pursuits.

We also thank E. Lindroth, A. F. Starace, C.-N. Liu, J. Xi and C. F. Fischer for fruitful discussions and for providing us with unpublished data. Financial support for this research has been obtained from the Swedish Natural Science Research Council (NFR). DJP acknowledge support from the Swedish Institute, The Royal Swedish Academy of Sciences and the US Department of Energy, Office of Basic Energy Sciences, Division of Chemical Sciences. UB acknowledge personal support from the Swedish Institute and, finally, IYK acknowledge support from the Wenner-Gren Center Foundation.

6. References

[1] For example, P. G. Hansen, A. S. Jensen and B. Jonson, Annual Review of Nuclear and Particle Science, **45**, eds. C. Quigg, V. Lueth and P. Paul (Palo Alto, CA, 1995), p.591.

[2] For example, P. G. Harris *et al.*, Phys. Rev. Lett. **65**, 309 (1990); H. C. Bryant *et al.*, Phys. Rev. Lett. **38**, 228 (1977); M. E. Hamm *et al.*, Phys. Rev. Lett. **43**, 1715 (1979); M. Halka *et al.*, Phys. Rev.A **44**, 6127 (1991).

[3] For example, H. R. Sadeghpour and C. H. Greene, Phys. Rev. Lett. **65**, 313 (1990); H. R. Sadeghpour, C. H. Greene and M. Cavagnero, Phys. Rev. A **45**, 1587 (1992); J. Z. Tang *et al.*, Phys. Rev. A **49**, 1021 (1994).

[4] See, for example, J. W. Cooper, U. Fano and F. Pratts, Phys. Rev. Lett. **10**, 518 (1963); D. R. Herrick, M. E. Kellman and R. D. Poliak, Phys. Rev. A **22**, 1517 (1980); C. D. Lin, Phys. Rev. Lett. **69**, 1633 (1992); C. D. Lin, Phys. Rev. A **29**, 1019 (1984); C. D. Lin, Adv. At. Mol. Phys. **22**, 77 (1986).

[5] C. Pan, A. F. Starace and C. H. Greene, J.Phys. B **27**, L137 (1994).

[6] E. Lindroth, Phys. Rev. **A 52**, 2737 (1995).

[7] S. L. Kaufman, Optics Comm. **17**, 309 (1976).

[8] For example, G. S. Hurst, M. G. Payne, S. D. Kramer and J. P. Young, Rev. Mod. Phys. **51**, 767 (1979).

[9] P. Balling *et al.*, J. Phys. B. **26**, 3531 (1993).

[10] U. Fano, Phys. Rev **124**, 1866 (1961).

[11] A. F. Starace, Phys. Rev. A **16**, 231 (1977).

[12] U. Ljungblad, Thesis at Göteborg University (1996).

[13] D. Hanstorp, J. Meas. Sci. Technol. **3**, 523 (1992).

[14] J. Dellwo, Y. Liu and D. J. Pegg, Phys. Rev. A **45**, 1544 (1992); Y. K. Bae and J. R. Peterson Phys. Rev. A **32**, 1917 (1985).

[15] For example, K. T. Chung and P. Fullbright, Phys. Scr. **45**, 445 (1992); C. Froese Fischer J. Phys. B. **26**, 855 (1993).

[16] For example, T. A. Patterson, H. Hotop, A. Kasdan, D. W. Norcross and W. C. Lineberger, Phys. Rev. Lett. **32**, 189 (1974).

[17] U. Berzinsh *et al.*, Phys. Rev. Lett. **74**, 4795 (1995).

[18] E. Lindroth, Phys. Rev. A **49**, 4473 (1994).

[19] S. Salomonson and P. Öster, Phys. Rev. A **40**, 5559 (1989).

[20] A. E. Klinkmüller *et al.*, Submitted to Phys. Rev. A.

[21] U. Ljungblad, D. Hanstorp, U. Berzinsh and D. J. Pegg, Phys. Rev. Lett. **77**, 3751 (1996).

[22] G. Haeffler *et al.*, Phys. Rev. A. **53**, 4127 (1996).

[23] S. J. Buckman and C. W. Clark, Rev. Mod. Phys. **66,** 539 (1994).

[24] J. Xi and C. Froese Fischer, Phys. Rev. A **53**, 3169 (1996).

[25] J. R. Peterson, Y. K. Bae and D. L. Huestis, Phys. Rev. Lett. **55**, 692 (1985).

[26] C. W. Walter, J. A. Seifert and J. R. Peterson, Phys. Rev. A. **50**, 2257 (1994).

[27] J. Xi and C. Froese Fischer, private communication.

[28] C. Pan, A. F. Starace and C. H. Greene, Phys Rev A **53**, 840 (1996).

[29] A. F. Starace and C.-N. Liu, private communication.

[30] E. P. Wigner, Phys. Rev., **73**, 1002 (1948).

[31] H. Hotop and W. C. Lineberger, J. Phys. Chem. Ref. Data **14**, 731 (1985).

[32] T. F. O´Malley, Phys. Rev. **137**, A1668 (1965).

[33] V. V. Petrunin *et al.* Phys. Rev. Lett.**75**, 1911 (1995).

[34] C. J. Sansonetti and B. Richou, Bull. Am. Phys. Soc. **40**, 1272 (1995).

Collinear Laser Spectroscopy on Radioactive Isotopes

J.R. Persson

School of Physics and Space Research
The University of Birmingham
Edgbaston
Birmingham B15 2TT
United Kingdom

Abstract.

The development of collinear laser spectroscopy has enabled measurements of isotope shifts and hyperfine structure over long chains of isotopes. This makes it possible to derive nuclear properties of isotopes far from stability. The experimental development is reviewed and some future prospects discussed.

1. Introduction.

At the end of the 19th century, the so called 'hyperfine' structure in spectral lines was discovered independently by Michelson [1], Fabry and Perot [2]. The origin of this structure, which consists of two parts, isotope shift (IS) and hyperfine structure (hfs), was in the case of IS explained partially by Bohr's atomic theory [3]. The first successful measurement of the IS, by Aronberg [4], in the 4058 Å line of lead, yielded such a large shift that it was soon realised that the isotope mass difference alone could not explain this. Ehrenfest [5] and Bohr [6] suggested that the larger shift could be due to differences in the nuclear structure of the isotopes, yielding a difference in the electric field of the nucleus, a field shift (FS). The first quantitative treatment of the field shift was made by Rosenthal and Breit [7] and Racah [8]. It was thus realised that measurements of the IS could give information on the size of the nucleus, i.e the nuclear charge distribution.

The other part of the 'hyperfine' structure was suggested by Pauli [9] to be the result of the magnetic interaction between the nucleus and the electrons in the atom. If hfs is due to only a magnetic interaction, the difference between two neighbouring hyperfine levels follows an interval rule. However, accurate measurements of Eu, performed by Schüler and Schmidt [10], showed a deviation from this rule. Theoretical calculations performed by Casimir [11] showed that the experimental result was in agreement with an electric quadrupole interaction with the nucleus, thus establishing a quadrupole moment for ^{151}Eu and ^{153}Eu.

ADVANCES IN QUANTUM CHEMISTRY, VOLUME 30

0065-3276/98 $25.00

While the early optical measurements suffered from limited resolution, the development of atomic beam methods provided a useful tool in the study of atomic and nuclear magnetic moments [12,13] (for a review see [14]) and it became possible to measure the nuclear magnetic moments (and nuclear spins) in a direct way for both stable and radioactive isotopes, by using a variety of methods [15]. The study of optical IS was, however, limited to Doppler-limited optical spectroscopy until the invention of the laser and the development of suitable high-resolution optical methods (a review can be found in [16]). It is also possible to obtain information on the nuclear charge distribution by electron scattering experiments and from muonic X-ray transitions and electron K X-ray IS [17], perhaps even with a higher accuracy than with optical spectroscopy.

However, optical (laser) spectroscopy offers high resolution and the possibility to measure isotope shifts using very small samples [16]. With the development of on-line mass separators, which provide long chains of isotopes extending far off stability, the interest and experimental work have increased. This article is focussing on the use of collinear laser spectroscopy in the study of nuclear properties, such as spin, radii and moments, but a short discussion on other methods will also be performed. In section 2, a short discussion on the origin of IS and hfs will be performed. Different experimental methods are presented in section 3. Results and special regions of interest are discussed in section 4 with concluding remarks in section 5.

2. Electron-nucleus interactions

The interaction between the nucleus and the electrons of an atom can conveniently be expanded in multipole terms

$$H_{hfs} = \sum_{k=0}^{\infty} \mathbf{T}^k \bullet \mathbf{M}^k \tag{1}$$

where $\mathbf{T}^k$ and $\mathbf{M}^k$ are tensor operators of rank k in the electronic and nuclear space, respectively.

The term $\mathbf{M}^0$ is the central part of the charge distribution of the nucleus. If we have two isotopes with different neutron number, the charge distribution will be slightly changed, thus changing the electric field inside the nucleus. For electrons with a non-zero wavefunction at the nucleus, this difference will cause a shift (field shift) in the energy levels between the isotopes. The field shift can in an approximate way be described as

$$\delta\nu_{FS} = \frac{Ze^2}{6h\varepsilon_0}\Delta|\Psi(0)|^2 \delta\langle r^2\rangle \tag{2}$$

Where $\Delta|\Psi(0)|^2$ is the change in the electron density at the nucleus, between the upper and the lower state in a transition, and $\delta\langle r^2\rangle$ is the change in the nuclear mean square charge radius between the isotopes. The electronic wavefunction is assumed to be constant inside the nucleus and equal to its value at r=0. This general approach is valid for a real system in such a way that the field shift can be separated into an atomic factor, depending on the electron density at the nucleus, and a nuclear factor, closely related to the difference in the mean square charge radii between the isotopes. For heavy isotopes the electron density can no longer be considered as constant and an expansion of the nuclear factor in to a power series of radial moments is required [18]. In addition to the field shift a mass dependent shift (mass shift), due to the fact that the nucleus is not infinitely heavy, is present. This mass shift consisting of two parts, the normal mass shift and the specific mass shift, must be evaluated in order to extract the field shift from experimental isotope shifts. The procedure is described in [19] and will not be presented here.

The nuclear tensor operators $\mathbf{M}^1$ and $\mathbf{M}^2$ are defined in terms of the nuclear magnetic dipole moment μ_I and the electric quadrupole moment Q

$$\left\langle II \middle| \mathbf{M}^1 \middle| II \right\rangle = \mu_I \tag{3}$$

$$\left\langle II \middle| \mathbf{M}^2 \middle| II \right\rangle = eQ/2 \tag{4}$$

The contributions of the magnetic octupole ($\mathbf{M}^3$) and the electric hexadecapole ($\mathbf{M}^4$) terms are always much smaller than the magnetic dipole and electric quadrupole terms and are therefore not considered here.

The magnetic dipole interaction and the electric quadrupole interaction, in an atom with nuclear spin I, will cause a level with angular momentum J to split into different energy levels denoted by the total angular momentum F (**F**=**I**+**J**), according to the Casimir formula [11],

$$E_F = \frac{1}{2}AC + B\frac{3/2\,C(C+1) - 2I(I+1)J(J+1)}{2I(2I-1)2J(2J-1)} \tag{5}$$

where $C = F(F+1) - I(I+1) - J(J+1)$. A and B are the hyperfine structure coupling constants, that can, as in the case of the field shift, be divided into two parts, one depending on electronic factors and one depending on nuclear properties (I, μ_I and Q) [20]. The electronic factors are thus the same for all isotopes of the same element, and therefore the ratio of A and B factors will give the nuclear moments for one isotope if the moment is known for the other. In reality, this is not the case since the s-electrons, which have a probability of being very close to or inside the nucleus, will not experience a point-like nucleus but the extended magnetisation will introduce an isotopical difference, the so-called hyperfine anomaly [21]. This effect is normally about 10^{-4} but has been found to be as large as 10%.

3. Experimental

3.1. Collinear Laser spectroscopy

The first experiment combining ion beams with laser excitation was performed by Andrä et al. [22] in 1973 in lifetime studies. Crossed-beam spectroscopy on ion-beams suffers from the short interaction time. This is why a collinear superposition of laser and ion beams is favourable. Another important advantage of a collinear geometry was proposed in 1976 by Kaufmann [23] and, independently, demonstrated by Wing et al. [24]. Since the kinetic energy spread(δE) of the ions remains constant under acceleration:

$$\delta E = \delta\left(\frac{mv^2}{2} \right) = mv\delta v = \frac{mc^2}{v^2} \Delta\nu_D \delta\nu_D \qquad (6)$$

it follows that the product of the average velocity(v) and velocity spread(δv) and the product of the Doppler shift($\Delta\nu_D$) and Doppler width($\delta\nu_D$) are constants of motion. Therefore, the acceleration of an ion beam (U), with a thermal velocity distribution (kT), gives rise to a reduced Doppler width equal to:

$$\delta\nu_D = \frac{1}{2}\sqrt{\frac{kT}{eU}}\,\delta\nu_D(0) \qquad (7)$$

A reduction of the Doppler width by a factor 10^3 is attainable. As an example, the barium resonance at 535 nm would, with an ion source temperature of 2000K and an acceleration of 60keV, have a residual Doppler width of about 1 MHz, compared with the natural linewidth of 19 MHz. In practice the experimental linewidths are 10-50 MHz depending on the natural linewidth, ion source, acceleration voltage stability and beam optics.

Another feature is the possibility of using a charge exchange cell producing a fast atomic beam. This is of importance as transitions in ions are in most cases in the UV or VUV region where single mode lasers are not readily available.

The standard method to obtain spectra is to lock the laser-frequency to a suitable atomic or molecular line, and apply a post-acceleration (or deceleration) voltage in the interaction region and Dopplertune the ions into resonance. The detection limit for this set-up is limited by the photo multiplier background, mainly due to scattered laser light and beam related light, still ion beams as low as 10^4 s^{-1} have given useful spectra.

It is of importance to improve the sensitivity when working on radioactive ion beams, where the beam intensity might be less than 100 s^{-1}. Rejection of background events becomes therefore important. One way of doing so is to only detect photons that come in coincidence with ions[25]. A schematic diagram of a set-up is shown in figure 1.

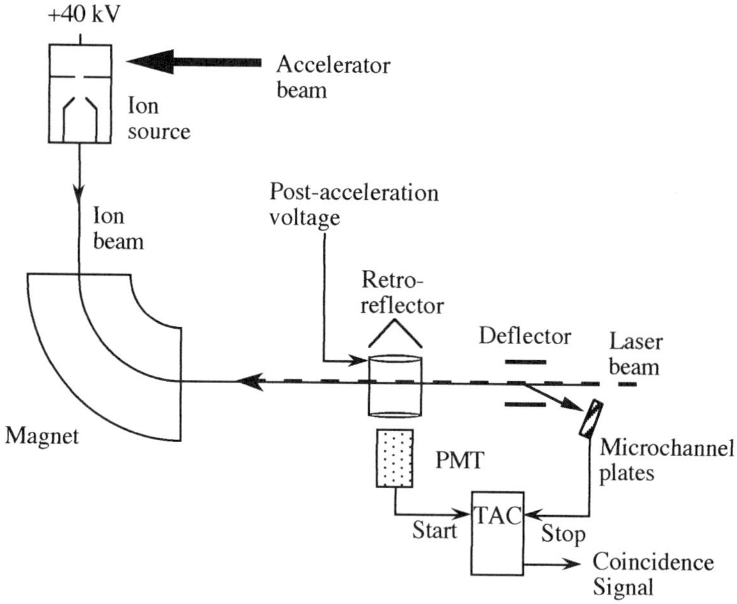

Figure 1. Schematic experimental set-up for coincidence collinear laser spectroscopy.

The ions are electrostatically deflected after passing the interaction/detection region, and detected by a microchannel plate. The time difference between the photon and the ion depends on the position of the ion when emitting the photon. An improvement by imaging the fluorescence on a position sensitive detector, yielding a determination of decay position to less than 1 mm and thus a time resolution of 20ns, made it possible to do measurements on ion beams of less than $100s^{-1}$. However, this method is limited by the maximal count-rate allowed on the channel plates. An alternative is to have a selective deflection, so that only when a photon has been detected the deflection is turned on and the ion detected.

3.2. Resonance Ionization Spectroscopy (RIS)

Resonance ionization of atoms and molecules by stepwise excitation is a well-established method[26], generally applied in trace element analysis [27]. Due to the high sensitivity, the method is well suited for applications in measurements of IS and hfs in radioactive isotopes. The IS and hfs is then probed in the first step of RIS, with a narrowband laser. For the other steps broadband lasers may be used. The first demonstration was performed at the St. Petersburg mass-separator on Eu isotopes [28] in 1983. Since then isotopic chains of other rare-earth elements have been investigated by the same group[29,30]. A collinear geometry has also been demonstrated by Schultz et al. [31] at ISOLDE on Yb, working on a charge exchanged atomic beam. Proposals for using RIS on ionic beams have been made, but no published results on such an experiment are available.

3.3. Ion Traps

The development of ion traps has yielded high-precision measurements of ground-state hfs using microwave-optical double resonance techniques [32]. It this way the magnetic octupole and electric hexadecapole hfs interaction constants have been determined in some cases[33]. Even if ionic traps open the possibility of precision measurements, difficulties arise when measuring IS as the velocity of the ions inside the trap might be very high, giving a large Doppler width. There are however different cooling techniques to overcome this, but it is doubtful if these can compete with collinear laser spectroscopy. It should be noted that ion traps are a very valuable tool for determining nuclear masses. The commission of the ISOLTRAP [34] at ISOLDE has given precise mass-measurements for a number of radioactive isotopes.

4. Results

Over the years substantial amounts of data on nuclear spins, moments and sizes have been gathered. Several compilations have been made of the data available [16,19,35-37]. The development of laser techniques has provided an increased sensitivity which together with the development of isotope separators on line (ISOL) has made measurements on nuclei far from stability possible[16]. The most important feature of the ion source, when studying nuclei far from stability, is that the release time for the element is short. In conventional ion sources this is possible for a limited number of elements.

The development of ion-guides has provided an ion-source that is both very fast and in principle independent of the chemical properties of the element under study [38]. This opens two regions of particular nuclear interest for laser spectroscopical studies: nuclear size and moments of isomers and the shape transition at N=59, showing a coexistence of spherical and deformed bands in for example ^{100}Sr and ^{100}Zr [39]. It is thus of interest to extend the systematic measurements performed in Rb and Sr [16] with measurements in Zr. As zirconium is a refractory element, it is not possible to obtain ion beams in conventional ion sources, so the ion-guide technique must be used. The combination of ion-guides and collinear laser spectroscopy was believed to suffer from a large energy spread of the ions, but measurements performed on Ba have shown that the energy spread can be reduced to the same order of magnitude as from conventional ion sources [40]. This makes studies in the unexplored Z=39 to 46 region, where no radioactive isotope shift data exist, feasible.

5. Conclusion

Laser spectroscopic studies of radioactive isotopes have proved to be a valuable source in obtaining nuclear properties. The continuing developments promise a way of meeting challenge in measuring nuclear properties far from stability. It may also be so that new isotopes are discovered by optical rather than nuclear methods. The extreme high precision measurements in ionic traps make rather small nuclear effects such as hyperfine anomaly interesting for both experimental and theoretical studies. In addition the theories of hyperfine structure and isotope shift are well understood, so that detailed information on nuclear properties can be extracted.

The illuminating discussions with Dr. Griffith and Prof. Hühnermann are kindly acknowledged.

References

[1] A. Michelson, Phil. Mag. **31**, 338 (1891)
[2] C.Fabry and A. Perot, Ann. Chim. et Phys. **10**, 457 (1903)
[3] N. Bohr, Phil. Mag. **26**, 1 (1913)
[4] L. Aronberg, Astrophys. J.**47**, 96 (1918)
[5] P. Ehrenfest, Nature **109**, 745 (1922)
[6] N. Bohr, Nature **109**, 746 (1922)
[7] J.E. Rosenthal and G. Breit, Phys. Rev. **41**, 459 (1932)
[8] G. Racah, Nature **129**, 723 (1932)
[9] W. Pauli, Naturwiss. **12**, 74 (1924)
[10] H. Schüler and T. Schmidt, Z. Phys. **94**, 457 (1935)
[11] H.B.G. Casimir, On the Interaction Between Atomic Nuclei and Electrons, (Teyler's Tweede Genootschap, Haarlem, 1936; new ed. W.H. Freeman Comp., San Francisco, 1963.)
[12] Gerlach and O. Stern, Ann. Physik, **74**, 673 (1924) and **76**, 163 (1925)
[13] I. I.Rabi, J. R. Zacharias, S. Millman, and P.Kusch, Phys. Rev. **53**, 318, (1938)
[14] N.F. Ramsey, Molecular Beams (Oxford University Press, 1956, new ed. 1985)
[15] W.A. Nierenberg and I Lindgren, in Alpha-, beta-, and gamma- ray spectroscopy, ed. K. Siegbahn, Vol 2. (North-Holland, Amsterdam, 1965) p.1263
[16] E. W. Otten, Treatise on Heavy-Ion Science Vol. 8, ed. D. Allan Bromley (Plenum, New York 1989)
[17] R. C. Barret and D. F. Jackson, Nuclear Sizes and Structure (Clearendon Press, Oxford 1977)
[18] E.C. Seltzer, Phys. Rev. **188**,1916 (1969)
[19] W.H. King, Isotope Shifts in Atomic Spectra, (Plenum, New York 1984)
[20] See for example. I. Lindgren and A. Rosen, Case Stud. in At. Phys. **4**, 93 (1974)
[21] A. Bohr and V.F. Weisskopf, Phys Rev. **77**, 94 (1950)

[22] H.J. Andrä, A. Gaupp, K. Tillman, W. Wittman, Nucl. Instr. Meth. **110**, 453 (1973)

[23] S. L. Kaufmann, Opt. Comm., **17**,309 (1976)

[24] W.H. Wing, G.A. Ruff, W.E. Lamb Jr, J.J. Spezeski, Phys. Rev. Lett. **36**, 1488 (1976)

[25] D.A. Eastham, A. Gilda, D.D. Warner, D.E. Evans, J.A.R. Griffith, J. Billows, M.P. Dancy, I.S. Grant, Opt. Comm. **82**, 23 (1991)

[26] V.S. Letokhov, Laser Photoionization Spectroscopy (Academic, New York 1987)

[27] G.S. Hurst, M.G. Payne (eds), Principles and Applications of Resonace Ionization Spectroscopy (Adam Hilger, Bristol 1988)

[28] G.D. Alkhazov, A.E. Barzakh, E.I. Berlovich, V.P. Denisov, A.J. Dernyatin, V.S. Ivanov, A.N. Zherikhin, O.N. Kompanets, V.S. Letokhov, V.I. Mishin, V.N. Fedoseyev, JETP lett. **37**, 274 (1983)

[29] G.D. Alkhazov, A.E. Barzakh, V.A. Bolshakov, V.P. Denisov, V.S. Ivanov, Yu.Ya. Sergeev, I.Ya. Chubukov, V.I. Tikhonov, V.S. Letokhov, V.I. Mishin, S.K. Sekatsky, V.N. Fedoseyev, Z. Phys. A **337**, 257 (1990)

[30] G.D. Alkhazov, A.E. Barzakh, V.P. Denisov, V.S. Ivanov, I.Ya. Chubukov, V.S. Letokhov, V.I. Mishin, S.K. Sekatsky, V.N. Fedoseyev, Z. Phys. A **337**, 367 (1990)

[31] Ch. Schulz, E. Arnold, W. Borchers, W. Neu, R. Neugart, M. Neuroth, E.W. Otten, M. Scherf, K. Wendt, P. Lievens, Yu.A. Kudryavtsev, V.S. Letokhov, V.I. Mishin, V.V. Petrunin, J.Phys. B. **24**, 4831 (1991)

[32] G. Werth, Physica. Scripta **T59**, 206 (1995)

[33] O. Becker, K. Enders, G.Werth, J. Dembczynski, Phys. Rev. A **48**, 3546 (1993)

[34] G. Bollen, Physica. Scripta. **T59**, 165 (1995)

[35] P. Aufmuth, K. Heilig, A Steudel, Atom. Data Nucl. Data **37**, 455 (1987)

[36] P. Raghavan, Atom. Data Nucl. Data **42**, 189 (1989)

[37] E.G. Nadjakov, K.P. Marinova, Yu.P. Gangrsky, Atom. Data Nucl. Data **56**, 133 (1994)

[38] J. Ärje, J.Äystö, P. Taskinen, J. Honkanen, K. Valli, Nucl. Instr. Meth. B **26**, 384 (1987)

[39] J.H. Hamilton, P.G. Hansen, E.F. Zganjar, Rep. Prog. Phys. **48**, 631 (1985)

[40] J.L. Cooke, J. Billowes, P. Campbell, E.C.A. Cochrane, T. Cooper, P. Dendooven, Da. E. Evans, J. A. R. Griffith, I.S. Grant, A. Honkanen, M. Huhta, J.M.G. Levins, M. Oinonen, M.R. Pearson, H. Pentillä, J.R. Persson,G Tungate, P. Wheeler, L Zybert and J.Äystö, to be published (1996)

Four Decades of Hyperfine Anomalies

Martin G. H. Gustavsson and Ann-Marie Mårtensson-Pendrill
Department of Physics, Chalmers University of Technology and Göteborg University, SE-412 96 Göteborg, Sweden

Isotopic differences in the distribution of nuclear charge and magnetization give rise to "hyperfine structure anomalies" which were observed already in the 1950s. More recently, the distribution of nuclear magnetization has been found to complicate the interpretation of the measured hyperfine splittings in highly charged hydrogen-like ions. In this paper, results of numerical calculations for a few hydrogen-like systems (^{133}Cs, ^{165}Ho, ^{185,187}Re and ^{209}Bi) of current experimental interest are presented in terms of moments of the nuclear charge and magnetization distribution, thereby displaying directly the sensitivity and emphasizing the need for a better understanding of nuclear wavefunctions. In addition, we also present results of many-body perturbation theory calculations for Cs hyperfine anomalies, in connection with experiments planned at ISOLDE.

I. INTRODUCTION

Impressively accurate hyperfine structure measurements were made possible in the 1950s through the development of radio frequency spectroscopy methods, including the atomic beam magnetic resonance (ABMR) technique developed by Rabi and coworkers [1–4] and introduced in Sweden by Ingvar Lindgren. The instrument built in Uppsala, nicknamed "The Hippopotamus", was used to determine a large number of spins and magnetic moments as discussed by Ekström [5,6]. The studies of hyperfine structure (hfs) became an early focus for the atomic theory work in the Göteborg group, starting with the development of self-consistent field methods, relativistic, as well as non-relativistic [7,8]. The development of atomic many-body theory calculations [9], and later coupled-cluster methods for open-shell systems [10] enabled relatively accurate predictions of hyperfine structure in alkali atoms. More recently, radiative corrections to the hyperfine structure have been calculated [11,12] in connection with experiments on highly-charged ions that have now reached sufficient accuracy to be sensitive to higher-order radiative effects [13,14].

Close investigations of measured hyperfine structures reveal small differences in the ratios between the "A-factors" for the hyperfine structure and the

ADVANCES IN QUANTUM CHEMISTRY, VOLUME 30

0065-3276/98 $25.00

nuclear g-factors, $g_I = \mu_I / I$. The resulting "hyperfine anomalies"

$$^1\Delta^2 = \left(\frac{A_1}{g_{I,1}} \frac{g_{I,2}}{A_2} - 1 \right)$$

between isotopes 1 and 2 have been listed for a number of stable isotope pairs by Arimondo *et al* in their 1977 review [15] of hyperfine structure data. Additional examples of measured hyperfine anomalies are given e.g. by Stroke *et al* [16]. The problem was analysed in pioneering work by Bohr and Weisskopf [17,18] and by Rosenthal, Breit [19] and others [20,21] emphasizing the importance of the distribution of nuclear magnetization and charge, respectively.

Nuclear charge distributions are often relatively well known [22,23] from electron scattering [23] and from energy levels in muonic atoms [24]. Isotopic differences in the charge distribution, also for radioactive isotopes, can be deduced from optical isotope shifts of spectral lines [25] and from K_α X-ray isotope shifts [26]. Much less is, however, known about magnetic moment distributions. One step into the investigation was taken in earlier work [27], where we combined many-body perturbation theory calculations for ^{203,205}Tl with hyperfine anomalies measured in the 1950s [28,29] and more recent experimental isotope shift data [30]. In this way it was possible to extract information about the magnetization distribution, which was found to change about twice as much as the charge distribution between the two isotopes.

Magnetic moment distributions, in fact, obscure the comparison between experimental and theoretical results for the hyperfine structure of highly charged hydrogen-like ions, such as $^{209}Bi^{82+}$ [13], $^{165}Ho^{66+}$ [14] and $^{185,187}Re^{74+}$ [31]. There is thus renewed interest in the analysis of the so-called "Bohr–Weisskopf" effect [17,18], calling for an improved theoretical description of the nucleus [32], and a suitable way to relate the electronic and nuclear parts of the calculation. The situation is similar to the studies of parity non-conservation, where the unknown neutron distribution would lead to an uncertainty in the interpretation of experiments for chains of isotopes [33,34], larger than the expected experimental uncertainty.

Series expansions of the electronic wavefunctions close to the nucleus can be obtained analytically for certain nuclear distributions, and were utilized already in the early days of hyperfine anomaly studies [17–19,21,16]. Shabaev [35] recently followed this approach to calculate the hyperfine structure for a number of hydrogen-like systems of experimental interest, using a homogeneous distribution of magnetization and charge, both with a radius R_0 given by $R_0^2 = \frac{3}{5}\langle r_c^2 \rangle$, using tabulated values for the charge radii, $\langle r_c^2 \rangle^{1/2}$ [36].

In this work, we use instead a direct numerical solution to extend that analysis to include changes in the charge distribution with respect to a reference distribution chosen as a two-parameter Fermi distribution, with the same value for $\langle r_c^2 \rangle$. Our parametrization also displays more clearly the sensitivity to uncertainties in the distribution of magnetization and the need for an improved understanding.

Section II describes the potential from Fermi and Fourier–Bessel distributions of nuclear charge. Section III shows the effect on the hyperfine structure of hydrogen-like systems, caused by the dependence of the wavefunction on the nuclear charge distribution — the Breit–Rosenthal effect. Section IV gives results for the Bohr–Weisskopf effect, due to the magnetization distribution. In Sec. V, we show how our calculations can be combined with known isotope shift data to analyse observed anomalies for a few Cs isotopes. Additional data are expected from experiments for radioactive Cs isotopes, where Ingvar Lindgren has been involved in the planning [37].

After four decades, the study of hyperfine anomalies may see a revival!

II. NUCLEAR CHARGE DISTRIBUTIONS

The approximation of the nucleus as an infinitely heavy point charge makes possible analytical solution of the Dirac equation for the hydrogen-like problem. The resulting orbitals are, however, too tightly bound and clearly unphysical within the nucleus. A homogeneously charged nucleus is a significant improvement and is sufficient for many applications. For more detailed studies of nuclear properties, it is, however, desirable to use a more physical nuclear distribution, such as the Fermi and the Fourier–Bessel distributions described below.

A. The two-parameter Fermi model

The two-parameter Fermi model gives a realistic description of the nuclear distribution [38,39], and at the same time provides considerable flexibility in the analysis:

$$\rho_{\mathrm{c}}(r) = \frac{\rho_0}{1 + \mathrm{e}^{(r-c)/a}}. \tag{1}$$

(The subscript, c, is used to distinguish the charge distribution from the magnetization distribution, discussed below.) Here, c is the half-density radius and a is related to the *skin thickness* t by $t = (4 \ln 3)a$, defined as the distance over which the charge falls from 90% of its central value to 10%. Experiments indicate that the value of the parameter a is roughly constant at 0.524 fm, corresponding to a skin thickness of about 2.30 fm for most nuclei. The case of a homogeneously charged nucleus with a radius $R_0 = c$ can be reproduced by choosing $a = 0$ in the Fermi distribution.

The shape of an arbitrary nuclear distribution can often be adequately described by the moments $\langle r^{2n} \rangle$ of the distribution. For the Fermi distribution above, these moments are given, to a good approximation [40], by the relations

$$\langle r_c^2 \rangle \approx \frac{3}{5}c^2 + \frac{7}{5}\pi^2 a^2,$$
$$\langle r_c^4 \rangle \approx \frac{3}{7}c^4 + \frac{18}{7}\pi^2 a^2 c^2 + \frac{31}{7}\pi^4 a^4,$$
$$\langle r_c^6 \rangle \approx \frac{3}{9}c^6 + \frac{11}{3}\pi^2 a^2 c^4 + \frac{239}{15}\pi^4 a^4 c^2 + \frac{127}{5}\pi^6 a^6.$$

(Exact relations are given in Appendix.) These formulæ are used to obtain the parametrization of numerical results, in Sec. III.

B. The potential from a Fermi distribution

The potential energy, $v(r)$, for an electron from an arbitrary, spherically symmetric, charge distribution, $\rho(r)$, can be written as

$$v(r) = -\frac{4\pi}{r}\int_0^r \rho(r')r'^2 dr' - 4\pi \int_r^\infty \frac{1}{r'}\rho(r')r'^2 dr'.$$

Outside the charge distribution, the potential energy reduces to the Coulomb expression, $v(r) = -Z/r$. (We use atomic units, where $e = m_e = 4\pi\epsilon_0 = \hbar = 1$.) For the special case of a homogeneous nucleus, the potential inside the nuclear radius, R_0, becomes $-Z\left[3 - (r/R_0)^2\right]/2R_0$, giving a less singular behaviour of the orbitals and thus, e.g. for the integrals of the hyperfine interaction. For the Fermi distribution (1), the electrostatic potential cannot be obtained analytically. However, convenient expressions for the evaluation of the potential have been given by Parpia and Mohanty [41]. These expressions are used in the procedure adopted in the general atomic structure package of computer programs GRASP^2 [42,43], and are summarized briefly below.

The radial integrals can be written in terms of infinite series, that converge to machine precision with just a few terms included. The potential energy for $r < c$ can be written as

$$v_{r<c}(r) = -\frac{N}{r}\left[\frac{3r}{2c} - \frac{r^3}{2c^3} + \frac{r\pi^2 a^2}{2c^3} - \frac{3ra^2}{c^3}S_2\left(\frac{r-c}{a}\right) + \frac{6a^3}{c^3}S_3\left(\frac{r-c}{a}\right) - \frac{6a^3}{c^3}S_3\left(-\frac{c}{a}\right)\right],$$

and for the case $r > c$ we have

$$v_{r>c}(r) = -\frac{N}{r}\left[1 + \frac{\pi^2 a^2}{c^2} + \frac{3ra^2}{c^3} S_2\left(-\frac{r-c}{a}\right) + \frac{6a^3}{c^3} S_3\left(-\frac{r-c}{a}\right) - \frac{6a^3}{c^3} S_3\left(-\frac{c}{a}\right)\right],$$

where S_k and N are given by

$$S_k(x) = \sum_{n=1}^{\infty}(-1)^n \frac{e^{xn}}{n^k}, \tag{2}$$

$$N = \frac{Z}{\left[1 + \frac{\pi^2 a^2}{c^2} - \frac{6a^3}{c^3} S_3\left(-\frac{c}{a}\right)\right]}.$$

Generalizations of the Fermi model to describe deformed nuclei have been applied e.g. in the study of energies for highly charged uranium [44,45]. The nuclear radius parameter c in the Fermi distribution (1) is then replaced by

$$R(\mathbf{r}) = c\left[1 + \beta_2 Y_{20}(\mathbf{r}) + \beta_4 Y_{40}(\mathbf{r})\right],$$

which includes an explicit dependence on the angular coordinates.

C. The Fourier–Bessel expansion

A more direct use of accurate experimental data is provided by analysis a model-independent approach, such as the Fourier–Bessel expansion of the nuclear charge distribution [46].

Within the interval $r \leq R$ the charge distribution is expanded into a Fourier–Bessel series and beyond the cutoff-radius R, the charge distribution is assumed to be zero:

$$\rho(r) = \begin{cases} \sum_{\nu} a_\nu j_0(\nu\pi r/R) & \text{for } r \leq R \\ 0 & \text{for } r > R, \end{cases}$$

where $j_0(x) = \sin(x)/x$ is the zeroth order spherical Bessel function. The first coefficients are deduced from experiments and tabulated, e.g. by de Vries *et al* [23] and Fricke *et al* [22].

The expressions for the total charge, Q, the potential energy $v(r)$ and the second moment, $\langle r_c^2 \rangle$, of the charge distribution are:

$$Q = 4\pi \sum_{\nu} a_\nu \left(\frac{R}{\nu\pi}\right)^3 \nu\pi(-1)^{\nu+1},$$

$$v(r) = \begin{cases} -4\pi \sum_{\nu} a_\nu \left(\frac{R}{\nu\pi}\right)^2 \left[j_0(\nu\pi r/R) - (-1)^\nu\right] & \text{for } r \leq R \\ -\frac{Z}{r} & \text{for } r > R, \end{cases}$$

$$\langle r_c^2 \rangle = \frac{4\pi}{Q} \sum_{\nu} a_\nu \left(\frac{R}{\nu\pi}\right)^5 \nu\pi(-1)^\nu \left[6 - (\nu\pi)^2\right].$$

This potential has been applied in the case of hydrogen-like Bi, as discussed in Sec. III.

D. Binding energies, nuclear distributions and isotope shifts

The change in binding energies due to changes in the nuclear charge distribution has been analysed in the context of studies of optical isotope shifts, with the aim of obtaining information about changes in the nuclear charge radii, as reviewed e.g. by King [47]. The "field" or "volume" isotope shifts is proportional to the change in electron density at the nucleus between the two states of a transition. In a first approximation only s electrons contribute and, following Kopferman [48], the size of the shift can be estimated from the hyperfine structure, since both effects arise within or close to the nucleus. This type of relation is useful also for an estimate of the leading terms in the Bohr–Weisskopf effect, discussed in Sec. IV. (The validity of this approach for many-electron systems is discussed in some detail in Refs. [49–51].) Quantitative studies by Bauche [52,53] using the Hartree–Fock approach, form the basis for the classic tabulations by Heilig and Steudel [25] (later extended by Aufmuth *et al* [54] to include more recent calculations).

Torbohm, Fricke and Rosén [55] used relativistic self-consistent field (RSCF) calculations to express the wavefunctions within the nucleus as a power series expansion, which was used to express the field isotope shift in terms of the moments $\langle r_c^{2n} \rangle$ of the charge distribution. The effect of the higher moments is commonly accounted for by introducing a parameter $\lambda_c = \kappa_c \langle r_c^2 \rangle$, where the correction $1 - \kappa$ is found to be about 3% for $Z \approx 55$, 5% for $Z \approx 70$, and 6% for $Z \approx 80$. For hyperfine structure, the higher moments are found to be considerably more important, as discussed below.

Isotope shifts in a spectral line arises not only through differences in the charge distribution, but also from differences in nuclear mass, M. The "normal" mass shift is simply obtained by replacing the electron mass m_e by the reduced mass $\mu = m_e/(1 + m_e/M)$, whereas the so-called "specific mass shift" (SMS) arising from a correlation of orbital momenta through the motion of the nucleus. The SMS is very sensitive to correlation effects, and poses a challenge to theoreticians. In earlier work [56] we performed lowest-order calculation for Cs, leading to revised values for the changes in nuclear charge radii for the Cs isotopes. These values are used in Sec. III to estimate the size of the "Breit–Rosenthal" effect.

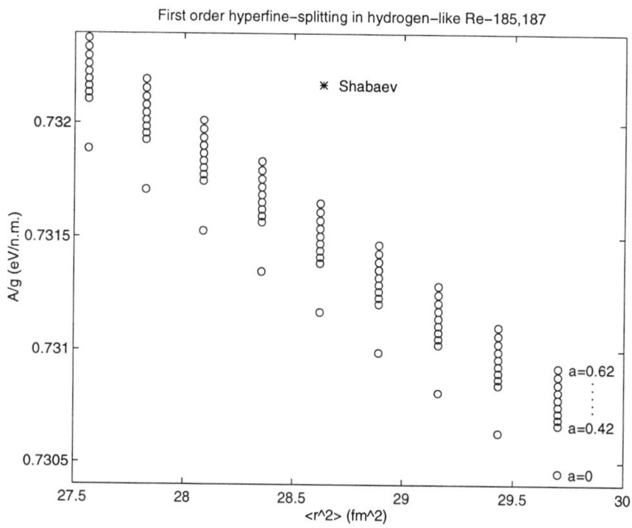

FIG. 1. The dependence of the hyperfine splitting in hydrogen-like ^{185,187}Re on the expectation value $\langle r_c^2 \rangle$ of the nuclear charge distribution. The hyperfine splitting constant A divided by the nuclear g-factor, g_I, is plotted as a function of $\langle r_c^2 \rangle$. The energy splitting has been calculated for several different values of the a-parameter in the two-parameter Fermi distribution Eq. (1), showing an essentially linear dependence of $\langle r_c^2 \rangle$ within the range studied.

III. HYPERFINE STRUCTURE AND NUCLEAR DISTRIBUTIONS

The magnetic moment, $\boldsymbol{\mu} = g_S \boldsymbol{S}_n + g_L \boldsymbol{L}_n$, of a nucleus (with $I \neq 0$) gives rise to a hyperfine interaction with the electrons, which can be described by a Hamiltonian

$$h_{\mathrm{p}}^{\mathrm{hfs}} = \hat{\boldsymbol{r}} \cdot \frac{\boldsymbol{\mu} \times \boldsymbol{\alpha}}{r^2} \tag{3}$$

for a point magnetic dipole. More general distributions of the magnetization lead to modifications of the radial dependence, $1/r^2$ in (3), as discussed in

Sec. IV. The hyperfine interaction can be expressed in terms of an "effective operator", $h^{\text{hfs}} = A\boldsymbol{I}\cdot\boldsymbol{J}$, and measured hyperfine structures are often analysed in terms of A factors. In the tables we present the results in terms of the purely electronic factor A/g_I.

The hyperfine interaction is sensitive to the wavefunction close to the nucleus and changes in the nuclear charge density leads to small changes in the hyperfine structure. The leading correction to the electronic wavefunction is proportional to the expectation value $\langle r_c^2\rangle$ of the nuclear charge distribution, so an uncertainty in $\langle r_c^2\rangle$ will lead to an uncertainty in the hyperfine structure. Figure 1 shows the sensitivity to changes in $\langle r_c^2\rangle$, for the case of Re ($Z = 75$). The dependence was investigated for a few different shapes of the charge distribution, obtained by varying the skin thickness. This effect is shown more directly in Fig. 2.

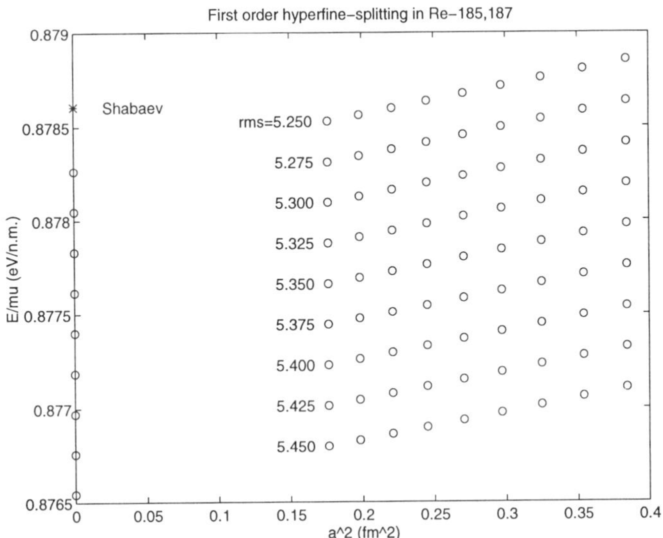

FIG. 2. The dependence of the hyperfine splitting in hydrogen-like ^{185,187}Re on the skin thickness parameter, a, of the nuclear charge distribution (1). The hyperfine splitting constant A divided by the nuclear g-factor, g_I, is plotted as a function of a^2 for several different values of the rms radius, $\langle r_c^2\rangle$.

The sensitivity to changes in the charge distribution is seen most easily by writing the hyperfine parameter in terms of changes in the moments, $\langle r_c^{2n}\rangle$, of the charge distribution, relative to a reference distribution, ρ_c^0:

$$\frac{A}{g_I} = \frac{A_0}{g_I}\Big[1 + x_2\,\delta\langle r_c^2\rangle + x_4\,\delta\langle r_c^4\rangle + x_6\,\delta\langle r_c^6\rangle\Big]. \tag{4}$$

For 0.42 fm $< a <$ 0.62 fm the relative error in the hyperfine structure obtained using this approximation is at most 10^{-5}. Alternatively, the calculated hyperfine splittings can be written in terms of $\delta\langle r_c^2\rangle$ and δa^2:

$$\frac{A}{g_I} = \frac{A_0}{g_I}\left[1 + x_r\ \delta\langle r_c^2\rangle + x_a\ \delta a^2\right]. \tag{5}$$

TABLE I. Nuclear properties for the systems studied, together with a comparison of the hyperfine structure parameters obtained for a homogeneous distribution and a Fermi distribution, respectively. In addition, the x-parameters in Eqs. (4) and (5) are given, showing the sensitivity to changes in the charge distribution. The numbers for ^{185}Re are also valid for ^{187}Re except for the magnetic moment, which is $\mu = 3.2197(3)\ \mu_N$[a]. (All values are obtained for a point distribution of the nuclear magnetization.)

System	$^{133}Cs^{54+}$	$^{165}Ho^{66+}$	$^{185}Re^{74+}$	$^{209}Bi^{82+}$
Z	55	67	75	83
μ (μ_N)	2.582024(34)[a]	4.132(5)[b]	3.1871(3)[a]	4.1106(2)[a]
I^π	$7/2^+$	$7/2^-$	$5/2^+$	$9/2^-$
$\langle r^2\rangle^{1/2}$ (fm)	4.807[c]	5.21[cd]	5.351[c]	5.519[d]
$\langle r^4\rangle^{1/4}$ (fm)[e]	5.145	5.578	5.700	5.849
$\langle r^6\rangle^{1/6}$ (fm)[e]	5.432	5.889	5.987	6.111
a (fm)[f]	0.524	0.57	0.524	0.468
A/g_I (eV/μ_N)[g]				
Homogeneous nucleus				
Shabaev[h]	0.224349	0.465024	0.732172	1.13797
This work	0.224277	0.464705	0.731160	1.13582
Fermi nucleus[e]				
This work	0.224330	0.464897	0.731500	1.13656
Parameters				
x_r (10^{-3}/fm^2)	−0.452	−0.683	−0.925	−1.25
x_a (10^{-3}/fm^2)	0.880	1.31	1.73	2.29
x_2 (10^{-3}/fm^2)	−0.710	−1.10	−1.49	−2.09
x_4 (10^{-5}/fm^4)	0.54	0.77	1.0	1.5
x_6 (10^{-8}/fm^6)	−2	−3	−3	−6

[a]Raghavan [57].
[b]Nachtsheim [58].
[c]Johnson and Soff [36].
[d]de Vries *et al* [23].
[e]The values for $\langle r^4\rangle^{1/4}$ and $\langle r^6\rangle^{1/6}$ are obtained for a Fermi distribution using tabulated values for $\langle r^2\rangle^{1/2}$ and a skin thickness parameters a as shown in the table. For the homogeneous nucleus with radius R the higher moments are given by $\langle r^{2n}\rangle = 3R^{2n}/(2n+3)$.
[f]For Cs and Re, the standard value has been used. The values for Ho and Bi are taken from the tabulations by de Vries *et al* [23].
[g]The hyperfine splitting is given by $\Delta E = A(I+1/2)$ and the nuclear g-factor is given by $g_I = \mu_I/I$.
[h]Shabaev [35]. For Bi, he used the value $\langle r^2\rangle^{1/2} = 5.531$ fm, taken from Ref. [36], slightly larger than the value given by de Vries *et al* [23].

A relation between the two parametrizations can be derived by expressing the hyperfine structure as

$$A = A(c, a) = A\left(\delta\langle r^2\rangle, \delta\langle r^4\rangle\, \delta\langle r^6\rangle\right).$$

This expression for A can then be differentiated with respect to c and a, using the above expressions between $\delta\langle r_c^{2n}\rangle$, in terms of c^2, and a^2.

Table I shows numerical results for $^{133}Cs^{54+}$, $^{165}Ho^{66+}$, $^{185,187}Re^{74+}$ and $^{209}Bi^{82+}$. The results for a homogeneous nuclear charge distribution are compared to the results given by Shabaev [35], who used analytical expressions and series expansions to evaluate the hyperfine structure. In that work, the effect of the nuclear charge distribution was treated only to lowest order in ZR_0/a_0 — the higher order effect may be responsible for the small discrepancy between our values and his, which, as seen in Table I, grows more important for higher Z. As seen below, this discrepancy between our numerical results and the values given by Shabaev is already larger than the uncertainty resulting from the uncertainty in $\langle r_c^2\rangle$.

The potential from a Fourier–Bessel expansion, discussed in Sec. II C, of the nuclear charge has been applied to the case of hydrogen-like Bi, using parameters given by [23]. For this distribution, $\langle r_c^4\rangle^{1/4} = 5.86060$ fm and $\langle r_c^2\rangle^{1/6} = 6.12840$ fm, and the hyperfine structure for a point magnetic dipole is given by $A/g_I = 1.13666$ eV$/\mu_N$, agreeing with the parametrization given by Eq. (4) and Table I with a relative error of about 1.3×10^{-5}.

IV. THE DISTRIBUTION OF NUCLEAR MAGNETIZATION

The calculations above, used a point nuclear magnetic moment, which is, of course, unrealistic. The magnetic moment distribution can, to a first approximation, be expected to follow the distribution of the unpaired nucleon(s), whereas *all* protons contribute to the charge distribution. We can thus expect $\langle r_m^2\rangle$ to be larger than $\langle r_c^2\rangle$ and also to find that the magnetization distribution is more sensitive to perturbations. Investigations of the "hyperfine anomalies" for the Tl isotopes showed that the difference in magnetization distribution between ^{203}Tl and ^{205}Tl is about twice as large as the change in the charge distribution [27].

In general the corrections for the distribution of nuclear magnetization are small and can be expressed in terms of a parameter ε, which relates the A-factors for point magnetization and distributed magnetization as

$$A_{\text{distributed}} = A_{\text{point}}(1 - \varepsilon).$$

As a starting point we consider a nucleus with the magnetization concentrated on a shell with radius, R. An arbitrary distribution $w(R)$ can, of course, be obtained as a superposition of contributions from different values R. In addition, a distinction must be made between different types of contributions to the nuclear magnetic moment μ since the interaction of an electron with a nuclear spin and orbital magnetic moment, respectively, have different radial dependence within the nucleus.

If the magnetic moment is due to the nuclear spin, the volume inside R gives no contribution, so the radial integral $\int fgdr$ should be corrected for the inner region, with a fraction

$$\kappa_S(R) = \frac{\int_0^R fgdr}{\int_0^\infty fgdr},$$

where f and g are the radial parts of the large and small component of the Dirac wavefunction, respectively.

If, instead, the magnetic moment arises from an *orbital* nuclear magnetic moment, the volume inside the shell leads to a correction

$$\kappa_L(R) = \frac{\int_0^R \left(1 - \frac{r}{R^3}\right) fgdr}{\int_0^\infty fgdr}.$$

These ratios are, in fact, essentially independent of the principal quantum number n. Both numerator and denominator receive their main contribution close to the nucleus, where the behaviour of the atomic wavefunction is determined mainly by the orbital angular momentum. The situation is analogous to the relation between field isotope shifts and hyperfine structures, commonly used to analyse experimental isotope shifts [25,48,53]

Just as in the case of isotope shifts discussed in Sec. II D, we expect the effect to scale essentially as R^2, but also higher moments of the magnetization distribution continue, and the integrals can be approximated by

$$\kappa_S = a_2R^2 + a_4R^4 + a_6R^6,$$
$$\kappa_L = b_2R^2 + b_4R^4 + b_6R^6.$$

The effect for an arbitrary charge distribution can be written in terms of

$$\langle\kappa_S\rangle = 4\pi \int_0^\infty \kappa_S(R) w_S(R) R^2 dR = a_2\langle R_S^2\rangle + a_4\langle R_S^4\rangle + a_6\langle R_S^6\rangle, \tag{6}$$

$$\langle\kappa_L\rangle = 4\pi \int_0^\infty \kappa_L(R) w_L(R) R^2 dR = b_2\langle R_L^2\rangle + b_4\langle R_L^4\rangle + b_6\langle R_L^6\rangle. \tag{7}$$

Here, we have allowed for the possibility that the orbital and spin part of the nuclear magnetization are described by different (normalized) distributions w_S

and w_L. The total correction due to the nuclear magnetization distribution can then be written as

$$\varepsilon = \alpha_S \left[\langle \kappa_S \rangle + \zeta \left(\langle \kappa_S \rangle - \langle \kappa_L \rangle \right) \right] + \alpha_L \langle \kappa_L \rangle,$$

where

$$\alpha_S = \frac{g_S}{g_I} \frac{g_I - g_L}{g_S - g_L}, \qquad \alpha_L = 1 - \alpha_S,$$

and the spin asymmetry parameter ζ is, in the extreme single particle model, given by [18]

$$\zeta = \begin{cases} \dfrac{2I - 1}{4(I+1)} & \text{for } I = l + \frac{1}{2} \\[2ex] \dfrac{2I+3}{4I} & \text{for } I = l - \frac{1}{2}. \end{cases}$$

In the extreme single particle model, the total nuclear angular momentum is considered to be due to one unpaired nucleon. The spin g-factor can then be determined from the relation [59]

$$\mu = I \left[g_L \pm (g_S - g_L) \frac{1}{2l+1} \right] \qquad \text{for } I = l \pm \frac{1}{2},$$

since g_L is commonly set to 1 and 0 for a proton and a neutron, respectively.

TABLE II. Sensitivity of the hyperfine structure to the magnetic moment distribution, in terms of parameters given in Eqs. (6) and (7). The values for the spin component, α_S and the spin asymmetry parameter, ζ, are taken from Shabaev [35]. The a- and b-parameters for ^{185}Re are also valid for ^{187}Re.

System	$^{133}Cs^{54+}$	$^{165}Ho^{66+}$	$^{185}Re^{74+}$	$^{209}Bi^{82+}$
Z	55	67	75	83
α_S	-0.506	0.281	0.372	-0.194
ζ	0.714	0.333	0.286	0.667
Parameters				
a_2 (10^{-4}/fm^2)	3.4	4.8	6.2	8.0
a_4 (10^{-6}/fm^4)	-1.5	-2.0	-2.5	-3.2
a_6 (10^{-8}/fm^6)	0.41	0.46	0.5	0.71
b_2 (10^{-4}/fm^2)	2.0	2.9	3.7	4.8
b_4 (10^{-6}/fm^4)	-0.64	-0.86	-1.1	-1.4
b_6 (10^{-8}/fm^6)	0.14	0.17	0.21	0.26

The values for these parameters and the ratios between them can be compared to those given by Shabaev (Eqs. (13)-(15) in [35]) from a series expansion of the wavefunctions for a homogeneously charged nucleus with radius R_0: The ratio between the parameters for κ_S and κ_L are then given by $b_2/a_2 = 3/5$, $b_4/a_4 = 3/7$ and $b_6/a_6 = 3/9$ (in reasonable agreement with the results from the fit). The importance of the higher correction from the series expansion is found by Shabaev [35] to be

$$\frac{a_4}{a_2} = -\frac{1}{2R_0^2}\left[\frac{1}{5} + \frac{3}{5}(Z\alpha)^2\right]$$

and

$$\frac{a_6}{a_2} = \frac{2}{3R_0^4}\left[\frac{19}{140}(Z\alpha)^2 + \frac{81}{1120}(Z\alpha)^4\right].$$

For Re, e.g., with $Z = 75$ and $R_0 = 6.91$ fm, the numerical values are $a_4/a_2 \approx -4.0 \times 10^{-3}$ fm^{-2} and $a_6/a_2 \approx 1.4 \times 10^{-5}$ fm^{-4} in reasonable agreement with the results from our fit, given in Table II.

For Re, the magnetic moment distribution leads to a reduction of the hyperfine structure of the order of 1.3% for a magnetic distribution with $\langle r_{\rm m}^2 \rangle \approx \langle r_{\rm c}^2 \rangle$. However, one might expect that the magnetic moment would essentially follow the unpaired proton, and should thus be more extended than the charge distribution. If, instead of a uniform distribution, the magnetization is distributed mainly on the surface, the correction to the hyperfine structure increases roughly by $\sqrt{5/3}$, giving a 2.2% "Bohr–Weisskopf" effect. The difference between the corrections due to the magnetizations is sufficient to explain the apparent discrepancy between the measured magnetic moment and the observed ground state hyperfine splitting of hydrogen-like Re [31].

V. HYPERFINE ANOMALIES IN CAESIUM ISOTOPES

The hyperfine structure for a few long-lived Cs isotopes were determined accurately in early radio-frequency measurement [60]: $A_{6s}(^{135}\mathrm{Cs}) = 2431.006(4)$ MHz and $A_{6s}(^{137}\mathrm{Cs}) = 2528.882(4)$ MHz. The hfs for the stable isotope, ^{133}Cs, is the SI definition of the second, corresponding to a value $A_{6s}(^{133}\mathrm{Cs}) = 2298.157943$ MHz. The magnetic moments for these isotopes are [57,61] $\mu(^{133}\mathrm{Cs}) = 2.5820246(34)\ \mu_{\rm N}$, $\mu(^{135}\mathrm{Cs}) = 2.7324(2)\ \mu_{\rm N}$, $\mu(^{137}\mathrm{Cs}) = 2.83413(1)\ \mu_{\rm N}$, and the hfs anomalies involving these isotopes become: $^{135}\Delta^{133} \approx -0.4(1) \times 10^{-3}$ and $^{137}\Delta^{133} \approx 2.5 \times 10^{-3}$. More recently, hyperfine structures were measured for a long chain of Cs isotopes [62]. However, the accuracy for the isotopes far from the stability line is, so far, in general insufficient to determine hfs anomalies.

The Bohr–Weisskopf effect for the ground state of Cs is found to be slightly increased by many-body effects, whereas the effect for the 6p states is changed considerably by the admixture of s orbitals. (Details results of the calculations will be published separately.) The Breit–Rosenthal correction also follows essentially the s orbitals, and modifications due to many-body effects for the 6s orbital should be negligible also for this effect. To analyse the ground state hfs anomaly, it would thus be sufficient to use the parameters given in Tables I and II.

The Breit–Rosenthal correction is found to account for a very small part of these observed anomalies: Based on changes in the charge distribution, $\delta\langle r_c^2\rangle$ deduced from optical isotope shifts [62,56], together with the parameter x_2 from Table I (and a factor of about 0.64 to account for the higher moments), we find that the Breit–Rosenthal contribution should be about -0.01×10^{-3} and -0.04×10^{-3}, respectively.

To evaluate the total Bohr–Weisskopf effect from these factors (together with the factors for the higher moments of the distribution obtained from the relations in Table II), it is necessary to assume some form of the magnetization distribution. In addition, the values for α_S, which is affected by the value of the nuclear magnetic moment, must be evaluated for each isotope, giving the α_S-values -0.506, -0.423 and -0.372, respectively for the isotopes ^{133}Cs, ^{135}Cs and ^{137}Cs. Assuming that the magnetization distribution for ^{133}Cs coincides with the charge distribution gives $\kappa_S = 0.0069$, $\kappa_L = 0.0042$ and a total Bohr–Weisskopf effect $\varepsilon = 0.0019$ for this isotope of which -0.0010 arises from the spin asymmetry term. For the other isotopes, the changes in α_S lead to the ε-values 0.0022 and 0.0025 for ^{135}Cs and ^{137}Cs, respectively. Additional contributions to the Bohr–Weisskopf effect can arise if the distribution of magnetization differs between the isotopes. As an illustration, it can be noted that if the magnetization were instead located on a shell with radius $R = 6.21$ fm, rather than following the Fermi distribution in Table I, the Bohr–Weisskopf effect for ^{133}Cs would instead be $\varepsilon = 0.0031$.

From these considerations, we conclude that the differences in the magnetization distribution must give relatively large positive contribution to $^{137}\Delta^{133}$, corresponding to a magnetization distribution closer to the nucleus for ^{137}Cs, than for ^{133}Cs.

VI. EPILOGUE

Atomic spectroscopy is a powerful tool to investigate also nuclear properties, and studies comparing theoretical and experimental results can provide a calibration of nuclear structure calculations, needed e.g., for the interpretation of experiments involving parity non-conservation [63], electric dipole moments [64] and tests of quantum-electrodynamics [65]. The tool of atomic spectroscopy is made sharper, not only through increasingly accurate experimental methods, but also through the development of atomic theory, enabling quantitative calculations for many-electron atoms. These calculations are possible through formal and numerical development, together with a stunning increase in computational capacity. The situation has, of course, changed drastically since the days in 1974 when Ingvar Lindgren helped one of us (A.-M. M.) prepare a deck of cards to run the self-consistent field program. Long gone are the days when a floating point multiplication lasted 5 μs and our pair program [66] was rejected because it needed more than the 1 MB available at the Göteborg University Computing Center IBM–360 mainframe. The problems to find room for the pair program led to the development, following a suggestion by Ingvar Lindgren, of a new numerical method to solve the "pair equation" [67], later replaced by a method to generate a numerical basis set, complete on the grid chosen [68,69]. These computer programs now form the basis for most calculations in our group, from non-relativistic cases to tests of QED — and have been used, somewhat extravagantly, to generate the 1s orbitals for the hydrogen-like systems studied in this work. Today, our calculations, even for relativistic many-electron systems [27,70,71] are performed using workstations, although the work on QED [65] is likely to bring us into problems where it will be of interest for us to ensure an adequate queueing situation at national computing facilities [72,73].

The work in the Göteborg theory group has focused on development of methods for calculations needed to interpret experimental data, combined with a strive to understand formal relations with other methods to treat atomic structure. Ingvar Lindgren's often impressive stamina in his efforts to go to the bottom of complicated problems can not fail to affect those who work with him!

ACKNOWLEDGMENTS

We would like to express our appreciation of stimulating helpful discussions with the atomic physics and subatomic physics groups at Göteborg. Financial support from the Swedish Natural Science Research Council (NFR) is gratefully acknowledged.

APPENDIX: SHAPE OF THE CHARGE DISTRIBUTION

We give here the exact expressions for the moments $\langle r_c^{2n} \rangle$ of the charge distribution using the parameters c and a in the Fermi distribution and the sums S_n given in Eq. (2):

$$\langle r^2\rangle = \frac{3}{5}c^2 + \frac{7}{5}\pi^2 a^2 + \frac{18a^3}{5c}\frac{\left[1+\dfrac{7\pi^2a^2}{3c^2}\right]S_3\left(-\dfrac{c}{a}\right) - 120\dfrac{a^2}{c^2}S_5\left(-\dfrac{c}{a}\right)}{1+\dfrac{\pi^2a^2}{c^2} - \dfrac{6a^3}{c^3}S_3\left(-\dfrac{c}{a}\right)},$$

$$\langle r^4\rangle = \frac{3}{7}c^4 + \frac{18}{7}\pi^2a^2c^2 + \frac{31}{7}\pi^4a^4 + \frac{18}{7}a^3c\frac{\left[1+\dfrac{6\pi^2a^2}{c^2}+\dfrac{31\pi^4a^4}{3c^4}\right]S_3\left(-\dfrac{c}{a}\right) - 840\dfrac{a^4}{c^4}S_7\left(-\dfrac{c}{a}\right)}{1+\dfrac{\pi^2a^2}{c^2} - \dfrac{6a^3}{c^3}S_3\left(-\dfrac{c}{a}\right)},$$

$$\langle r^6\rangle = \frac{3}{9}c^6 + \frac{11}{3}\pi^2a^2c^4 + \frac{239}{15}\pi^4a^4c^2 + \frac{127}{5}\pi^6a^6 + 2a^3c^3\frac{\left[1+\dfrac{11\pi^2a^2}{3c^2}+\dfrac{239\pi^4a^4}{5c^4}+\dfrac{381\pi^6a^6}{5c^6}\right]S_3\left(-\dfrac{c}{a}\right) - 60480\dfrac{a^6}{c^6}S_9\left(-\dfrac{c}{a}\right)}{1+\dfrac{\pi^2a^2}{c^2} - \dfrac{6a^3}{c^3}S_3\left(-\dfrac{c}{a}\right)}.$$

The last term in these three relations fall rapidly with c/a and can be neglected if $c/a << 1$. Already for ^{12}C the last terms gives a correction to $\langle r^2\rangle$ found to be only 0.003%.

---

[1] I. I. Rabi, J. R. Zacharias, S. Millman, and P. Kusch, Phys. Rev. **53**, 318 (1938).
[2] N. F. Ramsey, Phys. Rev. **76**, 996 (1949).
[3] N. F. Ramsey, Phys. Rev. **78**, 695 (1950).
[4] N. F. Ramsey, Phys. Rev. **84**, 506 (19451).
[5] C. Ekström, Adv. Quantum Chem., this volume, 1997.
[6] C. Ekström and I. Lindgren, in *Atomic Physics*, edited by R. Marrus, M. Prior, and H. Shugart (Plenum, New York, 1977), Vol. 5.
[7] I. Lindgren and A. Rosén, Case Studies in Atomic Physics **4**, 93 (1974).
[8] I. Lindgren and A. Rosén, Case Studies In Atomic Physics **4**, 150 (1974).
[9] I. Lindgren and J. Morrison, *Atomic Many-Body Theory, Series on Atoms and Plasmas*, 2nd ed. (Springer-Verlag, New York Berlin Heidelberg, 1986).
[10] I. Lindgren, Int. J. Quantum Chem. S **12**, 33 (1978).
[11] S. M. Schneider, W. Greiner, and G. Soff, Phys. Rev. A **50**, 119 (1994).

[12] H. Persson *et al.*, Phys. Rev. Lett. **76**, 1433 (1996).
[13] I. Klaft *et al.*, Phys. Rev. Lett. **73**, 2425 (1994).
[14] J. R. C. Lopez-Urrutria, P. Beiersdorfer, D. W. Savin, and K. Widmann, Phys. Rev. Lett. **77**, 826 (1996).
[15] E. Arimondo, M. Inguscio, and P. Violino, Rev. Mod. Phys. **49**, 31 (1977).
[16] H. H. Stroke, R. J. Blin-Stoyle, and V. Jaccarino, Phys. Rev. **123**, 1326 (1961).
[17] A. Bohr and V. F. Weisskopf, Phys. Rev. **77**, 94 (1950).
[18] A. Bohr, Phys. Rev. **81**, 331 (1951).
[19] J. E. Rosenthal and G. Breit, Phys. Rev. **41**, 459 (1932).
[20] M. F. Crawford and A. K. Schawlow, Phys. Rev. **76**, 1310 (1949).
[21] H. J. Rosenberg and H. H. Stroke, Phys. Rev. A **5**, 1992 (1972).
[22] G. Fricke *et al.*, At. Data Nucl. Data Tables **60**, 177 (1995).
[23] H. de Vries, C. W. de Jager, and C. de Vries, At. Data Nucl. Data Tables **36**, 495 (1987).
[24] R. Engfer *et al.*, At. Data Nucl. Data Tables **14**, 509 (1974).
[25] K. Heilig and A. Steudel, At. Data Nucl. Data Tables **14**, 613 (1974).
[26] F. Boehm and P. L. Lee, At. Data Nucl. Data Tables. **14**, 509 (1974).
[27] A.-M. Mårtensson-Pendrill, Phys. Rev. Lett. **74**, 2184 (1995).
[28] G. Gould, Phys. Rev. **101**, 1828 (1956).
[29] A. Lurio and A. G. Prodell, Phys. Rev. **101**, 79 (1956).
[30] M. Grexa, G. Hermann, G. Lasnitschka, and B. Fricke, Phys. Rev. A **38**, 1263 (1988).
[31] J. R. C. Lopez-Urrutria, P. Beiersdorfer, D. W. Savin, and K. Widmann, private communication, 1996.
[32] M. Tomaselli, D. Herold, and L. Grünbaum, Il. Nuovo Cimento **46A**, 468 (1978).
[33] E. N. Fortson, Y. Pang, and L. Wilets, Phys. Rev. Lett. **65**, 2857 (1990).
[34] S. J. Pollock, E. N. Fortson, and L. Wilets, Phys. Rev. C **46**, 2587 (1992).
[35] V. M. Shabaev, J. Phys. B **27**, 5825 (1994).
[36] W. R. Johnson and G. Soff, At. Data Nucl. Data Tables **33**, 405 (1985).
[37] H. T. Duong *et al.*, Nucl. Instrum. Methods A **325**, 465 (1993).
[38] R. Hofstadter, H. Fechter, and J. A. McIntyre, Phys. Rev. **92**, 978 (1953).
[39] B. D. Hahn, D. G. Ravenhall, and R. Hofstadter, Phys. Rev. **101**, 1131 (1956).
[40] M. G. H. Gustavsson, Realistic Nuclear Charge Distributions, Master's thesis, Dept. of Physics, Göteborg University and Chalmers University of Technology, 1995.
[41] F. A. Parpia and A. K. Mohanty, Phys. Rev. A **46**, 3735 (1992).
[42] K. G. Dyall *et al.*, Computer Physics Communications **55**, 425 (1989).
[43] F. A. Parpia and I. P. Grant, J. Phys. (Paris) Colloq. **1**, C1 (1991).
[44] P. Indelicato and E. Lindroth, Phys. Rev. A **46**, 2426 (1992).
[45] A. Ynnerman *et al.*, Phys. Rev. A **50**, 4671 (1994).
[46] B. Dreher *et al.*, Nucl. Phys. A **235**, 209 (1974).

[47] W. H. King, *Isotope shifts in atomic spectra* (Plenum Press, New York, 1984).
[48] H. Kopfermann, *Nuclear Moments* (Academic Press, New York, 1958).
[49] J. Bauche, Comments At. Mol. Phys. **10**, 57 (1981).
[50] A.-M. Mårtensson-Pendrill *et al.*, J. Phys. B **23**, 1749 (1990).
[51] A.-M. Mårtensson-Pendrill *et al.*, Phys. Rev. A **45**, 4675 (1992).
[52] J. Bauche and M. Klapisch, J. Phys. B **35**, 29 (1972).
[53] J. Bauche, J. Physique (Paris) **35**, 19 (1974).
[54] P. Aufmuth, K. Heilig, and A. Steudel, At. Data Nucl. Data Tables **37**, 455 (1987).
[55] G. Torbohm, B. Fricke, and A. Rosén, Phys. Rev. A **31**, 2038 (1985).
[56] A. Hartley and A.-M. Mårtensson-Pendrill, J. Phys. B **24**, 1193 (1991).
[57] P. Raghavan, At. Data Nucl. Data Tables **42**, 189 (1987).
[58] G. Nachtsheim, Ph.D. thesis, Universität Bonn, 1980.
[59] A. Bohr and B. R. Mottelson, *Nuclear Structure, I: Single-Particle Motion* (W. A. Benjamin, New York, 1969).
[60] H. H. Stroke, V. Jaccarino, J. D. S. Edmonds, and R. Weiss, Phys. Rev. **105**, 590 (1957).
[61] M. Winter, *WebElements - the periodical table on the WWW* http://www.shef.ac.uk/ chem/web-elements/, 1997.
[62] C. Thibault *et al.*, Nucl. Phys. A **367**, 1 (1981).
[63] S. A. Blundell *et al.*, Theor. Chim. Acta **80**, 257 (1991).
[64] A.-M. Mårtensson-Pendrill, in *Methods in Computational Chemistry*, edited by S. Wilson (Plenum Press, New York, 1992), Vol. 5, pp. 99–156.
[65] H. Persson, S. Salomonson, P. Sunnergren, and I. Lindgren, Phys. Rev. Lett. **76**, 204 (1996).
[66] S. Garpman, I. Lindgren, J. Lindgren, and J. Morrison, Phys. Rev. A **11**, 758 (1975).
[67] A.-M. Mårtensson, J. Phys. B **12**, 3995 (1979).
[68] S. Salomonson and P. Öster, Phys. Rev. A **40**, 5559 (1989).
[69] S. Salomonson and P. Öster, Phys. Rev. A **40**, 5548 (1989).
[70] E. Lindroth and A. Ynnerman, Phys. Rev. A **47**, 961 (1993).
[71] A.-M. Mårtensson-Pendrill, D. S. Gough, and P. Hannaford, Phys. Rev. A **49**, 3351 (1994).
[72] A.-M. Mårtensson-Pendrill, Int. J. Supercomp. Appl. and High-Performance Computing **9**, 312 (1995).
[73] A.-M. Mårtensson-Pendrill, Queueing Statistics from a User's Perspective, Poster presentation, Supercomputing '96, http://www.supercomp.org/sc96/posters.html, 1996.

SPINS AND MOMENTS OF NUCLEI FAR FROM STABILITY DETERMINED BY ON-LINE ATOMIC-BEAM TECHNIQUES

Curt Ekström

The Svedberg Laboratory, Box 533, S-751 21 Uppsala, Sweden

Abstract

Hyperfine structure (hfs) measurements using on-line atomic-beam techniques are of great importance in the systematic study of spins and moments of nuclei far from stability. I will discuss the atomic-beam magnetic resonance (ABMR) method, and laser spectroscopy methods using crossed-beam geometry with a collimated thermal atomic-beam, and collinear geometry with a fast atomic-beam. The advantages of the different methods will be compared, and selected results from the extensive measurements at the ISOLDE facility, CERN, will be presented.

1. Introduction

Nuclear spins and moments of ground and isomeric states are basic properties probing the structure and shape of atomic nuclei. The systematic experimental study of these quantities along isotopic and isotonic chains thus allows a mapping of the nuclear behaviour, to be compared with the predictions of different nuclear models.

The main experiments specifically aimed at revealing the nuclear structure by measuring nuclear spins and moments are those investigating the hyperfine structure (hfs) of free atoms using radio-frequency and optical spectroscopy techniques. Although a large number of data has been obtained in off-line experiments, it is only through the introduction of on-line techniques that short-lived nuclides far from stability, and thus long isotopic chains, have come within reach for study. The experiments performed at the ISOLDE facility, CERN, by the Mainz, Göteborg-Uppsala and Orsay groups constitute a main effort in this direction.

Here, I will concentrate on the different on-line atomic-beam techniques used in hyperfine structure measurements at ISOLDE, and on selected experimental results showing the importance of systematic spin and moment measurements.

2. Nuclear quantities accessible in hyperfine structure experiments

In hfs experiments on free atoms the small energy splitting of atomic levels due to the interaction between the nuclear multipole moments and the corresponding field quantities from the electrons are investigated. We may here distinguish between two classes of experiments: radio-frequency spectroscopy studying the hfs within an atomic level, generally the atomic ground state, and optical spectroscopy in which

ADVANCES IN QUANTUM CHEMISTRY, VOLUME 30

0065-3276/98 $25.00

the hyperfine structures of the initial and final states are superimposed on the optical line.

The narrow line-widhs obtained in rf-experiments constitute the main advantage compared to optical experiments, and allow high-precision measurements to be made. The Doppler broadening, limiting the precision in optical experiments, can however be strongly reduced by using crossed-beam geometry with a collimated thermal atomic-beam or collinear geometry with a fast atomic-beam.

Direct measurements of nuclear spins I are made by rf-experiments in weak external magnetic fields by studying the transitions between different Zeeman-levels in which the production and detection of the orientation may be done by several methods as will be seen below. The nuclear spins may also be determined in optical experiments by the number of hf-components and their intensity ratios. Whereas the latter method has been shown to be of limited validity, a combined analysis of the hfs of two optical lines in the same nuclide provides an unambiguous determination of the nuclear spin.

The energies of the zero-field hyperfine levels, characterized by the quantum number F, ($|I-J| \leq F \leq I+J$), may be written as a linear combination of the hyperfine coupling constants and known functions including the quantum numbers I, J and F:

$$W^F_0 = A\, f^F_1 + B\, f^F_2 \qquad (1)$$

neglecting the small energy contributions from higher order terms. The quantities

$$A \sim \mu \langle B \rangle / IJ \qquad I, J \geq 1/2$$

$$B \sim Q^S_2 \langle \partial^2 V / \partial z^2 \rangle \qquad I, J \geq 1 \qquad (2)$$

are the coupling constants for the dipole and quadrupole interactions, respectively, with the given restrictions on I and J. The quantities B and V are the magnetic field and the electric potential, respectively, due to the electronic shell.

Experimental data on the hyperfine separations thus give values of the coupling constants, from which the corresponding nuclear moments; magnetic dipole moment μ, and spectroscopic electric quadrupole moment Q^S_2, may be evaluated.

The evaluation of the magnetic dipole moments from the measured A-factors is generally made through a direct comparison with a stable isotope in which the values of I, μ and A are known. Neglecting the hyperfine anomaly (cf. below), the following relation holds within a given level of an element:

$$(A / g_I)_1 = (A / g_I)_2 \qquad (\mu = I\, g_I) \qquad (3)$$

The evaluation of the spectroscopic quadrupole moment from the measured B-factors has to rely to a large extent on theoretical calculations of the electronic part of the quadrupole interaction. From the study of muonic and pionic atoms, a number of accurate quadrupole moments have been determined [1]. These values serve as reference data in the same way as for the dipole interaction.

In addition to the nuclear spins and moments discussed above, hfs experiments may also provide a direct measure of nuclear g-factors through their interaction with an external magnetic field. Combined with experimental A-factors, one gets in this case information on the differential hyperfine anomaly Δ, defined by:

$$(A / g_I)_1 = (A / g_I)_2 \times (1 + \Delta) \qquad (4)$$

The anomaly, probing the difference in distribution of nuclear magnetism between two isotopes is generally small, but may reach a 10% level in $^2S_{1/2}$ states in heavy elements.

Here, I will not discuss the hyperfine anomaly nor the additional important nuclear information on changes of mean square charge radii obtained from isotope shift (IS) data in optical experiments.

3. Atomic-beam experiments at ISOLDE

In this section we will discuss briefly the different atomic-beam methods used in hfs measurements at ISOLDE, and give references to the isotopic chains studied. The description, although classified according to the technique used, follows to a large extent a chronological order.

3.1 Experiments with thermal atomic beams

Thermal atomic beams have been used extensively to determine nuclear spins and moments by investigations of the atomic hyperfine structure. The atomic-beam magnetic resonance (ABMR) method has already become classical [2]. More recent efforts include laser spectroscopy in a crossed-beam geometry, in which a large supression of the Doppler width is obtained by collimation of the atomic beam.

Several off-line experiments have been made with the specific aim of revealing nuclear properties in long isotopic chains. We mention here the systematic investigations of the rare-earth region by the Göteborg-Uppsala group [3, 4] using ABMR methods, and the laser spectroscopy study of IS and hfs of the sequence of barium isotopes $^{124-138}$Ba by the Karlsruhe group [5]. The introduction of on-line techniques, however, has extended considerably the field of application. Below, we will discuss the Göteborg-Uppsala ABMR experiment and the Orsay laser spectroscopy experiment, both conneced on-line to the ISOLDE mass separator.

3.1.1 Atomic-beam magnetic resonance (ABMR)

The ABMR apparatus connected on-line to the ISOLDE isotope separator is given schematically in Fig. 1. A detailed description of the design and operation may be found in Refs. [6, 7]. The apparatus consists of an oven section where the atomic beam is formed, a main section with focusing magnets of 6-pole and 4-pole type acting as polarizer and analyzer of the atomic beam, respectively, the rf-transitions being induced in the intermediate homogeneous field of the dipole magnet, and

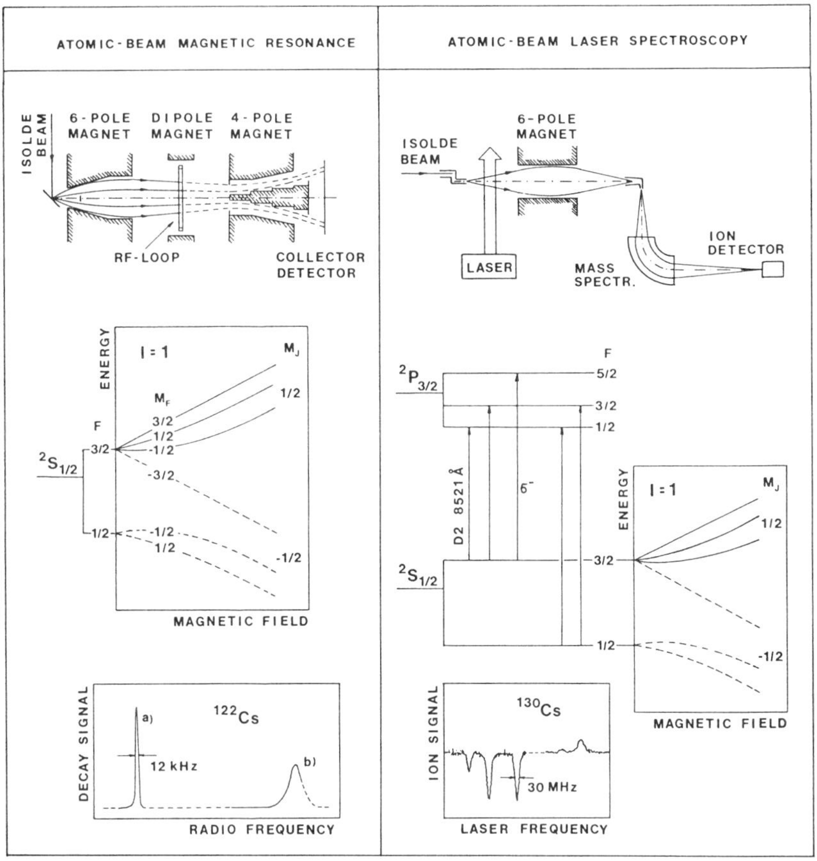

Fig. 1 Schematic view of the atomic-beam magnetic resonance and atomic-beam laser spectroscopy set-ups at ISOLDE. In both experiments, the 60 keV ion beam from the ISOLDE mass separator is converted to give a continuous flow of thermal atoms through the machines. Resonances are observed in the ABMR experiments when rf-transitions are induced in the atomic ground state hfs, changing the atoms from a focusing ($M_J > 0$) to a defocusing ($M_J < 0$) state. The $\Delta F = 1$ resonance transitions in ^{122}Cs ($I = 1$) are given at the bottom left; a) ($F = 3/2$, $M_F = 1/2$) --> (1/2, -1/2) and (3/2, -1/2) --> (1/2, 1/2), b) (3/2, 3/2) --> (1/2, 1/2). The small line-widths, determined mainly by the transit time of the atoms through the rf-field, should be noted. They give the A-factor in this particular case with a precision of $5x10^{-6}$. In the atomic-beam laser experiment, the resonances are recorded after state selection with a six-pole magnet; negative signals are obtained when the population of states with $M_J > 0$ in the strong-field limit is reduced by optical pumping with laser light via an excited state, and positive signals when the population is increased (cf. the laser-frequency scan in ^{130}Cs ($I = 1$) at the bottom right).

finally a collector and detector section. In the on-line experiments, the 60 keV ion beam from ISOLDE is deflected into the oven section and focused to a small spot at the source position from which it is evaporated in the form of thermal atoms. Atoms with positive effective magnetic moments (those in states with $M_J = 1/2$ in Fig. 1) are focused by the 6-pole magnet into the central region. When rf-transitions are induced changing the sign of the magnetic moment (to states with with $M_J = -1/2$ in Fig. 1), the atoms are defocused in the 4-pole magnet and transmitted through the apparatus. The resonances are recorded by exposing collectors to the atomic beam for certain periods of time and counting the activity in radiation detectors.

The ABMR method has been applied to several different elements at ISOLDE for measurements mainly of nuclear spins and magnetic dipole moments. The main results have been obtained in the alkali elements $_{37}$Rb [8, 9], $_{55}$Cs [9-11] and $_{87}$Fr [12, 13], and in $_{79}$ Au [14, 15]. Shorter sequences have been studied in $_{35}$Br [16], $_{49}$In [17], $_{53}$I, $_{63}$Eu [18, 19], $_{65}$Tm [19], and $_{81}$Tl [17].

3.1.2 Atomic-beam laser spectroscopy

A considerable extension of the ABMR measurements in the alkali elements, discussed above, was obtained by the introduction of the atomic-beam laser experiments at ISOLDE. With this method the nuclear quadrupole moments could be reached by studying the hfs of the $^2P_{3/2}$ excited state, as well as the IS in the isotopic sequences studied. A number of nuclear spins and magnetic dipole moments was also added. Of particular importance was the discovery of the D_2 optical line in francium which opened the way to hfs and IS measurements in this element. The atomic-beam laser spectroscopy works at ISOLDE on the alkali elements $_{37}$Rb, $_{55}$Cs and $_{87}$Fr have been presented in Refs. [20-25].

The atomic-beam laser method with non-optical detection, schematically given in Fig. 1, was developed at Orsay and first applied to the sequence of sodium isotopes ^{21,22,24,25}Na [26]. Later, long isotopic chains of $_{11}$Na and $_{19}$K were studied extensively [27-29] with an improved version of the apparatus put on-line at the proton synchrotron at CERN. The apparatus and the different oven systems used are described in Refs. [30, 31].

In the on-line experiments at ISOLDE, the 60 keV ion beam is converted to a thermal atomic beam, which after collimation is illuminated at right angle with the light from a tunable dye laser. The laser light induces optical pumping when tuned to eg. the resonances of the D_2 line (cf. Fig. 1), causing a change in the population of the ground state hyperfine levels. This change is monitored by a detector system consisting of a state selecting 6-pole magnet, a mass spectrometer and an electron multiplier. The resonances obtained in a laser frequency scan give the hfs of the ground state and the excited state. In double-resonance experiments including rf-transitions in weak external magnetic fields, nuclear spins and accurate values of the hyperfine separations in the ground and excited states have been determined.

In an effort to determine differential hyperfine anomalies (Eq. 4) by high-precision measurements of dipole constants and nuclear g-factors, the two atomic-beam groups above have installed an upgraded atomic-beam apparatus at ISOLDE [32].

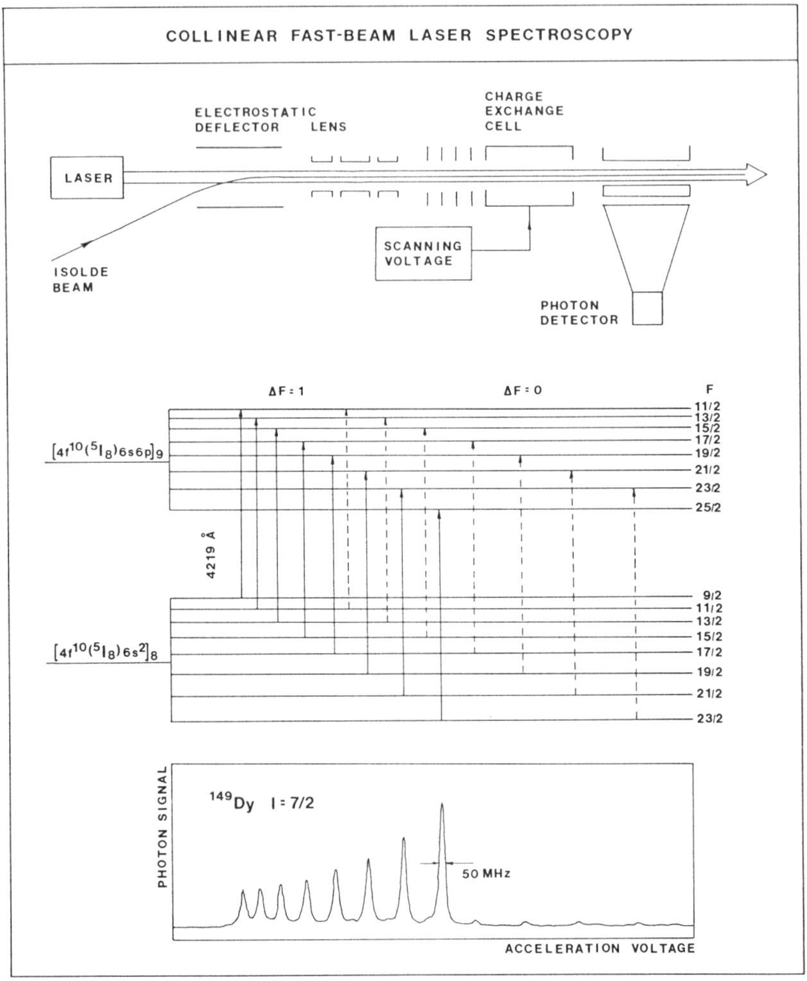

Fig. 2 Simplified drawing of the original version of the collinear laser experiment at ISOLDE. The ion beam from the ISOLDE mass separator is deflected into the the apparatus and passes, superimposed on a laser beam, along the optical axis. The ions are neutralized in a charge-exchange cell to give a beam of fast atoms. These are excited by the laser light and the detection of the resonances are made by recording the emitted fluorescence photons in a photomultiplier. The resonances are scanned by keeping the laser frequency at a fixed value and varying the Doppler shift of the absorption line by an additional voltage put on the retardation and charge-exchange systems. The photon spectrum of ^{149}Dy ($I = 7/2$) as a function of acceleration voltage is given at the bottom. The $\Delta F = 1$ and $\Delta F = 0$ resonances observed in the 4212 Å transition are indicated in the energy level diagram.

3.2 Experiments with fast atomic beams

The thermal atomic-beam experiments, discussed so far, are performed with samples obtained by collecting the mass-separated ion beam and a subsequent re-evaporation in the form of an atomic beam. Although this procedure may work continuously there are large losses, e.g. a factor of 10^4 because of the low transmission in the atomic-beam experiments. The collinear fast-beam laser spectroscopy method offers an elegant solution to this problem by working directly with the mass-separated beam from the separator. Combined with the large reduction in velocity spread, and thus in Doppler width, of the electrostatically accelerated ion beams [33], assuring a high sensitivity and resolution, the collinear method represents a major step towards the ideal spectroscopic tool for systematic hfs and IS measurements.

The collinear fast-beam laser spectroscopy method [34] was developed at Mainz, tested with stable beams of $_{11}$Na and $_{55}$Cs [35], and subsequently used for on-line measurements at the mass separator connected to the Mainz reactor. Here, neutron-rich fission products of the alkali elements $_{37}$Rb and $_{55}$Cs were investigated [36-38]. Full advantage of the potential of the method was taken by connecting an improved version of the apparatus on-line to the ISOLDE isotope separator, where a wide range of elements are available in high production yields. The "collinear" system has made a kind of revolution in systematic hfs and IS measurements, producing a vast amount of data on spins, moments and mean square charge radii.

The high sensitivity and resolution, due mainly to the large reduction in Doppler width in the accelerated ion beam, is fully utilized only if the acceleration voltage is sufficiently stable. At ISOLDE, a high-voltage stabilization system gives an absolute accuracy of ±0.1 V per day. This deviation is small compared to the energy spread of about 1 eV of the ions from a surface ionization source. The latter gives a Doppler broadening of about 5 MHz for a medium-mass nuclide and an optical wave-length of 6000 Å. This is to be compared to a typical value of 10 MHz for the natural line-widths in strong optical transitions.

In a measuring cycle, the nuclide under investigation and reference nuclides are directed alternatively through the apparatus. The absorption lines are simultaneously Doppler shifted into reasonable ranges by an offset voltage put on the retardation and charge-exchange systems. The resonances are subsequently scanned by applying an additional sweep voltage.

In the original version of the apparatus [34], shown in Fig. 2, the ions are neutralized in a charge-exchange cell and the fast atoms excited by laser light. The detection of the resonances is made by monitoring the fluorescence photons in a system consisting of cylindrical optics, a light-guide and a photomultiplier. The first measurements at ISOLDE were made in a long sequence of $_{56}$Ba isotopes [39-41] covering both sides of the N = 82 neutron-shell closure. Similarly, the chemical analogue $_{88}$Ra has been studied on both sides of N = 126 [42-46]. A main effort has been put into a systematic investigation of the neutron-deficient rare-earth nuclides, with special emphasis on those in the transitional region above N = 82 and with extensions into the region with strong nuclear deformation. The measurements include the elements $_{63}$Eu [47], $_{64}$Gd, $_{66}$Dy [48], $_{67}$Ho [49], $_{68}$Er [48], $_{70}$Yb [50] and $_{71}$Lu [51]. Furthermore, long isotopic chains of the elements $_{49}$In [52] and $_{80}$Hg [53] have been studied with the original technique, as have the elements $_{79}$Au [54], $_{81}$Tl [55] and $_{87}$Fr [24].

In order to improve the sensitivity of the collinear method new efficient detection schemes based on particle counting have been introduced. The first scheme, particularly applicable to alkaline-earth ions, is utilizing the state selectivity of the charge-transfer neutralization process and atom/ion counting [56]. The technique has been used in systematic investigations of $_{20}$Ca [57-60] and $_{38}$Sr [61-65].

The second scheme is based on the state-selective reionization of the optically pumped neutral atomic beam, followed by ion counting [66]. The method has been used successfully to study long isotopic chains of the noble gases $_{18}$Ar [67], $_{36}$Kr [68], $_{54}$Xe [69] and $_{86}$Rn [49, 70].

The previous experiments on ytterbium [50] have been extended using resonance ionization spectroscopy on a fast atomic ytterbium beam [71].

The combination of collinear fast-beam laser spectroscopy and β-RADOP (radiation-detected optical pumping) has been used to measure nuclear spins and moments of neutron-rich isotopes of the light alkali elements $_{3}$Li [72-74] and $_{11}$Na [75]. Here, the optically pumped fast atomic beam is implanted into a single crystal placed in a static magnetic field. The NMR signal is destroying the nuclear polarization detected by measuring the β-decay asymmetry.

4. Discussion of selected results

During the last decades, a tremendous progress has been made in the study of spins and moments of nuclei far from stability. The vast majority of results has been obtained using atomic-beam hfs techniques at the ISOLDE facility. Below, the experimental results in selected nuclear regions will be presented.

4.1 The rare-earth region

In the years around 1970, a systematic study of nuclear spins and moments of neutron-deficient rare-earth nuclides, produced by (p,xn)-rections in the Uppsala synchrocyclotron, were made by off-line ABMR techniques [3, 4]. At ISOLDE, these experiments have been extended by spin measurements in europium and thulium [18, 19]. The by far most important extension, however, has been made by the method of collinear fast-beam laser spectroscopy [47-51, 71]. Here, the method has shown its strength in high sensitivity, selectivity and resolution. A two-dimensional mapping of the (N,Z)-plane has been made, covering not only scattered isotopic sequences but wide isotopic chains of a range of neighbouring elements. The physical quantities studied include spins, moments and mean square charge radii.

The physics interest in the rare-earth region stems from the fact that the region is intercepted by the N = 82 neutron-shell closure. Thus, one is expected here to encounter spherical, transitional as well as strongly deformed nuclei. The main goals of the experiments is to obtain information on the nuclear single-particle structure and on the development of nuclear deformation away from the N = 82 shell closure, and to investigate the influence of the proton-subshell closure at Z = 64.

The information on the nuclear deformation, obtained mainly from the data on changes in mean square charge radii, indicates a smooth gradual increase on both sides of N = 82, followed by, on the heavy side, a sudden onset of strong deformation between N = 88 and N = 90 for the elements close to Z = 64. In the lighter and heavier elements, the step becomes less pronounced, to vanish completely in barium (Z = 56) and ytterbium (Z = 70). This behaviour supports the picture of a Z = 64 subshell closure, counteracting a smooth increase in deformation in the full sequence between N = 82 and N = 90 by its stabilizating effect for spherical shapes in gadolinium (Z = 64) and the neighbouring elements.

Within this general picture, the measured spins and moments give complementary information on the nuclear single-particle structure. Here, I will briefly discuss the ground-state structure of odd-neutron nuclei in the rare-earth region, and, as an example of odd-proton nuclei, the sequence of europium isotopes.

The odd-neutron nuclei close to the N = 82 neutron-shell closure are shown to be well described by available shell-model neutron states. The N = 81 isotones below the shell closure exhibit I = 3/2 ground states in the elements up to samarium (Z = 62), where there is a change to I = 1/2 ground states. This behaviour is explained by the lowering of the $s_{1/2}$ shell-model state with respect to $d_{3/2}$ with increasing proton number. The I = 11/2 isomeric states in the full range are due to $h_{11/2}$. The measured moments also support the assignments above.

The N = 83, 85 and 87 isotones above the neutron-shell closure at N = 82 are described essentially by the $f_{7/2}$ shell-model state. Calculations withing the particle-rotor model of Larsson et al. [77], assuming small nuclear deformations of the core, account qualitatively for the measured nuclear moments (cf. Fig. 3). The variation in the spectroscopic quadrupole moments reflects the successive filling of the $f_{7/2}$ neutron shell. This is realized from the formula [78]:

$$Q_2^s = Q_2^o \frac{3\sum_{\nu}(a_{\Omega}^{I\nu}\,\Omega)^2 - I(I+1)}{(I+1)\,(2I+3)} \tag{5}$$

relating the spectroscopic and intrinsic quadrupole moments, and the fact that the occupation of sublevels ranges from those with low-Ω to medium-Ω in the sequence N = 83 - 87 (cf. Fig. 3).

The nuclear moments of the I = 3/2 ground states of the N = 89 isotones ^{153}Gd and ^{157}Er [48] indicate permanently deformed nuclear shapes and a description by a mixture of the Nilsson orbitals [521 3/2] and [532 3/2]. For the I = 5/2 nuclides ^{145}Ba [39] and ^{159}Yb [71], on the other hand, there seems to be no straightforward explanation of the measured nuclear moments.

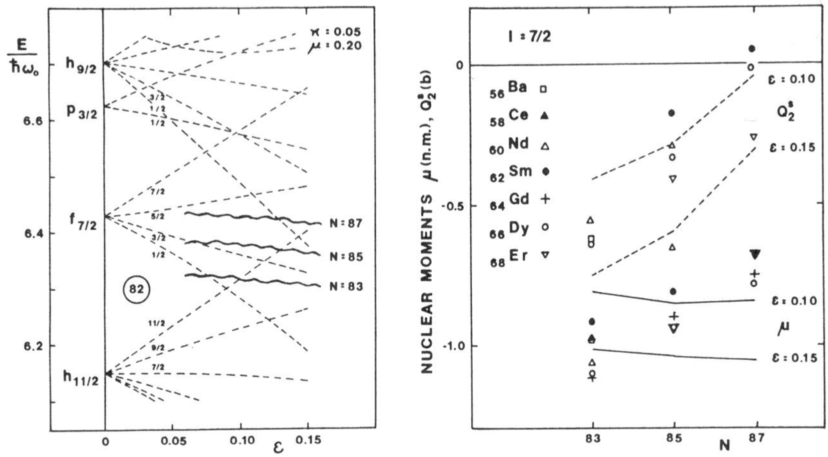

Fig. 3 Nilsson diagram for odd neutrons close to N = 82, calculated with the parameter set $\kappa = 0.05$ and $\mu = 0.20$ valid in the spherical limit [76]. The Fermi levels for N = 83, 85 and 87 indicate a successive filling of the $f_{7/2}$ neutron shell. In the right part, experimental data on magnetic dipole moments and spectroscopic quadrupole moments of I = 7/2 ground states in N = 83, 85 and 87 nuclei (Ba [39]; Ce, Nd and Sm [1]; Gd, Dy and Er [48]) are compared with the results from the particle-rotor calculations, assuming deformations of $\varepsilon = 0.10$ and 0.15.

The nuclear spins and moments of the strongly deformed rare-earth nuclides have been discussed in detail previously within the Nilsson model in connection with the ABMR experiments mentioned above [79]. The addition of the data from collinear fast-beam laser spectroscopy [48, 50, 71], the new reference values on spectroscopic quadrupole moments from muonic and pionic hfs [1], and the refined calculations within the partice-rotor model, including a number of orbitals close to the Fermi surface, have however resulted in a more complete picture and a better understanding of the nuclear single-particle structure in this region.

In Fig. 4 we compare the nuclear moments of the odd-neutron nuclei in the range N = 91 - 97 with the results from particle-rotor calculations using the "A = 161" potential parameters $\kappa = 0.0637$ and $\mu = 0.425$, and the g-factors $g_R = Z/A$, $g_s = 0.6\ g_s^{free}$ and gl = 0. The experimental data are shown to be well accounted for by the theory.

Among the odd-proton rare-earths, the element europium (Z = 63) has been thoroughly investigated at ISOLDE by collinear fast-beam laser spectroscopy in the range $^{140-153}$Eu [47]. The ABMR works include spin measurements in $^{143,145-150,156}$Eu [3, 18-19]. From the point of nuclear physics the element europium, being the immediate neighbour to the semi-magic (Z = 64) element gadolinium, is of particular interest.

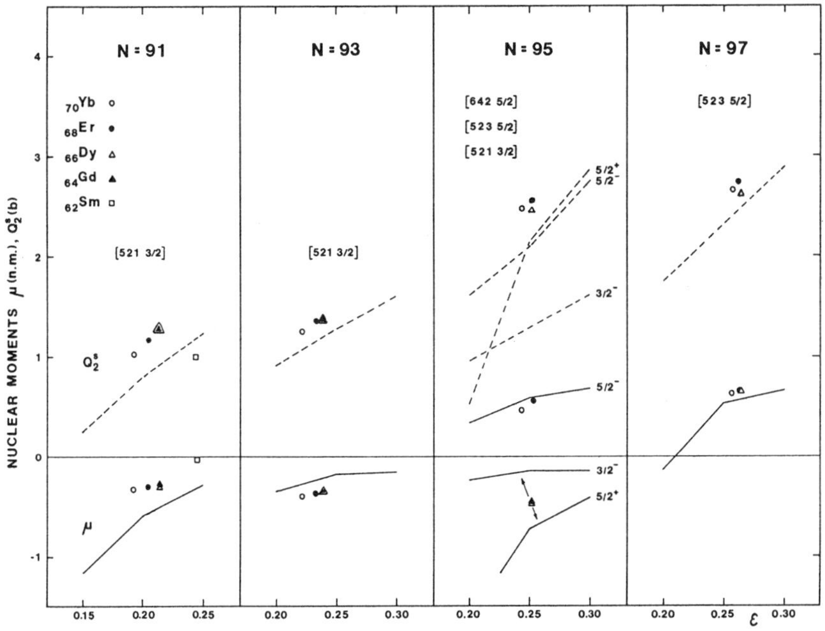

Fig. 4 Comparison between experimental [1, 48, 50, 71] and calculated magnetic dipole moments and spectroscopic quadrupole moments of odd-neutron nuclei in the range N = 91 - 97. The experimental data are given at the theoretical deformation positions [79]. The dominant Nilsson orbitals are indicated; in the case of the N = 95 isotones the orbital [521 3/2] is associated with ^{159}Gd, [642 5/2] with ^{161}Dy and [523 5/2] with ^{163}Er and ^{165}Yb.

The experimental magnetic dipole moments and spectroscopic quadrupole moments of the odd-A I = 5/2 europium isotopes in the mass range A = 141-153 are given in Fig. 5. The data indicate a gradual slight increase in nuclear deformation on both sides of the N = 82 neutron-shell closure, followed by an abrupt onset of strong deformation in going from N = 88 to N = 90. In fact, the onset of deformation occurs already between the ground states of ^{151}Eu and ^{152}Eu, i.e. between N = 88 and N = 89. The maximum deformation in europium is reached in this single step, as evidenced also by the data on the heavier isotopes up to ^{156}Eu, obtained by Dörschel et al. [80].

As to the ground-state configurations of the odd-A europium isotopes, the predominant part of the wave function in $^{141\text{-}151}$Eu is due to the shell model proton state $d_{5/2}$, whereas in the strongly deformed ^{153,155}Eu, there is a change to the Nilsson orbital [413 5/2]. The increasing spectroscopic quadrupole moments on each side of N = 82 indicate a positive projection factor in Eq. 5, and thus, for positive deformations, an association with the $(d_{5/2})_{5/2}$ state, which in the strong-coupling limit equals the Nilsson orbital [402 5/2]. The latter assignment has been proposed by Dörschel et al. [80] in $^{147-151}$Eu.

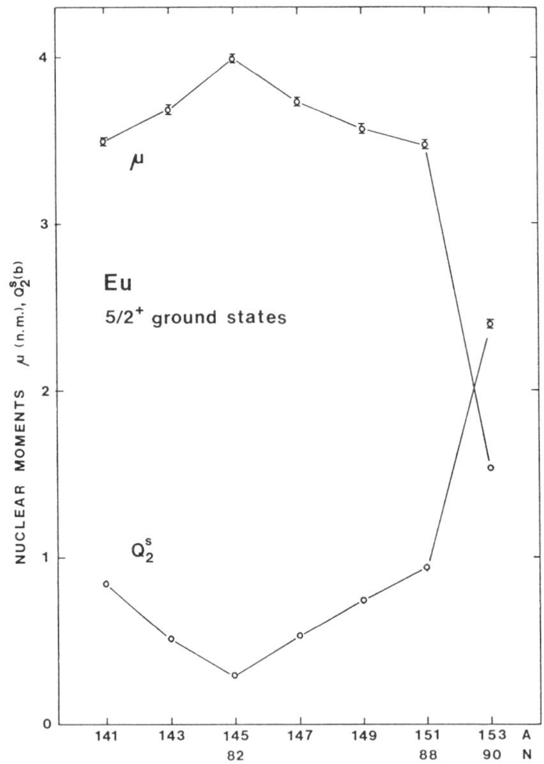

Fig. 5 Experimental nuclear moments of the sequence of I = 5/2 europium isotopes in the range $^{141-153}$Eu [1, 47]. The neutron-shell closure at N = 82 and the shape transition between N = 88 and N = 90 are well reflected in the data.

The doubly-odd europium isotopes in the range $^{140-150}$Eu are well described by the $d_{5/2}$ proton state coupled to the different neutron-shell model states, discussed above in connection with the odd-neutron nuclei; $d_{3/2}$ in $^{140-144}$Eu, $h_{11/2}$ in ^{142m}Eu and $f_{7/2}$ in $^{146-150}$Eu. By using g-factors of the neighbouring odd-A nuclei and the additivity theorem, the magnetic moments of these doubly-odd europium isotopes are well reproduced. The strongly deformed ^{152,154} Eu and ^{156}Eu are shown to be due to the configurations 3^-(p[413 5/2] n[505 11/2]) and 0^+(p[413 5/2] n[642 5/2]), respectively.

4.2 The francium isotopes

The isotopic sequence of the heaviest alkali element, francium, has been extensively studied at ISOLDE by the three atomic-beam techniques: ABMR, atomic-beam laser spectroscopy and collinear fast-beam laser spectroscopy.

The first hfs measurements in francium were made with the ABMR method, giving nuclear spin values for the sequence of isotopes $^{208\text{-}213,220\text{-}222}$Fr [12]. These experiments were followed by atomic-beam laser experiments [23] which, after the identification of the D_2 optical line, gave results on the hyperfine structure constants in the 7s $^2S_{1/2}$ ground state and the 7p $^2P_{3/2}$ excited state as well as isotope shifts in $^{208\text{-}213}$Fr. Further measurements in francium include the nuclear spin of ^{207}Fr, the nuclear g-factor of ^{211}Fr and the electronic g-factor of francium by the ABMR method [13], the identification of the D_1 optical transition by atomic-beam laser spectroscopy [24], and the 7s $^2S_{1/2}$ --> 8p $^2P_{1/2}$ and 7s $^2S_{1/2}$ --> 8p $^2P_{3/2}$ transitions together with hfs measurements on the heavy francium isotopes by collinear fast-beam laser spectroscopy [24], and a comprehensive study in $^{207\text{-}213,220\text{-}228}$Fr by atomic-beam laser spectroscopy [25].

The measurement of the nuclear g-factor of ^{211}Fr, g_I = 0.888(17), [13] has made possible a separation of the hyperfine constants into their nuclear and electronic parts, and to make comparisons with theoretical predictions. The magnetic moments in the range $^{207\text{-}213,220\text{-}228}$Fr have been evaluated from Eq. 3 using the measured dipole constants [25] and the g_I-factor of ^{211}Fr.

Furthermore, the works above have initiated several theoretical groups to calculate the electronic part of the magnetic dipole interaction [13, 81-84], obtained from the experimental dipole constant and g_I-factor in ^{211}Fr (cf. Eq. 2). The good agreement generally obtained for the magnetic hyperfine interaction indicate that the electronic wave functions obtained, reliably may be used in the evaluation of the spectroscopic quadrupole moments from the measured quadrupole constants. The value $Q^s_2(^{211}\text{Fr})$ = -0.19 b, obtained from many-body calculations of the electronic part of the hyperfine interaction [83], has been used as a reference value in the evaluation spectroscopic quadrupole moments.

The spectroscopic quadrupole moments of the odd-A francium isotopes below the neutron-shell closure at N = 126 show a characteristic behaviour, observed also in the bismuth isotopes below N = 126, and in iodine and cesium close to N = 82. As evidenced in Fig. 6, the largest moments appear at the neutron-shell closures at N = 82 and N = 126, with decreasing values as the distance from the shell closures increases. This behaviour is readily interpreted as an effect of the increasing deformation on each side of the shell closures. Assuming prolate deformation of the nuclei under consideration, the Fermi levels are shown to be located close to the low-Ω sublevels from the "high-spin" shell model states $g_{7/2}$, $d_{5/2}$ ($_{53}$I and $_{55}$Cs) and $h_{9/2}$ ($_{83}$Bi and $_{87}$Fr) as indicated in the Nilsson diagram in Fig. 7. With increasing deformation, the occupation of the low-Ω sublevels are favoured, resulting in decreasing spectroscopic quadrupole moments. This feature follows from Eq. 5.

The results from the quantitative calculations of Q^s_2 within the $d_{5/2}$ and $g_{7/2}$ states in iodine and cesium, and the $h_{9/2}$ state in bismuth and francium are given to the right

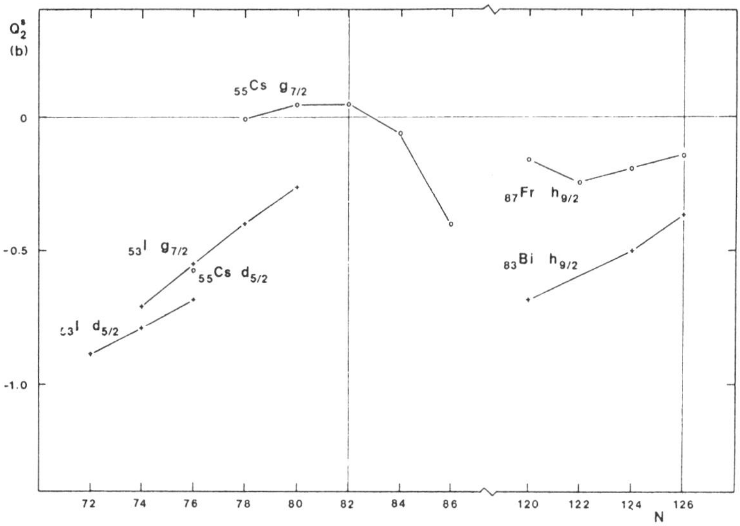

Fig. 6 Experimental spectroscopic quadrupole moments of some weakly deformed odd-proton nuclei appearing close to the neutron-shell closures at N = 82 and N = 126.

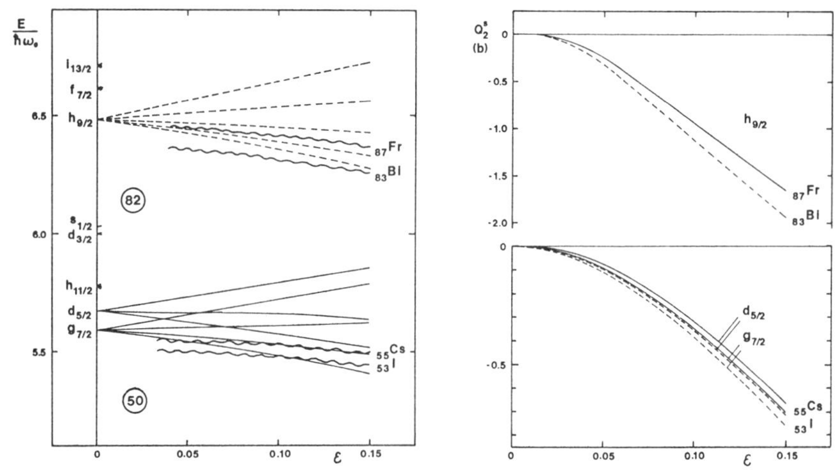

Fig. 7 Nilsson diagram for odd protons, including sublevels of the shell model states $g_{7/2}$, $d_{5/2}$ and $h_{9/2}$. The Fermi levels of $_{53}I$, $_{55}Cs$ and $_{83}Bi$, $_{87}Fr$ are shown to be located at the lower parts of the $g_{7/2}$, $d_{5/2}$ and $h_{9/2}$ sublevels, respectively. The theoretical results on the spectroscopic quadrupole moments as a function of deformation are given to the right. The potential parameters used are $\kappa = 0.066$, $\mu = 0.575$ for I and Cs, and $\kappa = 0.060$, $\mu = 0.630$ for Bi and Fr.

in Fig. 7. The trend and magnitude of the experimental moments are shown to be well reproduced, as well as the lower values of iodine and bismuth as compared to those of cesium and francium. The latter effect is related to the lower Fermi levels and the correspondingly higher occupancy of low-Ω states.

It may be noted that similar theoretical results are obtained by assuming triaxial or oblate shapes of the nuclei. This follows qualitatively from Eq. 5. For oblate shapes, the intrinsic quadrupole moments are negative, and with increasing deformation, the occupancy of high-Ω sublevels is favoured, resulting again in decreasing moments. The sign of the nuclear deformation may thus not be inferred from the spectroscopic quadrupole moments of weakly deformed nuclei, but has to be deduced from eg. the decoupled or coupled structure of the quasi-rotational bands built on the "high-spin" states.

There is shown to be a relatively small mixing for deformations $\varepsilon < 0.15$ between, on one hand the $g_{7/2}$ and $d_{5/2}$ states, and on the other the $h_{9/2}$ and $f_{7/2}$ states. Furthermore, the spin I = 5/2, 7/2 and 9/2 states discussed here are built up mainly of sublevels from $d_{5/2}$, $g_{7/2}$ and $h_{9/2}$, respectively. One may therefore conclude that the nuclear structure of these weakly deformed nuclei are described by relatively pure shell-model states, as evidenced also by the measured spins and magnetic moments, and that the variations in occupation of the corresponding sublevels with increasing deformation account for the observed changes in the spectroscopic quadrupole moments.

The short half-lives of the francium isotopes just above the N = 126 shell closure have prevented hfs measurements in the nuclides $^{214\text{-}219}$Fr. The structure of the odd-A isotopes, however, is most probably similar to that just below N = 126, here with increasing deformation with increasing mass number.

The nuclear single-particle structure of the francium isotopes $^{220\text{-}228}$Fr has been discussed in some detail within the particle-rotor model in Ref. [13]. The nuclear spin I = 5/2 and the negative spectroscopic quadrupole moment in ^{221}Fr indicate a decoupled structure in this transitional nucleus with components from the low-Ω orbitals of the $h_{9/2}$ and $f_{7/2}$ systems, the main component being $(h_{9/2})_{3/2}$. This gives a negative projection factor in Eq. 5. The odd-A nuclei ^{223}Fr and ^{225}Fr, both having a ground state spin I = 3/2, are well described by the Nilsson orbital [532 3/2] indicating increasing nuclear deformation. Going to ^{227}Fr, the intruder state assignment [400 1/2] is given to the nuclear ground state on the basis of the experimental spin and magnetic moment. Among the doubly-odd francium isotopes, the configurations of ^{220}Fr and ^{222}Fr are not evident because of the transitional character of these nuclei. The heavier ones, ^{224,226,228}Fr, however, are qualitatively well accounted for by combining the main proton and neutron orbitals of the neighbouring odd-A francium and radium isotopes. A particularly interesting case is ^{226}Fr, given by the combination (p[532 3/2] n[631 3/2]) with I = 1 and K = 0, explaining the negative spectroscopic quadrupole moment of this strongly deformed nucleus.

The measured nuclear spins and moments of the francium isotopes have thus given detailed information on the single-particle structure and deformation along the long isotopic chain.

5. Conclusion

The on-line atomic-beam techniques used at the ISOLDE facility at CERN have been extremely successful in measuring systematically nuclear spins and moments through hyperfine-structure studies along extended isotopic chains in wide regions of the nuclear chart. The measured nuclear spins and moments of the long isotopic sequences have given detailed information on the interplay between single-particle structure and collective effects, an interplay showing a large variation along the isotopic chains, particularly when passing neutron-shell closures.

The collinear fast-beam laser spectroscopy, ideally adapted to experiments directly on ion beams from the mass separator or, after charge exchange, on fast atomic beams, has shown to be the most versatile method for on-line hfs measurements. The apparatus of the Mainz group is currently connected to the general-purpose separator at the ISOLDE PS Booster Facility. To the same separator, the Orsay and Göteborg/Uppsala groups have connected an upgraded atomic-beam apparatus [32] intended for high-precision measurements of dipole constants and nuclear g-factors to determine differential hyperfine anomalies in long isotopic chains.

Acknowledgements

I am grateful to Prof. Ingvar Lindgren, founder of the Swedish atomic-beam group, for initiating the on-line application of the method at ISOLDE and for continuous support of the project. The fruitful cooperation with theoreticians of the Lund group, in particular with Dr. Ingemar Ragnarsson, in the interpretation of the experimental data is gratefully acknowledged. We also sincerely thank members of the different experimental hfs groups at ISOLDE, in particular Dr. Rainer Neugart, for a stimulating cooperation.

References

1. P. Raghavan, Table of Nuclear Moments, Atomic and Nuclear Data Tables 42(1989)189.
2. N.F. Ramsey, Molecular Beams, Oxford University Press, 1956.
3. C. Ekström, Thesis, University of Göteborg, 1972.
4. A. Rosén, Thesis, University of Göteborg, 1973.
5. K. Bekk et al., Z. Phys. A291(1979)219.
6. C. Ekström et al., Nucl. Instr. and Meth. 148(1978)17.
7. C. Ekström et al., Nucl. Instr. and Meth. 186(1981)261.
8. C. Ekström et al., Nucl. Phys. A311(198)269.
9. C. Ekström et al., Phys. Scr. 19(1979)516.
10. C. Ekström et al., Nucl. Phys. A292(1977)144.
11. C. Ekström et al., Phys. Lett 76B(1978)565.
12. C. Ekström et al., Phys. Scr.18(1978)51.
13. C. Ekström et al., Phys. Scr. 34(1986)624.
14. C. Ekström et al., Phys. Lett. 60B(1976)146.
15. C. Ekström et al., Nucl. Phys. A348(1980)25.
16. C. Ekström et al., Phys. Scr. 22(1980)344.

17. C. Bengtsson et al., Phys. Scr. 30(1984)164.
18. C. Ekström et al., Z. Phys. A302(1981)181.
19. C. Ekström et al., Z. Phys. A316(1984)239.
20. C. Thibault et al., Phys. Rev. C23(1981)2720.
21. G. Huber et al., Phys. Rev. Lett. 41(1978)459.
22. C. Thibault et al., Nucl. Phys. A367(1981)1.
23. S. Liberman et al., Phys. Rev. A22(1980)2732.
24. F. Touchard et al., 7th Int. Conf. on Atomic Masses and Fundamental Constants, Darmstadt-Seeheim, Germany, 1984, p. 353.
25. A.Coc et al., Phys. Lett. 163B(1985)66.
26. G. Huber et al., Phys. Rev. Lett. 34(1975)1209.
27. G. Huber et al., Phys. Rev. C18(1978)2342.
28. F. Touchard et al., Phys. Rev. C25(1982)2756.
29. F. Touchard et al., Phys. Lett. 108B(1982)169.
30. C. Thibault et al., Nucl. Instr. and Meth. 186(1981)193.
31. F. Touchard et al., Nucl. Instr. and Meth. 186(1981)329.
32. H.T. Duong et al., Nucl. Instr. and Meth. A325(1993)465.
33. S.L. Kaufman, Opt. Comm. 17(1976)309.
34. R. Neugart, Nucl. Instr. and Meth. 186(1981)165.
35. K.R. Anton et al., Phys. Rev. Lett 40(1978)642.
36. W. Klempt et al., Phys. Lett. 82B(1979)47.
37. B. Schinzler et al., Phys. Lett. 79B(1978)209.
38. J. Bonn et al., Z. Phys. A289(1979)227.
39. A.C. Mueller et al., Nucl. Phys. A403(1983)234.
40. K. Wendt et al., Z. Phys. A318(1984)125.
41. K. Wendt et al., Z. Phys. A329(1988)407.
42. S.A. Ahmad et al., Phys. Lett. 133B(1983)47.
43. K. Wendt et al., Z. Phys. D4(1987)227.
44. E. Arnold et al., Phys. Rev. Lett. 59(1987)771.
45. S.A. Ahmad et al., Nucl. Phys. A483(1988)244.
46. W. Neu et al., Z. Phys. D11(1989)105.
47. S.A. Ahmad et al., Z. Phys. A321(1985)35.
48. R. Neugart, private communication.
49. R. Neugart et al., 5th Int. Conf. on Nuclei far from Stability, Rosseau Lake, Canada, 1987, p. 161.
50. F. Buchinger et al., Nucl. Instr. and Meth. 202(1982)159.
51. W. Borchers et al., 6th Int. Conf. on Nuclear far from Stability and 9th Int. Conf. on Atomic Masses and Fundamental Constants, Bernkastel-Kues, Germany, 1992, p. 213.
52. J. Ebertz et al., Nucl. Phys. A464(1987)9.
53. G. Ulm et al., Z. Phys. A325(1986)247.
54. G. Passler et al., Nucl. Phys. A580(1994)173.
55. R. Neugart et al., Phys. Rev. Lett. 55(1985)1559.
56. R.E. Silverans et al., Nucl. Instr. and Meth. B26(1987)591.
57. R.E. Silverans et al., Z. Phys. D18(1991)351.
58. A.-M. Mårtensson-Pendrill et al., Phys. Rev. A45(1992)4675.
59. L. Vermeeren et al., Phys. Rev. Lett. 68(1992)1679.
60. L. Vermeeren et al., J. Phys. G22(1996)1517.
61. F. Buchinger et al., Z. Phys. A327(1987)361.
62. R.E. Silverans et al., Phys. Rev. Lett. 60(1988)2607.

63. F. Buchinger et al., Phys. Rev. C41(1990)2883.
64. P. Lievens et al., Phys. Lett. B256(1991)141.
65. P. Lievens et al., Phys. Rev. C46(1992)797.
66. R. Neugart et al., Nucl. Instr. and Meth. B17(1986)354.
67. A. Klein et al., Nucl. Phys. A607(1996)1.
68. M. Keim et al., Nucl. Phys. A586(1995)219.
69. W. Borchers et al., Phys. Lett. B216(1989)7.
70. W. Borchers et al., Hyp. Int. 34(1987)25.
71. Ch. Schultz et al., J. Phys. B24(1991)4831.
72. E. Arnold et al., Phys. Lett B197(1987)311.
73. E. Arnold et al., Z. Phys. A331(1988)295.
74. E. Arnold et al., Phys. Lett B281(1992)16.
75. M. Keim et al., Hyp. Int. 97/98(1995)543.
76. J. Rekstad and G. Løvhøiden, Nucl. Phys. A267(1976)40.
77. S.E. Larsson et al., Nucl. Phys. A307(1978)189.
78. C. Ekström, 4th Int. Conf on Nuclei far from Stability, Helsingør, Denmark,1981, CERN 1981-09, p. 12.
79. C. Ekström and I.-L. Lamm, Phys. Scr. 7(1973)31.
80. K. Dörschel et al., Z. Phys. A317(1984)233.
81. M. Vajed-Samii et al., Phys. Rev. Lett 48(1982)1330 and Erratum, Phys. Rev. Lett 49(1982)1466.
82. J.L. Heully and A.M. Mårtensson-Pendrill, Phys. Scr. 27(1983)29.
83. J.L. Heully and A.M. Mårtensson-Pendrill, Phys. Rev. A27(1983)3332.
84. V.A. Dzuba et al., J. Phys. B17(1984)1953.

Regularization Corrections to the Partial-Wave Renormalization Procedure *

Hans Persson, Sten Salomonson and Per Sunnergren
Department of Physics, Chalmers University of Technology and Göteborg University, S-412 96 Göteborg, Sweden

A description of the partial-wave renormalization (PWR), used for calculating the first-order self energy and certain higher-order effects for the energy levels in highly charged ions, is presented. We put special emphasis on correction terms which have to be considered due to the use of the non-covariant subtraction scheme used in PWR.

I. INTRODUCTION

Theoretical investigations of the α^2 quantum electrodynamical (QED) corrections have recently been motivated through the experimental success in accurately measuring energy levels in highly-charged ions. These corrections have to be calculated non-perturbatively in the electron-nucleus coupling parameter $Z\alpha$, which for high Z approaches unity. A variety of corresponding calculations have been performed during the last years for different first- and second-order QED corrections [1–6,9–14]. This subject is extensively covered in an overview given by G. Soff *et al* [15] in this Volume.

In this paper we are dealing with the fundamental aspects of the self energy renormalization procedure. In particular, we present a direct method of calculating self energy corrections which differs from the standard way of expanding the intermediate bound-electron propagator in the nuclear potential, in order to analytically isolate and subtract the mass divergency. This method, which is called *partial-wave renormalization* (PWR), was developed by Lindgren, Persson and Salomonson [5] and by Quiney and Grant [6] and appears to be rather promising for the higher-order QED calculations. Specifically, it has been applied on the electron-screened self energy (SE) and on

*This contribution is dedicated to Prof. I. Lindgren on the occasion of his 65th birthday.

ADVANCES IN QUANTUM CHEMISTRY, VOLUME 30

0065-3276/98 $25.00

some of the combined self energy - vacuum polarization (SEVP) two-photon diagrams [7,8]. However, because of the non-covariant regularization scheme used in PWR additional correction terms have to be considered. To identify these terms a covariant regulator scheme has to be employed.

In this paper, we investigate the correction terms obtained when applying a covariant Pauli-Villars regulator [16] when formulating the PWR procedure. In Section II we consider the first-order self energy and the Coulomb screened self energy is discussed in Section III. Finally, we derive an expression suitable for numerical evaluation of the correction term in the Coulomb screened case.

II. FIRST-ORDER SELF ENERGY

The renormalized first-order bound-electron self energy shift for a state $\Phi_a(\boldsymbol{x}) = \langle \boldsymbol{x}|a\rangle$ can formally be written as the real part of the bound self energy with the mass counter term subtracted

$$\begin{aligned}\Delta E^{\rm SE}_{\rm bound}(a) = \lim_{\Lambda\to\infty} \mathrm{Re}\Big\{ & ie^2 \int d^3x_2 \int d^3x_1 \int \frac{dz}{2\pi} \\ & \times \Phi^\dagger_a(\boldsymbol{x}_2)\alpha^\nu S_{\rm F}(\boldsymbol{x}_2,\boldsymbol{x}_1,E_a-z)\alpha^\mu\Phi_a(\boldsymbol{x}_1)D^\Lambda_{{\rm F}\nu\mu}(\boldsymbol{x}_2-\boldsymbol{x}_1,z) \\ & -\delta m(\Lambda)\int d^3x\, \Phi^\dagger_a(\boldsymbol{x})\beta\Phi_a(\boldsymbol{x})\Big\}\end{aligned} \tag{1}$$

where $S_{\rm F}$ denotes the time-independent electron propagator. $D^\Lambda_{\rm F}$ is the covariantly regulated (following the Pauli-Villars prescription [16]) photon propagator

$$\begin{aligned}D^\Lambda_{{\rm F}\nu\mu}(\boldsymbol{x}_2-\boldsymbol{x}_1,z) = & -g_{\nu\mu}\int \frac{d^3k}{(2\pi)^3} e^{i\boldsymbol{k}\cdot(\boldsymbol{x}_2-\boldsymbol{x}_1)} \\ & \times \left\{\frac{1}{z^2-\boldsymbol{k}^2+i\epsilon} - \frac{1}{z^2-\boldsymbol{k}^2-\Lambda^2+i\epsilon}\right\}\end{aligned} \tag{2}$$

which is introduced in order to define a rigorous subtraction scheme which yields the correct invariant shift. Schematically Eq. (1) can be written as

$$\Delta E^{\rm SE}_{\rm bound}(a) = \lim_{\Lambda\to\infty} \mathrm{Re}\{(B-B_\Lambda)-(M-M_\Lambda)\} \tag{3}$$

where B and B_Λ are the two parts which result from a separation of the first term in Eq. (1) into a Λ-independent part and a Λ-dependent part of the regulated photon propagator in Eq. (2). The $(M - M_\Lambda)$ term is a similar decomposition of the regulated mass term given in Eq. (1).

Consider first the B part. According to Eqs. (1) and (3) the corresponding expression reads

$$B = ie^2 \int d^3x_2 \int d^3x_1 \int \frac{dz}{2\pi} \times \Phi_a^\dagger(\boldsymbol{x}_2)\alpha^\nu S_F(\boldsymbol{x}_2, \boldsymbol{x}_1, E_a - z) D_{F\nu\mu}(\boldsymbol{x}_2 - \boldsymbol{x}_1, z)\alpha^\mu \Phi_a(\boldsymbol{x}_1) \quad , \tag{4}$$

where the electron propagator S_F is given by

$$S_F(\boldsymbol{x}_2, \boldsymbol{x}_1, z) = \sum_n \frac{\Phi_n(\boldsymbol{x}_2)\,\Phi_n^\dagger(\boldsymbol{x}_1)}{z - E_n(1 - i\eta)} = \sum_n \frac{\langle \boldsymbol{x}_2|n\rangle\langle n|\boldsymbol{x}_1\rangle}{z - E_n(1 - i\eta)} \quad . \tag{5}$$

The sum over n denotes a summation over positive and negative energy states. Evaluating the z-integration by means of complex contour integration yields

$$B = -\frac{\alpha}{4\pi^2} \int d^3k \, \frac{1}{k} \sum_n \frac{\langle a|\alpha_\mu e^{i\boldsymbol{k}\cdot\boldsymbol{x}_2}|n\rangle\langle n|e^{-i\boldsymbol{k}\cdot\boldsymbol{x}_1}\alpha^\mu|a\rangle}{E_a - E_n - \mathrm{sign}(E_n)k} \quad , \tag{6}$$

where k denotes $|\boldsymbol{k}|$. By performing the angular part of the $\boldsymbol{k}$ integration and employing the standard spherical wave expansion

$$\frac{\sin(k|(\boldsymbol{x}_2 - \boldsymbol{x}_1)|)}{k|(\boldsymbol{x}_2 - \boldsymbol{x}_1)|} = \sum_{l=0}^{\infty}(2l+1)j_l(kr_1)j_l(kr_2)\mathbf{C}^l(1)\cdot\mathbf{C}^l(2) \quad , \tag{7}$$

where the dot denotes the scalar product between the angular tensors

$$\mathbf{C}^l(2)\cdot\mathbf{C}^l(1) = \frac{4\pi}{2l+1}\sum_{m=-l}^{l} Y_{lm}(\Omega_2)\,Y_{lm}^*(\Omega_1) \quad , \tag{8}$$

we end up with the partial wave decomposition for B

$$B = -\frac{\alpha}{\pi}\int \frac{dk\,k^2}{k}\sum_{l=0}^{\infty}(2l+1)\sum_n \frac{\langle a|\alpha_\mu j_l(kr_2)\mathbf{C}^l|n\rangle\cdot\langle n|j_l(kr_1)\mathbf{C}^l\alpha^\mu|a\rangle}{E_a - E_n - \mathrm{sign}(E_n)k} \quad . \tag{9}$$

Here, the k in the denominator comes from the photon propagator and the k^2 in the numerator from the volume element. For later convenience, these are kept separate. In the following we will omit the dot for the scalar product.

The B_Λ term can be worked out in an analogous way

$$B_\Lambda = -\frac{\alpha}{\pi}\int \frac{dk\,k^2}{k'}\sum_{l=0}^{\infty}(2l+1)\sum_n \frac{\langle a|\alpha_\mu j_l(kr_2)\mathbf{C}^l|n\rangle\langle n|j_l(kr_1)\mathbf{C}^l\alpha^\mu|a\rangle}{E_a - E_n - \mathrm{sign}(E_n)k'} \quad , \tag{10}$$

where $k' = \sqrt{k^2 + \Lambda^2}$. Since $(B - B_\Lambda)$ is convergent for each Λ, we can change the order of the k-integration and the l-summation. Note that in Eq. (6) there is a logarithmic divergence in k. In Eq. (9), if we switch the order of the summation and the integration, the divergency is moved from the k-integration to the outer sum over partial waves. For each l-value the k-integration is finite.

We now turn to the second part of Eq. (1), i.e. the bound-state mass counter term. This term is defined as the free-electron self energy, in Feynman gauge, calculated for a momentum distribution determined by the bound state (French and Weisskopf [17]) and can be expressed in the following *symmetric* partial-wave form which is analogous to Eq. (9)

$$\begin{aligned} M = & -\frac{\alpha}{\pi} \sum_{r,r',s} \int d^3p \int d^3p' \int d^3q \int \frac{dk\, k^2}{k} \sum_{l=0}^{\infty} (2l+1) \\ & \times \frac{1}{2} \left(\frac{1}{E_{p,r} - E_{q,s} - \mathrm{sign}(E_{q,s})k} + \frac{1}{E_{p',r} - E_{q,s} - \mathrm{sign}(E_{q,s})k} \right) \\ & \times \langle a|\boldsymbol{p}, r\rangle \langle \boldsymbol{p}, r|\alpha_\mu j_l(kr_2)\mathbf{C}^l|\boldsymbol{q}, s\rangle \langle \boldsymbol{q}, s|j_l(kr_1)\mathbf{C}^l \alpha^\mu|\boldsymbol{p}', r'\rangle \langle \boldsymbol{p}', r'|a\rangle \quad , \end{aligned} \tag{11}$$

where $|\boldsymbol{p}, r\rangle$ denotes a free electron state with a specific momentum in a given spinor state. The M_Λ term can be written down directly by replacing k by k' in the denominators in Eq. (11) in the same way as for B_Λ in Eq. (10).

By using the above derived expressions in the partial-wave form we can write the self energy as

$$\Delta E^{\mathrm{SE}}_{\mathrm{bound}}(a) = \lim_{\Lambda \to \infty} \mathrm{Re} \left\{ \sum_{l=0}^{\infty} \left(B^l - B^l_\Lambda - M^l + M^l_\Lambda \right) \right\} \quad , \tag{12}$$

or since every term is finite we can group them to yield

$$\Delta E^{\mathrm{SE}}_{\mathrm{bound}}(a) = \mathrm{Re} \left\{ \sum_{l=0}^{\infty} \left(B^l - M^l \right) - \lim_{\Lambda \to \infty} \sum_{l=0}^{\infty} \left(B^l_\Lambda - M^l_\Lambda \right) \right\} \quad . \tag{13}$$

The Λ independent part can be treated using techniques described in earlier works [5] and will not be discussed here.

Now focus on the correction term, the Λ dependent part in the above expression, which is the main subject of this paper. Since we are interested in the correction term in the limit $\Lambda \to \infty$, we can assume Λ to be very large in our considerations. To get a non-vanishing contribution from the k integration, k must be of the same order as k'. This implies that only divergent terms, i.e. terms which contribute for extremely large momenta, can contribute to this correction. To be more specific, we can expand the bound electron propagator, given in the operator form, into a free propagator plus higher-order nuclear Coulomb corrections

$$\begin{aligned} \frac{1}{E_a - z - h_{bou}} &= \frac{1}{E_a - z - (h_{free} + V_{nuc})} \\ &= \frac{1}{E_a - z - h_{free}} + \frac{1}{E_a - z - h_{free}} V_{nuc} \frac{1}{E_a - z - h_{free}} + \ldots \end{aligned} \tag{14}$$

It can be shown from power counting arguments that only the first two terms in this expansion, the zero-potential and the one-potential terms, lead to divergent contributions and hence only these terms need to be considered in the Λ dependent part. In addition also the mass counter term has to be considered.

We will show below that this correction term vanishes for the first-order self energy and cancels for the Coulomb-screened self energy. This is, however, not always the case for higher-order effects. As an example, for the self energy in an external magnetic potential the correction term gives rise to a finite shift [17].

III. COULOMB SCREENED SELF ENERGY

To calculate the screening corrections of the self energy due to an external Coulomb potential, $V_c(r)$, we treat this potential as a perturbation of the first-order self energy [13]. This perturbation will affect the binding energy of the bound electron, the wave function and the bound-electron propagator

$$E_a \rightarrow E_a + \langle a|V_c(r)|a\rangle + \dots \,, \tag{15}$$

$$|a\rangle \rightarrow |a\rangle + \sum_{m\neq a} \frac{|m\rangle\langle m|V_c(r)|a\rangle}{E_a - E_m} + \dots = |a\rangle + |\delta a\rangle + \dots \,, \tag{16}$$

$$\begin{aligned} S_F(\boldsymbol{x}_2, \boldsymbol{x}_1; z) \rightarrow S_F(\boldsymbol{x}_2, \boldsymbol{x}_1; z) + \\ + \int \mathrm{d}^3 x_3\, S_F(\boldsymbol{x}_2, \boldsymbol{x}_3; z) V_c(r_3) S_F(\boldsymbol{x}_3, \boldsymbol{x}_1; z) + \dots \,. \end{aligned} \tag{17}$$

The wave-function modification term, originating from the replacement Eq. (16), takes the form

$$\begin{aligned} E^{\mathrm{wf}}(a) = -\frac{\alpha}{\pi} \sum_{l=0}^{\infty} (2l+1) \int \frac{dk\, k^2}{k} \\ \times \sum_n \frac{\langle a|\alpha_\mu j_l(kr_3) \mathbf{C}^l |n\rangle\langle n| j_l(kr_2) \mathbf{C}^l \alpha^\mu |\delta a\rangle}{E_a - E_n - \mathrm{sign}(E_n)k} \quad . \end{aligned} \tag{18}$$

There is also a corresponding mass counter part $E^{mc}(a)$. These both terms, which are shown in Fig. 1, are infrared finite, but ultraviolet mass divergent.

The vertex correction, replacement Eq. (17), leads to the following expression

$$\begin{aligned} E^{\mathrm{ve}}(a) = -\frac{\alpha}{\pi} \sum_{l=0}^{\infty} (2l+1) \int \frac{dk\, k^2}{k} \\ \times \sum_{m,n} \frac{\langle a|\alpha_\mu j_l(kr_3) \mathbf{C}^l |m\rangle\langle m|V_c(r_2)|n\rangle\langle n| j_l(kr_1) \mathbf{C}^l \alpha^\mu |a\rangle}{(E_a - E_m - \mathrm{sign}(E_m)k)(E_a - E_n - \mathrm{sign}(E_n)k)} \times F \,, \end{aligned} \tag{19}$$

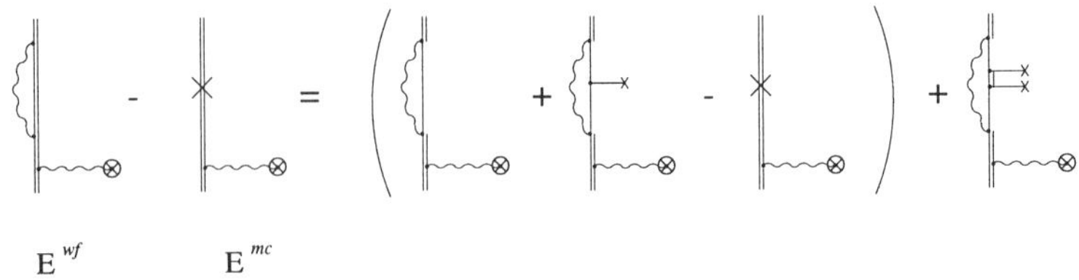

FIG. 1. Feynman graphs representing the mass renormalized wavefunction correction from an external Coulomb-like potential. The external potential is denoted by a ring with a cross. In order to isolate divergences, the internal self-energy electron propagator is expanded in the nuclear potential which is denoted by a horizontal line with a cross. The divergent zero-potential, the one-potential terms and the mass counter term are ultraviolet divergent and are grouped together. The remaining many-potential term is finite.

where we have introduced the function F defined by

$$F = 1 + [\mathrm{sign}(E_m) - \mathrm{sign}(E_n)] \frac{k}{E_m - E_n} \ . \tag{20}$$

For the binding energy term, replacement Eq. (15), the formula reads as follows

$$E^{\mathrm{be}}(a) = \frac{\alpha}{\pi} \sum_{l=0}^{\infty} (2l+1) \int \frac{dk\, k^2}{k} \times \langle a|V_c(r)|a\rangle \sum_m \frac{\langle a|\alpha_\mu j_l(kr_2)\mathbf{C}^l|m\rangle\langle m|j_l(kr_1)\mathbf{C}^l\alpha^\mu|a\rangle}{(E_a - E_m - \mathrm{sign}(E_m)k)^2} \ . \tag{21}$$

The vertex and the binding terms, which are shown in Fig. 2, are both infrared divergent and ultraviolet charge divergent. The infrared divergences can explicitly be shown to cancel between these terms [18]. Furthermore, the ultraviolet charge divergences will also cancel due to the Ward identity. However, it is essential to use a covariant regularization scheme to isolate the divergences. We use a Pauli-Villars prescription similar to the first-order case discussed above. To cancel the ultraviolet divergences in the screening case it is appropriate to organize the four different terms, depending on if they are mass divergent or charge divergent, in the following way

$$E^{\mathrm{wf-mc}}(a) = \mathrm{Re}\left\{ 2\sum_{l=0}^{\infty}(B^{wf,l} - M^{wf,l}) - \lim_{\Lambda\to\infty} 2\sum_{l=0}^{\infty}(B_\Lambda^{wf,l} - M_\Lambda^{wf,l}) \right\} , \tag{22}$$

and

$$E^{\mathrm{ve+be}}(a) = \mathrm{Re}\left\{ \sum_{l=0}^{\infty}(B^{ve,l} + B^{be,l}) - \lim_{\Lambda\to\infty} \sum_{l=0}^{\infty}(B_\Lambda^{ve,l} + B_\Lambda^{be,l}) \right\} \ . \tag{23}$$

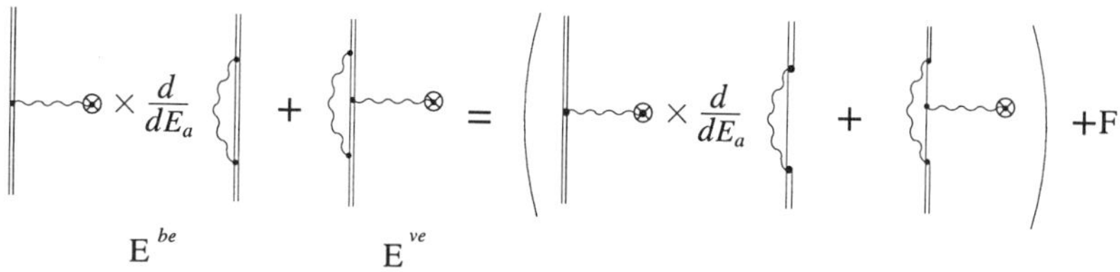

FIG. 2. Feynman graphs representing the binding energy and vertex correction. In this part the divergences occur only in the zero-potential terms, which are grouped together. F denotes the finite remainder.

The factor of two in the $E^{\mathrm{wf-mc}}$ term occurs since the wave-function correction can be situated both on the incoming and outgoing electron line. The $E^{\mathrm{wf-mc}}$ term and the $E^{\mathrm{ve+be}}$ term is divided up into Λ dependent and Λ independent parts. Note that it is essential to perform the k-integration and the l-summation in the same order in both these parts. In the PWR method the k-integration is performed first. The Λ independent parts can be calculated using standard procedures described in earlier works. We now study the Λ dependent parts in more detail.

A. Wavefunction Regularization Correction

Consider the Λ dependent parts in the wave-function term (Eq. 22)

$$C_\Lambda^{wf-mc} = 2\sum_{l=0}^{\infty}\left(B_\Lambda^{wf,l} - M_\Lambda^{wf,l}\right) = 2\frac{\alpha}{\pi}\sum_{l=0}^{\infty}(2l+1)\left(\tilde{B}_\Lambda^{wf,l} - \tilde{M}_\Lambda^{wf,l}\right) \quad . \tag{24}$$

The expression for a particular l is

$$\begin{aligned}\left(\tilde{B}_\Lambda^{wf,l} - \tilde{M}_\Lambda^{wf,l}\right) &= -\int \frac{dk\,k^2}{k'}\sum_{p,p'} \\ &\times \left\{\sum_n \frac{\langle a|p\rangle\langle p|\alpha_\mu j_l(kr_2)\mathbf{C}^l|n\rangle\langle n|j_l(kr_1)\mathbf{C}^l\alpha^\mu|p'\rangle\langle p'|\delta a\rangle}{E_a - E_n - \mathrm{sign}(E_n)k'}\right. \\ &- \sum_q \frac{1}{2}\left(\frac{1}{E_p - E_q - \mathrm{sign}(E_q)k'} + \frac{1}{E_{p'} - E_q - \mathrm{sign}(E_q)k'}\right) \\ &\times \left.\langle a|p\rangle\langle p|\alpha_\mu j_l(kr_2)\mathbf{C}^l|q\rangle\langle q|j_l(kr_1)\mathbf{C}^l\alpha^\mu|p'\rangle\langle p'|\delta a\rangle\right\} \quad ,\end{aligned} \tag{25}$$

where we have introduced the free-electron ket $|p\rangle$ as a short-hand notation for $|\boldsymbol{p}, r\rangle$ and replaced the integration over $\boldsymbol{p}$ and summation over r by a summation over p.

As in the first-order self energy case, we expand the bound propagator (Eq. (14)) and consider the zero-potential and the one-potential terms together with the mass counter part (see Fig. 1). Furthermore, we know that k is very large and that the free momenta k, p and q obey triangular conditions. Since the p momentum is cut off by the bound state $|a\rangle$ we can, by neglecting the rest mass energy, assume that $E_q \approx \pm k$.

Consider now the denominator difference between the zero-potential term and the mass counter term, which we can express as

$$\begin{aligned} &\frac{1}{2}\left(\frac{1}{E_a - E_q - \mathrm{sign}(E_q)k'} - \frac{1}{E_p - E_q - \mathrm{sign}(E_q)k'}\right) = \\ &= \frac{1}{2}\frac{(E_p - E_a)}{(E_a - E_q - \mathrm{sign}(E_q)k')(E_p - E_q - \mathrm{sign}(E_q)k')} \approx \\ &\approx \frac{1}{2}\frac{(E_p - E_a)}{(k + k')^2} \end{aligned} \tag{26}$$

plus a similar term with p replaced by p'. In the last step, we have neglected E_p and E_a in relation to k and k' in the denominators. Using this expression for the denominator difference, we can write the zero-potential term and the mass counter term as

$$\begin{aligned} -\int \frac{dk\,k^2}{k'(k+k')^2}\Big\{&\sum_p \langle a|p\rangle\langle p|\frac{1}{2}(E_p - E_a)(-2j_l^2(kr))|\delta a\rangle \\ &+ \sum_{p'} \langle a|\frac{1}{2}(-2j_l^2(kr))(E_{p'} - E_a)|p'\rangle\langle p'|\delta a\rangle\Big\} \end{aligned} \tag{27}$$

where we have summed over the intermediate free-electron spectrum. We have also used $\mathbf{C}^l(\mathbf{r}) \cdot \mathbf{C}^l(\mathbf{r}) = 1$ and the identity $\alpha_\mu \alpha^\mu = -2$. The one-potential term

$$-\int \frac{dk\,k^2}{k'} \sum_{p,p'} \frac{\langle a|\alpha_\mu j_l(kr_3)\mathbf{C}^l|p\rangle\langle p|V_{nuc}(r_2)|p'\rangle\langle p'|j_l(kr_1)\mathbf{C}^l\alpha^\mu|\delta a\rangle}{(E_a - E_p - \mathrm{sign}(E_p)k')(E_a - E_{p'} - \mathrm{sign}(E_{p'})k')} \times F\,, \tag{28}$$

where F is defined by

$$F = 1 + [\mathrm{sign}(E_p) - \mathrm{sign}(E_{p'})]\,\frac{k'}{E_p - E_{p'}}\,. \tag{29}$$

can be treated in a similar way as the above terms to yield

$$\int \frac{dk\,k^2}{k'(k+k')^2} \sum_{p,p'} \langle a|\alpha_\mu j_l(kr_3)\mathbf{C}^l|p\rangle\langle p|V_{nuc}(r_2)|p'\rangle\langle p'|j_l(kr_1)\mathbf{C}^l\alpha^\mu|\delta a\rangle \times G \tag{30}$$

where the factor G differs from the factor F by including the sign from the denominators. For positive-positive and negative-negative intermediate states we have $G = 1$. By using projection operators for positive-negative and negative-positive intermediate states we get

$$\begin{aligned}\Lambda_p^+ V_{nuc} \Lambda_{p'}^- &= \sum_{p,p'} \frac{1}{4} \left[\left(1 + \frac{h_{free}}{|E_p|}\right) |p\rangle\langle p|V_{nuc}|p'\rangle\langle p'| \left(1 - \frac{h_{free}}{|E_{p'}|}\right) \right] \\ &\approx \frac{1}{4} \left[\left(1 + \frac{\boldsymbol{\alpha} \cdot \hat{\boldsymbol{p}} + \beta m}{|E_k|}\right) V_{nuc} \left(1 - \frac{\boldsymbol{\alpha} \cdot \hat{\boldsymbol{p}} + \beta m}{|E_k|}\right) \right] \\ &\approx \frac{1}{4} \left(1 - \frac{\hat{\boldsymbol{p}}^2}{E_k^2}\right) V_{nuc} \approx \frac{1}{4} \left(1 - \frac{\boldsymbol{k}^2}{E_k^2}\right) V_{nuc} = 0 \quad , \end{aligned} \tag{31}$$

where we have used the fact that for large momenta h_{free} commutes with V_{nuc} and also that the rest mass part is negligible. Thus, after a summation over the intermediate states and by commuting the α matrices, the one-potential part can be written as

$$\int \frac{dk\, k^2}{k'(k+k')^2} \langle a|V_{nuc}(-2j_l^2(kr))|\delta a\rangle \ . \tag{32}$$

Finally, by adding these parts (zero + one + mass counter) we obtain

$$\begin{aligned}(\tilde{B}_\Lambda^{wf,l} - \tilde{M}_\Lambda^{wf,l}) = -\int \frac{dk\, k^2}{k'(k+k')^2} \Big\{ &\sum_p \langle a|p\rangle\langle p|(E_p - E_a + V_{nuc}) j_l^2(kr)|\delta a\rangle \\ &+ \sum_{p'} \langle a|j_l^2(kr)(E_{p'} - E_a + V_{nuc})|p'\rangle\langle p'|\delta a\rangle \Big\} \end{aligned} \tag{33}$$

or by using $h_{bou} = h_{free} + V_{nuc}$ we get

$$\begin{aligned}(\tilde{B}_\Lambda^{wf,l} - \tilde{M}_\Lambda^{wf,l}) = -\int \frac{dk\, k^2}{k'(k+k')^2} \{ &\langle a|(h_{bou} - E_a) j_l^2(kr)|\delta a\rangle \\ &+ \langle a|j_l^2(kr)(h_{bou} - E_a)|\delta a\rangle \} \ . \end{aligned} \tag{34}$$

The only difference between the first-order self energy correction and this wave function correction term is the appearance of $|\delta a\rangle$ instead of $|a\rangle$. Thus, from this expression we can conclude that in the case of the first-order self energy the correction term vanishes. For the screening case only the first term vanishes and we are left with

$$(\tilde{B}_{\Lambda}^{wf,l} - \tilde{M}_{\Lambda}^{wf,l}) = -\int \frac{dk\, k^2}{k'(k+k')^2} \langle a|j_l^2(kr)(h_{bou} - E_a)|\delta a\rangle$$
$$= -\int \frac{dk\, k^2}{k'(k+k')^2} \sum_{t\neq a} \langle a|j_l^2(kr)(h_{bou} - E_a)|t\rangle \frac{\langle t|V_c|a\rangle}{E_a - E_t}$$
$$= +\int \frac{dk\, k^2}{k'(k+k')^2} \langle a|j_l^2(kr)(1 - |a\rangle\langle a|)V_c|a\rangle \quad . \tag{35}$$

By summing over l we finally get

$$C_{\Lambda}^{wf-mc} = +2\frac{\alpha}{\pi} \sum_{l=0}^{\infty} (2l+1) \int \frac{dk\, k^2}{k'(k+k')^2} \langle a|j_l^2(kr)(V_c - \overline{V}_c)|a\rangle \quad , \tag{36}$$

where $\overline{V}_c$ denotes the expectation value $\langle a|V_c|a\rangle$.

B. Vertex and Binding Energy Regularization Correction

Consider next the Λ dependent part originating from the vertex term (Eq. (19)). Making a potential expansion of the bound propagators shows that only the free propagator term leads to divergences and hence needs to be considered. By using similar approximations as for the one-potential part in the wave function term we get

$$\tilde{B}_{\Lambda}^{ve,l} = -2\int \frac{dk\, k^2}{k'(k+k')^2} \langle a|j_l^2(kr)V_c|\delta a\rangle \quad . \tag{37}$$

The binding energy correction (see Eq. (21) can be considered in a similar manner (also here it is only the intermediate free electron propagator which gives rise to divergences) and yields

$$\tilde{B}_{\Lambda}^{be,l} = -\int \frac{dk\, k^2}{k'} \sum_p \langle a|V_c|a\rangle \frac{\langle a|\alpha_\mu j_l(kr_2)\mathbf{C}^l|p\rangle\langle p|j_l(kr_1)\mathbf{C}^l\alpha^\mu|a\rangle}{(E_a - E_p \mp ck)^2}$$
$$= 2\int \frac{dk\, k^2}{k'(k+k')^2} \langle a|V_c|a\rangle\langle a|j_l^2(kr)|\delta a\rangle \tag{38}$$

Putting the vertex and binding energy terms together we obtain

$$C_{\Lambda}^{ve+be} = \frac{\alpha}{\pi} \sum_{l=0}^{\infty} (2l+1)(\tilde{B}_{\Lambda}^{ve,l} + \tilde{B}_{\Lambda}^{be,l})$$
$$= -2\frac{\alpha}{\pi} \sum_{l=0}^{\infty} (2l+1) \int \frac{dk\, k^2}{k'(k+k')^2} \langle a|j_l^2(kr)(V_c - \overline{V}_c)|a\rangle \tag{39}$$

which, apart from the sign, is the same contribution as we obtained from the wave function correction (Eq. (36)). Thus, these extra contributions cancel each other when added. However, to be able to compare diagram by diagram with a similar calculation using dimensional regularization, these correction terms are needed. In the next section we derive an expression for these terms in a numerically suitable form.

C. Evaluation of the Correction Term

The correction term, Eqs. (36,39), can easily be seen to vanish if the l-summation is performed before the k-integration. This follows by using the identity

$$\sum_{l=0}^{\infty}(2l+1)j_l^2(kr) = 1 \quad . \tag{40}$$

In the PWR procedure, however, the k-integration is performed before the l summation in the Λ-independent terms and thus the same order of integration has to be used in the correction terms. The correction term can be written as

$$2\frac{\alpha}{\pi}\sum_{l=0}^{\infty}(2l+1)\lim_{K\to\infty}\langle a|(V_c(r)-\overline{V}_c)\int_0^K \frac{dk\,k^2}{k'(k+k')^2}j_l^2(kr)|a\rangle \quad , \tag{41}$$

where the expectation value over r is performed before $K\to\infty$. Since the bound state $|a\rangle$ effectively cuts off large r and also very small r, we can assume the range of the r-integration to be finite and with a non-zero lower limit. If, for each r, the substitution $kr \to x$ is made we get

$$\begin{aligned}
&\sum_{l=0}^{\infty}(2l+1)\lim_{K\to\infty}\\
&\langle a|(V_c(r)-\overline{V}_c)\int_0^{Kr}\frac{dx\,x^2\,j_l^2(x)}{\sqrt{x^2+(\Lambda r)^2}\left(x+\sqrt{x^2+(\Lambda r)^2}\right)^2}|a\rangle =\\
&\sum_{l=0}^{\infty}(2l+1)\lim_{K\to\infty}\\
&\left\{\int_0^{K\overline{r}}dx\,x^2\,j_l^2(x)\langle a|(V_c(r)-\overline{V}_c)\frac{1}{\sqrt{x^2+(\Lambda r)^2}\left(x+\sqrt{x^2+(\Lambda r)^2}\right)^2}|a\rangle+\right.\\
&\left.\langle a|(V_c(r)-\overline{V}_c)\int_{K\overline{r}}^{Kr}\frac{dx\,x^2\,j_l^2(x)}{\sqrt{x^2+(\Lambda r)^2}\left(x+\sqrt{x^2+(\Lambda r)^2}\right)^2}|a\rangle\right\}
\end{aligned} \tag{42}$$

where we have divided the x-range into an r-independent range, by using some arbitrary average $\overline{r}$, and the remaining large x range. The last term vanishes when $K \to \infty$ since for each l the x-integrand decreases as x^{-3} for large x. In the remaining term one is allowed to change the order between integration over x and summation over l. The reason for this is that the r-independent part of the x-integrand does not contribute to the $\langle a||a\rangle$ expectation value over r and that the r-dependent part has an extra factor $(\Lambda r/x)^2$). The convergence of the x-integral then does not depend on j_l^2.

Summing over l gives with the use of Eq. (40)

$$\lim_{K\to\infty} \langle a|(V_c(r) - \overline{V}_c) \int_0^{K\overline{r}} \frac{dx\, x^2}{\sqrt{x^2+(\Lambda r)^2}\left(x+\sqrt{x^2+(\Lambda r)^2}\right)^2}|a\rangle =$$

$$\lim_{K\to\infty} \langle a|(V_c(r) - \overline{V}_c) \int_0^{K\overline{r}/r} \frac{dk\, k^2}{\sqrt{k^2+\Lambda^2}\left(k+\sqrt{k^2+\Lambda^2}\right)^2}|a\rangle \quad . \tag{43}$$

As discussed above, an r-independent contribution from the k-integral does not contribute to the expectation value over r. Thus we can shift the lower integration limit to some large value. For large k the integrand goes as $(4k)^{-1}$ and only the r-dependent contribution -$ln(r)/4$ from the upper limit needs to be considered. We then finally get the contribution from the regulator correction term (Eq. (36)) as

$$-\frac{1}{2}\frac{\alpha}{\pi}\langle a|(V_c(r) - \overline{V}_c)ln(r)|a\rangle \quad . \tag{44}$$

In this form the correction term can easily be evaluated numerically.

IV. CONCLUSION

In QED certain Feynman diagrams contain divergences. When adding a group of Feynman diagrams with cancelling divergences one obtains a well defined finite remainder if a covariant regularization scheme is used. However, using a non-covariant regularization scheme, one can obtain additional finite correction terms. Since the correction terms are dependent on the regularization scheme, one has to investigate these corrections within the scheme used.

In the partial-wave renormalization approach, which we consider in this paper, we have shown that these correction terms vanish for the first-order self-energy and also for Coulomb screened self-energy. However, in the last case we obtain non-zero correction terms for the different subgroups of diagrams, but they cancel when added. This is not always the case and e.g. for the Breit screened self-energy one gets a remaining non-vanishing correction.

In conclusion, if a non-covariant regularization scheme is used for evaluating e.g. higher-order self energies, correction terms from the regularization have to be considered.

ACKNOWLEDGMENTS

We want to express our deepest gratitude to Professor Ingvar Lindgren. He has participated actively in our research and has been, and still is, a rich source for new ideas and inspiration. Apart from this, we have also become very good friends. We hope that this successful collaboration will continue for years to come !

---

[1] K T Cheng, W R Johnson and J Sapirstein, Phys. Rev. Lett **66**, 2960 (1991).
[2] N. J. Snyderman, Ann. Phys. **211**, 43 (1991).
[3] S. A. Blundell and N. J. Snyderman, Phys. Rev. **A44**, 1427 (1991).
[4] S. Blundell, Phys. Rev. A **46**, 3762 (1992).
S. Blundell, Phys. Rev. A **47**, 1790 (1993).
[5] H. Persson, I. Lindgren, and S. Salomonson, Phys. Scr. **T46**, 125 (1993).; I. Lindgren, H. Persson, S. Salomonson and A. Ynnermann, Phys. Rev. **A47**, 4555 (1993).
[6] H. M. Quiney and I. P. Grant, J. Phys. **B27**, L199 (1994).
[7] H. Persson, I. Lindgren, S. Salomonson, and P. Sunnergren, Phys. Rev. A **48**, 2772 (1993).
[8] H. Persson, I. Lindgren, L. Labzowsky, G. Plunien, T. Beier, and G. Soff, Phys. Rev. A **54**, 2805 (1996).
[9] S. Blundell, P. J. Mohr, W. R. Johnson and J. Sapirstein, Phys. Rev. A **48**, 2615 (1993).
[10] P. J. Mohr and G. Soff Phys. Rev. Lett. **70**, 158 (1993).
[11] I. Lindgren, H. Persson, S. Salomonson and L. Labzowsky, Phys. Rev. A **51**, 1167, (1995).
[12] A. Mitrushenkov, L. Labzowsky, I. Lindgren, H. Persson and S. Salomonson, Phys. Lett. **A200**, 51 (1995).
[13] H. Persson, S. Salomonson, P. Sunnergren, and I. Lindgren, Phys. Rev. Lett. **76**, 204 (1996).

[14] H. Persson *et al.*, Phys. Rev. Lett. **76**, 1433 (1996).
[15] G. Soff, T. Beier, M. Greiner, H. Persson and G. Plunien, This Volume, (1997).
[16] W. Pauli and F. Villars, Rev. Mod. Phys.**21**, 434 (1949).
[17] J. B. French and V. Weisskopf, Phys. Rev. **75**, 1240 (1949).
[18] P. Sunnergren, Licentiate Thesis, Göteborg University (1996).

The Reference State Coulomb-Breit QED Corrections for the Few–Electron Highly Charged Ions

L.N.Labzowsky and M.A.Tokman

St. Petersburg State University Department of Physics
198904 Petrodvorets St. Petersburg Russia

The QED Coulomb-Breit reference state corrections (RSC) for the $2p_{1/2} - 2s$ shift of Li-like ions are investigated in detail. The numerical calculations of the RSC-box and RSC-cross for the two- and three-electron configurations for Z from 1 to 92 are given.

I. INTRODUCTION

The interest to the Quantum Electrodynamic (QED) calculations of the energy levels of the few-electron highly charged ions can be explained by recent accurate measurements of the energy shifts in these ions [1]- [3]. In particular it is necessary to take into account all the second order in α (α is the fine structure constant) corrections to the energy (see general discussion in [4]). Up to now not all the α^2 one-electron corrections are calculated. The recent status of the one-electron α^2 QED corrections is discussed in detail in [5]. There is also the possibility to measure the energy difference between two- and one-electron ions, i.e. directly the two-electron contributions [3]. The full calculation of the two-electron α^2 QED corrections was given recently in [6] for the ground state of the two-electron ions (see also [7]- [11]).

In this paper we investigate the so called "reference state" Coulomb-Breit QED corrections. These corrections are the non-trivial parts of the two-photon exchange corrections. They are absent in the ordinary many-body perturbation theory (MBPT) and arise in QED due to the dependence of the effective potentials on the energy. These corrections are connected with the singularities caused by the presence of the reference state in the sums over the intermediate states. More exactly, all the states degenerate with the reference state will contribute to these singularities. The singularities should be removed by some procedure and the finite remainder is called reference state corrections (RSC). The Feynman graphs, corresponding to the two-photon exchange between the

ADVANCES IN QUANTUM CHEMISTRY, VOLUME 30

0065-3276/98 $25.00

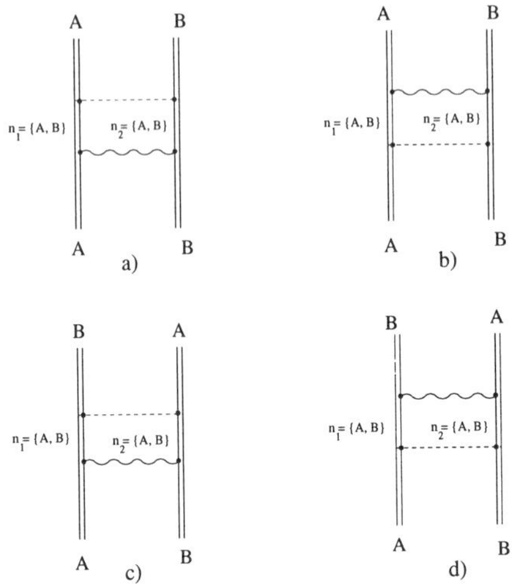

FIG. 1. The Feynman two photon box Coulomb-Breit graphs, giving rise to the RSC. The solid vertical lines correspond to the electron in the field of the nucleus. The dashed horizontal lines denote the Coulomb interaction and the wavy line denotes the Breit interaction. The graphs a), b) are the "direct graphs", and c), d) are the "exchange graphs". The notations $n_1 = \{A, B\}, n_2 = \{A, B\}$ mean that the summation over n_1, n_2 in the electron propagators are restricted to the subset of states, degenerate with the reference state $|AB>$

electrons and containing the singularities that give rise to RSC are shown in Fig. 1. We use the Coulomb gauge, distinguishing the Coulomb-Breit (CB) graphs given in Fig. 1 and the Breit-Breit (BB) given in Fig. 2. In this work we restrict ourselves to the Coulomb-Breit RSC. The reason is that the CB and BB RSC have different scaling with Z: the corrections $\Delta E^{(CB)(RSC)}$ scale like $m\alpha^2(\alpha Z)^3$ and the corrections $\Delta E^{(BB)(RSC)}$ scale like $m\alpha^2(\alpha Z)^5$. The difference in scaling is important even for high Z values. For example the total two-photon corrections for the ground state of He-like uranium U^{90+} look like [10]: $\Delta E^{(CB)} = -0.217$ a.u.; $\Delta E^{(BB)} = -0.024$ a.u.

Two different procedures were used for the removal of the reference state singularities and the derivation of RSC. As it was shown in [12], the Coulomb-Coulomb contribution to RSC is absent. The RSC were first introduced in QED in [13] in the frames of the Green function formalism in the Feynman gauge [13,14]. The expressions for RSC are much simplified in the equal energies case ($E_A = E_B$), i.e. for the ground state of the two-electron ion. The explicit expression for the equal energies correction $\Delta E^{(CB)}$ was derived in [15] with the use of the adiabatic S-matrix approach [16]- [19]. In principle, the

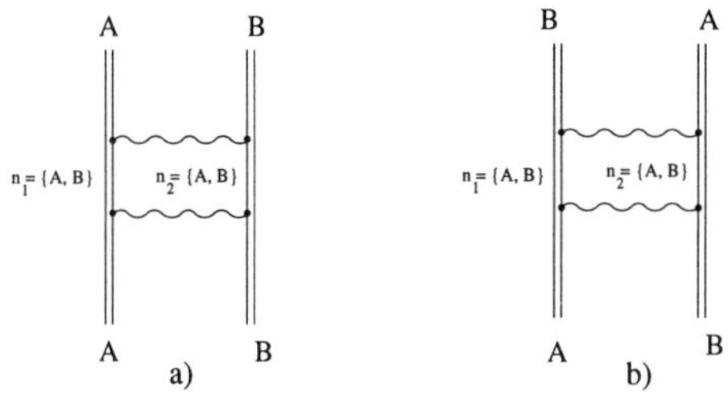

FIG. 2. The Feynman two-photon box Breit-Breit graphs, giving rise to RSC. The notations are the same as in Fig. 1. The graph a) is the "direct" graph and the graph b) is the "exchange" graph.

reference state contributes not only to the "box" graphs, given in Fig. 1, but also to the "cross" graphs, given in Fig. 3. Unlike the "box" graphs, the "cross" graphs do not have the singularities for the reference states. Hence the "cross" RSC do not require any special consideration. However it became traditional in the literature [9,10] to combine the RSC-box and RSC-cross contributions and we shall follow this tradition in this work.

In the equal energies case the Coulomb-Breit RSC with all the "direct" and "exchange" terms taken into account looks as [15]

$$\Delta E_{AB}^{(CB)}(box) = \frac{2e^4}{\pi}\left(\frac{1}{r_{12}}\right)_{AB;AB} (\vec{\alpha}_1\vec{\alpha}_2\, ln\, r_{12})_{AB;AB} \tag{1.1}$$

where e is the electron charge, $r_{12} = |\, r_1 - r_2\,|$, $\vec{\alpha}_i$ are the Dirac matrices for different variables and

$$(T_{12})_{AB;AB} = (T_{12})_{ABAB} - (T_{12})_{ABBA} \tag{1.2}$$

The "cross" Coulomb-Breit contribution to RSC in the equal energies case is [20]:

$$\Delta E_{AB}^{(CB)}(cross) = -\Delta E_{AB}^{(CB)}(box), \tag{1.3}$$

so the total Coulomb-Breit RSC is zero.

The expression (1.1) was later derived also by the "line profile" approach in [4].

In the Feynman gauge the equal-energy RSC expressions have been obtained in [21] by the modified Green functions approach [22] and in [9,10] in the framework of S-matrix adiabatic approach. The unequal energies ($E_A \neq E_B$) Coulomb-Breit RSC (box) were derived in [20] by the line profile approach and RSC (cross) were derived for this case by the same method in [23].

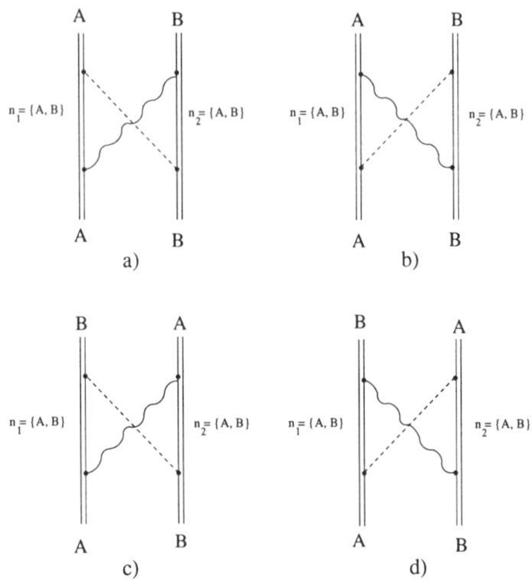

FIG. 3. The Feynman two-photon cross Coulomb-Breit graphs, that give contributions to RSC. The notations are the same as in Fig. 1.

The unequal-energies RSC in the Feynman gauge have been derived by the modified Green function approach in [24].

The numerical calculations for the ground state (equal energies case) of the He-like heavy ions were given in [9,10]. The Feynman gauge calculations include also the Breit-Breit contributions. Since for these contributions the Eq. (1.3) does not hold, the total RSC was nonzero.

The numerical calculation of the unequal energies Coulomb-Breit "box" RSC was performed in [20] for the $(1s_{1/2})^2 2s_{1/2}$ and $(1s_{1/2})^2 2p_{1/2}$ states of Li-like heavy ions. The same calculation for the "cross" RSC was given in [23].

In this paper we give the Coulomb-Breit RSC numerical calculations for $(1s_{1/2})(2l_{1/2})$ two-electron configurations and $(1s_{1/2})^2(2l_{1/2})$ three-electron configurations with $l = 1, 2$. The results are given for the nuclear charge Z from 1 to 92. We also obtain the analytic formulae for the low-Z limit that help to check the numerical results. The scaling law for the Coulomb-Breit RSC: $\Delta E^{(CB)} \sim m\alpha^2(\alpha Z)^3$ is confirmed. Few improvements in the formulae for the "box" and "cross" Coulomb-Breit RSC given in [20] and [23] that change the previous numerical results are also made.

II. NONEQUAL ENERGIES "BOX" AND "CROSS" RSC CORRECTIONS

In the case of $(1s_{1/2})^2(2l_{1/2})$ configurations we can write the total Coulomb-Breit RSC as the sum of the following contributions

$$\Delta E^{(CB)}_{(1s_{1/2})^2 2l_+} = \Delta E^{(CB)}_{1s_+,1s_-} + \Delta E^{(CB)}_{1s_+,2l_+} + \Delta E^{(CB)}_{1s_-,2l_+} \tag{2.1}$$

where $|nl\frac{1}{2}m> \equiv |nl\pm>$. The first term in Eq.(2.1) is zero for the total "box"+"cross" RSC due to Eq.(1.3). RSC for For high Z values the difference in the RSC for three-electron and two-electron ions is of the order $1/Z$ and is negligible.

Considering the configuration $(1s_{1/2})^2 2l_+$ we should take into account also the intermediate one-electron states $2l_-$, degenerate with $2l_+$. Part of the intermediate states, containing $2l_-$, is prohibited by symmetry. In the Table 1 we list all the intermediate states, contributing to the graphs Fig. 1 and Fig. 3. Note, that it is enough to consider only the contributions Fig. 1 a), c) and multiply the results by factor 2.

The nonequal-energies Coulomb-Breit "box" RSC with all the "direct" and "exchange" terms taken into account looks as [20]:

$$\Delta E^{(CB)}_{AB}(box) = \frac{e^4}{\pi} Re \sum_{i=1}^{4} \Delta E^{(box)}_i \tag{2.2}$$

where

$$\Delta E^{(box)}_1 = 2\left(\frac{1}{r_{12}}\right)_{AB\tilde{A}\tilde{B}} (\vec{\alpha}_1\vec{\alpha}_2 \ln r_{12})_{\tilde{A}\tilde{B}AB} \,, \tag{2.3}$$

$$\Delta E^{(box)}_2 = -2\left(\frac{1}{r_{12}}\right)_{BA\tilde{A}\tilde{B}} (\vec{\alpha}_1\vec{\alpha}_2 \ln r_{12})_{\tilde{A}\tilde{B}AB} \,, \tag{2.4}$$

$$\Delta E^{(box)}_3 = -\left(\frac{1}{r_{12}}\right)_{AB\tilde{A}\tilde{B}} \left(\frac{\vec{\alpha}_1\vec{\alpha}_2}{i\, r_{12}} I'((E_B-E_A), r_{12}) + \right.$$
$$\left. + \frac{(E_B-E_A)^2}{i\, r_{12}} K'((E_B-E_A), r_{12})\right)_{\tilde{A}\tilde{B}BA} \,, \tag{2.5}$$

$$\Delta E^{(box)}_4 = \left(\frac{1}{r_{12}}\right)_{AB\tilde{B}\tilde{A}} \left(\frac{\vec{\alpha}_1\vec{\alpha}_2}{i\, r_{12}} I'((E_B-E_A), r_{12}) + \right.$$
$$\left. + \frac{(E_B-E_A)^2}{i\, r_{12}} K'((E_B-E_A), r_{12})\right)_{\tilde{B}\tilde{A}AB} \,, \tag{2.6}$$

and

$$I(E,r) = \int_{-\infty}^{\infty} \frac{e^{i|\omega|r}}{\omega + E - i0} d\omega \tag{2.7}$$

$$K(E,r) = \int_{-\infty}^{\infty} \frac{e^{i|\omega|r} - 1 - i|\omega|r}{(\omega + E - i0)\omega^2} d\omega \tag{2.8}$$

By I' and K' we denote the derivatives of I and K with respect to E. We suppose that $E_B > E_A$. $\tilde{A}$, $\tilde{B}$ denote intermediate states degenerate with the state A, B and contributing to ΔE_i^{box} according to the Table 1.

Performing explicitly the integration over ω in Eqs.(2.7), (2.8) we obtain:

$$I(E,r) = 2i\frac{E}{|E|}G(|E|r) + \pi i(1 - \frac{E}{|E|})e^{i|E|r} \tag{2.9}$$

$$K(E,r) = \frac{\pi i}{E^2}(1 - \frac{E}{|E|})(e^{i|E|r} - 1 - i|E|r) +$$

$$+\frac{2i}{E^2}\frac{E}{|E|}(G(|E|r) - |E|r\, ln\,(|E|r) + (1-C)|E|r - \frac{\pi}{2}) \tag{2.10}$$

where

$$G(z) = sin\, z\, ci\, z - cos\, z\, si\, z\,, \tag{2.11}$$

$si(z)$ and $ci(z)$ are the integral sine and cosine functions and C is the Euler's constant. The term with $-\pi/2$ in the last line of Eq.(2.10) was omitted in [20].

The nonequal energy Coulomb-Breit "cross" RSC with all the "direct" and "exchange" terms is [23]

$$\Delta E_{AB}^{(CB)}(cross) = -\frac{e^4}{\pi}\, Re \sum_{i=1}^{4} \Delta E_i^{(cross)} \tag{2.12}$$

where

$$\Delta E_1^{(cross)} = 2\left(\frac{1}{r_{12}}\right)_{A\tilde{B}\tilde{A}B} (\vec{\alpha}_1\vec{\alpha}_2\, ln\, r_{12})_{\tilde{A}BA\tilde{B}}\,, \tag{2.13}$$

$$\Delta E_2^{(cross)} = \frac{1}{2i\,(E_B - E_A)}\left(\frac{1}{r_{12}}\right)_{A\tilde{A}\tilde{B}B}\left(\frac{\vec{\alpha}_1\vec{\alpha}_2}{r_{12}}[\,I((E_B - E_A), r_{12}) - \right.$$

$$\left. -I((E_A - E_B), r_{12})\,] + \frac{(E_B - E_A)^2}{r_{12}}[\,K((E_B - E_A), r_{12}) - K((E_A - E_B), r_{12})\,]\right)_{\tilde{B}BA\tilde{A}}, \tag{2.14}$$

$$\Delta E_3^{(cross)} = \frac{1}{i\,(E_B - E_A)}\left(\frac{1}{r_{12}}\right)_{B\tilde{B}\tilde{A}B}\left(\frac{\vec{\alpha}_1\vec{\alpha}_2}{r_{12}}[\,\pi i - I((E_B - E_A), r_{12})\,]\right)_{\tilde{A}AA\tilde{B}}, \tag{2.15}$$

$$\Delta E_4^{(cross)} = \frac{1}{i\,(E_B - E_A)}\left(\frac{1}{r_{12}}\right)_{B\tilde{A}\tilde{B}B}\left(\frac{\vec{\alpha}_1\vec{\alpha}_2}{r_{12}}[\,\pi i - I((E_B - E_A), r_{12})\,]\right)_{\tilde{B}AA\tilde{A}} \tag{2.16}$$

In the limit $E_B - E_A \to 0$ the formulae (2.2) and (2.12) come over to the Eq. (1.1) and (1.3).

III. NUMERICAL RESULTS AND DISCUSSION

In the Table 2 the RSC-box corrections for the ground state are represented. Though these corrections are canceled by the RSC-cross contributions according to Eq. (1.3), we give these values for completeness. In some numerical methods (e.g. in the Coulomb Green function approach) it is more convenient not to separate out RSC-cross. In the case of two-electron configuration we give the contributions of triplet and singlet states. The first one coincides with the contribution of configuration $1s_{1/2+}2l_{1/2+}$ (second term in Eq. (2.1)). The contribution of the third term in Eq.(2.1) is given separetely. In the Tables 3-6 we give the numerical values for the "box" and "cross" $\Delta E^{(CB)}$ RSC for two- and three-electron configurations, mentioned in Sec. 1. The last term from Eq. (2.1) is demonstrated in Tables 3-4.

According to the scaling of $\Delta E^{(CB)}$ RSC with Z we can express these corrections for the different configurations in the form

$$\Delta E^{(CB)} = \frac{m\alpha^2}{\pi}(\alpha Z)^3 F(\alpha Z) \tag{3.1}$$

where $F(\alpha Z)$ should be a smooth function of αZ. This is confirmed by the results given in Tables 2-6.

The low Z limit of the F function in Eq. (3.1) $F(\alpha Z) \to F_0$, can be evaluated analytically. For this evaluation we use the expansion for G-function in Eq.(2.11)

$$G(z) = \frac{\pi}{2} + (C-1)z + z\,ln\,z - \frac{\pi}{4}z^2 - \frac{C}{6}z^3 - \frac{1}{6}z^3\,ln\,z + \frac{11}{36}z^3 + \frac{\pi}{48}z^4 + ... \tag{3.2}$$

and the known expressions for $1s_{1/2}, 2s_{1/2}, 2p_{1/2}$ Dirac functions for the Coulomb field. The latter should be also expanded in αZ. The values of F_0 for the different configurations are given in Tables 7,8. We write F_0 as

$$F_0 = A_1 + A_2\,(C + ln(\alpha Z)) + A_3\,ln2 + A_4\,ln3 + A_5\,ln7 \tag{3.3}$$

where A_i are the fractional numbers. The numerical results for the low Z values coincide with the analytic results obtained by Eq.(3.1) with $F(\alpha Z) = F_0$.

The relativistic and QED corrections of the order $m\alpha^5 = \alpha^3$ a.u. for the low Z limit were obtained by Araki [25] and Sucher [26] (see also [27]). The Coulomb-Breit RSC also should be among these corrections. However it looks nonevident how to separate out the RSC from the total contribution of the "box" and "cross" Coulomb-Breit graphs, given in [25]- [27].

Finally, we give the Table 9, showing the various corrections to the $(1s_{1/2})^2$ $(2p_{1/2}) - (1s_{1/2})^2(2s_{1/2})$ shift for the Li-like uranium ion. This Table coincide with the corresponding Table in [5], but RSC values are changed, according to the improved calculations given in the present paper.

[1] J. Schweppe, A. Belkacem, L. Blumenfeld, N. Claytor, B. Feinberg, H. Gould, V. Kostroun, L. Levy, S. Misawa, J. Mowat and M. Prior, Phys. Rev. Lett. **66**, 1434 (1991)

[2] H. Beyer GSI-94-12 Report Nov 1994, ISSN 0171-4546

[3] R. Marrs, S. Schneider and Th. Stöhlker, Phys. Rev. A (submitted)

[4] L. Labzowsky, V. Karasiev, I. Lindgren, H. Persson and S. Salomonson, Phys. Scr. **T46**, 150 (1993)

[5] H. Persson, I. Lindgren, L. Labzowsky, G. Plunien, Th. Beier and G. Soff, Phys. Rev. A**54**, 2805 (1996)

[6] H. Persson, S. Salomonson, P. Sunnergren and I. Lindgren, Phys. Rev. Lett. **76**, 204 (1996)

[7] S. Blundell, Phys. Rev. A**46**, 3762 (1992)

[8] S. Blundell, Phys. Rev. A**47**, 1790 (1993)

[9] S. Blundell, P. Mohr, W. Johnson and J. Sapirstein, Phys. Rev A**48**, 2615 (1993)

[10] I. Lindgren, H. Persson, S. Salomonson and L. Labzowsky, Phys. Rev. A**51**, 1167 (1995)

[11] V. Yerokhin, and V. Shabaev, Phys. Lett. A**207**, 274 (1995)

[12] L. Labzowsky, Sov. Phys. JETP**32**, 94 (1970)

[13] M. Braun and A. Shirokov, Izv.Akad. Nauk USSR, ser. fiz.**41**, 2585 (1977)

[14] M. Braun, A. Gurchumelia and U. Safronova, *Relativistic Theory of Atoms*, Moscow, Nauka 1984 (in Russian)

[15] T. Timofeeva and L. Labzowsky, Izv. Akad. Nauk USSR, ser. fiz. **45**, 2390 (1981)

[16] M. Gell-Mann and F. Low, Phys. Rev. **84**, 350 (1951)

[17] J. Sucher, Phys. Rev. **107**, 1448 (1957)

[18] L. Labzowsky, J. Phys. B **26**, 1039 (1993)

[19] L. Labzowsky, G. Klimchitskaya, Yu. Dmitriev, *Relativistic Effects in the Spectra of Atomic Systems* IOP. Publishing 1993

[20] I. Lindgren, H. Persson, S. Salomonson, V. Karasiev, L. Labzowsky, A. Mitrushenkov, and M. Tokman, J. Phys. B **26**, L503 (1993)

[21] V. Shabaev, Sov. Phys. Journ. **33**, 60 (1990)

[22] V. Shabaev, Teor. Mat. Fiz. **82**, 83 (1990)

[23] L. Labzowsky and M. Tokman, J. Phys. B **28**, 3717 (1995)

[24] V. Shabaev and I. Fokeeva, Phys. Rev. A **49**,4489 (1994)

[25] H. Araki, Progr. Theor. Phys. **17**, 619 (1957)

[26] J. Sucher, Phys. Rev. **109**, 1010 (1958)

[27] M. Braun and L. Labzowsky, Zh. Eksp. Teor. Fiz. **53**, 1776 (1967)

[28] I. Lindgren, H. Persson, S. Salomonson, and A. Ynnerman, Phys. Rev. A **47**,

R4555 (1993)
[29] H. Persson, I. Lindgren, S. Salomonson, and S. Sunnergren, Phys. Rev A **48**, 2772 (1993)
[30] A. Mitrushenkov, L. Labzowsky, I. Lindgren, H. Persson, and S. Salomonson, Phys. Lett. A **200**, 51 (1995)
[31] S. Schneider, W. Greiner, and G. Soff, J. Phys. B **26**, L529 (1993)
[32] G. Plunien, B. Müller, W. Greiner, and G. Soff, Phys. Rev. A **43**, 5853 (1991)

Tables

Table 1. Intermediate states for the graphs Fig. 1 and Fig. 3

Fig. 1 a), c)

i / AB	1	2	3	4
$1s_+2s_+$	$1s_+2s_+$	$1s_+2s_+$	$1s_+2s_+$	$1s_+2s_+$
$1s_-2s_+$	$1s_-2s_+$	$1s_+2s_-$	$1s_-2s_+$	$1s_+2s_-$
$1s_+2p_+$	$1s_+2p_+$	$1s_+2p_+$	$1s_+2p_+$	$1s_+2p_+$
$1s_-2p_+$	$1s_-2p_+$	$1s_-2p_+$ $1s_+2p_-$	$1s_-2p_+$	$1s_-2p_+$ $1s_+2p_-$

Fig. 3 a), c)

i / AB	1	2	3	4
$1s_+2s_+$	$1s_+2s_+$	$1s_+2s_+$	$1s_+2s_+$	$1s_+2s_+$
$1s_-2s_+$	$1s_-2s_+$	$1s_+2s_-$	$1s_+2s_+$	$1s_+2s_+$
$1s_+2p_+$	$1s_+2p_+$	$1s_+2p_+$ $1s_-2p_-$		
$1s_-2p_+$	$1s_-2p_+$	$1s_+2p_-$		

Table 2. The RSC-box corrections for the ground state. The values of ΔE are given in eV. The numbers in brackets denote multiplication by powers of 10.

	$(1s_{1/2})^2$	
Z	ΔE	F
1	0.129501[-5]	0.3846
10	0.129776[-2]	0.3856
20	0.104318[-1]	0.3874
30	0.354945[-1]	0.3906
40	0.851255[-1]	0.3952
50	0.168885	0.4014
60	0.297784	0.4096
70	0.485097	0.4202
80	0.747757	0.4339
82	0.805626	0.4345
90	1.108887	0.4519
92	1.195700	0.4566

Table 3. The RSC-box corrections for the two-electron configurations. The values of ΔE are given in eV. The numbers in brackets denote multiplication by powers of 10.

	$1s_{1/2+}2s_{1/2+}$ $2S_1$		$1s_{1/2-}2s_{1/2+}$	$1s_{1/2-}2s_{1/2+}$ $2S_0$	
Z	ΔE	F	ΔE	ΔE	F
1	0.207089[-7]	0.6150[-2]	0.251716[-7]	0.458998[-7]	0.1363[-1]
10	0.149120[-4]	0.4443[-2]	0.313841[-4]	0.578985[-4]	0.1720[-1]
20	0.609436[-4]	0.2263[-2]	0.315142[-3]	0.589841[-3]	0.2190[-1]
30	0.101399[-3]	0.1115[-2]	0.118473[-2]	0.226807[-2]	0.2495[-1]
40	0.111603[-3]	0.5179[-3]	0.310308[-2]	0.609455[-2]	0.2828[-1]
50	0.132308[-3]	0.3144[-3]	0.634071[-2]	0.125491[-1]	0.2982[-1]
60	0.175351[-3]	0.2410[-3]	0.113999[-1]	0.226245[-1]	0.3111[-1]
70	0.208556[-3]	0.1806[-3]	0.189363[-1]	0.376641[-1]	0.3261[-1]
80	0.196586[-3]	0.1140[-3]	0.298430[-1]	0.595020[-1]	0.3452[-1]
82	0.212735[-3]	0.1146[-3]	0.325190[-1]	0.648252[-1]	0.3492[-1]
90	0.131273[-3]	0.5348[-4]	0.453066[-1]	0.904820[-1]	0.3686[-1]
92	0.105086[-3]	0.4008[-4]	0.490965[-1]	0.980879[-1]	0.3741[-1]

Table 3. (*Continued*)

	$1s_{1/2+}2p_{1/2+}$ $2P_1$		$1s_{1/2-}2p_{1/2+}$	$1s_{1/2-}2p_{1/2+}$ $2P_0$	
Z	ΔE	F	ΔE	ΔE	F
1	0.523004[-7]	0.1553[-1]	0.394531[-7]	0.725792[-7]	0.2158[-1]
10	0.805326[-4]	0.2392[-1]	0.167703[-4]	0.310561[-4]	0.9225[-1]
20	0.781583[-3]	0.2901[-1]	0.118334[-3]	0.224253[-3]	0.8325[-1]
30	0.289344[-2]	0.3183[-1]	0.448325[-3]	0.868191[-3]	0.9515[-1]
40	0.908988[-2]	0.4218[-1]	0.516341[-3]	0.101522[-2]	0.4711[-1]
50	0.216398[-1]	0.5142[-1]	0.69532[-2]	0.137789[-1]	0.3274[-1]
60	0.499611[-1]	0.6869[-1]	0.796903[-2]	0.158367[-1]	0.2177[-1]
70	0.742032[-1]	0.6425[-1]	0.734175[-2]	0.146162[-1]	0.1266[-1]
80	0.102469	0.5944[-1]	0.531247[-2]	0.106037[-1]	0.6151[-2]
82	0.108390	0.5839[-1]	0.527631[-2]	0.103459[-1]	0.5573[-2]
90	0.116320	0.4739[-1]	0.454122[-2]	0.907011[-2]	0.3695[-2]
92	0.120998	0.4615[-1]	0.310120[-2]	0.619868[-2]	0.2364[-2]

Table 4. The RSC-cross corrections for the two-electron configurations. The values of ΔE are given in eV. The number in brackets denote multiplication by powers of 10.

	$1s_{1/2+}2s_{1/2+}$ $2S_1$		$1s_{1/2-}2s_{1/2+}$	$1s_{1/2-}2s_{1/2+}$ $2S_0$	
Z	ΔE	F	ΔE	ΔE	F
1	0.105567[-7]	0.3136[-2]	-0.425868[-7]	-0.733711[-7]	-0.2179[-1]
10	0.111354[-4]	0.3363[-2]	-0.436388[-4]	-0.759834[-4]	-0.2257[-1]
20	0.104447[-3]	0.3877[-2]	-0.319769[-3]	-0.568842[-3]	-0.2112[-1]
30	0.359543[-3]	0.3955[-2]	-0.991575[-3]	-0.178083[-2]	-0.1959[-1]
40	0.104871[-2]	0.4866[-2]	-0.223410[-2]	-0.407740[-2]	-0.1892[-1]
50	0.227668[-2]	0.5409[-2]	-0.436878[-2]	-0.805113[-2]	-0.1913[-1]
60	0.424544[-2]	0.5837[-2]	-0.775435[-2]	-0.143779[-1]	-0.1977[-1]
70	0.726228[-2]	0.6288[-2]	-0.129192[-1]	-0.240158[-1]	-0.2079[-1]
80	0.117657[-1]	0.6825[-2]	-0.206174[-1]	-0.383562[-1]	-0.2225[-1]
82	0.128729[-1]	0.6933[-2]	-0.225255[-1]	-0.419344[-1]	-0.2259[-1]
90	0.183640[-1]	0.7481[-2]	-0.319181[-1]	-0.595048[-1]	-0.2424[-1]
92	0.200218[-1]	0.7637[-2]	-0.347491[-1]	-0.648226[-1]	-0.2472[-1]

Table 4. (*Continued*)

	$1s_{1/2+}2p_{1/2+}$ $2P_1$		$1s_{1/2-}2p_{1/2+}$	$1s_{1/2-}2p_{1/2+}$ $2P_0$	
Z	ΔE	F	ΔE	ΔE	F
1	-0.372257[-7]	-0.1106[-1]	-0.559210[-8]	-0.888606[-8]	-0.2639[-2]
10	-0.412548[-4]	-0.1225[-1]	-0.506995[-5]	-0.889481[-5]	-0.2642[-2]
20	-0.384521[-3]	-0.1428[-1]	-0.409453[-4]	-0.733274[-4]	-0.2722[-2]
30	-0.131932[-2]	-0.1451[-1]	-0.159823[-3]	-0.290028[-3]	-0.3190[-2]
40	-0.473932[-2]	-0.2199[-1]	-0.428352[-3]	-0.785627[-3]	-0.3646[-2]
50	-0.999683[-2]	-0.2375[-1]	-0.940721[-3]	-0.173572[-2]	-0.4124[-2]
60	-0.189366[-1]	-0.2603[-1]	-0.163551[-2]	-0.302811[-2]	-0.4163[-2]
70	-0.299929[-1]	-0.2596[-1]	-0.154117[-2]	-0.286341[-2]	-0.2479[-2]
80	-0.452906[-1]	-0.2628[-1]	-0.143342[-2]	-0.267384[-2]	-0.1551[-2]
82	-0.488411[-1]	-0.2631[-1]	-0.116024[-2]	-0.217192[-2]	-0.1170[-2]
90	-0.556353[-1]	-0.2266[-1]	-0.981263[-3]	-0.184726[-2]	-0.7525[-3]
92	-0.583745[-1]	-0.2228[-1]	-0.821263[-3]	-0.155900[-2]	-0.5946[-3]

Table 5. The RSC-box corrections for the three-electron configurations. The values of ΔE are given in eV. The numbers in brackets denote multiplication by powers of 10.

	$(1s_{1/2})^2 2s_{1/2+}$		$(1s_{1/2})^2 2p_{1/2+}$		$(1s_{1/2})^2 2p_{1/2+} - (1s_{1/2})^2 2s_{1/2+}$	
Z	ΔE	F	ΔE	F	ΔE	F
1	0.134088[-5]	0.3982	0.138676[-5]	0.4119	0.458799[-7]	0.1363[-1]
10	0.134406[-2]	0.3993	0.139506[-2]	0.4144	0.510023[-4]	0.1515[-1]
20	0.108079[-1]	0.4012	0.113317[-1]	0.4207	0.523815[-3]	0.1945[-1]
30	0.367806[-1]	0.4046	0.388362[-1]	0.4272	0.205559[-2]	0.2261[-1]
40	0.883402[-1]	0.4099	0.947317[-1]	0.4396	0.639153[-2]	0.2966[-1]
50	0.175358	0.4166	0.197478	0.4692	0.221201[-1]	0.5255[-1]
60	0.309359	0.4254	0.355714	0.4891	0.463549[-1]	0.6373[-1]
70	0.504242	0.4366	0.566642	0.4906	0.624003[-1]	0.5403[-1]
80	0.777803	0.4512	0.855554	0.4963	0.777352[-1]	0.4509[-1]
82	0.838358	0.4516	0.919293	0.4963	0.809354[-1]	0.4360[-1]
90	1.154329	0.4703	1.229748	0.5010	0.754192[-1]	0.3072[-1]
92	1.244902	0.4748	1.319789	0.5021	0.748974[-1]	0.2856[-1]

Table 6. The RSC-cross corrections for the three-electron configurations. The values of ΔE are given in eV. The number in brackets denote multiplication by powers of 10.

	$(1s_{1/2})^2 2s_{1/2+}$		$(1s_{1/2})^2 2p_{1/2+}$		$(1s_{1/2})^2 2p_{1/2+} - (1s_{1/2})^2 2s_{1/2+}$	
Z	ΔE	F	ΔE	F	ΔE	F
1	-0.132704[-5]	-0.3941	-0.133782[-5]	-0.3973	-0.107882[-7]	-0.3204[-2]
10	-0.133034[-2]	-0.3951	-0.134415[-2]	-0.3993	-0.138314[-4]	-0.4099[-2]
20	-0.106471[-1]	-0.3953	-0.108571[-1]	-0.4031	-0.210032[-3]	-0.7796[-2]
30	-0.361115[-1]	-0.3972	-0.369736[-1]	-0.4067	-0.862120[-2]	-0.9485[-2]
40	-0.863109[-1]	-0.4005	-0.902928[-1]	-0.4190	-0.398192[-2]	-0.1848[-1]
50	-0.170977	-0.4062	-0.179823	-0.4272	-0.884613[-2]	-0.2102[-1]
60	-0.301293	-0.4143	-0.318356	-0.4377	-0.170634[-1]	-0.2346[-1]
70	-0.490754	-0.4249	-0.516631	-0.4473	-0.258805[-1]	-0.2241[-1]
80	-0.756609	-0.4389	-0.794481	-0.4608	-0.378720[-1]	-0.2197[-1]
82	-0.815279	-0.4391	-0.855628	-0.4609	-0.403487[-1]	-0.2174[-1]
90	-1.122441	-0.4573	-1.165504	-0.4748	-0.430625[-1]	-0.1754[-1]
92	-1.210427	-0.4617	-1.254896	-0.4786	-0.444691[-1]	-0.1696[-1]

Table 7. The F_0 coefficients for the different two-electron configurations.

Configurations	A_1	A_2	A_3	A_4	A_5
$(1s_{1/2})^2$(box)	$\frac{5}{13}$				
$1s_{1/2+}2s_{1/2+}\ 2S_1$(box)	$\frac{5}{729}$	$-\frac{548}{254421}$	$-\frac{275}{68283}$	$-\frac{411}{62208}$	
$1s_{1/2+}2s_{1/2+}\ 2S_1$(cross)	$\frac{335}{78732}$	$\frac{67}{34992}$	$-\frac{201}{148899}$	$\frac{536}{106106}$	$\frac{268}{203391}$
$1s_{1/2-}2s_{1/2+}\ 2S_0$(box)	$\frac{103}{6561}$	$\frac{7}{2187}$	$-\frac{283}{75816}$	$\frac{86}{6561}$	
$1s_{1/2-}2s_{1/2+}\ 2S_0$(cross)	$-\frac{164}{6561}$	$-\frac{97}{13122}$	$-\frac{64}{2187}$	$\frac{5}{1701}$	$-\frac{359}{59049}$
$1s_{1/2+}2p_{1/2+}\ 2P_1$(box)	$\frac{9275}{531441}$	$-\frac{4667}{708588}$	$-\frac{4667}{419904}$	$-\frac{74672}{3601989}$	
$1s_{1/2+}2p_{1/2+}\ 2P_1$(cross)	$-\frac{2388}{177147}$	$-\frac{82}{19683}$	$-\frac{1910}{52488}$	$\frac{19}{2187}$	

Table 7. (*Continued*)

Configurations	A_1	A_2	A_3	A_4	A_5
$1s_{1/2-}2p_{1/2+}\ 2P_0$(box)	$\frac{19525}{767637}$	$\frac{2423}{314928}$	$-\frac{1471}{236196}$	$\frac{2423}{78732}$	
$1s_{1/2-}2p_{1/2+}\ 2P_0$(cross)	$-\frac{98}{34992}$	$-\frac{142}{209952}$	$-\frac{205}{96228}$	$-\frac{331}{279936}$	

Table 8. The F_0 coefficients for the different three-electron configurations.

Configurations	A_1	A_2	A_3	A_4	A_5
$(1s_{1/2})^2 2s_{1/2+}$(box)	$\frac{34144}{85293}$	$-\frac{248}{763263}$	$-\frac{439207}{63912888}$	$\frac{653}{559872}$	
$(1s_{1/2})^2 2s_{1/2+}$(cross)	$-\frac{404593}{1023515}$	$-\frac{85}{34992}$	$-\frac{1883684}{108547371}$	$\frac{561974}{90243153}$	$-\frac{3943}{1830519}$
$(1s_{1/2})^2 2p_{1/2+}$(box)	$\frac{2884412}{6908733}$	$-\frac{5339}{2834352}$	$-\frac{66595}{3779136}$	$-\frac{27665}{14407956}$	
$(1s_{1/2})^2 2p_{1/2+}$(cross)	$-\frac{4910197}{12282192}$	$-\frac{2815}{610173}$	$-\frac{2720}{72171}$	$\frac{743}{93312}$	

Table 9. $2p_{1/2} - 2s_{1/2}$ shift in Li-like uranium.

Correction	Order of magnitude and scaling	Numerical value (eV)	Reference
RMBPT[a)]	$m\alpha(\alpha Z)$	322.33(15)	[28]
Electron self-energy (SE)	$m\alpha(\alpha Z)^4$	-55.87	[28]
Vacuum polarization (VP)	$m\alpha(\alpha Z)^4$	12.94	[29]
SE screening	$m\alpha^2(\alpha Z)^3$	1.55	[28]
VP screening	$m\alpha^2(\alpha Z)^3$	-0.39	[29]
SESE[b)]	$m\alpha^2(\alpha Z)^5$	0.09	[30]
VPVP[c)]	$m\alpha^2(\alpha Z)^4$, $m\alpha^2(\alpha Z)^5$	0.13	[31], [5]
SEVP[d)]	$m\alpha^2(\alpha Z)^4$	-0.21	[20], [5]
RSC-box	$m\alpha^2(\alpha Z)^3$	0.075	This work
RSC-cross	$m\alpha^2(\alpha Z)^3$	-0.045	This work
Recoil	$m\frac{m}{M}(\alpha Z)^2$	-0.08	[7]
Nuclear polarization		0.03	[32], [20]
Total (theory)		280.55(15)	
Experiment		280.59(9)	[1]

[a)] RMBPT-relativistic many-body perturbation theory; differs from the full QED calculation by the absence of the retardation, negative energy states and cross photon contributions.

[b] SESE-second order electron self energy; this calculation is up to now incomplete.

[c] VPVP-second order vacuum polarization.

[d] SEVP-mixed second order self energy-vacuum polarization correction.

Some Aspects on the Bloch-Lindgren Equation and a Comparison with the Partitioning Technique.

By Per-Olov Löwdin*
Department of Quantum Chemistry, Uppsala University,
Box 518, S-75120, Uppsala, Sweden, and
Florida Quantum Theory Project,
Departments of Physics and Chemistry, 362 Williamson Hall,
University of Florida, Gainesville, FL 32611-8435, USA.
*Professor Emeritus at Uppsala University, and Graduate Research Professor Emeritus at the University of Florida.

Abstract: After a brief introduction of the intuitive idea of replacing the ordinary Schrödinger equation with a simpler relation involving a *model Hamiltonian* and a set of model functions, a strict derivation of the non-linear form of the Schrödinger equation is given by using a skew-projector Θ associated with two linear subspaces, which justifies this approach. In the case when the Hamiltonian has the form $H = H_0 + V$, the Bloch and Bloch-Lindgren equations are derived, and their solutions in terms of power series in V are discussed. Since the model Hamiltonian does not depend explicitly on the energy E, the attempts to use this approach to build a bridge between the semi-empirical and *ab-initio* methods in quantum chemistry are briefly reviewed. Finally a comparison of the treatment of similar systems by means of conventional partitioning technique is given, and the problem of the derivation of the wave and reactions operator in a generalized closed form of perturbation theory is discussed.

List of contents:

1. Introduction.
2. The Energy-Independent Wave Operators and the Non-Linear Form of the Schrödinger Equation.
3. The Bloch and Bloch-Lindgren Equations.
4. On the Bridge between Semi-Empirical and *Ab-Initio* Methods in Atomic and Molecular Quantum Theory.
5. A Comparison with the Conventional Multi-Dimensional Partitioning Technique.
   The reduced resolvent.
   Connection with perturbation theory; the wave and reaction operators in the general case.
   Connection with conventional perturbation theory and the Bloch method.
6. References.

ADVANCES IN QUANTUM CHEMISTRY, VOLUME 30

0065-3276/98 $25.00

1. Introduction.

A wave operator W is an operator which maps a selected reference function φ into an exact eigenfunction Ψ to the Hamiltonian H, or a set of such reference functions into a set of eigenfunctions. Such wave operators are fairly easily derived by means of partitioning technique [1], in which case they contain the associated energy eigenvalues E. In the beginning of the 1960's, there was an interest in nuclear physics [2] in trying to find energy-independent wave operators in order to construct a model relation, which was a simplification of the conventional Schrödinger equation $H\,\Psi = E\,\Psi$. In order to briefly review this development, let us start from a reference set $\boldsymbol{\varphi} = \{\varphi_1, \varphi_2, \varphi_3, \ldots \varphi_p\}$ of order p, which set is linearly independent and characterized by a self-adjoint projector [3] $O = |\boldsymbol{\varphi}\rangle\langle\boldsymbol{\varphi}|\boldsymbol{\varphi}\rangle^{-1}\langle\boldsymbol{\varphi}| = |\boldsymbol{\varphi}\rangle\langle\boldsymbol{\varphi}_r| = |\boldsymbol{\varphi}_r\rangle\langle\boldsymbol{\varphi}|$, where $\boldsymbol{\varphi}_r = \boldsymbol{\varphi}\langle\boldsymbol{\varphi}|\boldsymbol{\varphi}\rangle^{-1}$ is the reciprocal set, which is bi-orthonormal to the original set $\boldsymbol{\varphi}$, so that $\langle\boldsymbol{\varphi}|\boldsymbol{\varphi}_r\rangle = \mathbf{1}$. In this paper, bold-face Roman fonts and Greek shadowed fonts will denote vectors and matrices.

Let further $\boldsymbol{\Psi} = \{\Psi_1, \Psi_2, \Psi_3, \ldots \Psi_p\}$ be a set of p exact eigenfunctions satisfying a system of p Schrödinger equations

$$H\,\boldsymbol{\Psi} = \boldsymbol{\Psi}\,\mathbf{E}, \qquad (1.1)$$

where $\mathbf{E}$ is a diagonal matrix with the eigenvalues $E_1, E_2, E_3, \ldots E_p$ on the diagonal. The projections $O\,\boldsymbol{\Psi} = \boldsymbol{\eta}$ of the eigenfunctions on the reference space are called the *model functions*, and it is explicitly assumed that they are linearly independent. It is further assumed that there exists a wave operator W, which maps the modelfuntions back on the eigenfunctions, so that $\boldsymbol{\Psi} = W\,\boldsymbol{\eta}$. However, since it is evident that every operator $W' = W + A(1-O)$ for an arbitrary operator A will serve the same purpose, it is customary to consider the product $\Theta = W'O = WO$, which is unambiguous. The starting relations which are essentially definitions are hence of the form

$$O\,\boldsymbol{\Psi} = \boldsymbol{\eta}, \qquad \boldsymbol{\Psi} = \Theta\,\boldsymbol{\eta}, \qquad \Theta O = \Theta. \qquad (1.2)$$

Of special interest is now the effective Hamiltonian

$$H_{eff} = O\,H\,\Theta, \qquad (1.3)$$

having the property $H_{eff}\boldsymbol{\eta} = O\,H\,\Theta\,\boldsymbol{\eta} = O\,H\,\boldsymbol{\Psi} = O\,\boldsymbol{\Psi}\,\mathbf{E} = \boldsymbol{\eta}\,\mathbf{E}$, or

$$H_{eff}\boldsymbol{\eta} = \boldsymbol{\eta}\,\mathbf{E}, \qquad (1.4)$$

where **E** consists of the exact eigenvalues. For this reason, the effective Hamiltonian H_{eff} is often referred to as the *model Hamiltonian*, and equation (1.4) is then the *model relation* which replaces the original Schrödinger equation. It is believed [4] that this relation forms a better basis for the semi-empirical formulation of quantum theory, and we will return to this question later.

Multiplying the equation $O\,\boldsymbol{\Psi} = O\,\Theta\,\boldsymbol{\eta} = \boldsymbol{\eta}$, to the right by $<\boldsymbol{\eta}_r|$, where $\boldsymbol{\eta}_r = \boldsymbol{\eta} < \boldsymbol{\eta}|\boldsymbol{\eta}>^{-1}$ is the reciprocal set to $\boldsymbol{\eta}$, one gets directly $O\Theta O = O$, which gives $O\Theta = O$. One has further $\Theta^2 = \Theta\,\Theta = \Theta\,O\,\Theta = \Theta\,O = \Theta$, which implies that the wave operator Θ must be idempotent, i.e. a projector. Since $\boldsymbol{\Psi} = \Theta\,\boldsymbol{\eta}$, one gets also $\Theta\,\boldsymbol{\Psi} = \Theta^2\,\boldsymbol{\eta} = \Theta\,\boldsymbol{\eta} = \boldsymbol{\Psi}$, which means that the eigenfunctions $\boldsymbol{\Psi}$ are situated in the range of the projector Θ. Multiplying the Schrödinger equation (1.1) to the left by Θ, one gets directly $\Theta\,H\,\boldsymbol{\Psi} = \boldsymbol{\Psi}\,\mathbf{E} = H\,\boldsymbol{\Psi}$, or $\Theta\,H\,\Theta\,\boldsymbol{\eta} = H\,\Theta\,\boldsymbol{\eta}$ and - multiplying this relation to the right by $<\boldsymbol{\eta}_r|$ and observing that $\Theta\,O = \Theta$ - one gets finally $\Theta\,H\,\Theta = H\,\Theta$, which is a non-linear form of the Schrödinger equation. The search for a model relation leads hence to the search for a wave operator Θ, which satisfies the relations

$$\Theta\,O = \Theta, \qquad O\,\Theta = O, \qquad \Theta^2 = \Theta, \qquad (1.5)$$

$$\Theta\,H\,\Theta = H\,\Theta, \qquad (1.6)$$

and the question is, of course, whether such an operator exists. A weak point in this approach is also that one starts from the explicit assumption that one knows p exact eigenfunctions to the Hamiltonian, from which one may derive the model functions.

2. The Energy-Independent Wave Operators and the Non-Linear Form of the Schrödinger Equation.

Let us in this section start from the assumption that one has selected a specific reference set $\boldsymbol{\varphi} = \{\varphi_1,\varphi_2,\varphi_3, \ldots\varphi_p\}$, which is linearly independent but not necessarily orthonormal and characterized by a self-adjoint projector $O = |\varphi\rangle\langle\varphi|\varphi\rangle^{-1}\langle\varphi| = |\varphi\rangle\langle\varphi_r| = |\varphi_r\rangle\langle\varphi|$. The set of Schrödinger equations (1.1) is solvable for every value of the parameter E - real or complex - and the physically important values of E are singled out by the boundary conditions selected. We will here use the reference set φ to specify the boundary conditions of our choice by demanding that - for the eigenfunctions Ψ under consideration - the matrix $\mathbf{C} = \langle\varphi|\Psi\rangle$ is non-singular, i.e. that the value of the determinant $|\langle\varphi|\Psi\rangle|$ is different from zero:

$$|\langle\varphi|\Psi\rangle| \neq 0. \tag{2.1}$$

This boundary condition will be automatically fulfilled for all eigenfunctions associated with discrete or continuous eigenvalues to the Hamiltonian, except for those eigenfunctions Ψ which accidentally lack projections in the reference space and then correspond to "lost eigenvalues". Since the matrix $\mathbf{C} = \langle\varphi|\Psi\rangle$ has an inverse, it may now be feasible to look at a skew projector [5], which is given by the formula:

$$\Theta = |\Psi\rangle\langle\varphi|\Psi\rangle^{-1}\langle\varphi|. \tag{2.2}$$

It is immediately clear that it satisfies the three relations (1.5), whereas it is a little bit more cumbersome to show that it satisfies (1.6). Multiplying the Schrödinger equation (1.1) to the left by $\langle\varphi|$, one gets $\langle\varphi|H|\Psi\rangle = \langle\varphi|\Psi\rangle\mathbf{E}$ and $\mathbf{E} = \langle\varphi|\Psi\rangle^{-1}\langle\varphi|H|\Psi\rangle$, and the Schrödinger equation then takes the form

$$H\Psi = \Psi\langle\varphi|\Psi\rangle^{-1}\langle\varphi|H|\Psi\rangle. \tag{2.3}$$

Multiplying this relation to the right by $\langle\varphi|\Psi\rangle^{-1}\langle\varphi|$, one gets the operator relation

$$H\Theta = \Theta H \Theta, \tag{2.4}$$

which is the non-linear form of the Schrödinger equation desired. Let us now consider the question whether Θ is really a wave operator. One gets directly $\Theta\varphi_r = \Psi\mathbf{C}^{-1}$, or $\Psi = \Theta\varphi_r\mathbf{C}$, and - if one introduces the model functions through the relation $\eta = \varphi_r\mathbf{C}$ - one obtains $\Psi = \Theta\eta$ as well as $O\Psi = \eta$, as in the introduction. Hence one has

$$\eta = \varphi_r \mathbf{C}, \quad \Psi = \Theta \eta, \tag{2.5}$$

where it remains to determine the matrix $\mathbf{C}$. Using the previous expression for the energy, one gets finally $\mathbf{E} = <\varphi|\Psi>^{-1}<\varphi|H|\Psi> = \mathbf{C}^{-1}<\varphi|H|\Theta\eta> = \mathbf{C}^{-1}<\varphi|H\Theta|\varphi_r>\mathbf{C}$, which means that the matrix $\mathbf{C}$ defines the similarity transformation that brings the fundamental matrix $\mathbf{H} = <\varphi|H\Theta|\varphi_r>$ to classical canonical form:

$$\mathbf{C}^{-1}<\varphi|H\Theta|\varphi_r>\mathbf{C} = \mathbf{E}. \tag{2.6}$$

We note that the matrix $\mathbf{H} = <\varphi|H\Theta|\varphi_r>$ is not hermitian, but that the classical canonical form is still diagonal with real eigenvalues, which is essential for the applications. Starting from an arbitrary reference set φ, one can hence solve the Schrödinger equation (1.1) provided that one can determine the skew projector Θ by solving the non-linear equation (2.4). Some aspects of this problem will be given below.

3. The Bloch and Bloch-Lindgren Equations.

Let us now consider the case when the Hamiltonian has the form $H = H_0 + V$, where the perturbation V is not necessarily small. The non-linear Schrödinger equation (2.4) takes then the form $\Theta(H_0 + V)\Theta = (H_0 + V)\Theta$, or

$$\Theta H_0 \Theta - H_0 \Theta = V\Theta - \Theta V \Theta, \tag{3.1}$$

where it is sometimes convenient to introduce an abbreviation for the right-hand member $V\Theta - \Theta V\Theta = V_\theta$. Let us further assume that the reference set φ is here chosen to consist of an orthonormal set of p eigenfunctions to H_0, in which case one has also $\varphi_r = \varphi$, $O = |\varphi><\varphi|$, and $H_0 O = |\varphi>_0<\varphi| = OH_0$. This gives further $\Theta H_0 \Theta = \Theta O H_0 \Theta = \Theta H_0 O \Theta = \Theta H_0 O$, and we note that this important simplification is possible whenever the projector O for the reference space commutes with the unperturbed Hamiltonian H_0.

Let us start with the simplest case, in which the reference space is associated with an eigenvalue E_0 which is degenerate of order p, so that $H_0 O = E_0 O = OH_0$. In such a case, one has $\Theta H_0 \Theta = \Theta H_0 O =$

$\Theta\, E_o\, O = E_o\, \Theta$, and instead of (3.1) one gets the relation

$$(E_o - H_o)\Theta = V\,\Theta - \Theta\, V\,\Theta, \qquad (3.2)$$

which is the famous *Bloch equation* [6] first derived in nuclear physics. To solve this equation, one needs the projector $P = 1 - O$ for the orthogonal complement to the reference space O and the reduced resolvent R_o to the unperturbed Hamiltonian defined through the relation

$$R_o = (E_o - PH_o)^{-1}P = P\,(E_o - H_oP)^{-1} = P\,(E_o - PH_oP)^{-1}P =$$

$$= \Sigma_{k\neq 0} \frac{O_o^k}{E_o - E_o^k}, \qquad (3.3)$$

where we note that all the terms associated with the degeneracy are omitted. It is easily shown that $PR_o = R_oP = R_o$, $(E_o - H_o)\, R_o = R_o\,(E_o - H_o) = P$, and $R_o\, \varphi_o = 0$. Multiplying (3.2) to the left by R_o and observing that $\Theta = (O+P)\,\Theta = O + P\Theta$, one obtains

$$\Theta = O + R_o\,(V\,\Theta - \Theta\, V\,\Theta.) = F(\Theta), \qquad (3.4).$$

which equation may be solved by iteration. Conventionally one expands the operator Θ in terms of powers of V, and in this way one gets Bloch's solution:

$$\Theta_o = O,\ \ \Theta_1 = R_o\, V\, O, \qquad \Theta_2 = R_o\, V\, R_o\, V\, O - R_o^2\, V\, O\, V\, O,$$
$$\Theta_3 = R_o\, V\, R_o\, V\, R_o\, V\, O - R_o\, V\, R_o^2\, V\, O\, V\, O - R_o^2 V\, O\, V\, R_o\, V\, O -$$
$$- R_o^2\, V\, R_o\, V\, O\, V\, O + R_o^3\, V\, O\, V\, O\, V\, O \ ,.... \qquad (3.5)$$

One may then substitute this expression for Θ into the secular equation (2.6), which gives the matrix **C**, and the model functions η and the eigenfunctions Ψ are then determined by the relations (2.5). By means of this approach, it was hence possible to generalize Schrödinger's perturbation theory to the case when the reference level E_o is degenerate of order p.

A more difficult problem occurs when the reference set φ is associated not with a pure degeneracy but with p very close-lying energy eigenvalues $\mathbf{E}_o$. This problem was first attacked by Ingvar Lindgren [7].

Instead of (3.1), one gets now the equation

$$(\Theta H_0 - H_0\Theta)\, O = V\Theta - \Theta V\Theta, \qquad (3.6)$$

which is known as the generalized Bloch equation or the *Bloch-Lindgren equation*. To solve this equation, Lindgren introduced as an orthonormal basis the complete set $(\varphi,\bar{\varphi})$ of all unperturbed eigenfunctions, where φ is the *reference set* with the projector $O = |\varphi\rangle\langle\varphi|$, and $\bar{\varphi}$ is the *virtual set* spanning the orthogonal complement with the projector $P = 1 - O = |\bar{\varphi}\rangle\langle\bar{\varphi}|$. One has also $H_0 O = |\varphi\rangle \mathbf{E}_0 \langle\varphi| = OH_0$ as well as $H_0 P = |\bar{\varphi}\rangle \bar{\mathbf{E}}_0 \langle\bar{\varphi}| = PH_0$. In the following it is convenient to label the reference functions by Roman indices k, l, ... and the virtual states by Greek indices α, β, ... In solving the equation (3.6), Lindgren has shown the general structure of the solution, and we will now use his results as a guide for a slightly different derivation. For this purpose, we will introduce the superoperator $\hat{A}$ which maps an arbitrary linear operator T on another operator $\hat{A}\,T$ defined by the relation

$$\hat{A}\, T = \Sigma_{\alpha,k} \frac{P_a\, T\, O_k}{(\alpha;k)}, \qquad (3.7)$$

where $P_\alpha = |\bar{\varphi}_\alpha\rangle\langle\bar{\varphi}_\alpha|$, $O_k = |\varphi_k\rangle\langle\varphi_k|$, and the symbol $(\alpha;k)$ is an abbreviation for the energy difference $(\alpha;k) = (\bar{E}_0^{\alpha} - E_0^{k})$. It follows from the definition that one has the four relations

$$\hat{A}\,(OT) = 0\,,\; \hat{A}\,(TO) = \hat{A}\,T\,,\; \hat{A}(PT) = \hat{A}\,T\,,\; \hat{A}\,(TP) = 0, \qquad (3.8)$$

and further that

$$\hat{A}\; T)H_0 - H_0(\hat{A}\; T) = \Sigma_{a,k}\,[P_a\, T\, O_k\, E_0^{k} - \bar{E}_0^{\alpha}\, P_a\, T\, O_k\,](\bar{E}_0^{\alpha} - E_0^{k}) =$$
$$= -\,\Sigma_{a,k}\; P_a\, T\, O_k = -\,P\,T\,O, \qquad (3.9)$$

which gives the fundamental identity

$$(\hat{A}\; T)H_0 - H_0(\hat{A}\,T) = -\,P\,T\,O, \qquad (3.10)$$

for an arbitrary linear operator T. One may now use this identity to transform the Bloch-Lindgren equation (3.6). Putting $\Theta' = P\Theta = P\Theta O$, and observing that $V_\theta = PV_\theta O$, one may write (3.6) on the form $\Theta' H_0 - H_0\Theta' = - PV_\theta O$, and - according to (3.10) - this equation is then satisfied by the expression $\Theta' = - \hat{\mathbb{A}} V_\theta$, which gives $\Theta' = \Theta - O = - \hat{\mathbb{A}} V_\theta$. Hence the wave operator Θ satisfies the implicit relation

$$\boxed{\Theta = O - \hat{\mathbb{A}} (V\Theta - \Theta V\Theta) \equiv F(\Theta)} \qquad (3.11)$$

from which the wave operator Θ may be solved by various iteration procedures. Starting from $\Theta_0 = O$, and observing the relations (3.8), one gets immediately

$$\Theta_1 = - \hat{\mathbb{A}}\, V, \;\; \Theta_2 = \hat{\mathbb{A}}\, [V(\hat{\mathbb{A}}\, V)] - \hat{\mathbb{A}}\, [(\hat{\mathbb{A}}\, V)V], \text{ etc.} \qquad (3.12)$$

and using (2.7) one finds that these expressions are in complete agreement with Lindgren's original formulas. This iteration process is easily continued to any order. Once one has found the wave operator Θ, one may then find the coefficient matrix **C** by solving (2.6) and the model functions η and the eigenfunctions Ψ by using (2.5).

Ingvar Lindgren and his group in Gothenburg have used their profound knowledge of perturbation theory to carry out many non-relativistic calculations in atomic and molecular physics and in some cases also relativistic corrections by means of quantum electrodynamics [8], and the importance of this work can hardly be overestimated.

4. On the Bridge between Semi-Empirical and *Ab-Initio* Methods in Atomic and Molecular Quantum Theory.

It has been mentioned in the introduction that many authors [4] believed that the model-Hamiltonian $H_{eff} = O\, H\, \Theta$ would give a better basis for the semi-empirical quantum theory than the derivations starting from e.g. the Hartree-Fock Hamiltonian. One has previously had the dilemma that the parameters in the semi-empirical approach determined from selected experiments were usually rather different from those calculated by means of the ab-initio methods. This applied e.g. to Slater's F- and G-integrals in the theory of atomic spectra, to Hückel's parameters α and β in the theory of conjugated systems, or to the γ parameter in the Pariser-Parr-Pople scheme. Careful studies by Karl Freed and his group [9] in Chicago have shown that the discrepancy between the two sets of parameters disappears, if one bases the semi-empirical methods on the model Hamiltonian, and in this way Freed has succeeded in building a bridge between the semi-empirical and *ab-initio* methods of fundamental importance. This work has been very cumbersome, since for each system under consideration the wave operator Θ had to be determined by the best ab-initio-methods available. One problem which remains in this approach is that the model Hamiltonian $H_{eff} = O\ H\ \Theta$ is not self-adjoint, but this difficulty seems to disappear at least in low-order perturbation theory. For a survey of the development in this field, the reader is referred to Ref. [10].

5. A Comparison with the Conventional Multi-Dimensional Partitioning Technique.

Many years ago the author [11] suggested that perhaps it would be convenient to treat degenerate perturbation theory by means of multi-dimensional partitioning technique, an approach later investigated in greater detail in a series of papers called "Studies in Pertubation Theory" [12]. An important feature of this method is that the characteristic equation for the Hamiltonian H was replaced by a *reduced* characteristic equation $E = f(E)$, which had only single roots, and this implied that all degeneracies had been removed, which is often a great deal of simplification. The solution of the eigenvalue problem $E = f(E)$ by means of various iteration methods usually was further simplified by introducing a complex variable z and by studying the analytic function $z_1 = f(z)$ and its "crossings" with the straight line $z_1 = z$. Since one has the theorem that, in the interval between z and z_1, there is at least one true eigenvalue E, the function $z_1 = f(z)$ is often referred to as the *bracketing function*. In this section, we will give some new aspects on this approach in view of the solutions to the Bloch and Bloch-Lindgren equations.

Let us for a moment assume that we start from an arbitrary reference set $\varphi = \{\varphi_1, \varphi_2, \varphi_3, \ldots \varphi_p\}$ of order p with the self-adjoint projector $O = |\varphi\rangle\langle\varphi|\varphi\rangle^{-1}\langle\varphi| = |\varphi\rangle\langle\varphi_r| = |\varphi_r\rangle\langle\varphi|$, and with the projector P $= 1 - O$ for its orthogonal complement. Instead of the original Schrödinger equation (1.1), we will now study the inhomogeneous equation

$$(H - z.1)\Psi_z = \varphi\, \mathbf{a}, \qquad (4.1)$$

where $\mathbf{a}$ is a quadratic matrix of order p to be determined partly by the condition that the matrix $\mathbf{C} = \langle\varphi|\Psi_z\rangle$ is non-singular, i.e. that $|\mathbf{C}| \neq 0$. Multiplying this relation to the left by $\langle\varphi_r|$, one obtains

$$\mathbf{a} = \langle \varphi_r | \, H - z.1 \, | \Psi_z \rangle. \tag{4.2}$$

and we note that the inhomogeneous equation goes over into the ordinary Schrödinger equation whenever **a** is a zero matrix: $\mathbf{a} = \mathbf{0}$. Multiplying (4.1) to the left by the projector P and observing that $P\varphi = \mathbf{0}$ and that $P\,\Psi_z = \Psi_z \, O\,\Psi_z = \Psi_z - \varphi\ \mathbf{C}$, one obtains $P(H - z.1)\Psi_z = PH\Psi_z - z\Psi_z + z\varphi\,\mathbf{C} = \mathbf{0}$, which gives the solution

$$\Psi_z = (1 - PH/z)^{-1}\varphi\ \mathbf{C}, \tag{4.3}$$

with

$$\mathbf{a} = [\,\langle \varphi_r | H\,(1 - PH/z)^{-1} - z.1 \,|\, \varphi \rangle]\ \mathbf{C}\,. \tag{4.4}$$

The two key quantities in the theory is the *wave operator* W and the so-called *bracketing operator* Ω defined by the relations

$$W(z) = (1 - PH/z)^{-1}, \tag{4.5}$$
$$\Omega(z) = HW = H(1 - PH/z)^{-1}. \tag{4.6}$$

Introducing the matrix $\boldsymbol{\Omega}(z) = \langle \varphi_r | H(1 - PH/z)^{-1} | \varphi \rangle$ and putting $\mathbf{a} = \mathbf{0}$, one finds that **C** brings this matrix to diagonal form by the similarity transformation

$$\mathbf{C}^{-1}\,\boldsymbol{\Omega}(z)\,\mathbf{C} = z.\,\mathbf{1}, \tag{4.7}$$

where both members depend on the complex variable z. At this point, it is convenient to decouple the two members by diagonalizing the matrix $\boldsymbol{\Omega}(z)$ for arbitrary values of z, which gives

$$\mathbf{C}^{-1}\,\boldsymbol{\Omega}(z)\,\mathbf{C} = z_1.\,\mathbf{1}. \tag{4.8}$$

This relation defines a p-fold function $z_1 = f(z)$, and one then looks for the "crossing" points with the straight line $z_1 = z$, which gives the eigenvalues $z_1 = z = E$. Substitution into the relation (4.3) gives then the exact wave functions Ψ_z. In comparison to the previous sections, this approach deals also with a secular equation of order p, but the wave operator now contains the energy E explicitly, and further all the degeneracies of the Hamiltonian H are removed. This means that the connection with the idea of the existence of a "model Hamiltonian" and a set of "model functions" is definitely lost. However, from the point-of-view of *ab-initio* applications this approach may offer other advantages. For instance, the methods for solving the equation $E = f(E)$ by iteration procedures of the first and second-order are highly developed [13], whereas the corresponding theory for the operator relation $\Theta = F(\Theta)$ remains still to be developed in greater detail. For this reason, it may be convenient to give a brief review of the current structure of the multi-dimensional partitioning technique.

The reduced resolvent.- Let us now consider the wave operator (4. 5) in somewhat greater detail. By using the first of the two well-known identities $(A-B)^{-1} = A^{-1} + A^{-1}(A-B)^{-1}B\,A^{-1} = A^{-1} + A^{-1}B(A-B)^{-1}$, one gets the expansion $W(z) = 1 + (1 - PH/z)^{-1}(PH/z) = 1 + (z-PH)^{-1}PH = 1 + T(z)H$, where $T(z) = (z-PH)^{-1}P$ is referred to as the *reduced resolvent*. By using the well-known identity $(1-RS)^{-1}R = R(1 - SR)^{-1}$, one gets immediately the various relations

$$T(z) = (z-PH)^{-1}P = P\,(z-HP)^{-1} = P(z-PHP)^{-1}P. \qquad (4.9)$$

where the last form contains the resolvent for the operator $\bar{H} = PHP$, which is referred to as the *outer projection* of the Hamiltonian H [14]. One knows that, if the Hamiltonian H is bounded from below, one has the inequality $\bar{E}_k > E_k$ in order from below, which means that T(z) as a rule does not become singular for the eigenvalues z = E. Instead of the relations (4.5) and (4.6), we will now use the expressions:

$$W(z) = 1 + T(z)H, \qquad \Omega\,(z) = H + H\,T(z)\,H, \qquad (4.10)$$

where we note that - for real z - the bracketing operator $\Omega(z)$ is self-adjoint, which is a convenient feature of this approach. In addition to the complex variable z, it is sometimes convenient to introduce a fixed parameter a and to write T(z) in the form $T(z) = P(z-HP)^{-1} = (a - PH + PQ)^{-1}P$. Cross multiplication gives $P(a - PH + PQ) = P(z-HP)$ or $PQ = P(z-a)$, and - putting $Q = (z-a)$ and $H' = H - (z-a)$ - one gets the alternative forms

$$T(z) = P(a-H'P)^{-1} = (a-PH')^{-1}P, \quad H' = H - (z-a), \qquad (4.11)$$

which will be used frequently in the following,

Connection with perturbation theory; the wave and reaction operator in the general case.- Let us now consider the case when the Hamiltonian H consists of two terms $H = H_0 + V$, where the perturbation V does not necessarily have to be small. The reference set φ is still arbitrary of order p. In addition to the reduced resolvent T(z), we will now introduce the reduced resolvent T_0 for the unperturbed Hamiltonian H_0 through the relation

$$T_0(a) = (a - PH_0)^{-1}P, \tag{4.12}$$

where a is a fixed parameter at our choice. Putting

$$V' = V - (z-a), \tag{4.13}$$

and using (4.11), one gets immediately the transformation

$$\begin{aligned} T(z) = (a-PH')^{-1}P= [(a - PH_0)- PV']^{-1}P = \\ [1- (a - PH_0)^{-1}PV']^{-1}(a - PH_0)^{-1}P \\ = [1 - T_0(a)V']^{-1}T_0(a), \end{aligned} \tag{4.14}$$

and this formula forms the essential basis for our approach. Using the second of the two identities $(1-B)^{-1} =1+ (1-B)^{-1}B = 1+B(1-B)^{-1}$, one gets immediately

$$\begin{aligned} T(z) &= T_0(a) + T_0(a)V'[1 - T_0(a)V']^{-1}T_0(a) \\ &= T_0(a) + T_0(a)\, t(a:z)\, T_0(a), \end{aligned} \tag{4.15}$$

where we have introduce the notation

$$t(a;z) = V'[1 - T_0(a)V']^{-1}, \tag{4.16}$$

which is obviously a generalization of the concept of the *reaction operator* [12], discussed in some previous papers. If there is no risk for misunderstanding, we will in the following often use the abbreviations $t(a:z) = t$, $T(z) = T$, and $T_0(a) = T_0$. Since one has also $t = [(V')^{-1} - T_0]^{-1}$, one gets the two-term formula

$$t^{-1}= (V')^{-1} - T_0, \tag{4.17}$$

where it is assumed that the inverse operator $(V')^{-1}$ exists. Expanding the relation (4.16) one step, one gets

$$t = V' + V' [1 - T_0V']^{-1} T_0V' = V' + V' T (z) V'. \tag{4.18}$$

Hence one has the two reciprocity formulas

$$T = T_0 + T_0\, t\, T_0 ; \quad t = V' + V' T V', \tag{4.19}$$

which are closed expressions which may be used instead of the conventional perturbation expansions. From the definitions, one gets also the two relations

$$V' T = t T_{o,} \qquad T V' = T_o t, \qquad (4.20)$$

which are useful in the following. The idea is now to try to express the wave operator W = ! + TH and the bracketing operator $\Omega = H + H T H$ in terms of the new fundamental quantities T_o and t. For the wave operator, one obtains

$$\begin{aligned} W &= 1 + T H = 1 + (T_o + T_o t T_o) H_{o,} + T V = \\ &= 1 + T_o H_o + T_o t T_o H_o + T_o t + T(z-a) = \\ &= (1 + T_o t)(1 + T_o H_o) + T(z-a). \end{aligned} \qquad (4.21)$$

At this point, it is convenient to introduce the *modified* reference set $\bar{\varphi} = (1 + T_o H_o)\varphi$, and - observing that $T\,\varphi = 0$ - one gets further according to (4.3) that

$$\Psi_z = (1 + T H)\, \varphi\, \mathbf{C} = (1 + T_o t)\, \bar{\varphi}\, \mathbf{C}, \qquad (4.22)$$

which is the explicit formula for the eigenfunctions in the case when z = E. For the bracketing operator, one gets further after some simple algebraic manipulations using (4.19) and (4.20):

$$\begin{aligned} \Omega &= \Omega_o + (V + VTV - t) + \\ &+ (1+H_o T_o)\, t\, (1+T_o H_o) + (z-a)(T H_o + H_o T). \end{aligned} \qquad (4.23)$$

Observing again that $T\,\varphi = 0$, one gets finally

$$\boldsymbol{\Omega}(z) = \langle \varphi_r | \Omega | \varphi \rangle = \langle \varphi_r | \Omega_o | \varphi \rangle + (z-a).\mathbf{1} + \langle \bar{\varphi}_r | t | \bar{\varphi} \rangle, \qquad .(4.24)$$

where $\bar{\varphi}_r$ is a symbol for the set $\bar{\varphi}_r = (1+T_{o,}H_o)\varphi_r$ which is usually not the reciprocal to the set $\bar{\varphi}$.. In the next section, we will further simplify this formula.

Connection with conventional perturbation theory and the Bloch method. - For the sake of simplicity we will now choose the reference set φ so that it consists of p eigenfunctions to the unperturbed

Hamiltonian, so that $H_o\varphi = \varphi E_o$. In such a case, one gets directly $\bar{\varphi} = (1 + T_o H_o)\varphi = \varphi$, and the fundamental formulas (4.22) and (4.24) take the simple form

$$\Psi_z = (1 + T_o t)\, \varphi\, \mathbf{C}, \tag{4.25}$$

$$\Omega(z) = E_o + (z\text{-}a).\mathbf{1} + \langle\varphi\,|\,t\,|\,\varphi\rangle. \tag{4.26}$$

In order to evaluate the reaction operator t, one may combine the two relations (4.19), which gives the formula

$$t = V' + V'\,T_o\,V' + V'\,T_o\,t\,T_o\,V', \tag{4.27}$$

leading to the power expansions in the perturbation V. At this point it should be mentioned that a different approach is possible, which is based on the concept of the *inner projection* [15]. If A is an arbitrary linear operator having an inverse A^{-1}, and $\mathbf{h} = \{h_1, h_2, h_3, \ldots .h_n\}$ is a linearly independent set of order n, then the operator A' defined by the relation

$$A' = |\mathbf{h}\rangle\langle\mathbf{h}|A^{-1}|\mathbf{h}\rangle^{-1}\langle\mathbf{h}|, \tag{4.28}$$

is an approximation to A, which converges to A when $n\rightarrow\infty$ and the set **h** becomes complete. If the operator A is positive definite or has only a finite negative part, the convergence is for sufficiently large values of n always from *below*. By using (4.17) one gets for the inner projection of the reaction operator

$$t' = |\mathbf{h}\rangle\langle\mathbf{h}|t^{-1}|\mathbf{h}\rangle^{-1}\langle\mathbf{h}| = |\mathbf{h}\rangle\langle\mathbf{h}|(V')^{-1} - T_o|\mathbf{h}\rangle^{-1}\langle\mathbf{h}|. \tag{4.29}$$

For p = 1, this approach has been used by the Uppsala and Florida groups to calculate true *lower bounds* to energy eigenvalues for some simple systems [16].

Let us now consider the connection with perturbation theory in somewhat greater detail. For the resolvent of the outer projection PH_oP, one obtains the conventional spectral resolution $(a - PH_oP)^{-1} = \Sigma_k |\varphi_k\rangle\langle\varphi_k|/a + \Sigma_\alpha |\varphi_\alpha\rangle\langle\varphi_\alpha|/(a\text{-}E_\alpha)$, where the first sum goes over the reference set and the second sum over the virtual set spanning the orthogonal complement. This gives for the reduced resolvent

$$T_o(a) = P(a\text{-}\,PH_oP)^{-1}P = \Sigma_\alpha\,|\varphi_\alpha\rangle\langle\varphi_\alpha|/(a\text{-}E_\alpha).. \tag{4.30}$$

Putting $a = E_0$, one get the Schrödinger resolvent R_0:

$$R_0 = \Sigma_\alpha \frac{|\varphi_\alpha\rangle\langle\varphi_\alpha|}{E_0 - E_\alpha}, \qquad (4.31)$$

which for p = 1 occurs in the ordinary Schrödinger perturbation theory and for p equal to the order of the degeneracy of E_0 in the solution of the Bloch equation (3.5). For the fundamental bracketing matrix (4.26) in our approach, one gets now

$$\boldsymbol{\Omega}(z) = z.\mathbf{1} + \langle\boldsymbol{\varphi}|t|\boldsymbol{\varphi}\rangle. \qquad (4.32)$$

where t is given by the relation (4.27)

$$t = V' + V' R_0 V' + V' R_0 t R_0 V'. \qquad (4.33)$$

It follows from the definition (4.16) that, for any real value of the parameter a, the operator t is self-adjoint and the matrix (4,32) is hermitean. It is evident that the matrix $\langle\boldsymbol{\varphi}|t|\boldsymbol{\varphi}\rangle$ in (4.32) for z = E in some way must be similar and sometimes identical to the Bloch matrix $\langle\boldsymbol{\varphi}|H\Theta|\boldsymbol{\varphi}\rangle$, which is easily verified by considering the power series expansions in V. It is then also clear that, as an alternative to the non-linear Schrödinger equation $H\Theta = \Theta H\Theta$, one may use multi-dimensional partitioning technique, which treats the eigenvalue problem from a rather different point of view. It is also applicable to the case when - instead of an exact degeneracy of order p - one has p close-lying unperturbed reference states $\boldsymbol{\varphi}$. It should also be remembered that, whereas the nonlinear form $H\Theta = \Theta H\Theta$ always corresponds to the full characteristic equation for the Hamiltonian H, the partitioning technique always gives the *reduced* characteristic equation, which means that the connection with the model Hamiltonian and the model functions is gone in the latter case. On the other hand, it may sometimes be simpler to solve the p-fold algebraic equation $E = f(E)$ by first- and second-order iteration procedures, than it is to solve the operator relation $\Theta = F(\Theta)$ iteratively. One may also be able to utilize the bracketing properties of the p-fold function $z_1 = f(z)$. A further comparison of the two approaches in the applications to particular systems may hence be of some interest.

This paper is dedicated to Professor Ingvar Lindgren in connection with his 65th anniversary in view of his many outstanding contributions to physics and particularly to the development of perturbation theory and its applications to the non-relativistc theory of atomic and molecular systems and in some cases also to relativistic corrections by means of quantum electrodynamics.

6. References.

1. P.O.Löwdin, Phys. Scrip. **21**, 229 (1980).
2. J. des Cloiseaux, Nucl. Phys. 20, 321)1960). Morita, Prog. Theor. Phys. **29**, 351 (1963); B.H. Brandow, Proc. Int. School Phys. "Enrico Fermi" **36**, 496 (1966); Rev. Mod. Phys. **39**, 771 (1967); P.G.H. Sandars, R. Lefebrand, C. Moses, Ed. (Interscience, London, 1969) p. 365; J.A.R. Coope, Mod.Phys. **18**, 571 (1970); V. Kvasnicka, Phys. Rev. **A12**, 1159 (1975); D. Mukherjee, R.K. Moitra, and A. Mukhopadhyay, Mol. Phys. **30**, 1861 (1975); **33**, 955 (1977); Pramana **9**, 6 (1977); and many others.
3. P.O. Löwdin, Phys.Rev. **139**, A 357 (1965), part. Appendix A, p.A370.
4. See e.g. B. H. Brandow, Adv. Quantum Chem. **10**, 187 (1977); Adv. Phys. **26**, 651 (1977); Int. J. Quantum Chem. **15**, 207 (1979); in "New Horizons of Quantum Chemistry", 15 (Eds. P.O. Löwdin and B. Pullman, Reidel, Dordrecht 1983).
5. P.O. Löwdin, J. Math. Phys. 24, 70 (1983), particularly p. 75-76.
6. C. Bloch, Nucl. Phys. **6**, 329 (1958).
7. I. Lindgren, J. Phys. **B 7**, 2441 (1974); Cf. also V. Kvasnicka, Czech. J. Phys. **B 24**, 605 (1974); see also S. Garpman, I Lindgren, J.Lindgren, and J. Morrison, Phys. Rev. **A. 11**, 758 (1974); Z. Phys. **A 276**, 167 (1976); I Lindgren, J. Lindgren, and A.-M. Mårtensson, Zs.Physik **A 279**, 113 (1976); I Lindgren, Int. J. Quantum Chem. Symp. **12**, 33 (1978); A.-M. Mårtensson, J. Phys. **B 12**, 24 (1979); I. Lindgren and S. Salomonson, Phys. Scripta **21**, 335 (1980); J. Morrison and S. Salomonson, Phys. Scripta **21**, 343 (1980); I. Lindgren and J. Morrison, in "Atomic Many-Body Theory" (Springer, Berlin 1982); for some further applications, see e.g. I. Hubac and P. Carsky, Phys. Rev. **A 22**, 2392 (1980); M. Urban, I. Hubac, V. Kellö, and J. Noga, J. Chem. Phys. **72**, 3378 (1980); I. Hubac, Int. J. Quantum Chem. **17**, 195 (1980); B. Jeziorski and H. J. Monkhorst, Phys. Rev. **A 24**, 1668 (1981); P. Carsky, I. Hubac, and V. Staemmler, Theor. Chim. Acta 60, 445 (1982); I. Hubac and P. Carsky, Int. J. Quantum Chem. **24**, 141 (1983), I. Lindgren, Rep. Prog. Phys. **47**, 345 (1984); I lindgren, Phys. Rev. A 31, 1273 (1985).
8. E. Lindroth, Phys. Scripta **36**, 485 (1987); I. Lindgren, Phys. Scripta **36**, 591 (1987); E. Lindroth, Phys. Rev. **A 17**, 316 (1988), I. Lindgren, Nucl. Instr. and Methods in Phys. Research. **B 31**, 102 (1988); in Proc. "Programs on Relativistic, Quantum Electro-dynamics, and Weak Interaction Effects", 371 (AIP Conference Series, 1989); S. Salomonson and P. Öster, Phys. Rev. **A 40**, 5548, 5559, (1989); **A 41**, 4670 (1990); E. Lindroth and S. Salomonson, Phys. Rev. **A 41**, 4659 (1990); S. Salomonson and A. Ynnerman,

Phys. Rev. **B 43**, 88 (1990); I. Lindgren, Phys. Scripta **T 34,** 36 (1991); E. Lindroth, H. Persson, S. Salomonson, and A.-M. Mårtensson-Pendrill, Phys. Rev. **A 45**, 1493 (1992); E. Lindroth and J. Hvarfner, Phys. Rev. **A 45**, 2771 (1992); I. Lindgren, in "Recent Progress in Many-Body Theories" **3**, 245 (Eds. T.L Aimsworth et. al., Plenum Press, New York 1992); E. Lindroth and A. Ynnerman, Phys. Rev. **A 47**, 961 (1993); L. Labzowsky, V. Karrsiev, I. Lindgren, H. Persson, and S. Salomonson, Phys. Scripta **T 46**, 150 (1993); I. Lindgren, H. Persson, S. Salomonson, and A. Ynnerman, Phys. Rev. **A 47**, R4555 (1993); H. Persson, I. Lindgren, S. Salomonson, and P. Sunnergren, Phys. Rev. **A 48**, 1 (1993); I. Lindgren, H. Persson, S. Salomonson, V. Karasiev, L. Labzowsky, A. Mitrushenkov, and M. Tokman, J. Phys. **B 26**, L503 (1993); E. Lindroth and P. Indelicato, Nucl. Instr. and Methods in Phys. Research. **B 87**, 222 (1994); I. Lindgren, H. Persson, S. Salomonson, and L. Labzowsky, Phys. Rev. **A 51,** 1167 (1995). I. Lindgren, Int. J. Quantum Chem. **57**, 683 (1996).

9. K.F. Freed, Chem. Phys. Lett. **15**, 331 (1972); in *Energy, Structure, and Reactivity*, 374 (ed. Darwin Smith et.al., John Wiley & Sons, New York 1973); Chem.P Phys. Lett. **24**, 275 (1974); Chem. Phys. **3**, 463; **4,** 80 (1974); J.Chem. Phys. **60**, 1765 (1974); S. Iwata and K. Freed, J. Chem. Phys. **61**, 1500 (1974); Chem. Phys. Lett. **28**, 176 (1974); S. Iwata and K. F. Freed, J. Chem. Phys. **65**, 1071 (1976); S. Iwata and K. F. Freed, Chem. Phys. Lett. **38**, 425 (1976); D. J. Yeager, H. Sun, and K.F. Freed, Chem. Phys. Lett. **57**, 490 (1978); M.G. Sheppard, K.F. Freed, M.F. Herman, and D. J. Yeager, Chem. Phys. Lett. **61**, 577 (1979); D. J. Yeager, M.G. Sheppard, and K.F. Freed, J. Amer.Chem. Soc. **102**, 1270 (1980); H. Sun, K. F. Freed, M.F. Herman, J. Chem. Phys. **72**, 4158 (1980); K. F. Freed and H. Sun, Israel J. Chem. **19**, 99 (1980); Y.S. Lee, H. Sun, M.G. Sheppard, and K.F. Freed, J. Chem. Phys. **73**, 1472 (1980); H. Sun, M.G. Sheppard, K.F. Freed and M.F. Herman, Chem. Phys. Lett. **77**, 555(1981); H. Sun and K.F. Freed, Chem. Phys. Lett. **78**, 531(1981); H. Sun, M.G. Sheppard, and K.F. Freed, J. Chem. Phys. **74**, 6942 (1981); M.G. Sheppard, and K.F. Freed, Chem. Phys. Lett. **82**, 235 (1981); J. Chem. Phys. **75**, 4507, 4525 (1981); M.G. Sheppard and K.F. Freed, Int. J. Quantum Chem. Symp. QC **15**, 21 (1981); H. Sun and K.F. Freed, J. Chem. Phys. **76**, 5051 (1982); J.J. Oleknik and K.F. Freed, J. Chem. Phys. **79**, 1396 (1983); X.-C. Wang and K.F. Freed, J. Chem. Phys. **91**, 1142, 1151, 3002 (1989); A.W. Kanzler, H. Sun and K.F. Freed, Int. J. Quantum Chem. **39**, 269 (1991); A.W. Kanzler and K.F. Freed, J. Chem. Phys. **94**, 3778 (1991); Richard L. Graham and K.F. Freed, J. Chem. Phys. **96**, 1304 (1992); A.W. Kanzler and K.F. Freed, J. Chem. Phys. **96**, 5245 (1992); V. Hurtubise, J. Chem. Phys. **99**,

265 (1993); C.H. Martin, R.L. Graham and K.F. Freed, J. Chem.Phys. **99**, 7833 (1993); V. Hurtubise and K.F. Freed, J. Chem. Phys. **99**, 7946 (1993); J. Chem. Phys.**100**, 4955 (1994); C.H. Martin and K.F. Freed, J. Chem. Phys.**100**, 7454 (1994); J. Chem. Phys.**101**, 4011, 5029 (1994); J.E. Stevens, K.F. Freed, M.F. Arendt and R.L. Graham, J. Chem. Phys. **101**, 4832 (1994); J.P. Finley, J. Chem. Phys. **102**, 1306 (1995); J.P. Finley, R.K. Chaudhuri and K.F. Freed, J. Chem. Phys. **103**, 4990 (1995); C.H. Martin and K.F. Freed, J. Phys. Chem. **99**, 2701 (1995); C.H. Martin, J. Phys. Chem. **100**, 14310 (1996); Chem. Phys. Lett. **257**, 239 (1996).

10. K.F. Freed, in *Structure and Dynamics of Atoms and Molecules; Conceptual Trends*, 25 (Eds.J.L. Calais,E.S.Kryachko, Kluwer, 1995).
11. P.O. Löwdin, J. Chem. Phys. **19**, 1323 (1951).
12. P.O. Löwdin, J. Mol. Spectrosc. **10**, 12 (1963); **13**, 326 (1964); **14**, 112 (1964); **14**, 119 (1964); **14**, 131 (1964); J. Math. Phys. **3**, 969 (1962); **3**, 1171 (1962); **6**, 1341 (1965); Phys. Rev. **139**, A357 (1965); J. Chem. Phys. **43**, S175 (1965); Int. J. Quantum Chem. **2**, 867 (1968); Int. J. Quantum Chem. **S4**, 231 (1971); **5**, 685 (1971) (together with O. Goscinski); Phys. Scrip. **21**, 229 (1980); Adv. Quantum Chem. (Academic Press, New York, 1980) **12**; Int. J. Quantum Chem. **21**, 69 (1982).. See also P.O. Löwdin, Int.J. Quantum Chem. **55**, 77 (1995).
13. E. Schröder, Math. Ann. 2, 317 (1870); see also D.R. Hartree, Proc. Cambridge Phil.Soc. 45, 230 (1949); P.O. Löwdin, J. Mol. Spectrosc. **10**, 12 (1963); particularly Appendix A, p. 26.
14. P.O. Löwdin, Phys. Rev. **139**, A357 (1965), particularly p. A 358.
15. See ref. 10, particularly p. A 359-360: for a review, see also P.O. Löwdin, Int. J. Quantum Chem. **4S**, 231 (1971).
16. J. Gay, Phys. Rev. A **135**, 1220 (1964); P.O. Löwdin, Phys. Rev. A **139**, 357 (1965); J. Chem. Phys. Suppl. **43**, S175 (1965); C. Reid, *ibid.* S 186; J.H. Choi and D.W. Smith, *ibid.* S 189: C.F. Bunge and A. Bunge, *ibid.* S 194; J.H. Choi, J.Chem. Phys. **45,** 442S (1966); J. Math. Phys. **10**, 2142 (1969); T.M. Wilson, J. Chem. Phys. **47**, 3912 (1967); T.M. Wilson and C.G. Reid, *ibid.* 3920 (1967); as well as O. Goscinski and E. Brändas, Int. J. Quantum Chem. **5**, 131 (1971); E. Brändas and R.J. Bartlett, Chem. Phys. Lett. **8**, 153 (1970); R.J. Bartlett and E. Brändas, Int. J. Quantum Chem. **5,** 151 (1971); E. Brändas and R.J. Bartlett, J. Chem. Phys. **56**, 5467 (1972): D.A. Micha and E. Brändas, J. Chem. Phys. **55**, 4792 (1971); E. Brändas and D. Micha, J. Math. Phys. **13**, 155 (1972).

WHAT IS THE FORCE BETWEEN ELECTRONS?

J. Sucher

Center for Theoretical Physics and Department of Physics
University of Maryland, College Park, MD 20742, USA

Contents

I. Introduction and historical review
II. Scattering potentials: One-photon exchange
A. Two spin-1/2 particles
B. Two spin-0 particles
C. A spin-1/2 and a spin-0 particle
D. Remarks on the orbit-orbit interaction
III. Scattering potentials: Two-photon exchange
A. Two spin-0 particles
B. A spin-1/2 and a spin-0 particle
C. Two spin-1/2 particles
IV. Scattering potentials and bound states

I. Introduction and historical review

It is great pleasure for me to be able to contribute to this volume in honor of my good friend, Ingvar Lindgren. A large part of Ingvar's illustrious career has been devoted to work in theoretical atomic physics. Since the structure and properties of atoms are largely determined by the nature of the interaction between its electrons, I thought this would be a nice occasion to present a discussion of this topic. I will focus on aspects which are important for the relativistic theory of atomic structure, to which Ingvar has made many important contributions. Because he has had an amazingly large number of graduate students (over 60!), there is a good chance that some of them will find something here which Ingvar would think they ought to know. (Note: The exclamation mark after 60 is to indicate astonishment, not "factorial" - even Ingvar has his limitations.)

The question posed in the title may be short, but it does not have a snappy answer. If one asks "What is the force between planets?", everybody knows that it is given with great accuracy by Newton's law of universal gravi-tation. But for electrons, the story is much more complicated and indeed the question itself must be recast. There are at least three reasons for this: (a) physical systems involving electrons must be described by quantum mechanics (QM) (b) in many such systems, the relative velocities are either not very small compared to *c* and/or high accuracy is of interest, so special relativity must be used, and (c) electrons have spin.

With regard to (a), recall that in nonrelativistic classical mechanics both the concept of force and the concept of the force between particles is well-defined; if the force is conservative, so is the concept of two-body potential U(1,2). So in this context the answer is of course the familiar one: the force is the Coulomb force, now known for almost exactly 200 years [1]. For two point particles "1" and "2", the associated potential is just the Coulomb potential $U_C(1,2)$,

$$U_C(1,2) = k_{12}/r_{12} = e_1e_2/4\pi r_{12}, \tag{1.1}$$

with $r_{12} = |\mathbf{r}_1-\mathbf{r}_2|$ and $k_{12} \to e^2/4\pi = \alpha \approx 1/137$ for two electrons. However, in quantum theory the concept of force plays a secondary role and the question should really be: "What is the potential

ADVANCES IN QUANTUM CHEMISTRY, VOLUME 30

0065-3276/98 $25.00

the force between particles is well-defined; if the force is conservative, so is the concept of two-body potential $U(1,2)$. Thus, in this context the answer is of course the familiar one. The force is the Coulomb force, known for 200 years, and for two point particles, "1" and "2", the associated potential is just the Coulomb potential $U_C(1,2)$,

$$U_C(1,2) = k_{12}/r, \quad [k_{12} \equiv e_1 e_2/4\pi] \tag{1.1}$$

with $r = |\mathbf{r}_1 - \mathbf{r}_2|$; for two electrons, $k_{12} \rightarrow \alpha \equiv e^2/4\pi \approx 1/137$. However, in quantum theory the concept of force plays a secondary role and the question should really be: "What is the potential acting between two electrons?" In the context of nonrelativistic quantum mechanics (NRQM) and a configuration space description of the system in question, the answer is again $U_C(1,2)$, now reinterpreted as a multiplicative operator acting on a many-body Schrödinger wave function.

With regard to (ii), when the effects of the finite speed of propagation of light are taken into account, things get more complicated, both on conceptual and technical grounds. Now even in classical electrodynamics the question becomes murky: While the meaning of the force on "1" is sharp, via the Lorentz force law, the force exerted by "2" on "1" depends on the previous history of "2" (and *vice versa*). The question arises: To what extent can one describe the force between "1" and "2" in terms of properties of their motion at a given instant? This question appears to have been first addressed by C.E. Darwin [1], the grandson of Charles Darwin. Darwin showed that, to order v^2/c^2, the effects of retardation can be taken into account by adding to the free Lagrangian for two point particles not only the Coulomb term $-U_C$ but also a term $-U_D$, with

$$U_D = -\tfrac{1}{2}(\mathbf{v}_1 \cdot \mathbf{v}_2 + \mathbf{v}_1 \cdot \hat{\mathbf{r}}\mathbf{v}_2 \cdot \hat{\mathbf{r}})U_C/c^2, \tag{1.2}$$

from which the corresponding force may be calculated. On passing to the Hamiltonian formalism and introducing canonical momenta one gets a momentum-dependent interaction term of the form

$$U_D = -\tfrac{1}{2}(\mathbf{p}_1 \cdot \mathbf{p}_2 + \mathbf{p}_1 \cdot \hat{\mathbf{r}}\mathbf{p}_2 \cdot \hat{\mathbf{r}})U_C/m_1 m_2 c^2. \tag{1.3}$$

This must have a counterpart in quantum theory. But how is one to order the factors, if $\mathbf{p}_i$ is replaced by $\mathbf{p}_i^{op}$? In whatever way this question is answered, it becomes clear that at this level any effective potential between electrons will not be strictly local.

Finally, with regard to (iii), after the discovery of the spin-1/2 character of the electron and the associated magnetic moment, it was obvious that spin-dependent terms must be added to U_D, corresponding to the interaction of the magnetic field produced by "1" with the magnetic moment of "2" and *vice versa*. This led to the now familiar spin-other-orbit and spin-spin potentials and gave rise to an operator $U_{fs}^{(2)}$ which could describe the effect of electron-electron interaction on atomic fine structure correctly to order α^2Ry (apart from a missing contact term). So one could say with some justification that, to order e^2 and v^2/c^2, the operator $U_C+U_{fs}^{(2)}$ describes the force between electrons.

After the development of the Dirac equation one might have guessed that, within a framework in which the state of a many-electron system is described by a multi-Dirac spinor, the velocity factors $\mathbf{v}_i$ in (1.2) should simply be replaced by their formal counterparts in Dirac theory, viz. $c\boldsymbol{\alpha}_i$. This yields the operator

$$U_B(1,2) = -\tfrac{1}{2}(\boldsymbol{\alpha}_1\cdot\boldsymbol{\alpha}_2+\boldsymbol{\alpha}_1\cdot\hat{\mathbf{r}}\boldsymbol{\alpha}_2\cdot\hat{\mathbf{r}})U_C, \tag{1.3}$$

known as the Breit operator. It was first obtained by Breit [2], but not in the way I have described. After the development of QED by Dirac, Breit studied the level shift in the helium atom arising from the exchange of a transverse photon between the two electrons, within a formally relativistic (but actually unsound) framework. He concluded that, to a good approximation, this shift is given by the expectation value of U_B with a Dirac-type wave function which includes the effect of U_C. From this one would expect that an effective interaction between electrons which takes into account items (i), (ii), and (iii) is given by the Coulomb-Breit potential,

$$U_{CB}(1,2) = U_C(1,2) + U_B(1,2). \tag{1.4}$$

However, although accepted for many years, this expectation turns out to be misleading at best and wrong at worst.

In the following sections I will consider the question of the electron-electron interaction from the perspective of modern QED. It will be useful to consider both spin-1/2 and spin-0 particles. Sec. II deals with the relatively simple case of one-photon exchange potentials, in the context of the scattering of two point particles. However, 200 years after Coulomb one ought to have some idea about the extent to which the exchange of two photons can also be represented by an effective potential. This is discussed Sec. III. In Sec. IV I turn to the question of the connection between these potentials and the properties of bound states.

II. Scattering potentials: One-photon-exchange

Within the framework of perturbative quantum field theory, there is a sharp contrast between the beautiful methods available for the calculation of collision amplitudes and those used in practice for the calculation of the properties of composite systems or "bound states," especially in the case of a gauge theory such as QED. As an example, consider the scattering of two particles, "1" and "2",

$$1 + 2 \rightarrow 1 + 2, \tag{2.1}$$

with initial and final four-momenta p_1, p_2 and p_1', p_2', respectively. In a Lorentz-invariant theory the transition amplitude has the form $t_{fi} = N_f M N_i$, where $N_f = (4E_1'E_2')^{-1/2}$ and $N_i = (4E_1E_2)^{-1/2}$ are the usual kinematic factors, and

$$M = M(s,t) \qquad [s \equiv P^2,\ t = Q^2] \tag{2.2}$$

is the invariant Feynman amplitude, with $\underline{s}$ and $\underline{t}$ the invariant squares of the total four-momentum $P = p_1+p_2$ and four-momentum transfer $Q = p_1-p_1'$. Let us first consider the effects arising from one-photon exchange between two distinct point-like spin-1/2 particles, such as an e^- and a μ^+.

A. Two spin-1/2 particles

Using Feynman gauge for the photon propagator one finds that the lowest-order contribution to $\underline{M}$, regarded as a power series in e_1e_2, is given by

$$M^{(2)} = \bar{u}_1'\bar{u}_2'F^{(2)}u_1u_2 \tag{2.3}$$

with the u_i's denoting lepton spinors normalized to $2m_i$ and

$$F^{(2)} = e_1e_2\gamma_1\cdot\gamma_2/t. \tag{2.4}$$

In the c.m. system, $Q \rightarrow (0,\mathbf{q})$, with $\mathbf{q}$ the three-momentum transfer, $\mathbf{q} = \mathbf{p}-\mathbf{p}'$. On taking the Fourier transform of $F_{cm}^{(2)}$ with a factor $\exp(-i\mathbf{q}\cdot\mathbf{r})$, one sees that the corresponding transition amplitude $t^{(2)}$ can be obtained by taking the matrix element between plane wave spinors of the operator

$$U_{CG}(1,2) = k_{12}(1-\boldsymbol{\alpha}_1\cdot\boldsymbol{\alpha}_2)/r = U_C(1,2)+U_G(1,2) \tag{2.5}$$

with $U_G \equiv -k_{12}(\boldsymbol{\alpha}_1\cdot\boldsymbol{\alpha}_2/r)$, the so-called Gaunt potential [3].

It is therefore tempting to say that, to leading order in e_1e_2, but regardless of the relative speed, the effective interaction between, "1" and "2" is given by (2.5). However, suppose that instead one uses the Coulomb gauge for the photon propagator. Then one finds that (2.4) is replaced by

$$F_T^{(2)} = e_1e_2[\gamma_1^0\gamma_2^0/\mathbf{Q}^2+(\gamma_1\cdot\gamma_2-\gamma_1\cdot\mathbf{Q}\ \gamma_2\cdot\mathbf{Q}/\mathbf{Q}^2)/Q^2]. \tag{2.6}$$

Fourier transformation of $F_{T;cm}^{(2)}$ yields

$$U_{CB}(1,2) = U_C+U_B, \tag{2.7}$$

where U_B is the operator

$$U_B = -k_{12}(\boldsymbol{\alpha}_1\cdot\boldsymbol{\alpha}_2+\boldsymbol{\alpha}_1\cdot\hat{\mathbf{r}}\ \boldsymbol{\alpha}_2\cdot\hat{\mathbf{r}})/2r \tag{2.8}$$

originally obtained by Breit in the context of a bound state problem, as mentioned above. By construction, the Coulomb-Gaunt potential U_{CG} and the Coulomb-Breit potential U_{CB} have the same matrix element between product plane wave spinors, provided these are on the energy shell ($\mathbf{p}^2 = \mathbf{p}'^2$), a minimal requirement for an effective potential. Thus even in the simplest circumstances, asking for "**the** effective electron-electron potential" is too naive.

More serious is the fact that neither U_{CB} or U_{CG} are permissible approximations to an effective potential, in the context of a Dirac-spinor description of spin-1/2 particle wave functions. This can be most easily seen by including a third particle "3". The use of a sum such as U(1,2)+U(1,3)+U(2,3) to describe the interaction is disastrous from the start, because of the problem of continuum dissolution [4,5]. Analysis shows that a theoretically well-founded choice for the second-order potential is neither U_{CG} nor U_{CB}, but either of the two operators V_{CG} or V_{CB}, defined by

$$V_{CG} = \Lambda_{++}U_{CG}\Lambda_{++}, \qquad V_{CB} = \Lambda_{++}U_{CB}\Lambda_{++}, \tag{2.9}$$

where $\Lambda_{++} = \Lambda_+(1)\Lambda_+(2)$ is the product of positive-energy Casimir-type projection operators for the leptons [5,6]. Since the spinors in (2.3) are eigenfunctions of the $\Lambda_+(i)$, these operators also reproduce the lowest-order amplitude.

All this is by now fairly well known. But a related question has received relatively little attention: What is an effective potential which describes the scattering amplitude correctly to fourth order? Before considering this, it is useful to study two simpler but instructive cases: two spin-0 particles, in scalar QED, and a spin-1/2 and a spin-0 particle, in spinor-scalar QED.

B. Two spin-0 particles

On using Feynman gauge to write down the one-photon exchange amplitude, one gets

$$M^{(2)} = e_1e_2(p_1+p_1')\cdot(p_2+p_2')/t = e_1e_2(2a+t)/t, \tag{2.10}$$

where $a = s-m_1^2-m_2^2$. Fourier transformation of the c.m. value of (2.10) yields a term proportional to U_C, with an energy-dependent coefficient, plus a contact term proportional to $\delta(\mathbf{r})$. Such a potential is not suitable for use in a Schrödinger type of equation. In second-order perturbation theory it would lead to an ultraviolet (UV) divergence. A potential which is iterable can be obtained by first writing $M^{(2)}$ in a different form (which does not change its value on the mass shell) and then finding an equivalent operator in $\mathbf{r}$-space, now involving derivative operators [7]. One is thereby led to what can be termed a Feynman-gauge-inspired (FGI) potential $V_{1\gamma}^{FGI}$, which in the c.m. system ($\mathbf{p}_1^{op}\to\mathbf{p}_{op}$, $\mathbf{p}_2^{op}\to-\mathbf{p}_{op}$) is

$$V_{1\gamma}^{FGI} = z'_{op}U_C z'_{op} + y_{op}(\mathbf{p}_{op}\cdot U_C\mathbf{p}_{op}/2m_A m_B)y_{op}, \tag{2.11}$$

with $z'_{op} \equiv (1+\mathbf{p}_{op}^2/2E_1^{op}E_2^{op})^{1/2}$ and $y_{op} = (m_1m_2/E_1^{op}E_2^{op})^{1/2}$. The corresponding Coulomb-gauge inspired (CGI) one-photon exchange potential $V_{1\gamma}^{CGI}$ is given by [8],

$$V_{1\gamma}^{CGI} = V_C+V_T \tag{2.12}$$

where, with a curly bracket denoting an anticommutator,

$$V_C = y_{op}[\{E_1^{op},\{E_2^{op},U_C\}\}]y_{op}/4m_1m_2 \tag{2.13}$$

is a relativistic version of the Coulomb interaction and

$$V_T = -(1/2)y_{op}\{\mathbf{p}_{1;i}^{op},\{\mathbf{p}_{2;j}^{op},(\delta_{ij}+\hat{r}_i\hat{r}_j)U_C\}\}]y_{op}/4m_1m_2 \tag{2.14}$$

is the potential arising from the exchange of a transverse photon.

C. One-photon exchange potential in spinor-scalar QED

To complete the one-photon story, consider the exchange of a photon when "1" has spin-0 and "2" has spin-1/2. The Feynman amplitude is then given by

$$M^{(2)} = -e_1e_2\bar{u}'\gamma_2\cdot(p_1+p_1')u/t \tag{2.15}$$

The corresponding FGI potential is [9]

$$V_{1\gamma}^{FGI} = y_1^{op}\Lambda_+^{op}(2)U^{(2)}\Lambda_+^{op}(2)y_1^{op} \quad [y_1^{op} \equiv (m_1/E_1^{op})^{1/2}] \tag{2.16}$$

where

$$U^{(2)} = (E_1^{op}U_C+U_C^{op}E_1^{op} - \boldsymbol{\alpha}\cdot\mathbf{p}_{op}U_C-U_C\boldsymbol{\alpha}\cdot\mathbf{p}_{op})/2m_1. \tag{2.17}$$

This is an analogue of $V_{1\gamma}^{FGI}$ for two spin-1/2 particles. There is a corresponding CGI potential which I do not write down. Use will be made of (2.16) in Sec. III.

D. Remarks on the orbit-orbit interaction

The difference between the two choices, $V_{1\gamma}^{FGI}$ and $V_{1\gamma}^{CGI}$ is connected with the form of the so-called orbit-orbit interaction. To see this, note that in the n.r. limit (2.11) yields as the leading correction to U_C an orbit-orbit interaction $U_{0\text{-}0}$ of the form

$$U_{0\text{-}0}^{FGI} = \{\mathbf{p}_i^{op},\{\mathbf{p}_j^{op},\delta_{ij}U_C\}\}/4m_A m_B, \tag{2.18}$$

whereas (2.12) yields

$$U_{0\text{-}0}^{CGI} = (1/2)\{\mathbf{p}_i^{op},\{\mathbf{p}_j^{op},(\delta_{ij}+\hat{r}_i\hat{r}_j)U_C\}\}/4m_A m_B. \qquad (2.19)$$

Note that (2.19) is a manifestly hermitian form of the orbit-orbit interaction familiar from atomic physics, usually described in texts as resulting from the reduction of the Breit operator (2.8) to n.r. form. This is unfortunate from a pedagogical point of view, since the Breit operator refers only to spin-1/2 particles. Spin has nothing to do with it! I will return to the difference between (2.18) and (2.19) shortly.

III. Scattering potentials: Two-photon exchange

A key point in the definition of any potential, often overlooked, is that one must specify in advance how it is to be used. Let us work in the c.m. system of the particles, scattering with total energy E. With $h_0 = h_0(1)+h_0(2)$ the free Hamiltonian for the particles, we will require that the transition amplitude computed from

$$T_{op}(E) = V_{eff} + V_{eff}(E-h_0-V_{eff}+i\epsilon)^{-1}V_{eff} \qquad (3.2)$$

reproduce the scattering amplitude obtained from field theory to a given accuracy. I consider first the case of two spin-0 particles, then the mixed case of a spin-0 and a spin-1 particle, and finally the case of two spin-1/2 particles.

A. Two spin-0 particles

In scalar QED there are five Feynman diagrams corresponding to two-photon exchange, which contribute to the fourth-order amplitude $M^{(4)}$: the two-rung ladder (or "box") graph, and the two-rung crossed-ladder graph, a pair of "single-seagull" graphs, and a "double-seagull" graph. The sea-gull graphs are ultraviolet (UV) divergent, but these divergences are taken care of by renormalization and do not affect the long-distance behavior. More serious is the fact that the "box" and "crossed box" graphs are infrared (IR) divergent. Thus it seems at first sight that one cannot even begin to talk about a two-photon exchange potential!

The resolution of the problem lies in the recognition that the iteration amplitude M_I arising from $V_{1\gamma}$ is now also infrared divergent. This divergence is just the field-theoretic counterpart of the fact that in NRQM the second (and higher) Born approximations to the scattering of a charged particle in a Coulomb field is divergent (for any value of the momentum transfer). It turns out that the difference $M_{2\gamma}-M_I$ is IR convergent, so that the associated potential $V_{2\gamma}$ is well-defined after all.

I will only state the form of the result at large $\underline{r}$ (that is, $\underline{r}$ large compared to the Compton wave length of either particle) and at low momenta. On using $V_{1\gamma}^{FGI}$ to compute M_I one finds that [7],

$$V_{2\gamma}^{FGI} = c_2^{FGI} r^{-2} + c_3^{FGI} r^{-3} + \cdots \qquad (3.2)$$

where, with $k \equiv e_1 e_2/4\pi$,

$$c_2^{FGI} = k^2/2(m_A+m_B), \quad c_3^{FGI} = -\ 7k^2/6\pi m_A m_B. \qquad (3.3)$$

In contrast, use of $V_{1\gamma}^{CGI}$ yields [8]

$$V_{2\gamma}^{CGI} = c_2^{CGI} r^{-2} + c_3^{CGI} r^{-3} + \ldots \qquad (3.4)$$

where

$$c_2^{CGI} = 0, \quad c_3^{CGI} = -\ 7k^2/6\pi m_1 m_2. \qquad (3.5)$$

The difference between the asymptotic forms (3.2) and (3.4) can be traced back the difference in the associated forms of the orbit-orbit interactions mentioned above. Thus we see that in the case of two charged particles the leading asymptotic behavior of $V_{2\gamma}$ depends on the precise definition of $V_{1\gamma}$. This observation resolves a long-standing puzzle concerning conflicting results for the value of c_2. Further, as was noted some time ago by L. Spruch, c_2^{FGI} is classical in character, i.e. if $\underline{h}$ and $\underline{c}$ are restored, c_2 turns out to be independent of $\underline{h}$. One should therefore try to understand the source of this term from classical electrodynamics. It turns out that this is indeed possible by a reexamination of the work of Darwin [1], but I will not enter into the details here [10].

B. A spin-1/2 and a spin-0 particle

A similar analysis can be carried out for the mixed case. If one uses $V_{1\gamma}^{FGI}$ to compute the iteration amplitude one finds that the spin-independent part of $V_{2\gamma}$ is

$$V_{2\gamma}^{s.i.} = k^2[2(m_1+m_2)]^{-1} r^{-2}, \qquad (3.6)$$

at large $\underline{r}$ and low energies, which coincides with that for two spin-zero particles. The spin-dependent part is essentially a spin-other-orbit type of interaction [9]

$$V_{2\gamma}^{s.o.} = -(e_1 e_2/4\pi)^2 (\boldsymbol{\sigma}\cdot\boldsymbol{\ell}/4m_2^2)\{[(3m_1+5m_2)/m_1(m_1+m_2)] r^{-4} \qquad (3.7)$$

The two-photon-exchange spin-orbit interaction therefore decreases as r^{-4}, or one power more rapidly than that arising from one-photon exchange. Applications of this result to a variety of exotic systems, such as pionium and muonic helium, are considered in Ref. [9].

C. Two spin-1/2 particles

The computation of $V_{2\gamma}$ for two spin-1/2 particles, even of its asymptotic form for large $\underline{r}$, turns out to be remarkably complex and has not yet been completed. However, some aspects are known and others can be guessed. For example, one can show that if $V_{1\gamma}^{FGI}$ is used for the lowest-order potential, then the spin-independent part of $V_{2\gamma}$ coincides, for large $\underline{r}$, with that for two spin-0 particles and the spin-other-orbit part coincides with that for a spin-0 particle "1" and a spin-1/2 particle "2", plus a similar term with "1" and "2" interchanged. At large $\underline{r}$ and low energies one expects that the spin-spin part $V_{2\gamma}^{s-s}$ has the form

$$V_{2\gamma}^{s-s} = k^2[A(r)\boldsymbol{\sigma}_1\cdot\boldsymbol{\sigma}_2 + B(r)\boldsymbol{\sigma}_1\cdot\hat{\mathbf{r}}\ \boldsymbol{\sigma}\cdot\hat{\mathbf{r}}\ ], \tag{3.8}$$

with $A(r) = ar^{-p}$ and $B(r) = br^{-q}$. The coefficients $\underline{a}$ and $\underline{b}$, and the exponents $\underline{p}$ and $\underline{q}$, remain to be determined. For equal-mass particles, such as two electrons, $a = a'/m^{p-1}$ where a' is dimensionless and likely to be of order unity. To guess the value of $\underline{p}$, note that in the large $\underline{r}$, low-energy limit the spin-spin part coming from $V_{1\gamma}$ has one more inverse power of $\underline{r}$ than the spin-other-orbit part. If the same relation holds for two-photon exchange, then $p = 5$.

IV. Scattering potentials and bound states

Let us now consider the possible relevance of all this to the computation of relativistic atomic structure, one of Ingvar Lindgren's keen interests.

The application of QED to compute such structure to an accuracy beyond order α^2Ry, for systems more complicated than the classic two-body systems (hydrogen, positronium, and muonium) began in the mid-fifties, with calculations of the fine structure of helium to an accuracy of order α^3Ry [11,12]. These calculations, and subsequent ones of spin-dependent effects of order α^4Ry [13], were based on a generalization of the two-body Bethe-Salpeter (BS) equation to include an external field. The generalized BS equation leads in a somewhat convoluted way to an equation of the form

$$H_{++}\psi = E\psi \tag{4.1}$$

where

$$H_{++} = h_D^{ext}(1) + h_D^{ext}(2) + L_{++}U_C(1,2)L_{++}, \tag{4.2}$$

with $h_D^{ext}(i) = \boldsymbol{\alpha}_i\cdot\mathbf{p}_i+\beta_i m-Z\alpha/r_i$ the usual external-field Dirac Hamiltonian and L_{++} the product of the associated external-field positive-energy projection operators. This equation, called the (external-field) no-pair Coulomb-ladder equation in [12], can be used as the starting point of a systematic perturbative approach to the energy levels of He and He-like systems. The corresponding equation with U_C replaced by U_C+U_B contains not only all α^2Ry corrections to fine-structure, but a number of corrections of order α^3Ry as well. (For large $\underline{Z}$ these can be expressed in analytic form. Recently, Lindgren and his collaborators [14] have developed methods for solving such equations numerically with sufficient accuracy to enable them to verify the analytic results.)

However, the further calculations needed to obtain the remaining α^3Ry and still higher-order corrections, such as those coming from two-photon exchange, are complicated, messy, and have all the appeal of a thorny, black box. They contrast strongly with the beauty of scattering-amplitude calculations, where among other things one can advantageously use a covariant propagator for the photon. It is an attractive idea to try to understand what part of the level structure can be thought of as arising purely from those forces which act when the electrons undergo free scattering. To put it another way, note that in NRQM one can describe both the scattering of two electrons and their interaction within a bound state with one and the same interaction potential U_C. The question then is to what extent one can, within the framework of QED, understand the bound states in terms of the same two-body effective potentials which describe the scattering accurately. It is clear that in QED there are effects which cannot be so described, e.g. the effects of the external potential provided by the nucleus during the exchange of photons between the electrons.

The availability of a lepton-lepton potential which reproduces the scattering amplitude exactly to order $(e_1e_2)^2$ would be a significant step in the direction of such a goal [15]. An early test of such a program would be provided, for example, by solution the "free no-pair ladder equation" [5]

$$h_{++}\psi = E\psi \tag{4.3}$$

where

$$h_{++} = h_+^{ext}(1) + h_+^{ext}(2) + V_{++}. \tag{4.4}$$

Here $h_+^{ext}(i) = \boldsymbol{\alpha}_i\cdot\mathbf{p}_i+\beta_i m + \Lambda_+(i)(-Z\alpha/r_i)\Lambda_+(i)$, and V_{++} is of the form $V_{++} = \Lambda_{++}U(1,2)\Lambda_{++}$, with Λ_{++} the product of free positive-energy projection operators, as in (2.9). The operator $U(1,2) = U^{(2)} + U^{(4)}$ must be chosen so that if the external potential is turned off, the one-photon and two-photon exchange scattering amplitudes are reproduced. (I note in passing that if one uses $U^{(2)} = U_{CC}$ rather that U_{CB} one must already include $U^{(4)}$ to get the α^2Ry level shift correctly!) On comparing the eigenvalues of (4.3) with those obtained from, e.g., the Bethe-Salpeter equation approach, one could disentangle those dynamical effects involved in the scattering from those which are peculiar to the bound-state situation. For $\underline{Z}$ not too large one would have an initial description in the language of configuration space which is not only quite accurate, but which also could actually be explained to students, without requiring that they become experts in the arcana of four-dimensional wave equations and all that. So it may well appeal to Ingvar Lindgren: As the photo-op at the end of this symposium showed, the results of his pedagogical skill and devotion could only be captured by a wide-angle lens!

Acknowledgements

This work was supported in part by the U.S.National Science Foundation.

References

1. C.E. Darwin, Philos. Mag. **39**, 537 (1920).
2. G. Breit, Phys. Rev. **34**, 553 (1929).
3. J.B. Gaunt, Proc. R. Soc. London, Ser. A **122**, 513 (1929).
4. G. Brown and D.G Ravenhall, Proc. R. Soc. London, Ser. **A 208**, 552 (1951).
5. J. Sucher, Phys. Rev. A **25**, 348 (1980); Phys. Rev. Lett. **55**, 1023 (1985).
6. For a review, see e.g. J. Sucher, in Program on Relativistic, Quantum Electrodynamic and Weak Interaction Effects in Atoms, edited by W.R. Johnson, P. Mohr, and J. Sucher (AIP, New York, 1989).
7. G. Feinberg and J. Sucher, Phys. Rev. D **38**, 3763 (1988).
8. J. Sucher, Phys. Rev. D **49**, 4284 (1994).
9. G. Feinberg and J. Sucher. Phys. Rev. D **45**, 2493 (1992).
10. J. Sucher, Comm. At. Mol. Phys. **30**, 129 (1994).
11. H. Araki, Prog. Theoret. Phys. **17**, 6191 (1957).
12. J. Sucher, Phys. Rev. **109**, 1010 (1958).
13. M.H. Douglas and N.M. Kroll, Ann. Phys. (N.Y.) **82**, 89 (1974)
14. I. Lindgren et al., Phys. Rev. **A51**, 1167 (1995).
15. There is an additional force between leptons, arising from the exchange of a neutrino-antineutrino pair, which is unfortunately too weak to be accessible experimentally. For a review, see J. Sucher, in Proceedings of the XIVth Moriond Workshop: Particle Astrophysics, Atomic Physics and Gravitation, Villars-sur-Ollon, January, 1994.

Atomic physics and the laser metrology of time and length

L R Pendrill[1] and L Robertsson[2]

[1]Swedish National Testing & Research Institute, Box 857, S-501 15 Borås, Sweden

[2]Bureau International des Poids et Mesures, Pavillon de Breteuil, F-92310 Sèvres, France

Abstract

Frequency measurements, which are amongst the most accurate measurements made, form a basis of, on the one hand, practical measurements of length as defined in terms of the SI metre for example, and on the other, determination of universal physical constants such as the Rydberg constant. Atomic physics lies at the heart of these measurements, providing the tools and methods of laser spectroscopy as well as the theory of atomic structure and atom-light interaction with which to interpret the results of measurements and relate them to other fields of physics.

ADVANCES IN QUANTUM CHEMISTRY, VOLUME 30

0065-3276/98 $25.00

1 Introduction

The lively atomic physics community in Sweden can be traced to a line of research, through the Siegbahns, Rydberg and Ångström, which has much to do with accurate measurement and fundamental metrology.

It seems fitting therefore to give the often-quoted statement of Maxwell [1]:

"... If we wish to obtain standards of length, time and mass which shall be absolutely permanent, we must seek them not in the dimensions, or the motion, or the mass of our planet, but in the wave-length, the period of vibration, and the absolute mass of these imperishable and unalterable and perfectly similar molecules"

Interpretation of the most precise measurements in atomic physics with the best *ab initio* theories leads to values of the fundamental constants, which reflect relations between phenomena in the different realms of physics, and which can be regarded as *the* ultimate base units of physical measurement. In the past decades, a "revolution" has occurred in our system of measurements, whereby the accuracy of many of the fundamental physical constants, especially of the "atomic" constants (such as the Rydberg constant), has caught up with our ability to realise the definitions of many of the SI units [2, 3].

In this paper, we review the advances made in the most accurate measurement of time, frequency and length with particular emphasis on the use of the laser, and on the relation of these to research in atomic physics.

2 Absolute and universal measures. Atomic hyperfine structure and Microwave frequency standards

Many will recognise the plaque (figure 1), copies of which have been sent on a number of spacecraft, such as Pioneer and Voyager, on the off-chance that some extra-terrestrial life will one day have the opportunity of learning some basic facts about Mankind. The wave-length of radiation of the ground state hyperfine structure in atomic hydrogen has been chosen as a length measure based on one of the most ubiquitous of measures, available throughout the Universe. Two stylised hydrogen atoms - with electron spin up and down - separated by a distance equal to the wave-length of 21 cm, are depicted in the upper part of the plaque.

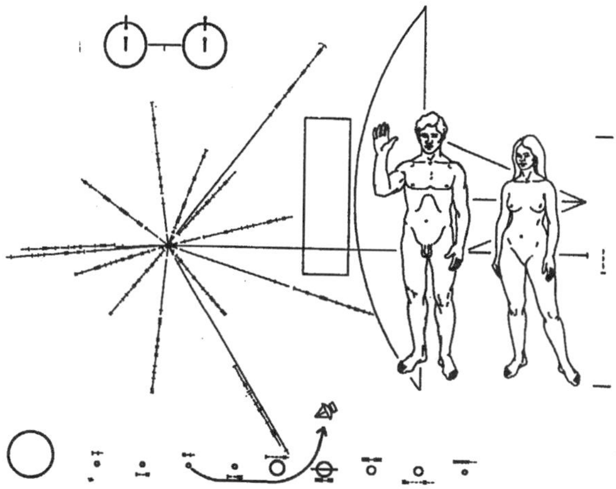

Figure 1 Plaque on board the Pioneer F spacecraft

The hydrogen maser, operating at the frequency of this ground state hyperfine interval, remains one of the most stable frequency standards, as shown in figure 2, especially at short time intervals. For practical reasons, one has chosen to define the unit of time, the second of the Système International, in terms of the analogous hyperfine structure interval, about 9 GHz, in the ground state of the ^{133}Cs atom [4]. This is due to the fact that at longer time intervals, the caesium atomic beam magnetic resonance clock is more accurate. For example, modern commercial Cs clocks have reported stabilities of about 3 parts in 10^{15} at 10 days (9×10^5 s) [5].

Although a mature technology, Cs atomic clocks continue to be steadily refined. In particular, the ability with laser light to probe and pump atoms to specific internal energy states as well to control the external motion of the atom (with laser cooling for example) has lead to further refinements in the classical ABMR atomic clocks [6] as well as to new clock designs. Recent results for a caesium fountain clock [7] based on an original idea of Zacharias [8] report a measured frequency stability of about $2 \times 10^{-13}\ \tau^{-1/2}$ and an accuracy estimated to be 3×10^{-15}.

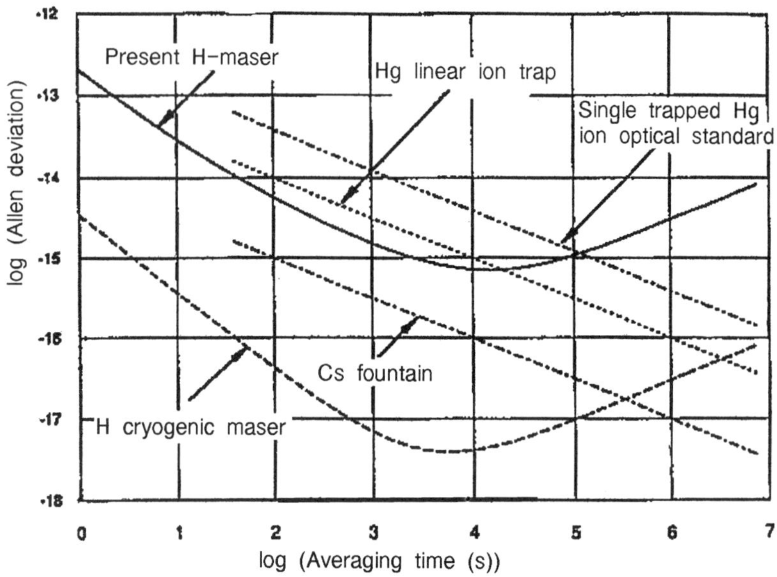

Figure 2 Successive sample deviations for various frequency standards [9]

Other stable frequency standards in the microwave region based on the ground-state hyperfine structure of hydrogenic atomic and ionic systems, include alkaline earth ions, such as Be^+ or Mg^+ often in electromagnetic (Penning or Paul) traps that can also make impressive claims to accuracy. Laser cooling is used to reduce uncertainties in these traps due to thermal or random motion of the ions to levels as low as 10^{-19} [10] but other, coherent motions, such as a rotation of a cloud of ions about the symmetry axis of a Penning trap or micromotion at the driving field frequency of a Paul trap, remain and limit accuracies at the 10^{-13} level [11] Linear traps [12] and ultimately a single string of ions, have been suggested [13] and realised [14, 15] to reduce some of these uncertainties in ion traps intended as frequency standards.

Applications of these stable microwave clocks, which drift by as little as some nanoseconds a day, are widespread, with the obvious example of the Global Positioning System which enables cellular navigation with unprecedented ease [16].

Is then the impressive accuracy of these, Mankind's most accurate standards of measurement, matched by a similarly impressive understanding of the underlying physics? That is, how well can theory be used to express the atomic energy level splittings in terms of the equations of physics and the fundamental constants of Nature?

A theoretical expression of the hyperfine structure of the ground state of atomic hydrogen is given in eqn (1) below. Despite the high experimental accuracy of frequency measurement, values of the hydrogen ground state hyperfine structure are principally limited in accuracy at the *ppm* level by incomplete knowledge of the structure of the finitely-sized proton, represented in eqn. (1) by one term in a sum of various corrections ε_i [17].

$$\Delta E_H = \frac{16R_\infty}{3} c\alpha^2 \frac{\mu_p}{\mu_B} [\frac{m_p}{(m_e + m_p)}]^3 (1 + a_e + \frac{3}{2}\alpha^2 + \sum_i \varepsilon_i) \qquad (1)$$

Studies of the ground-state hyperfine structures of hydrogen-like heavy ions, such as Bi^{82+} [18], Ho^{66+} and Re^{74+} [19], reveal effects of nuclear size, magnetic moment and QED. With the latest atomic theory [20], these hyperfine structures can be described with an accuracy of about 0,1%, mainly limited by knowledge of the nuclear magnetic moment.

Muonium has a simpler, point-like muon nucleus and recent improvements in the theory [21, 22, 23] yield results which compare favourably with measurements of the muonium ground state hyperfine structure [24, 25] at a level of 0,13 ppm, mainly determined by uncertainties in the measured muon mass.

Trapped ion spectroscopy does enable tests of fundamental symmetries, such as of the equivalence principle, made for example by comparing the frequency of a $^9Be^+$ "clock" with the frequency of a hydrogen maser [26]. In another comparison of microwave atomic clocks, based on Cs and Mg^+, Godone *et al.* [27] were able to set new experimental limits on the time variation of various fundamental constants.

3. Stable optical frequency standards

3.1 Lasers as optical frequency standards

The higher frequencies of visible light (in the range 10^{13} Hz - 10^{15} Hz) offer potentially higher accuracy than microwave standards because of the larger number of oscillations able to be measured in a given time interval. Progress in optical frequency standards requires the tackling of natural line-width broadening which increases, with the rate of spontaneous emission, with the cube of the frequency.

The laser of course is a key enabling technology for optical frequency standards. Firstly, it is instrumental in preparing atomic samples as references. In particular, transitions to meta-stable levels may be selected which lead to narrow line-widths. Laser-based methods such as laser-cooling, saturated absorption or two-photon spectroscopy enable the effects of atomic velocities to be reduced, thereby minimising shifts and broadening due to the Doppler effect, collisions, transit-time effects, etc. The laser secondly provides the source of highly monochromatic and coherent light to match the ultra-narrow reference lines. Normally one reckons to be able to lock an oscillator to within 10^{-4} of the line-width with active electronics, where the line-width is determined by a convolution of the oscillator and reference line-widths.

3.2 Trapped ions as optical frequency standards

A state-of-the-art example [28] for trapped-ion optical frequency standards is the case of a laser with a line-width of less than 25 Hz locked to a electric quadrupole transition at 282 nm in a single laser-cooled Hg^+ ion. The inherent stability of this trap based on the radiative lifetime of the metastable upper level of this transition is calculated to be about $1{,}5 \times 10^{-15}\ \tau^{-1/2}$ [29]. This is an exceptional example - in most cases as yet, the lasers used as oscillators for optical frequency standards based on ion traps often do not have a stability which matches the spectral sharpness of the trapped ion reference resonance.

As shown in figure 2, examples of a stable optical frequency standard of similar performance to the best microwave oscillators mentioned above are those with Ba^+ [30] and Hg^+ [31] trapped in Paul and Penning traps. In many cases, uncertainties due to coherent ion motion present limitations similar to those of the microwave standards discussed above.

3.3 Trapped atoms as optical frequency standards

Laser cooling works by exchanging momentum between light and atom. The last 10 years or so have seen a spectacular development in neutral atom manipulation with laser light. "Doppler" cooling, explainable in terms of the simplest atomic structure, lead to "optical molasses" with a combination of laser beams in three dimensions with temperatures in the range of a few hundred microkelvin [32]. Even lower temperatures were observed shortly after, called by some "Sysiphus" cooling [33], and it is this extra cooling, down to a couple of μK for the caesium atom, based on optical pumping amongst an underlying structure of Zeeman sub-levels, which is playing an important role in neutral atom frequency standards. A particular arrangement of magnetic fields together with laser cooling, the so-called magneto-optic trap (MOT), has been highly successful in the trapping neutral atoms. Experiments with trapped neutral atoms can be said to have reached a culmination with the observation of Bose-Einstein condensation in MOTs loaded with some of the alkali atoms such as Rb [34] and Na [35], with the extra cooling down to the few nanokelvin required in these experiments being provided by so-called "evaporative" cooling [36]. With this, the field of laser cooling and trapping of atoms has expanded to include the study of atomic motion under well-controlled conditions, including quantum mechanical collective movement [37].

Where does this spectacular development lead concerning frequency standards? Traps for neutral atoms unfortunately perturb the atomic energy levels; both the magnetic fields used as well as the laser light for cooling. As an example of what can be achieved, mention may be made of a tuneable dye laser, locked to the inter-combination line at 657 nm of Ca atoms, either in an atomic beam or in a magneto-optic trap, has an estimated relative uncertainty of below 10^{-12} [38].

3.4 The hydrogen atom as an optical frequency standard

The optical spectrum of the hydrogen atom, with its simple and symmetric structure, has always attracted physicists searching for ultimate units of measurement [39, 40, 41, 42]. The frequencies of different transitions in hydrogen scale to high precision as the square of the principal quantum numbers in a rational way. If one then makes measurement and can interpret the results accurately enough, then the hydrogen atom provides a complete set of frequency standards spanning the complete spectral range, from the microwave to the x-ray region. In recent years intensive research has come to the point where both measurement of the optical frequencies of transitions and theory of the energy levels in H are now known to about the same accuracy, that is, parts in 10^{11}. Most experiments have been performed with classical Doppler-free methods such as crossed beams and saturation and two-photon laser spectroscopy [41, 42]. The binding energy of a hydrogen energy level is expressed as:

$$E_{nj} = -hcR_H \frac{1}{n^2}\left[1+\frac{\alpha^2}{n^2}\left(\frac{n}{j+\frac{1}{2}}-\frac{3}{4}\right)\right]+\cdots$$

including the effects of relativity, to which should be added QED corrections and nuclear size effects. A measure of the success of recent hydrogen spectroscopy can be had by noting that the Rydberg constant R_H is now the most accurately determined fundamental constant.

Mention should be made here of a new experiment using millimetre radiation (about 256 GHz) to determine the Rydberg constant from the energy spacings of neighbouring "circular" Rydberg states of atomic hydrogen [43] in a way essentially free of QED and the Lamb shift.

3.5 Standards of length

Primary length measurements are these days based on optical frequency standards. If one needs a unit of length, for example, a wave-length for interferometric measurement, then one divides the optical frequency by the value of the speed of light (299 792 458 m.s^{-1}) as defined in the SI metre. The *mise en pratique* of the metre [44] lists a number of frequency-stabilised lasers at various wavelengths in the visible and near infrared spectral regions.

The close relationship between the field of laser spectroscopy of atoms and molecules and the techniques used for laser frequency stabilisation in precision length metrology might be worth underlining. Developments in laser spectroscopy for pure atomic physics research, such as various techniques of eliminating the Doppler effect (such as saturation absorption), have been instrumental in the evolution of the frequency-stabilised reference laser systems used today in length metrology.

Beside the purely spectroscopic part there has also been important development of "active" optical components such as electro-optic and acousto-optic modulators. Such components allowed the use of modulation schemes earlier used in the microwave region to be used in optical spectroscopy. In the early 80's the FM-sideband technique both for stabilisation to molecular lines as well as to passive cavities were demonstrated [45, 46]. The developing field of non-linear optics and the wave-mixing picture established a framework in which line shapes of different modulation schemes in saturation spectroscopy could be understood [47, 45, 46, 48, 49].

One more general consideration could be useful to have in mind before entering into details. For more or less complex systems such as stabilised lasers, a choice has often to be made as to which properties of the system should be optimised, be it portability, stability, cost, accessibility, wavelength etc. Often these properties are complementary so improving one is made at the cost of another. Comparing systems optimised in largely different ways has therefore limited relevance.

For many applications involving the wavelength of the light the uncertainty in the determination of the index of refraction puts a limit on the relative uncertainty at a level around 10^{-8} - 10^{-9}. Applications in vacuum can surpass this level by some few orders of magnitude but become finally limited by thermal and material instability together with laser field aberrations and diffraction at a level of about 10^{-11} - 10^{-12}. In the foreseeable future it would thus hardly be meaningful to improve further the frequency stability and reproducibility of laser systems in this group far beyond the level 10^{-12} - 10^{-13}. Here rather, considerations like reliability, portability, wavelength, lifetime etc. would be more important.

The workhorse of modern optical metrology is the He-Ne laser, emitting light at 472 THz (633 nm) and locked for the highest accuracy to the hyperfine structure of the iodine diatomic molecule with Doppler-free saturated absorption spectroscopy. Even though it is more than 25 years since the red He-Ne laser was first stabilized by intracavity spectroscopy to I_2 [50], it is today still the reference system used almost exclusively in national metrological laboratories to realize the metre. In the last update of the "Mise en pratique" [44] for the definition of the Metre the relative uncertainty for this wavelength was given as 2,5 x 10^{-11} (about 10 kHz), as demonstrated in international comparisons such as [51]. The reason for the durable success of the He-Ne laser could be that this laser is in many respects a good compromise of different properties like stability, size, cost etc. which to a large extent provide the features requested in the field of length metrology. Other traditional gas laser systems are He-Ne lasers locked to various molecular lines at 515 nm, 543 nm, 612 nm and 640 nm as well as the CO_2 laser with absorbing molecules like CO_2, SF_6, SiF_4, OsO_4 etc. [52, 53].

Semiconductor lasers are now readily available over a wide spectral range extending from the infrared to the visible with high efficiency and reliability, thanks to developments spurred by the requirements of applications such as optical communications. The small size, relatively high power and large tuneability of these diode lasers could enable this type of laser to play an important role in the future. For instance, in the proximity of the 633 nm He-Ne transition there are several stronger lines in iodine that easily can be reached with diode lasers [54]. Recent techniques using optical comb-generators could provide such reference transition in iodine over THz ranges by making the connection to the well established P(127)11-5 transition used by the He-Ne/I_2 laser. Such red diode laser based systems are important because they could be seen as a refinement and a more modern version of the old "work-horse" He-Ne/I_2 whose wavelength will certainly be used for years to come.

One laser that has the potential to gain the same practical importance in the near future as the 633 nm He-Ne laser is the frequency doubled YAG laser with the wavelength 532 nm [55, 56]. This wavelength coincides with several strong transitions in iodine. A stability of 10^{-13} in one second reaching around 3 x 10^{-14} in 30 s has been demonstrated [57, 58]. This stability appears to be satisfactory for most applications involving the wavelength of the laser. Today, still more expensive, more cumbersome but with superior performance there are thus reasons to believe that this green wavelength will be used more and more for the next years to come.

Diode laser systems near 780 nm locked to Rb two-photon transitions [59] have recently been developed. The $5^2S_{1/2}$-$5^2D_{3/2}$ transition can be excited by two photons from a 778 nm diode laser [60]. This transition has shown good signal to noise ratio giving a short term stability of 6×10^{-13} in one second reaching 2×10^{-14} in 1000 sec (61, 62].

Another notable system is the CH_4-stabilised He-Ne laser at 3,39 μm which has recently been reported to have stabilities (10^{14}) approaching that of the better microwave clocks mentioned above. Here, the approach is to use selectively only those molecules with very low speed. This work has been pioneered by the Novosibirsk group [63, 64]. By using very low light powers only the molecules with very low transverse speeds will experience the light field long enough to get excited and contribute to the absorption. This technique can also give very narrow line widths but at the cost of extremely weak signals. Using this technique, excellent results have been obtained with the system He-Ne/CH_4. Their estimated reproducibility was as low as 3×10^{-13} with a stability of 5×10^{-15} in 10 seconds [65, 66].

Ro-vibrational transitions in molecules provide an enormous number of narrow absorption lines in the infrared region. By using overtones of these transitions a large number of visible and near visible absorption lines could be provided. However, the weakness of these lines require very sensitive spectroscopic techniques. An example is the NICE OHMS technique [58, 67] where a gas sample of C_2HD contained in a high finesse cavity absorbs at the $(\nu_2 + 3\nu_3)$ overtone at a 1064 nm. In beat experiments between this system and a iodine-stabilized YAG-laser, stabilities of below 10^{-14} in 800 s have been observed. Especially interesting, is the demonstration of how this technique could be combined with the selective cold-atom technique to obtain line widths as narrow as 20 kHz.

From the brief discussion above it can be seen that there are many roads to follow in the search for the reference systems that will cover the needs of standards in the future. However, it appears as if the wavelengths of largest practical importance for the next 5 years to come will continue to be 633 nm or wavelengths in its vicinity excited by diode lasers, together with the green 532 nm frequency doubled YAG-laser.

4. Frequency chains linking microwave and optical standards

In order to measure the frequency of a standard with traceability to the definition of the SI second entails for the optical standards a technologically demanding set-up consisting of a chain (or series of chains) of stable oscillators and means of producing and determining the difference frequency at each step. (Heterodyne (or beat) spectroscopy is the usual method of comparing frequencies. The reason for this is that the highest frequency which can be measured with a detector is at several hundred GHz and the large range of frequencies - from MHz to about 500 THz - has to be bridged in small steps.

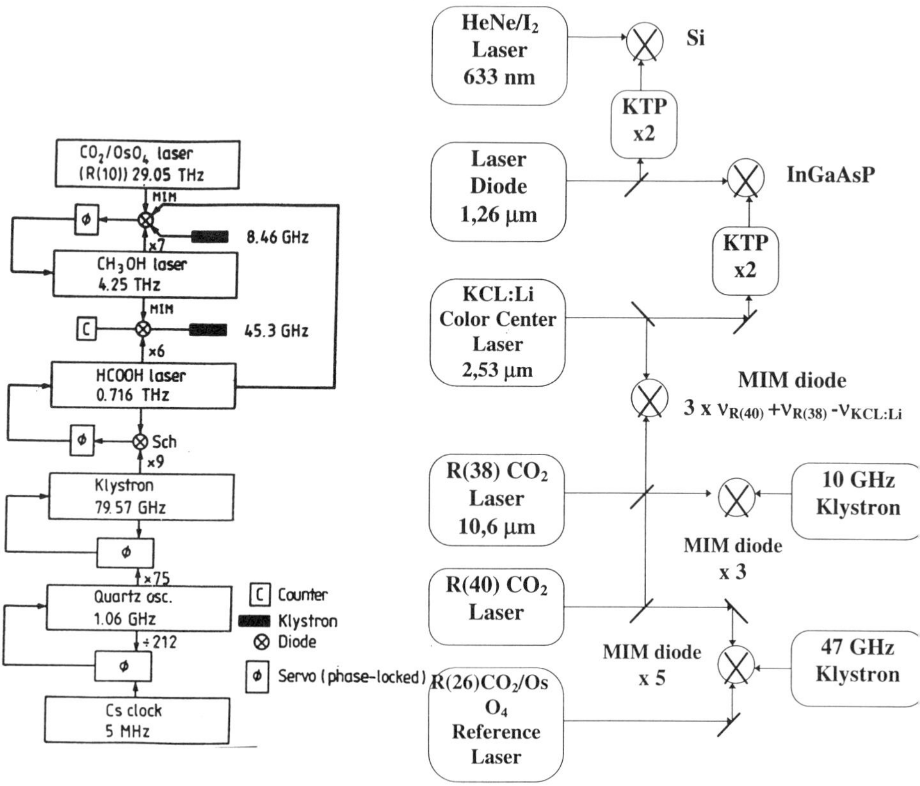

(a) (b)

Figure 3 Frequency synthesis chains linking a Cs clock and an I_2-stabilised He-Ne laser
(a) Clairon *et al.* [68] (b) Acef *et al.* [69]

An example of such a chain links a 5-MHz frequency derived from a primary Cs clock to the iodine-stabilised He-Ne laser at 473 THz and is shown in figure 3. The first part of the chain (figure 3 (a)) was used to measure the frequency of a CO_2 laser, locked at 29 THz to the OsO_4 molecule, with an accuracy of 10^{-12} [68]. The second stage links this laser to the I_2-stabilised He-Ne laser at 473 THz with four intermediate laser lines in the infra-red [69].

Another impressive example is work made at Physikalisch-Technische Bundesanstalt in Braunschweig [38]. The optical frequency of the intercombination line, $^3P_1$-$^1S_0$, at 657 nm in ^{40}Ca has been measured, both in an atomic beam as well as with trapping/cooling techniques. This transition was connected all the way to the Cs frequency by a chain giving a total uncertainty of the absolute frequency for their measurements as low as 10^{-12} which is the lowest uncertainty in the visible domain. This transition has a linewidth $\leq$ 400 Hz which gives good prospects of considerably improving the results.

Limitations on the highest frequencies which may be counted with available photodetectors and electronics motivates the search and exploitation of fortunate coincidences of sum and differences of optical frequencies [70]. As more optical reference systems are established in the visible and neighbouring spectral regions, especially using tuneable lasers, such as Ti:Sapphire and the semiconductor diodes, the chances increase of finding coincidences, such as:

$$2\text{x } f(633 \text{ nm}) = f(780 \text{ nm}) + f(532 \text{ nm}) + \Delta f$$

where Δf is a manageable frequency (in this example, - 263 GHz).

5. Wave-length measurement and the speed of light. Metre definition. Optical interferometry

Optical interferometry is the technique of choice for the most accurate measurement of length, in which light is divided into separate beams which are subsequently re-combined after travelling different optical paths. Analysis of the interference pattern at re-combination yields the path difference in terms of the wave-length of the light.

While the frequency measurements described above are ultimately limited in accuracy by the finite time interval available in which an oscillation of a particular frequency may be followed, the accuracy of the best wave-length measurements is limited by finite spatial apertures in the experimental apparatus through the effects of diffraction. Diffraction leads to light scattered at angles away from the optical axis, and this lack of collimation results in interference occurring at path-lengths leading to errors which include the so-called Abbé cosine error. This limits wave-length accuracy to the level of 10^{11} at best.

This fundamental limit in accuracy of optical wave-length interferometry caused by diffraction (as well as other sources of uncertainty discussed below) was one of the principal reasons behind the most recent change in definition of the SI metre in terms of the distance travelled by light (in vacuum) in a certain fraction of a second [CIPM 1984]. In order to benefit fully from the much higher accuracy of frequency measurements, the wave-length of any stable oscillator may be calculated from the frequency of that oscillator (determined for example in a frequency synthesis chain of the type described in §4) using the defined value of the speed of light, c = 299 792 458 m. s^{-1}. This value of the speed of light was obtained prior to the re-definition of the metre from combinations of optical frequency measurements (from synthesis chains) and the best optical interferometry.

Optical interferometers fall into two categories - the *two-pass*, such as the Michelson, interferometer and the *multi-pass*, such as the Fabry-Perot, interferometer, distinguished by the number of reflections of light between mirror pairs. Both types of interferometer form a Haidinger interference ring pattern in the image of a spatially extended light source, where constructive interference, marked by transmission maxima for light of wavelength λ, occurs at an integer number (m) of half wavelengths:

$$m\frac{\lambda}{2} = t\cos\theta$$

where t is the length of the interferometer.

The sharpness of the interference fringes, which determines the ultimate accuracy in terms of the ability to separate neighbouring fringes as well to determine the exact location of each fringe in the wave-number or length scale, is much greater in the multi-pass interferometer compared with the two-pass, simply owing to the greater number of reflections. Traditionally, this advantage has been offset by transmission loss and asymmetry of interference fringes due to maladjustment or optical defects in multipass interferometers, so that two-pass interferometers have being used as often as multipass in the most accurate length measurement [71]. It is not uncommon in the most demanding interferometry (such as in the detection of gravity waves with ultra-stable lasers [72] to use combinations of the two types of interferometer.

Figure 4 **Fabry-Perot interferometer spacer (of Zerodur) and mirrors** [73]

Laser light has normally to be spatially prepared in order to form the extended light source required for interferometry. In addition to beam expansion with high-quality optics, it is also necessary to eliminate the transverse coherence of the laser light (for example, by passing the light through a rotating matt disc) in order to avoid the disturbing non-uniform illumination of laser speckle. Mirror optics is usually employed, since it is easier to manufacture with the required optical flatness (normally $< \lambda/100$ for visible wavelengths) at a mirror surface rather than in transmission through a lens, especially when one considers the requirement for large-diameter (typically 50 mm) optics to avoid diffraction.

Modern mirror coatings can have reflectivities very close to 100 % but often only over a limited range of wave-lengths. Reflectivity may in general be described in terms of complex numbers, where the imaginary component is associated with phase changes which need to be accounted for in precision wave-length interferometry. Rapid and complicated changes in mirror reflectivity as a function of wave-length are often accompanied by similar variations in phase shift at reflection. It is common therefore to accept a lower finesse in order to provide an interferometer of wider spectral coverage with a monotonic phase shift dispersion.

The distance between the interferometer optics (usually some decimetres) should be constant, both in the long and short term. Ultra-low thermal expansion materials (such as ULE, Zerodur, etc) have coefficients of $< 10^{-8}$, but in some cases, long-term creep has been observed which depends on treatment and preparation (annealing) of the material on manufacture. If measurements are made with the interferometer in atmosphere of varying gas pressures, then account should be taken not only of refractive index but also of the compressibility of the interferometer parts [73]. Contamination, especially by water which may be adsorbed on the various surfaces of the interferometer, can affect the optical path length directly on the mirrors or indirectly through changes in the (background) gas refractivity. The refractivity of moist air has been recently re-measured [73, 74, 75].

Analysis of the interferogram can either be performed by registering the Haidinger ring system at different angles (with a CCD camera, for example) or by placing a pin-hole aperture at the centre of the system and recording successive constructive and destructive interference as the interferometer is swept (for example with pressure). The interferometer is normally illuminated simultaneously with light from a laser of known wave-length and the laser to be determined.

Despite all of the above-mentioned limitations in accuracy of optical interferometry, it is still widely used in the determination of the wave-numbers of atomic transitions, since optical frequency metrology (synthesis chains, optical frequency combs, etc, §4) does not yet have the wide spectral coverage provided by the broad-band interferometers. As an example, a recent absolute wave-number determination of the Cs D_2 resonance line at 852 nm is with a Fabry-Perot interferometer, saturated absorption and a grating-cavity semiconductor laser [76]. These results are of interest to various Cs atomic fountain measurements and lead to better determinations of fundamental constants, such as h/m_p and α, [77] as well as of the acceleration due to gravity, g [78, 79].

Optical interferometry has an interest in its own right in the determination of the refractive indices of gases. For example, precise laser refractometry of pure helium gas can lead to a better determination of the Boltzmann constant [80] when combined with the latest *ab initio* theory of atomic polarisability [81].

6. Conclusion

Both microwave and optical frequency standards have benefited greatly from the development of the laser and the methods of laser spectroscopy in atomic physics. In particular, the ability to determine both the internal and external (that is, motion) atomic states with laser light - by laser cooling for example - has opened up the prospect of frequency standards with relative uncertainties below 10^{-15}, for example, the Cs atomic fountain clock. The best atomic theories in some cases at starting to match in accuracy that of measurement, providing thereby refined values of the fundamental, so-called atomic constants. Even quite practical measurements (such as used in GPS navigation and primary standards of length) have advanced in recent years.

References

1 Maxwell J C 1870 Presidential address, British Association for the Advancement of Science

2 Petley B W 1992 *Metrologia* **29**, 95

3 Pendrill L R 1995 *"Some comments on Fundamental Constants and Units of Measurement related to Precise Measurements with Trapped Charged Particles"*, Nobel Symposium NS91 "Trapped Charged Particles and Fundamental Physics", Lysekil, August 19 - 25, 1994 Phys. Scripta **T59**, 46 - 52 (1995) and World Scientific Publishing, ed. I Bergström, C Carlberg & R Schuch, ISBN 981-02-2481-8 (1995)

4 BIPM 1992 '*Le Système International d'Unités*', Bureau International des Poids et Mesures, BIPM, Pavillon de Breteuil, F-92319 SÈVRES, France

5 Allan D W, Lepek A, Cutler L, Giffard R and Kusters J 1995 *Proc. Symposium on Frequency Standards and Metrology*, Woods Hole, MA (USA), (World Scientific Publishing, ed. J. C. Bergquist), 97 - 104

6 de Clercq E and Makdissi A 1996 "Current status of the LPTF optically pumped Cs beam standard", *Proc. Symposium on Frequency Standards and Metrology*, Woods Hole, MA (USA), (World Scientific Publishing, ed. J. C. Bergquist), pp. 409 - 10

7 Ghezali S, Laurent Ph, Lea S N, Santarelli G, Bahoura M, Simon E, Weyers S, Szymaniec K, Cognet L and Clairon A 1996 "About the accuracy evaluation of the LPTF caesium fountain frequency standard", *Conference on Precision Electromagnetic Measurements*, Braunschweig (DE), TH1B-2

8 Zacharias J R 1953 unpublished (see Ramsey [1985]) Ramsey N F 1985 "Molecular beams", Oxford University Press, Oxford

9 Vessot R F C, Mattison E M, Levine M W and Walsworth R L 1992 "*Status of local oscillators for operating ultra-high resolution frequency discriminators as frequency standards*", 24th Annual PTTI Applications and Planning Meeting, VA (USA), Dec. 1-3

10 Diedrich F, Bergquist J C, Itano W M and Wineland D J 1989 *Phys. Rev. Lett.* **62**, 403

11 Bollinger J J, Heinzen D J, Itano W M, Gilbert S L, and Wineland D J 1991 *IEEE Trans. Instru. Meas.* **40**, 126

12 Prestage J D, Dick G J and Maleki L 1989 *J. Appl. Phys.* **66**, 1013

13 Dehmelt H G 1989 in "*Frequency Standards and Metrology*", Proc. 4th Symposium, ed. A deMarchi, Ancona (IT) (Springer Verlag, Berlin, Heidelberg), p. 289

14 Raizen M G, Gilligan J M, Bergquist J C, Itano W M, and Wineland D J 1992 *Phys. Rev.* **A45**, 6493

15 Poitzcsh M E, Bergquist J C, Itano W M and Wineland D J 1994 *Proc. IEEE Int. Freq. Contr. Symp.*, p. 744

16 Getting I A 1994 "The Global Positioning System", *IEEE Spectrum* 36 - 47 (Dec.)

17 Gidley DW, Rich A, Zitzewitz P W and Paul D A L 1978 *Phys Rev Letts* **40**, 737

18 Klaft I *et al.* 1994 *Phys Rev Lett* **73**, 2425

19 Crespo López-Urrutia, J R, Beiersdorfer P, Savin D W and Widmann K 1996 "Hyperfine transitions in ground state hydrogen-like $^{165}Ho^{66+}$ and $^{185,187}Re^{74+}$" DAMOP WL05

20 Persson H *et al.* 1996 "Self-energy correction to the hyperfine structure splittings of hydrogen-like atoms" *Phys Rev Lett* **76**, 1433

21 Kinoshita T and Nio M 1994 "Improved theory of the muonium hyperfine structure",
Phys Rev Lett **72**, 3803 - 6

22 Kinoshita T 1994 "Improved determination of fine structure constant based on the electron g - 2 and muonium hyperfine structure", *Conference Precision Electromagnetic Measurements Digest*, WE2B-1, Boulder CO (USA), (July 1994)

23 Pachucki K 1996 "$\alpha(Z\alpha)^2E_F$ correction to the hyperfine splitting in hydrogenic atoms", *Phys. Rev.* **A 54**, 1994 - 8

24 Mariam F G *et al* 1982 *Phys Rev Lett* **49,** 993

25 Klempt E *et al* 1982 *Phys Rev D* **25**, 652

26 Prestage J D, Bollinger J J, Itano W M and Wineland D J 1985 , "Limits for spatial anisotropy by use of nuclear-spin-polarized $^9Be^+$ ions", *Phys Rev Lett* **54**, 2387 - 90

27 Godone A, Novero C, Tavella P and Rahimullah "New experimental limits to the time variations of g_p, (m_e/m_p) and α", *Phys Rev Lett* **71**, 2364 (1993)

28 Bergquist J C, Itano W M and Wineland D J 1994 in "*Frontiers in Laser Spectroscopy*", Proc. Int. Sch. Phys <<E. Fermi>>, CXX, eds. T W Hänsch and M Inguscio (North Holland, Amsterdam) pp. 359 - 76

29 Wineland D J *et al.* 1996 "Application of laser-cooled ions to frequency standards and metrology", *Proc. Symposium on Frequency Standards and Metrology*, Woods Hole, MA (USA), (World Scientific Publishing, ed. J. C. Bergquist), pp. 11 - 9

30 Janik G, Nagourney W and Dehmelt H G 1985 *Opt. Commun.* **79**, 176

31 Bergquist J C, Itano W M and Wineland D J 1987 *Phys Rev* **A36**, 428

32 Chu S, Hollberg L, Bjorkholm J, Cable A and Ashkin A 1985 *Phys. Rev. Lett.* **55**, 48

33 Cohen-Tannoudji C and Phillips W D 1990 *Physics Today* **43**, 33

34 Anderson M H, Ensher J R, Matthews M R, Wieman C R and Cornell E A 1995 *Science* **269**, 198

35 Davis K B, et al. 1995 *Phys. Rev. Lett.* **75**

36 Hess H F 1986 *Phys Rev* **B34**, 3476

37 Metcalf H 1997 "Laser cooling as a form of optical pumping in the quantum domain of atomic motion", in "George W. Series Memorial Essays", *Phys. Scripta* **T70** Topical Issue (ed. A Corney, J N Dodd, L R Pendrill and D N Stacey) in press

38 Riehle F, Schnatz H, Lipphart B, Zinner G, Kersten P and Helmcke J 1996 "Optical frequency standard based on laser-cooled Ca atoms", *Proc. Symposium on Frequency Standards and Metrology*, Woods Hole, MA (USA), (World Scientific Publishing, ed. J. C. Bergquist), pp.277 - 82

39 Series G W 1957 "Spectrum of atomic hydrogen", Oxford Univ. Press

40 McIntyre D H and Hänsch T W 1988 "Precision measurements of the Rydberg constant", *Metrologia* **25**, 61 - 6

41 Cagnac B 1993 "Progress on the Rydberg constant: the H atom as a frequency standard", *IEEE Trans. Instrum. Meas.* **42**, 206 - 12

42 Cagnac B 1997 "Hydrogen Metrology: Up to what limit?", in "George W. Series Memorial Essays", *Phys. Scripta* **T70** Topical Issue (ed. A Corney, J N Dodd, L R Pendrill and D N Stacey) in press

43 Lutwak R, Holley J, de Vries and Kleppner D 1996 "Millimetre-wave measurement of the Rydberg frequency", *Proc. Symposium on Frequency Standards and Metrology*, Woods Hole, MA (USA), (World Scientific Publishing, ed. J. C. Bergquist), pp. 259 - 63

44 Quinn T J 1993/4 "*Mise en Pratique* of the Definition of the Metre (1992)" *Metrologia* **30**, 523 - 41

45 Bjorklund G. C., *Opt. Lett. Vol.* 5, 1980, pp.15-17.

46 Hall. J. L., Hollberg L., Baer. T. and Robinson H. G., *Appl. Phys. Lett.* Vol. 39,1981, pp. 680-682.

47 Raj R. K., Bloch D., Snyder J. J., Camy G. and Ducloy M., " High-Frequency Optically Heterodyned Saturation Spectroscopy Via Resonant Degenerate Four-Wave Mixing", *Phys. Rev. Lett.* Vol. 44, 1980, pp. 1251-1254.

48 Schenzle A., DeVoe R. and Brewer R. G., "Phase-modulation laser spectroscopy", *Phys. Rev. A.*, Vol. 25, 1982, pp. 2606-2621.

49 Shirley J. H., "Modulation transfer processes in optical heterodyne saturation spectroscopy", *Opt. Lett.*, Vol. 7, 1982, pp. 537-539.

50 Hanes G.R. and Dahlstrom C.E., 1969 "Iodine hyperfine structure observed in saaturated adoption at 633 nm", App. Phys. Lett. 14, 362

51 Chartier JM, Darnedde H, Frennberg M, Henningsen J, Kärn U, Pendrill L R, Hu Jianpei, Petersen JC, Poulsen O, Ramanujam P S, Riehle F, Robertsson L, Ståhlberg B and Wahlgren H 1992 *Metrologia*, **29**, 331 - 9

52 Clarion A. Acef. O. Chardonnet C. and Bordé C. J. 1989, in "Frequency Standards and Metrology" edited by A. de Marchi (Springer-Verlag, Heidelberg,1989), p. 212ff.

53 Bobin B, Bordé C. J. Bordé J and Bréant C. 1987, *J. Mol. Spec.* **121**, p.91ff

54 Åman J, Hammersberg M, Pendrill L R, Talvitie H, Zarka A and Chartier J-M 1996 "Laser spectroscopy of molecular iodine with a tuneable semiconductor laser around 633 nm", *Proc. Symposium on Frequency Standards and Metrology*, Woods Hole, MA (USA), (World Scientific Publishing, ed. J. C. Bergquist), 437 - 40

55 Arie A., Schiller S., Gustavson E. and Byer R. L. 1993, "Absolute frequency stability of diode-pumped Nd:YAG lasers to hyperfine transitions in molecular iodine", *Opt. Lett.* **17**, 1204

56 Arie A and Byer R L 1994 *J. Opt. Soc. Am.* **B11**, 866

57 Eickhoff M. and Hall J. L. 1995, "Developing an optical frequency standard at 532 nm", *IEEE Trans. Instrum. Meas.* **44**, 155-158

58 Hall J. L., Jun Ye, Long-Sheng Ma, Swartz S., Jungner P and Waltman S., "Optical Frequency Standards; some improvements, some measurements, and some dreams", Proceedings of the 5:th Symposium on Frequency Standards and Metrology, Woods Hole, Massachusetts, Oct. 1995, pp. 267-276.

59 Millerioux Y, Touhari D, Hilico L, Clairon A, Felder R, Biraben F, and de Beauvoir B 1994 *Opt. Commun.* **108**, 91

60 Ryan R. E., Westling L. A. and Metcalf H. J. 1993, "Two-photon spectroscopy in rubidium with a diode laser", *J. Opt. Soc. Am. B.* **10**, 1643-1648

61 Nez F., Biraben F., Felder R. and Millerioux Y. 1993, "Optical frequency measurement of the $5^2S_{1/2}$-$5^2D_{3/2}$ two-photon transition in Rubidium", *Opt. Comm.* **102**,. 432

62 Zondy J.-J., Touahri D., Acef O., Clarion A., Felder R., Nez F. and Hilico L. 1996, "Toward the Frequency Measurement of a Laser Diode Locked to 5S-5D Rubidium Two-Photon Transition (2 x 385 THz), Proceedings of the 5th Symposium on Frequency Standards and Metrology, Woods Hole, Massachusetts, Oct. 1995, pp.310-315

63 Bagayev S N, and Chebotaev V P 1990, " High-Stability - He-Ne/CH_4 Laser for Precision Frequency Measurements", *Opt. Spectrosc.*(USSR), **69**, 406-407.

64 Bagayev S N, Chebotaev V P, Dmitriyev A K, Om A, Nekrasov Yu V and Skvortsov B N 1991, *Appl. Phys.* **B 52**, 63

65 Bagayev S. N., Dmitriyev A. K., Pokasov P. K. and Skvortsov B. N 1996., "He-Ne/CH_4 Laser Frequency Standard for Precise Measurements", Proceedings of the 5th Symposium on Frequency Standards and Metrology, Woods Hole, Massachusetts, Oct. 1995, pp.289-296

66 Tyurikov D A, Gubin M A, Shelkovnikov A S and Koval'chuk E V 1995 *IEEE Trans Instrum Meas* **44**, 166

67 Ye J, Ma L-S and Hall J L 1996 "Sub-Doppler optical frequency reference at 1,064 mm by means of ultrasensitive cavity-enhanced frequency modulation spectroscopy of a C_2HD overtone transition", *Opt. Lett.*, **21**, 1000 - 2

68 Clairon *et al.* 1985, *IEEE Trans. Instrum. Meas.*, **IM-34**, 260 - 265

69 Acef O, Zondy J -J, Abed M, Rovera D G, Gérard A H, Clairon A, Laurent Ph, Millerioux Y and Juncar P 1994 *Opt. Commun* **97**, 29

70 van Baak, D A and Hollberg L 1994 "Proposed sum-and-difference method for optical-frequency measurement in the near infrared", *Opt. Lett.* **19**, 1586 - 8

71 Rowley W R C 1984 "Laser wavelength measurements and standards for the determination of length", in *Precision Measurement and Fundamental Constants II*, B N Taylor & W D Phillips, eds., *N.B.S.(U.S.), Spec. Publ.* **617** 57 - 64

72 Stephens M 1993 "A sensitive interferometric accelerometer", *Rev Sci Instrum* **64**, 2612 - 4 *and references therein*

73 Andersson M, Eliasson L and Pendrill L R 1987 "Compressible Fabry-Perot refractometer", *Appl. Opt.* **26**, 4835 - 40

74 Birch K P and Downs M J 1994 "Correction to the up-dated Edlén equation for the refractive index of air", *Metrologia* **31**, 315 - 6

75 Bönsch G 1994 "Revidierte Edlén-Formel", PTB (DE) private communication

76 Carlsson G, Kastberg A and Pendrill L R 1996 "Absolute wavelength measurement of Cs D_2 resonance line", submitted for publication

77 Weiss D S, Young B C and Chu S 1994 *Appl Phys* **B 59**, 217 - 56

78 Kasevich M and Chu S 1992 "Measurement of the gravitational acceleration of an atom with a light-pulse atom interferometer", *Appl. Phys.* **B54**, 321 - 32

79 Peters A, Chung K-Y and Chu S 1996 private communication

80 Pendrill L R 1996 "Macroscopic and microscopic polarisabilities of helium gas", *J. Phys. B: Atom. Molec Opt. Phys.* **29**, 3581 - 6

81 Johnson W R and Cheng K T 1996 "Relativistic configuration-interaction calculation of the polarizabilities of helium-like ions", *Phys. Rev.* **A 53**, 1375 - 8

Optical Trapping of Absorbing Particles

H. Rubinsztein-Dunlop, T.A. Nieminen, M.E.J. Friese and N.R. Heckenberg

Department of Physics, The University of Queensland, Brisbane, Queensland 4072, Australia

Abstract

Radiation pressure forces in a focussed laser beam can be used to trap microscopic absorbing particles against a substrate. Calculations based on momentum transfer considerations show that stable trapping occurs before the beam waist, and that trapping is more effective with doughnut beams. Such doughnut beams can transfer angular momentum leading to rotation of the trapped particles. Energy is also transferred, which can result in heating of the particles to temperatures above the boiling point of the surrounding medium.

1 Introduction

The availability of lasers has enabled the observation of forces due to light interacting with microscopic objects. In 1970, Ashkin [1] reported optical trapping of micrometre sized spheres using two opposing laser beams, and by 1980 [2] had proposed many experiments using focussed laser beams, and discussed widely varying possible applications, including automatic force measurement, particle size measurement using surface wave resonances in scattering particles, a modified Millikan experiment, measurement of radiometric forces, and separation and manipulation of biological particles. In 1986 [3] the single-beam gradient optical trap, or *optical tweezers* was first demonstrated. Optical tweezers can be used for three-dimensional manipulation of transparent particles around 1-100 μm in diameter, or other small particles which behave as reactive dipoles.

The single-beam gradient optical trap consists of a single laser beam, tightly focussed to create a very strong field gradient both radially and axially, which acts on polarisable particles to cause a dipole force. Polarisable particles are attracted to the strongest part of the field, at the beam focus, due to the gradient force. A scattering force results from momentum transfer to the particle when light is scattered by it. Under the right conditions the gradient force can balance the scattering and gravity forces, to trap particles three-dimensionally in the laser beam. If the laser beam is not tightly focussed, the axial component of the gradient force will be weak, and only radial trapping will be possible [1,4].

ADVANCES IN QUANTUM CHEMISTRY, VOLUME 30

0065-3276/98 $25.00

A practical optical tweezers setup usually consists of a laser beam with a power of a few hundred mW, introduced into a microscope and focussed using a high numerical aperture 100× objective lens. Both conventional upright and inverted microscopes are used, the inverted microscope providing stronger axial trapping due to the scattering force opposing gravity.

The ability to manipulate transparent particles of size 1-100 μm in a closed sterile environment has been exploited by researchers in the biological sciences, to move [5], isolate [6,7], cut (using UV beams, where biological specimens are highly absorbing) [8], and perform surgery on [9] cells and to do various forms of analysis. The very high intensity region at the beam focus can be used for two photon spectroscopy, which enables analysis of very thin sections of a sample, since the high intensity region is extremely localised [10]. Using calibrated forces from optical tweezers, it is also possible to measure physical properties of specimens, such as the compliance of bacterial flagella [11], mechanical properties of a single protein motor [12] and tube-like motion of a single polymer chain [13].

Although the particles normally trapped with optical tweezers are highly transparent, there is usually some absorption occurring. Studies have been made of the wavelength dependence of the heating of specimens and the ability of live specimens to remain viable after having been trapped [14,15,16].

Svoboda and Block [17] reported that small metallic Rayleigh particles with sizes on the order of tens of nm can be trapped using dipole forces. However when the absorptivity of micron-sized particles becomes high enough, radiation pressure becomes much greater than the gradient force, and they can no longer be trapped using dipole forces. These absorbing particles are then affected primarily by radiation pressure, whereby the momentum of the absorbed light is transferred to the particle. The momentum of the light field is normal to the wavefronts, so if the curvature of the wavefront is such that the resulting force can be resolved into a radially inward force and a force in the direction of beam propagation, then particles constrained in the direction of beam propagation (e.g. by a glass microscope slide as in Figure 1) can be trapped two-dimensionally.

Trapping of micron-sized reflective particles has been modelled in a similar way [18]. On the basis of ray optics, a totally reflective particle can be shown to be radially constrained when located before the beam focus. However, the forces produced by reflection are dependent on the angle at which the laser light impinges on the particle. Trapping of absorbing particles is much less dependent on particle geometry as the radiation force is perpendicular to the wavefront and independent of the orientation of the particle surface. In experiments on the trapping of microscopic reflective particles, heating effects such as bubble formation and radiometric forces were observed [18], indicating a noticeable absorption effect,

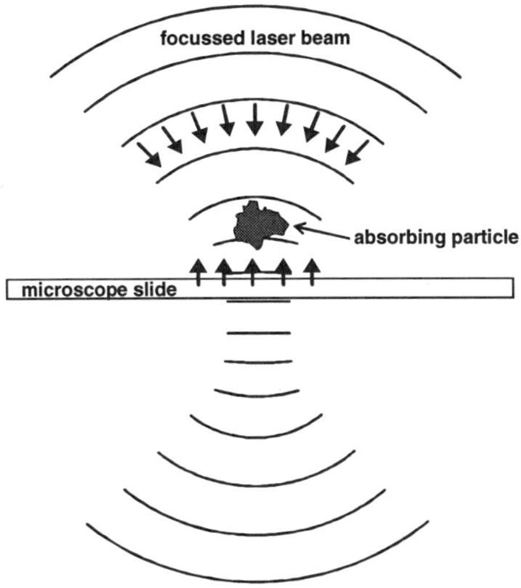

Figure 1: Schematic diagram of an absorbing particle in a focussed laser beam
The curvature of the wavefront is such that the resulting force can be resolved into a radially inward force and a force in the direction of the beam propagation. The particles are constrained in the vertical direction by the glass microscope slide, and hence trapped.

so absorption of linear momentum also contributes to the trapping forces acting on these particles.

Using such transfer of linear momentum, absorbing particles such as zinc dust [19], CuO particles [20] and ceramic powder [21,22] have been trapped and manipulated. The laser beam used in most of these experiments was equivalent to a Gauss-Laguerre LG_{03} mode, a *doughnut* beam. When a doughnut beam is used for trapping, the laser intensity is concentrated in a ring of light: this means that the radiation force along the beam axis is less than for a TEM_{00} mode, and the radiation trapping is comparatively stronger.

As well as linear momentum, angular momentum is transferred from light to absorbing particles during trapping. Both *orbital* angular momentum due to the helical wavefront of a LG_{03} mode [19,21,23] and *spin* angular momentum due to the polarisation of the light [19,23,24] have been observed to set particles into rotation.

In this paper we review the work performed on absorbing particles. Both Gaussian and Laguerre-Gauss modes are used. We outline a model for the trapping of absorbing particles for these beam types based on the effects of transfer of momentum from the beam to the particle. We also consider in some detail the heating effects.

2 Laser Tweezers Experiments

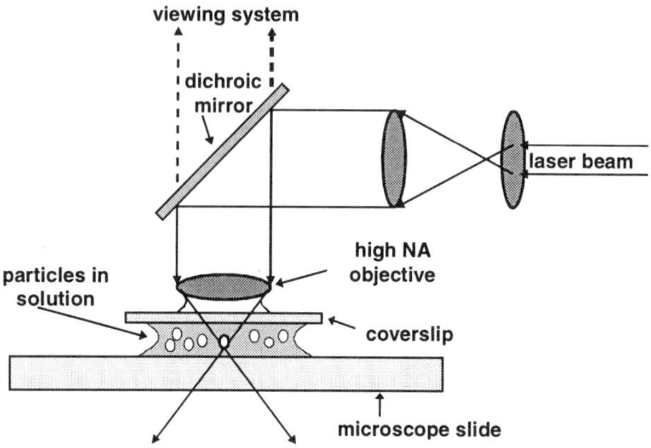

Figure 2: Basic experimental setup for laser tweezers

A dichroic mirror is used to direct the beam into the objective. The particles to be trapped are suspended in a solution, between a glass microscope slide and a coverslip. An oil-immersion objective is used for low distortion and tight focussing of the beam. The laser beam is expanded to fill the objective in order to produce the smallest spot and largest intensity gradient to ensure optimal trapping.

The most commonly used experimental set up (depicted in Figure 2) for the single beam optical trap is optimised for trapping transparent particles and consists of a laser light source directed into a port of an optical microscope [25]. A variety of lasers can be used for trapping, ranging from He-Ne lasers, CW NdYAG lasers, Ti Sapphire, Ar ion lasers to diode laser sources [26-33]. A single mode laser beam is introduced into the microscope in such a way as not to interfere with normal microscope function. It is brought to a tight focus at (or near) the specimen plane, usually with a high numerical aperture (NA $\geq$ 1) oil immersion 100$\times$ objective lens. As microscope optics are designed to minimise aberrations near the specimen plane, arranging the optics in such a way that the trap is parfocal with the specimen allows trapped objects to be visualised and improves the quality of the trap. High numerical aperture is essential to maximise the light intensity gradient near the focal plane and ensure stable trapping in the axial direction. The diameter of the laser beam is normally expanded just before being introduced into the microscope using a set of appropriate lenses. This is done in order to ensure exactly filling, or somewhat over-filling the back pupil of the objective lens. In this way minimum focal spot size and maximum intensity gradient, and hence the strongest trapping force on an object to be trapped and manipulated is achieved. The expanded beam is deflected to the objective by a dichroic mirror. The spot size of the laser beam is of the order of 1-2 μm at the focal plane of the microscope.

Particles to be trapped are placed between a microscope slide and coverslip. Depending on the type of particle, the solution in which they are kept varies (water, kerosene etc). The optical trapping is observed using a CCD camera mounted on the microscope.

In most experiments the trap has to be moved with respect to the specimen. This can be done by either moving the specimen or by moving the beam [34]. A specimen can be positioned in the x-y plane by moving the microscope stage in the conventional way, leaving the trap fixed on the optic axis. When small precise displacements are needed, the sample can be mounted on an x-y piezoelectric stage which is computer controlled. Movement of the trap in the z-direction (depth) is achieved by focusing the microscope up or down, taking advantage of the parfocality of the trap and specimen. The z-direction movement can also be made in extra fine steps if either the sample or the objective is placed on a vertical piezoelectric element. A change in the vertical position of the trap with respect to the specimen plane can be achieved by moving an external lens which controls the beam divergence. A large number of experiments involving the trapping of transparent objects with refractive indices higher than the surrounding medium have been performed using the above described set-up [for reviews, see 35-38].

For transparent (non-absorbing) objects, the strongest trapping is achieved at the beam waist which is near the object plane of the microscope lens and therefore trapped particles are in focus when viewed. The forces acting on the trapped particle have been studied by calibrating against viscous drag exerted by fluid flow using escape-force methods [25,26,29,39-41]. The force can also be measured as a function of displacement from the trap centre. In this way, the trap stiffness, when the particle in a single-beam gradient optical trap is modelled as a mass in a 3-D (ellipsoidal) harmonic potential well [42-44], can be determined.

2.1 Absorbing Particles

It was considered for some time, that a strongly absorbing particle with a high complex index of refraction should be impossible to trap using a Gaussian beam. Such a particle would get pushed out of the beam. Therefore, in a number of experiments hollow beams of different types have been used to achieve trapping of absorbing or reflecting particles. In one of the experiments the trapping was achieved by scanning a beam in a circle using galvanometer mirrors and in this way producing a hollow beam - an intensity minimum surrounded by bright circle. Successful trapping was obtained with a particle being confined in the intensity minimum region of the beam [45]. It has been also demonstrated that optical *levitation* of metal particles could be obtained using a TEM^*_{01} mode laser beam [4].

Recently, we have shown in a series of experiments that two-dimensional optical trapping of highly absorbing particles can be achieved using a Gauss-Laguerre LG_{03} mode laser beam [20-22,44]. In these experiments, particles were trapped against a microscope slide in the converging beam before the waist where the radial component of momentum at any position is directed radially inwards.

2.2 Doughnut Beams

Beams of this type contain a phase singularity which is defined as a point in an optical field around which the phase of the field changes by an integer multiple of 2π. The integer is denoted by l and called the topological *charge* of the phase singularity. The charge l is the *azimuthal mode index* of a Gauss-Laguerre beam. At the singularity the phase is undefined and it appears as a dark spot on a bright background. Beams containing phase singularities can be produced in a variety of ways such as by transformation of Hermite-Gaussian modes, or using cooperative mode locking of a laser. We have shown earlier that beams containing phase singularities can be conveniently produced using computer generated holograms with high efficiency [22]. A phase singularity hologram is similar to a grating except that it has a defect in the centre of the pattern. Using computer graphic techniques, a blazed interference pattern with a central defect can be produced. This is followed by photoreduction of the pattern onto a film which results in an amplitude hologram. The photoreduced patterns are contact printed onto a holographic plate and the developed plate is bleached to produce a phase hologram of high efficiency. In this way we can produce holograms with different order phase singularities. Angular momentum is associated with the helical structure of the wave surrounding a singularity so that a linearly polarised beam with a charge l singularity will carry angular momentum $\hbar l$ per photon.

2.3 Transfer of Angular Momentum

In our experiments using an optical tweezers set-up, only slightly modified compared to the conventional system described above, we have shown that using phase singular fields we can not only trap absorbing particles but also set them into rotation. This demonstrates the transfer of the angular momentum from the light beam to the particles.

The purpose of our experiments was to unambiguously demonstrate transfer of angular momentum, evaluate the resulting rotation speed of the particles and investigate the relationship of the angular momentum associated with the helical structure of the beam to that associated with circular polarisation.

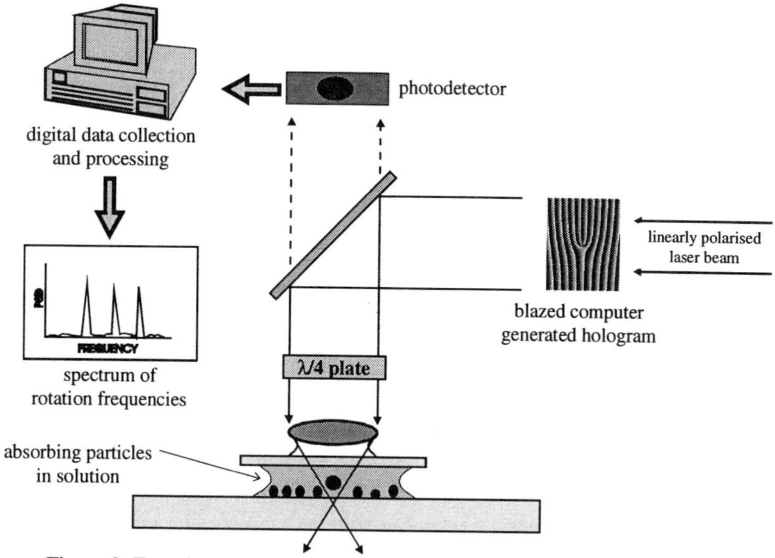

Figure 3: Experimental arrangement for measurement of rotation frequency of absorbing particles

Scattered light from rotating particles is measured with a photodiode. The photodiode signal is digitally sampled, then this data is processed using Fourier analysis to produce spectra of rotation frequencies.

The transfer of angular momentum from light to an absorbing particle can be understood by considering that the light torque arises from the angular momentum carried by photons. The Gauss-Laguerre modes can be seen as eigenmodes of the angular momentum operator L_z and as such have an *orbital* angular momentum $\hbar l$ per photon. If the Gauss-Laguerre mode is circularly polarised we can then also assign it *spin* angular momentum of $\sigma_z\hbar$ per photon, where σ_z is ± 1 (or zero for linear polarisation). So in a paraxial approximation, considering the total number of photons absorbed per second, the torque due to both the polarisation and the helical Poynting vector of the Gauss-Laguerre mode is given by

$$\Gamma = \frac{P_{abs}}{\omega}\left(l + \sigma_z\right) \tag{1}$$

where P_{abs} is the power absorbed by the particle and ω is the frequency of light. Barnett and Allen [46] have developed a general nonparaxial theory according to which the torque will be given by

$$\Gamma = \frac{P_{abs}}{\omega}\left\{\left(l+\sigma_z\right)+\sigma_z\left(\frac{2kz_r}{2p+l+1}\right)^{-1}\right\} \tag{2}$$

where k is the wave number, p and l are mode indices and z_r is a length term, which in the paraxial limit is associated with the Rayleigh range. However it can be shown that the mixed term in the above equation becomes significant only when the beam is very strongly focussed. Even in the experimental situation, when the Gauss-Laguerre LG_{03} mode is used and is focussed to approximately 2 μm size waist, the contribution of this term can be neglected.

The experimental set-up we used for studies of trapping of absorbing particles is similar to the one described in the earlier part of this paper with only a few modifications : A Gaussian TEM_{00} beam from a 17 mW He-Ne laser is passed through a computer generated phase hologram, which produces a Gauss-Laguerre LG_{03} mode in the far field. This plane polarised, helical doughnut beam is then introduced in the usual way into the microscope (see Figure 3).

The trapping of absorbing particles was performed using irregular black ceramic particles and CuO particles dispersed in kerosene. We have also trapped slightly absorbing latex spheres which were in clumps so that their rotation could be more easily observed.

We found that most efficient trapping was achieved with the particles slightly above the focal plane of the doughnut beam. A variety of experiments was performed. Firstly, trapping and rotation could be observed using a CCD camera fitted to the microscope. In this set of studies the absorbing particles were trapped when the hologram of charge 3 was placed in the beam. A particle was trapped and observed to rotate. The rotation could be observed for a long period of time. Subsequently the hologram was moved sideways so that the particle was illuminated by a Gaussian beam and stopped rotating. As the hologram used here is blazed the sign of the doughnut can be simply reversed by turning the hologram around. In principle, on the reversal of the hologram, the particle should rotate in the opposite direction. However, in the process of turning the hologram, the particle is no longer trapped and so on the reversal of the hologram, it is not certain that the retrapped particle is the same as before. It is, however, certain that on the reversal of the hologram, all of the trapped particles rotate in the opposite direction. Another method of reversing the helicity of the doughnut beam is to introduce a Dove prism into the beam path between the hologram and the microscope. As the beam undergoes one reflection in the prism, its helicity is reversed and the prism can be adjusted in such a way so that the beam will be undeviated. With the Dove prism in the beam path, we now trap the particle, detect its direction of rotation and then quickly remove the Dove prism so that the same particle is still trapped. The direction of rotation is reversed [19,47]. The sequence of events is recorded and reproduced frame by frame (see Figure 4).

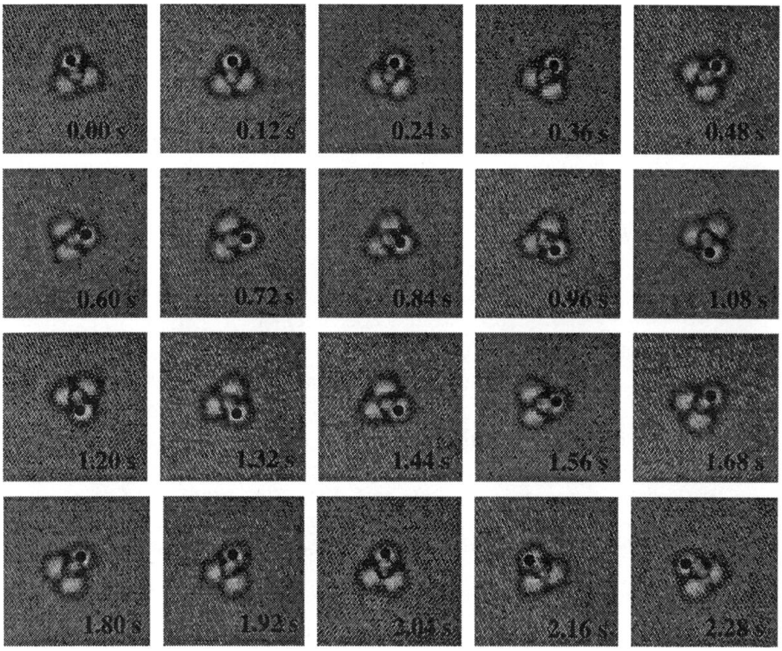

Figure 4: Reversal of rotation of particles trapped in a helical beam

The helicity of the beam is reversed at time 1.20s by removing a Dove prism from the beam path, causing the rotation of the clump of weakly absorbing polystyrene spheres to be reversed.

This method of detection has serious limitations, as particles move out of clear focus when they are trapped and so their images are not well defined on video. In the second set of experiments the experimental set-up was refined in two ways. Firstly, a quarter wave plate was introduced into the beam path directly before the 100× microscope objective (NA = 1.3) which, as before, focussed the doughnut beam to a waist size of approximately 2 μm in diameter. The purpose of this experiment was to allow a precise measurement of the rotation speed and also to study the influence of the added circular polarisation of the beam by which the theoretical prediction of the total transfer of angular momentum could be verified. A second refinement was made to allow for precise measurement of the rotation frequency. The rotation was this time measured using a photodetector positioned off centre to detect a portion of the light reflected from the rotating particle. The particles are irregularly shaped and the protruding parts of the particles reflect a "flash" of light onto the displaced detector. The signal of intensity fluctuations over time obtained from the photodetector is Fourier transformed yielding the rotation frequency of the particle. The modified experimental set-up for the above experiments is shown in Figure 3. The sequence of the experiment is as follows: The LG_{03} beam enters the microscope. It passes through a $\lambda/4$ plate and its focus is adjusted to be slightly below the focal point of the microscope. The polarisation state is checked after the objective lens by an analyser and the power of the beam is monitored for each position of the plate (left and right-circularly polarised light and linearly polarised light). It was established that the power output varied less than 1% between different polarisations. Care was taken to position the $\lambda/4$ plate in such a way as not to cause any deflection of the beam when the plate was rotated. The sample of absorbing CuO particles of sizes up to 20 μm in kerosene was placed between a glass microscope slide and coverslip. A particle was trapped into the linearly polarised helical doughnut and the $\lambda/4$ plate rotated firstly to right circular polarisation, back to linear and then to left circular polarisation. The signal was sampled at 20 Hz for a period of 100 s. The sequence was repeated and the reflected light monitored on the photodetector. An example of the spectrum resulting from this experiment is shown in Figure 5. It can be seen from this figure that rotation frequency increases when the helicity

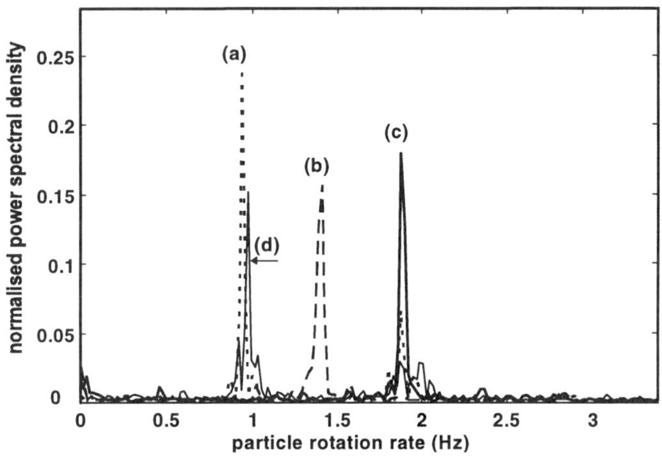

Figure 5: Rotation speeds for absorbing particles trapped by a helical doughnut beam when the polarisation state of the trapping beam is varied

Power spectra obtained by Fourier analysis of the photodiode signal obtained from light scattered off rotating CuO particles. The peaks at (a), (b), and (c) represent the measured rates of rotation of CuO particles trapped in a left-circularly polarised right-helical, plane polarised right-helical, and right-circularly polarised right-helical doughnut beam respectively. The peak at (d) is the 4th set of data, taken with left-circularly polarised right-helical light, in order to verify that this rotation rate did not vary during the experiment. The particle's rotation rates are approximately in the ratio 2:3:4 .

of the electric field vector has the same direction as the helicity of the Poynting vector and decreases when these directions are opposite to each other.

Using an LG_{03} doughnut the theoretical prediction is that the frequencies should scale as 2:3:4 when going from opposite helicity Poynting vector and spin to linearly polarised helical wave front through to the same helicities. This was verified experimentally (see Figure 5).

Recently a somewhat similar experiment has been performed by Simpson et al [23] using a charge one Gauss-Laguerre laser mode produced by operating a laser in a Hermite-Gaussian mode and converting the output into the corresponding Gauss-Laguerre mode using a cylindrical lens mode converter. A quarter-wave plate was used to change polarisation. The authors showed that spin angular momentum of $\pm\hbar$ per photon associated with circularly polarised light can add to, or subtract from, the orbital angular momentum and observed the mutual cancellation of the spin and orbital angular momentum. Their results confirm the

(a)

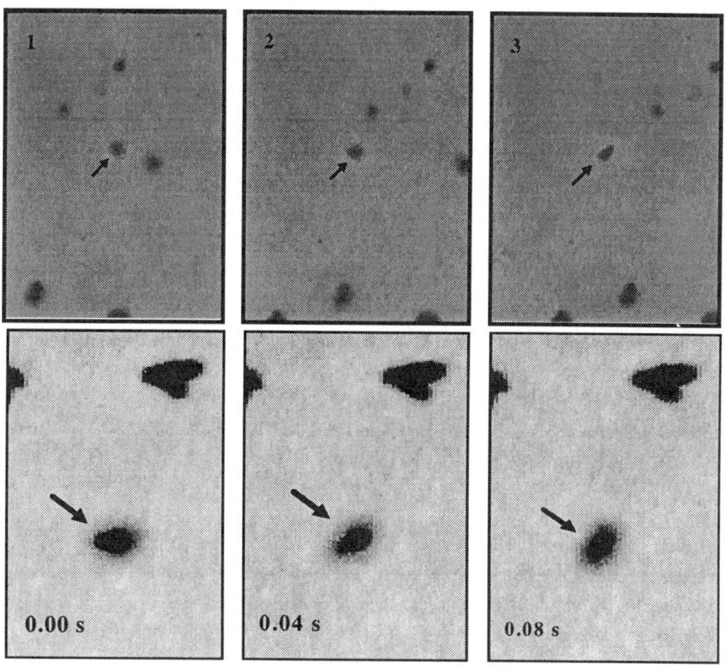

(b)

Figure 6: CuO particle trapped in a Gaussian beam

Frames from a video recording CuO particles being trapped, moved and rotated using a Gaussian beam. The arrow points to the trapped particle. (a) shows motion with respect to the surrounding particles and (b) shows rotation produced by circular polarisation of the beam.

results obtained by our group and show that the LG_{01} mode of charge 1 has a well defined orbital angular momentum corresponding to $\hbar$ per photon.

In our most recent experiments we have shown that an absorbing particle can in fact be trapped by a Gaussian beam. Two-dimensional trapping can be achieved before the waist of a tightly focussed Gaussian beam, where the spot size is rapidly changing, against a surface such as a microscope slide, using radiation pressure. The particle was first trapped using linearly polarised Gaussian beam and subsequently, by turning a $\lambda/4$ plate, a circularly polarised Gaussian beam was used to trap the particle, resulting in a rotation of this particle. When the $\lambda/4$ plate was rotated to produce circularly polarised light of the opposite sense, the particle reversed its direction of rotation. Successive frames from a video demonstrating trapping are shown in Figure 6.

3 Theory of Trapping of Absorbing Particles

The physical basis of the trapping of absorbing particles can be investigated in terms of the interaction between radiation and matter. For highly absorbing particles, effects such as induced polarisation within the particle, refraction and Doppler shifts which are important for transparent particles and atoms can be neglected. The properties of the trap depend mainly on the properties of the trapping beam, which will be assumed to be paraxial. While a non-paraxial beam is necessary for laser trapping, the use of the paraxial approximation does not seem to introduce excessive error. Errors due to aberration within the optical system are expected to be greater.

The theory will be developed in terms of the two most common types of beams used for trapping particles, namely TEM_{00} Gaussian beams (denoted (G)) and Laguerre-Gaussian "doughnut" beams (denoted (LG)) described by a radial mode index p and an azimuthal mode index l.

3.1 The Transfer of Linear Momentum and Orbital Angular Momentum

We begin with consideration of the transfer of linear momentum to an absorbing infinitesimal element. This microscopic behaviour can then be extended to the particle as a whole. In the course of this, it can be seen that the transfer of orbital angular momentum from a helical beam to a particle is a simple radiation pressure process involving the transfer of linear momentum to different portions of the particle. Due to the spatial structure of the helical beam, this results in a transfer of angular momentum.

3.2 Microscopic transfer of momentum

Due to the cylindrical symmetry of the beam, it is convenient to use a cylindrical coordinate system, with radial, azimuthal and axial coordinates r, ϕ and z and corresponding unit vectors $\hat{\mathbf{r}}$, $\hat{\phi}$, and $\hat{\mathbf{z}}$. The linear momentum flux of a laser beam is given by the time-averaged Poynting vector **S** (in cylindrical coordinates) [48]:

$$\mathbf{S} = \frac{c\varepsilon}{2} E_0^{\,2} \left(\frac{zr}{z_r^{\,2} + z^2} \hat{\mathbf{r}} + \hat{\mathbf{z}} \right) \text{ (G)} \tag{3}$$

$$\mathbf{S} = \frac{c\varepsilon}{2} E_0^{\,2} \left(\frac{zr}{z_r^{\,2} + z^2} \hat{\mathbf{r}} + \frac{l}{kr} \hat{\phi} + \hat{\mathbf{z}} \right) \text{ (LG)} \tag{4}$$

where z_r is the Rayleigh range, k is the wavenumber of the beam and E_0 is the amplitude of the beam, given by [49]

$$E_0 = \sqrt{\frac{c\varepsilon}{2}} \sqrt{\frac{2P}{\pi}} \frac{1}{w(z)} e^{\frac{-r^2}{w^2(z)}} \text{ (G)} \tag{5}$$

$$E_0 = \sqrt{\frac{c\varepsilon}{2}} \sqrt{\frac{2p!P}{\pi(l+2+p)!}} \left(\frac{r\sqrt{2}}{w(z)} \right)^l L_p^l \left(\frac{2r^2}{w^2(z)} \right) \frac{1}{w(z)} e^{\frac{-r^2}{w^2(z)}} \text{ (LG)} \tag{6}$$

where P is the beam power, $w(z)$ is the beam width, L_p^l is the generalised Laguerre polynomial, c is the speed of light and ε is the permittivity. The Rayleigh range z_r and the beam width $w(z)$ are related to each other and the beam waist spot size w_0 by:

$$w^2(z) = \frac{2(z_r^{\,2} + z^2)}{kz_r} \tag{7}$$

$$w(z) = w_0 \sqrt{1 + z^2/z_r^{\,2}} \tag{8}$$

$$z_r = kw_0^{\,2}/2 \tag{9}$$

For most cases of interest involving doughnut beams, the radial mode index $p = 0$ and the Laguerre polynomial $L_p^l = 1$. If an area element **dA** is highly absorbing, the rate of momentum transfer to it will be given by the Poynting vector:

$$\mathbf{F} = \frac{1}{c} \mathbf{S} \cdot (-\mathbf{dA}) \frac{\mathbf{S}}{|\mathbf{S}|} \tag{10}$$

The transfer of momentum from the beam to an absorbing particle is therefore straightforward compared to other cases, such as transparent particles and atoms. For transparent particles, refraction and induced polarisation must be taken into account. For an atom, the frequency dependence of the absorption and spontaneous emission must be considered, while for an absorbing particle, the absorption can be assumed to be independent of frequency, and inter-atomic collision rates within the particle can be assumed to be high enough to cause de-excitation without re-emission.

3.3 Macroscopic effects

As particles typically are sufficiently large that the Poynting vector of the laser beam is not constant over their absorbing surfaces, it is necessary to integrate

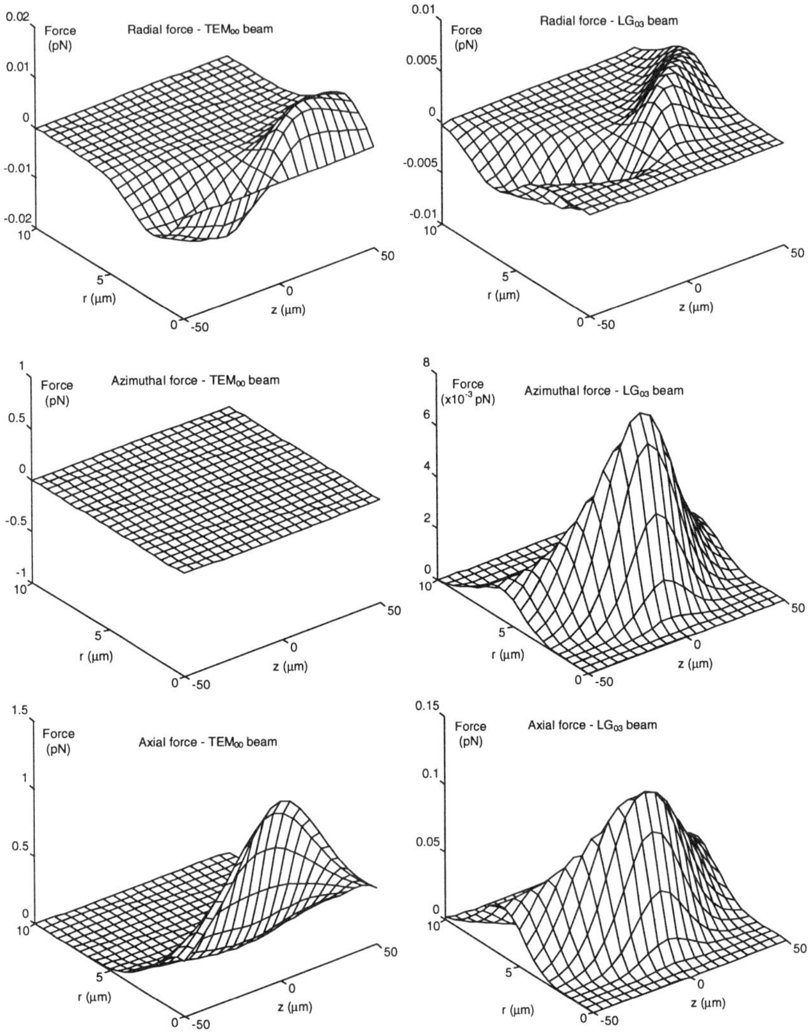

Figure 7: Spatial dependence of optical force on an absorbing particle

The radial and axial variation of the optical force is shown for both a TEM$_{00}$ Gaussian beam and an LG$_{03}$ Laguerre-Gaussian beam. Both beams have the same power (1 mW), spot size (2 μm) and wavenumber (free space wavelength 632.8 nm). The particle has a circular cross-section of radius 1 μm. Due to the cylindrical symmetry, there is no azimuthal variation of the force. The beam is propagating in the +z direction, with the beam waist at z = 0.

the Poynting vector over the surface. This requires knowledge of the geometry of the particle and the beam. This calculation can be readily performed numerically by dividing the particle into a number of small area elements **dA** over which the Poynting vector is approximately constant, and integrating equation (10). Where the product $\mathbf{S}\cdot(-\mathbf{dA}) < 0$, the particle will not be illuminated, so such regions can be neglected. A useful and simple approximation is to represent the particle as a disk with the same cross-section (essentially applying the paraxial approximation to the illumination of the particle). The resultant force on a particle is shown in Figure 7.

Equilibrium points in these force fields exist only where the force is zero, and a particle can be trapped only at a stable equilibrium point. Thus, it can be readily seen that a particle cannot be axially trapped without an external force (such as gravity, viscous drag due to convection, or a reaction force due to the particle resting on the bottom of the trapping cell) acting on it. The particle can be trapped radially in the portion of the beam which is converging towards the beam waist. The particle cannot be trapped after the beam waist, as the convergence of the beam required for trapping no longer exists. The absorbing particle trap can therefore be considered to be a two-dimensional trap. Also, it should be noted that in the case of Laguerre-Gaussian beams, angular momentum is transferred to the particle as a whole, although microscopically only linear momentum is seen to be present.

The relative efficiencies of trapping absorbing particles using Gaussian beams and doughnut beams can be measured in terms of the ratio of the radial force to the axial force. As can be seen from Figure 8, the doughnut beam trap has smaller axial forces for trapped particles.

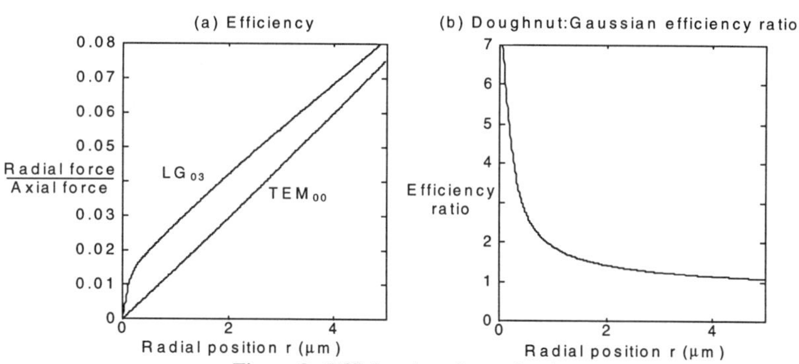

Figure 8: Efficiencies of trapping beams

The efficiency of the trap can be measured by the ratio of the radial force to the axial force. These ratios are calculated here for beams of waist size 2 μm and power of 1mW. The particle has a radius of 1 μm and is in a plane 30 μm before the waist.

As the force acting on a particle at any position within the beam can be calculated, the overall properties of the trap, such as the trapping force, or the motion of trapped particles can be determined. If particles are confined to a two-dimensional plane in which they are radially trapped, particles trapped in Gaussian beams are pushed into the centre of the trap, while particles trapped in Laguerre-Gaussian beams also tend to orbit about the beam axis due to the azimuthal component of the force. This orbital motion can lead to instability in the trap, as a particle undergoing such motion will only remain within the trap if the radial force can provide a centripetal force sufficient to force the path to become circular. If the viscosity of the surrounding medium is too low, the terminal orbit speed will be high enough so that the radial force will be unable to provide the necessary centripetal acceleration and the particle will escape from the trap. Particles trapped in fluids such as water or kerosene, however, have very low terminal speeds and can be trapped.

4 Transfer of Spin Angular Momentum

As seen above, any orbital angular momentum present in a helical laser beam (for example, a Laguerre-Gaussian beam) is transferred to a particle by the same mechanisms as linear momentum. All laser beams, however, can carry angular momentum if the beam is circularly polarised, with each photon having angular momentum of magnitude $\hbar$. The rate of absorption of angular momentum by a small section of the particle is given in terms of the Poynting vector **S** and the wavenumber k by

$$\tau = \frac{\sigma_z}{k}\mathbf{S}\cdot(-\mathbf{dA})\frac{\mathbf{S}}{|\mathbf{S}|} \tag{11}$$

where $\sigma_z = \pm 1$ for left- and right-circular polarisation and $\sigma_z = 0$ for plane polarisation.

The total torque on the particle due to polarisation can then be found by integrating equation (11) over the absorbing surface of the particle (see Figure 9). The rotational behaviour of particles trapped in a Gaussian beam is straightforward. However, that of a particle illuminated by a Laguerre-Gaussian beam is more complex, due to the presence of the orbital angular momentum. In practice a particle is trapped on the beam axis. The total torque due to polarisation on a particle of radius r trapped on the beam axis is given by

$$\tau_p = \frac{\sigma_z P}{\omega}\left(1 - e^{\frac{-2r^2}{w^2(z)}}\right) \quad \text{(G)} \tag{12}$$

$$\tau_p = \frac{\sigma_z P}{\omega}\left(1 - e^{\frac{-2r^2}{w^2(z)}}\sum_{k=0}^{l+2}\frac{1}{k!}\left(\frac{2r^2}{w^2(z)}\right)^k\right) \quad \text{(LG)} \tag{13}$$

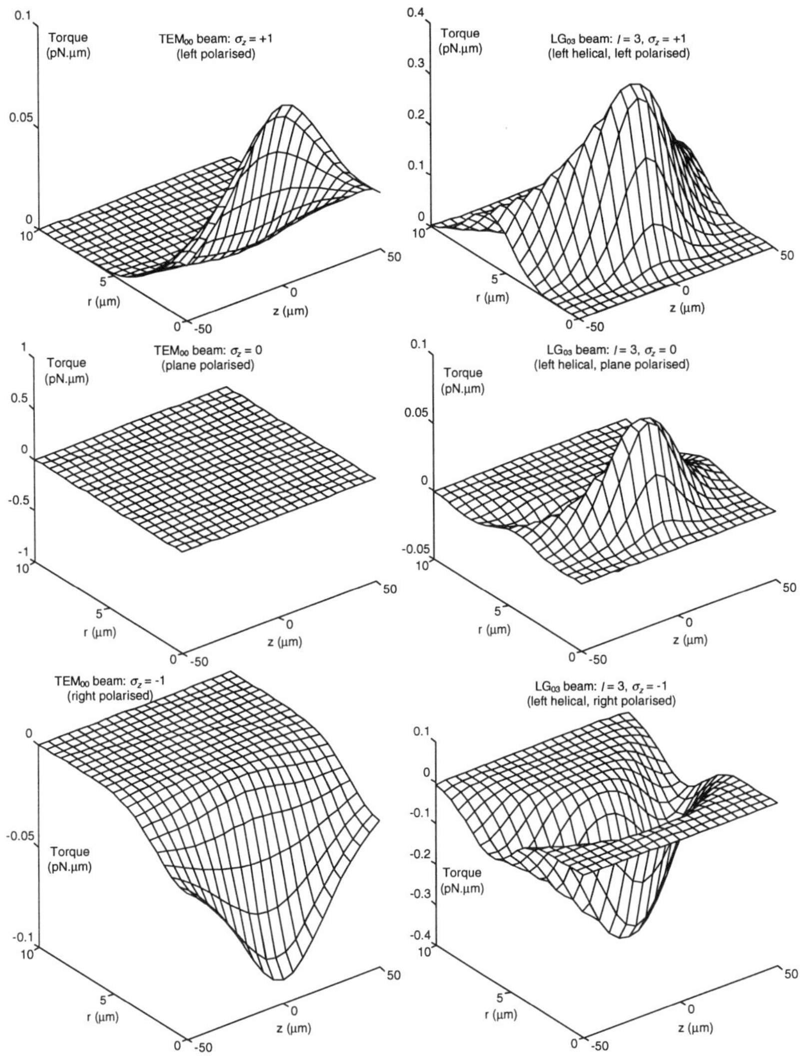

Figure 9: Spatial dependence of torque parallel to beam axis on absorbing particle

The torque experienced by a 1 μm radius particle through absorption of a combination of spin and orbital angular momentum is shown for both a Gaussian TEM_{00} beam and a left-helical LG_{03} doughnut beam. The beams have a waist size of 2 μm, a power of 1 mW, and a free space wavelength of 632.8 nm. Note that the torque acting on a particle trapped on the axis of the LG_{03} beam is positive for all cases, but is small compared to the torque on a particle in the high intensity region of the beam.

where ω is the angular frequency of the light. The total torque can be found by combining this with the torque due to the orbital angular momentum carried by the helicity of the beam

$$\tau_o = 0 \text{ (G)} \tag{14}$$

$$\tau_o = \frac{lP}{\omega}\left(1 - e^{\frac{-2r^2}{w^2(z)}} \sum_{k=0}^{l+2} \frac{1}{k!}\left(\frac{2r^2}{w^2(z)}\right)^k\right) \text{ (LG)} \tag{15}$$

The angular velocity Ω in a viscous medium will depend on the drag torque, which for a smooth spherical particle of radius r in a medium of viscosity η is given by [50]

$$\tau_d = -8\pi\eta r^3\Omega \tag{16}$$

and the optical torque. The resultant spin rates for particles trapped on the beam axis are

$$\Omega = \frac{\sigma_z P}{8\pi\eta r^3\omega}\left(1 - e^{\frac{-2r^2}{w^2(z)}}\right) \text{ (G)} \tag{17}$$

$$\Omega = \frac{(\sigma_z + l)P}{8\pi\eta r^3\omega}\left(1 - e^{\frac{-2r^2}{w^2(z)}} \sum_{k=0}^{l+2} \frac{1}{k!}\left(\frac{2r^2}{w^2(z)}\right)^k\right) \text{ (LG)} \tag{18}$$

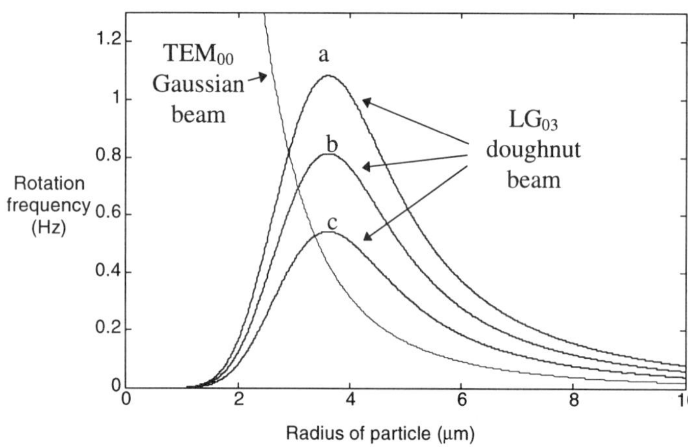

Figure 10: Spin rates of trapped spherical absorbing particles in kerosene

A (two-dimensionally) trapped particle will experience a torque dependent on its size and the beam power. This torque will depend on the position along the beam axis where the particle is trapped and on the direction of circular polarisation of the beam. The beams here are of width 2 μm and power 1 mW. The three cases for particles trapped by a LG_{03} doughnut beam are (a) polarisation and helicity in the same direction, (b) plane polarisation, and (c) polarisation and helicity in opposite directions.

The rotation rate for particles trapped in a Gaussian beam rises as the particles become small. When the spin rate becomes large, equation (16) will cease to be applicable. For particles trapped in the doughnut beam, the rotation rate is maximum for a particle large enough to absorb most of the beam, but small enough to keep the drag surface area small.

In a typical experiment, ceramic particles with radii of about 2 μm, suspended in kerosene, were trapped in a beam of 4 mW power and a beam width of 1.75 μm. Rotation rates of 1-2 Hz were observed (see Figure 5), consistent with the predictions of equation (18) and Figure 10.

5 Absorption of Energy

The energy density in a focussed beam of even modest power can be very large. For example, in a 1 mW beam focussed to a 1 μm waist the irradiance will exceed 10^9 W/m^2. The transparent objects conventionally trapped by optical tweezers evidently absorb very little, partly because they are very thin, and they are surrounded by a conducting medium so experience only modest rises in temperature. Indeed it has been shown that bacteria will survive and even reproduce while trapped.

The situation for strongly absorbing particles is very different. A typical 1 μm radius particle absorbing 1 mW will, in the absence of any losses, experience a temperature rise of order 10^8 K/s. Clearly the temperature of a trapped particle will rise rapidly and it is necessary to consider what losses will limit the rise.

Since the surface area is small, even neglecting local absorption in the surrounding medium, the amount of heat which can be radiated is limited. Stefan's Law shows that a particle of 1 μm radius needs to have a surface temperature in excess of 1000 K before the power radiated exceeds 1 μW. The small surface area also severely limits the rate at which heat can be exchanged with the surrounding fluid.

If we consider a spherical particle in equilibrium losing the heat absorbed from the beam by isotropic conduction into a surrounding medium of thermal conductivity k, the radial variation of the temperature T in the surrounding medium can easily be shown to be given by

$$T = \frac{P_a}{4\pi rk} + T_0 \qquad (19)$$

where P_a is the total power absorbed by the particle from the beam and T_0 is the ambient temperature. The absorbed power for a particle with absorption coefficient α trapped on the beam axis of a TEM_{00} Gaussian or LG_{0l} Laguerre-Gaussian beam of power P is given by

$$P_a = \alpha P\left(1 - e^{\frac{-2r^2}{w^2(z)}}\right) \quad \text{(G)} \qquad (20)$$

$$P_a = \alpha P\left(1 - e^{\frac{-2r^2}{w^2(z)}} \sum_{k=0}^{l+2} \frac{1}{k!}\left(\frac{2r^2}{w^2(z)}\right)^k\right) \text{(LG)} \tag{21}$$

Calculated equilibrium temperatures reached by typical particles trapped in a 1 mW beam are shown in Figure 11.

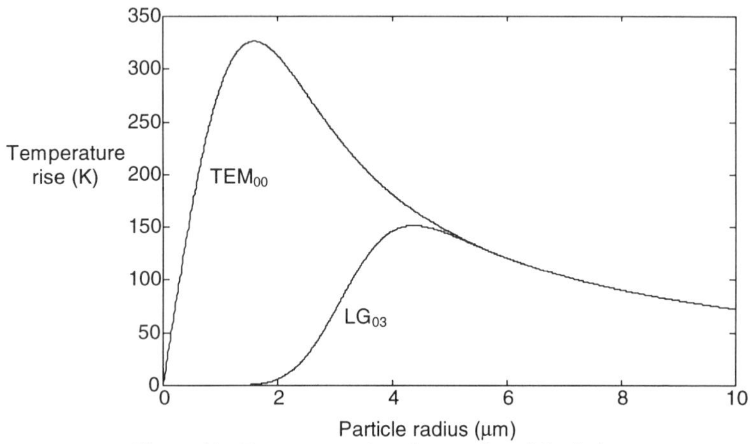

Figure 11: Temperature rise of trapped particles in kerosene

The temperature rise due to power absorbed from a laser beam is shown as a function of particle size. The trapping beams have a power of 1 mW and a width of 2 μm where the particle is trapped. Small particles absorb little energy from the beam while very large particles lose significantly more heat through conduction due to their large surface area.

Quite high temperatures are predicted, especially for particle sizes on the order of the beam spot size.

Convection is another possible means of heat loss. The importance of convection can be found in terms of the mean Nusselt number $\overline{Nu}$, which, at low flow rates, is given in terms of the Prandtl number Pr (the ratio between the momentum and thermal diffusivities) and the Grashof number Gr (the ratio between buoyant and viscous forces) by [51]

$$\overline{Nu} = 2 + Gr + Gr^2\left(0.139 - 0.419\,Pr + 1.1902\,Pr^2\right) \tag{22}$$

The Grashof number is given by [52]

$$Gr = g\left(1 - \rho_T/\rho_\infty\right)D^3/v^2 \tag{23}$$

and the Prandtl number by

$$Pr = c_p\eta/k \tag{24}$$

where c_p is the specific heat of the fluid, η is the dynamic viscosity and ν is the kinematic viscosity.

For a temperature difference of 100K, an ambient temperature of 300K, and a particle diameter of D = 3 μm (at which size high temperatures are reached for a typical beam), $\overline{Nu} = 2 + 3\times10^{-4}$ for water and $\overline{Nu} = 2 + 7\times10^{-5}$ for kerosene, values which are very close to $\overline{Nu} = 2$ for pure conduction. The convective contribution to the cooling is therefore negligible.

We are therefore driven to the conclusion that even in the presence of a liquid medium, an absorbing particle will experience a large rise in temperature, probably limited by thermal conduction. Note that due to the $1/r$ dependence of the temperature rise in that case, the volume where elevated temperatures will be found is quite small which means in turn that the thermal relaxation time will be short, so that the temperature rises will be difficult to measure.

However, we do have evidence of such large temperature rises. In experiments where we trapped small particles of photocopy toner, we observed that initially jagged fragments almost instantly became smooth when irradiated by a beam with a power of a few mW. According to manufacturer's specifications, the fusing point of the toner is 185°C. These results are at first rather surprising as it might have been expected that boiling of the liquid would enhance cooling or disrupt the trapping process. Sometimes, with beam powers well in excess of that required to melt toner, a bubble is generated but it is not clear if this is vapour or merely absorbed gases being evolved from the particle. The boiling process in general is a somewhat mysterious one and it is widely accepted that bubble formation requires some seed to overcome surface tension forces which scale inversely as the radius. Therefore it cannot be taken for granted that normally observed behaviour will scale to very small dimensions. If so, further study of superheated particles may help shed new light on the fundamentals of the boiling process.

6 Interaction with the Medium

Interactions with the surrounding transparent medium include the heating of the medium through conduction from the absorbing particle, and the generation of fluid motions through stirring and convection currents.

Temperature changes in the medium will be localised to within 100 μm of the particle but could affect properties such as the refractive index and viscosity. It is also possible that thermophoretic effects may occur in the strong temperature gradients formed.

Stirring and convection effects have both been observed when CuO particles were trapped, using a doughnut beam, in water to which a little detergent had been added. This greatly reduced the tendency of particles to stick to the slide and generally freed up their motion so that Brownian motion was more evident.

When a particle was trapped, others in its vicinity began to move toward it, presumably as a result of a convection currents, and began to circle around the trapped particle in the same direction as its rotation, presumably carried around by a flow generated by the rotation of the trapped particle. The fluid velocities of the order of 10 μm/s are too small to significantly affect heat transport.

7 Conclusion

It has been shown that the converging section of a focussed laser beam can radially trap absorbing particles against a supporting substrate using a Gaussian or doughnut beam. Doughnut beams with phase singularities carry angular momentum which is also transferred to the trapped particle causing it to rotate. Angular momentum resulting from circular polarisation of the light can also be transferred. Energy absorbed from the beam can lead to rapid heating and high equilibrium temperatures, even well in excess of the normal boiling point of the surrounding medium.

References

[1] A. Ashkin *Physical Review Letters* **24**, 156 (1970)
[2] A. Ashkin *Science* **220**, 4474 (1980)
[3] A. Ashkin, J.M. Dziedzic, J.E. Bjorkholm and S. Chu *Optics Letters* **11**, 288 (1986)
[4] G. Roosen and C. Imbert *Optics Communications* **26**, 432 (1978)
[5] W.H. Wright, G.J. Sonek, Y. Tadir and M.W. Berns *IEEE Journal of Quantum Electronics* **26**, 2148 (1990)
[6] J.G. Mitchell, R. Weller, M. Beconi, J. Sell and J. Holland *Microbial Ecology* **25**, 113 (1993)
[7] R. Huber, S. Burggraf, T. Mayer, S.M. Barns, P. Rossnagel and K.O. Stetter *Nature* **376**, 57 (1995)
[8] K. Schütze *Nature* **368**, 667 (1994)
[9] K. Schütze, A. Clementsengewald and A. Ashkin *Fertility and Sterility* **61**, 783 (1994)
[10] Y. Liu, G.J. Sonek, M.W. Berns, K. Konig and B.J. Tromberg *Optics Letters* **20**, 21 (1995)
[11] S.M. Block, D.F. Blair and H.C. Berg *Nature* **338**, 514 (1989)
[12] T. Nishizaka, H. Miyata, H. Yoshikawa, S. Ishiwata and K. Kinosita Jr *Biophysical Journal* **68**, 75s (1995)
[13] T.T. Perkins, D.E. Smith and S. Chu *Science* **264**, 819 (1994)
[14] Y. Liy, D.K. Cheng, G.J. Sonek, M.W. Berns, C.F. Chapman and B.J. Tromberg *Biophysical Journal* **68**, 2137 (1995)
[15] H. Liang, K.T. Vu, T.C. Trang, D. Shin, S. Kimel and M.W. Berns *Biophysical Journal* **70**, 1529-1533 (1996)

[16] K. König, H. Liang, M.W. Berns and B.J. Tromberg *Nature* **377**, 20 (1995)
[17] K. Svoboda and S.M. Block *Optics Letters* **19**, 930 (1994)
[18] S. Sato, Y. Harada and Y. Waseda *Optics Letters* **19**, 1807 (1994)
[19] M.E.J. Friese, H.He, N.R. Heckenberg and H. Rubinsztein-Dunlop *Transfer of angular momentum to absorbing particles from a laser beam with a phase singularity* in N.B. Abraham and Y.I. Kanin (eds) *Laser Optics '95: Nonlinear Dynamics in Lasers* Proc SPIE 2792 190 -195(1996)
[20] M.E.J. Friese, J. Enger, H. Rubinsztein-Dunlop and N.R. Heckenberg *Physical Review A* **54**, 1593-1596 (1996)
[21] H. He, M.E.J. Friese, N.R. Heckenberg and H. Rubinsztein-Dunlop *Physical Review Letters* **75**, 826 (1995)
[22] H. He, N.R. Heckenberg and H. Rubinsztein-Dunlop *Journal of Modern Optics* **42**, 217 (1995)
[23] N.B. Simpson, K. Dholakia, L. Allen and M.J. Padgett (Submitted to *Optics Letters*)
[24] T. Sugiura, S. Kawata and S. Minami *Journal of the Spectroscopical Society of Japan* **39**, 342 (1990)
[25] A. Ashkin, J.M. Dziedzic and T. Yamane *Nature* **330**, 769 (1987)
[26] S. Sato, M. Ohyumi, H. Shibate and H. Inabe *Optics Letters* **16**, 282 (1991)
[27] L. Malmqvist and H.M. Hertz *Optics Communications* **94**, 19 (1992)
[28] I.A. Vorobjev, H. Liang, W.H. Wright and M.W. Berns *Biophysical Journal* **64**, 533 (1993)
[29] C. D'Helon, E.W. Dearden, H. Rubinsztein-Dunlop and N.R. Heckenberg *Journal of Modern Optics* **41**, 595 (1993)
[30] S. Sato and H. Inabe *Optics Letters* **19**, 927 (1994)
[31] T.N. Buican, M.J. Smyth, H.A. Crissman, G.C. Salzman, C.C. Stewart and J.C. Martin *Applied Optics* **26**, 5311 (1987)
[32] L.P. Ghislain, N.A. Switz and W.W. Webb *Review of Scientific Instruments* **65**, 2762 (1994)
[33] H. Liang, K.T. Vu, P. Krishnan, T.C. Trang, D. Shin, S. Kimel and M.W. Berns *Biophysical Journal* **70**, 1529 (1996)
[34] E. Fällman and O. Axner, to appear in *Applied Optics* (1996)
[35] M.W. Berns, W.H. Wright and R.W. Steubing *International Reviews of Cytology* **129**, 1 (1991)
[36] S.M. Block *Nature* **360**, 493 (1992)
[37] S.C. Kuo and M.P. Sheetz *Trends in Cell Biology* **2**, 116 (1992)
[38] G. Weber and K.O. Greulich *International Reviews of Cytology* **133**, 1 (1992)
[39] A. Ashkin, K. Schütze, J.M. Dziedzic, U. Euteneuer and M. Schliwe *Nature* **348**, 346 (1990)
[40] S.M. Block, D.F. Blair and H.C. Berg *Cytometry* **12**, 492 (1990)
[41] S.C. Kuo and M.P. Sheetz *Science* **260**, 232 (1993)

[42] L.P. Ghislain and W.W. Webb *Optics Letters* **18**, 1678 (1993)

[43] K. Svoboda, C.F. Schmidt, B.J. Schnapp and S.M. Block *Nature* **365**, 721 (1993)

[44] M.E.J. Friese, H. Rubinsztein-Dunlop, N.R. Heckenberg and E.W. Dearden, to appear in *Applied Optics*

[45] K. Sasaki, M. Koshioka, H. Misawa and N. Kitamura *Applied Physics Letters* **60**, 807 (1992)

[46] S.M. Barnett and L. Allen *Optical Communications* **110**, 670 (1994)

[47] http://www.physics.uq.oz.au:8001/lp/tweezers.html

[48] M.J. Padgett and L. Allen *Optics Communications* **121**, 36 (1995)

[49] M.W. Beijersbergen, L. Allen, H.E.L.O van der Veen and J.P.Woerdman *Optics Communications* **96**, 123-132 (1993)

[50] L.D. Landau and E.M. Lifshitz *Fluid Mechanics* 2nd ed., Pergamon Press (1987)

[51] R.C. Armstrong *Heat Exchanger Design Handbook Vol 2* Hemisphere Publishing Corporation (1989)

[52] F.P Incropera and D.P. De Witt *Fundamentals of Heat and Mass Transfer* 4th ed., Wiley (1996)

Index

A

Ab initio methods, in quantum theory, 422–423
Absorbing particles
  laser tweezing, 473–474
  optical trapping, 471
    energy absorption, 487–489
    interaction with medium, 489–490
    linear and orbital angular momentum transfer, 480
    macroscopic momentum transfer, 481–484
    microscopic momentum transfer, 480–481
    momentum transfer from light, 471
    spin angular momentum transfer, 484–487
Absorption, energy in optical trapping, 487–489
Absorption spectroscopy, boron quartet system, 301–302
Abundance spectroscopy, clusters, 245–246
Alkali elements, atomic beam laser spectroscopy, 365–366
Amplification systems, in high-power laser, 210–214
Argon, cluster structure, 247
Argon laser, for optical trapping, 472
Atomic beam laser spectroscopy, experiments at ISOLDE, 365–366
Atomic beam magnetic resonance
  experiments at ISOLDE, 363–365
  Fr isotopes, 373–375
  in hyperfine structure measurements, 343
Atomic beams, experiments at ISOLDE
  atomic beam laser spectroscopy, 365–366
  atomic beam magnetic resonance, 363–365
  fast atomic beams, 367–368
  thermal atomic beams, 363
Atomic ions
  negative, photodetachment, role in physics, 312–313
  trapped, coherent quantum state
    conditional dynamics and single ion controlled not logic gate, 56–58
    entangled states, 45
    Jaynes–Cummings-type coupling between internal and motional states, 46–47
      for trapped $^{9}Be^{+}$ ion, 47–50
    laser cooling to zero-point state, 46
    motional state reconstruction, 53–55
    quantum computation, 58–59
    quantum logic, 55–56, 58–59
    quantum logic application to spectroscopy, 59–60
    quantum registers, 58
    Schrödinger cats, 58
    and Schrödinger cat states of motion, 50–53
    trap types, 42–45
Atomic many-body theory, development, 343
Atomic nuclei, properties, 237
Atomic physics
  Ingvar Lindren's career, 1–4
  nuclear spins and hyperfine interaction constants, 237–238
Atomic quantum theory, semi-empirical and *ab initio*
methods, 422–423
Atomic spectroscopy, for nuclear properties studies, 356–357
Atomiziation, in pulsed laser excitation, 290–291
Atoms
  alkali, hyperfine structure, 343
  alkali clusters, shell closing, 247–249
  electron–nucleus interaction, 336–337

Atoms *(continued)*
excitation state, resonance ionization, 315–316
focusing and trapping, in spectroscopy, 15–16
noble gas, cluster structure, 246–247
in optical lattices, 24–28
Bragg scattering, 28–34
Doppler cooling, 19–20
driven motion, 34–35
sub-Doppler cooling, 20–23
radiative lifetimes, pulsed laser spectroscopy
atomization, 290–291
error sources, 294
excitation, 285
experimental methods, 285
fluorescence decay, 290
short pulse generation, 286–287
transition probabilities, 284
tunable laser radiation generation, 288–289
resonance ionization spectroscopy, 339–340
role of electric hexadecapole, 337
role of electric quadrupole, 337
role of magnetic dipole, 337
role of magnetic octupole, 337
trapped, as optical frequency standard, 451

B

Baryons
collisions, 203
in lead energy levels, 199
Beam–foil studies, boron quartet system, 304–305
Beryllium
hyperfine structure, 448
trapped, Jaynes–Cummings-type coupling, 47–50
Beutler–Fano profile, fitting of resonance energy, 324
Big bang nucleosynthesis model, boron quartet system in, 302
Bismuth ion
effect of magnetic moment, 344
ground-state hyperfine structures, 449
hyperfine structure splitting, 126, 152–157
Bloch equation
approach to solution, 423
derivation, 419–422
and perturbation theory, 427–428
Bloch–Lindgren equation
approach to solution, 423
derivation, 419–422
Boron, quartet system
absorption spectroscopy, 301–302
in astrophysics, 302
beam–foil studies, 304–305
in big bang nucleosynthesis model, 302
flash pyrolysis, 302
multiconfiguration Hartree–Fock approach, 301–302
theoretical calculations, 305–306
in thermonuclear fusion research, 304
Bracketing function, in multi-dimensional partitioning technique, 423
Bragg scattering, from optical lattices, 28–34

C

C_{60}
early proposed structure, 251–252
NMR characterization, 253
production by laser vaporization method, 252
Carbon clusters
C_{60}
early proposed structure, 251–252
NMR characterization, 253
production by laser vaporization method, 252
cyanopolyynes, production, 251
geometrical structure, 254
mass spectrometry, 250–251
–metal clusters, production, 242–243
nanotubes, 255
production by laser vaporization method, 250
Catalysts, clusters in, 260
Ceramic particles, optical trapping, 471, 476
Cesium
isotopes, hyperfine anomalies, 355–356
in microwave atomic clock, 449

Charge, nuclear distribution
  Fourier–Bessel expansion, 347–348, 352
  and hyperfine structure, 349–352
  potential from Fermi distribution, 346–347
  relationship to electron scattering, 344
  role in binding energies, 348–349
  shape, 357–358
  two-parameter Fermi model, 345–346
Chirped pulse amplification system, in high-power lasers, 210–214
Clusters
  alkali atoms, shell closing, 247–249
  behavioral periodicities, 241–242
  carbon, *see* Carbon clusters
  in catalysts, 260
  copper, energy dependencies, 240
  electronic structure, 256–257
  energy dependencies, 240
  expansion
    non-perturbative single reference coupled cluster theory, 165–169
    single reference many-body perturbation theory, 165–169
  geometrical structure, 256
  gold, energy dependencies, 240
  niobium, theoretical approach, 255–256
  noble gas atoms, structure, 246–247
  noble metals, periodicities, 249–250
  in Periodic Table of Elements, 239
  potentials
    Jellium model, 257–258
    3D spherical potentials, 258–259
  in powder technology, 260
  production methods, 242–244
  properties, 245–246
  sodium clusters, 248–249
  surface area and volume, 239
  transition metal, production by laser vaporization method, 250
  2D electron gas, 261
Cluster theory
  single reference coupled, non-perturbative, 165–169
  state-specific coupled
    with contracted reference function, 176–188
    with decontracted reference function, 169–176
Coherency, in spectroscopy with highly coherent radiation, 13–14
Collinear fast beam laser spectroscopy
  development, 367
  rare earth region, 368
Collinear laser spectroscopy, radioactive isotopes, 338–339
Complete active space approach, in MCHF calculations, 169–176, 305–306
Computation, quantum, in coherent state manipulation, 58–59
Conditional dynamics, in logic gate, 56–58
Cooling, laser, to zero-point state, 46
Copper
  cluster energy dependencies, 240
  as ion–atom collision target, 278–279
Copper oxide particles, optical trapping, 471, 476
Correction terms
  Coulomb–Breit QED
    highly charged ions, 393–396
    and nonequal energy box, 397–398
    numerical results, 399
  Coulomb screened, self energy, 383–385
  evaluation, 389–390
  in QED
    higher-order radiative corrections, 89–90
    radiative-recoil corrections, 90
    Reduced-mass corrections, 90
  vertex and binding energy regularization correction, 388–389
  wavefunction regularization correction, 385–388
Correlation diagram, time-dependent ion–atom collisions, 274–275
Coulomb–Breit QED corrections, few-electron highly charged ions
  and nonequal energy box, 397–398
  numerical results, 399
  procedures, 393–396
Coulomb potential
  correction term evaluation, 389–390
  between electrons, 433–434
  screened self energy, 383–385
  vertex and binding energy regularization correction, 388–389
  wavefunction regularization correction, 385–388
Coupled-cluster methods, development, 343
Cyanopolyynes, production, 251

D

Degrees of freedom, internal nuclear, role in QED, 144–149
Density functional theory, for cluster electronic structure, 256–257
Diode laser sytem, development, 455
Dipoles
  measurements, 237–238
  role in interactions in atoms, 337
Dirac eigenvalue, in QED, 86
Dirac Hamiltonian
  and Green's function, 131
  spin-orbit force, 197–198
Direct amplification system, in high-power laser, 214
Doppler cooling, atoms in optical lattices, 19–20
Doppler effect, in resonance fluorescence of single ion, 67–73
Dysprosium, fast atomic beam experiments, 367–368

E

Electric dipole, role in interactions in atoms, 337
Electric hexadecapole
  role in interactions in atoms, 337
  role of ion traps, 340
Electron affinity, clusters, 245–246
Electron dynamics, photodetachment
  definition, 313–315
  resonance structures, 315
Electronic shell
  alkali atom clusters, closing, 247–249
  geometrical, noble gas atom clusters, 246–247
Electronic structure, clusters, 256–257
Electrons
  force between, 433–435
  hyperfine interactions, effect of magnetic moment, 349–352
  Li^-, cross section for photodetachment, 319–321
  –nucleus interaction, in atom, 336–337
  one-photon exchange potential, 438
  one-photon exchange scattering potentials, 435–536
  orbit–orbit interaction, 438–439
  scattering potentials
    and bound states, 441–443
    spin-0 particle, 440
    spin-1/2 particle, 440
    two spin-0 particles, 439–440
    two spin-1/2 particles, 441
  2D, gas clusters, 261
  two-photon exchange scattering potentials, 439
  two spin-0 particles, 437–438
  two spin-1/2 particles, 436–437
Electron scattering, relationship to nuclear charge distributions, 344
Electrostatic quadrupole deflectors, effect on negative ion beam, 317–319
Energy
  absorption, in optical trapping, 487–489
  binding
    role of nuclear charge distribution, 348–349
    term in regularization correction, 388–389
  dependencies of clusters, 240
  Dirac, eigenvalue, as Lamb shift, 127–128
  independent wave operators, 417–419
  levels in lead, 199
  nonequal, and RSC, 397–398
  resonance, fitting to Beutler–Fano profile, 324
  self, *see* Self energy
Erbium, fast atomic beam experiments, 367–368
Europium
  fast atomic beam experiments, 367–368
  spin measurements, 368
Exchange potential, one-photon, in spinor–scalar QED, 438
Excitation, atoms and ions, pulsed laser spectroscopy, 285

F

Faraday cup, for monitoring negative ions, 317–318
Fermi model
  potential from nuclear charge distribution, 346–347

two-parameter, for nuclear charge distribution, 345–346
Feshback resonance, in partial cross section measurements, 321
Fields
oscillatory, successive, in spectroscopy, 13–14
strong, *see* Strong fields
Fine structure
muonium in n = 2 state, 117–119
positronium in n = 2 state, 104–106
Fine-structure constant, in quantum electrodynamics
equations, 83–85
recent progress, 80–81
Flash pyrolysis, boron quartet system, 302
Fluorescence, resonance, single ion, 67–73
Fluorescence decay, in pulsed laser excitation, 290
Focusing, atoms and ions, in spectroscopy, 15–16
Foldy–Wouthuysen transformation, for Dirac Hamiltonian spin-orbit force, 197–198
Fourier–Bessel expansion, for nuclear charge distribution, 347–348, 352
Fragmentation
clusters, 245–246
molecular, from high-power lasers, 222–223
Francium
fast atomic beam experiments, 368
isotopic sequence, 373–375
Frequency, *see* Optical frequency standard
Fundamental constants, QED
1986 adjustment, 79
1990 update, 79–80
background, 78–79
fine-structure constant, 83–85
future adjustments, 92–93
Planck constant, 81–83
recent progress, 80–81
Rydberg constant, 85–92

G

Gadolinium, fast atomic beam experiments, 367–368
Gas, as ion–atom collision target
MO X-rays, 277
triple differential X-ray cross sections, 276
Global positioning system, microwave clock applications, 448
Gold
cluster energy dependencies, 240
fast atomic beam experiments, 368
Green's function, and Dirac Hamiltonian, 131

H

Harmonics, high, in high-power laser
harmonic generation and applications, 219–221
overtone generation and characterization, 215–219
Helium ion, doubly excited states, 321–324
Helium–neon laser
as mobile system, 455
optical metrology in, 454
for optical trapping, 472
Holmium ion
effect of magnetic moment, 344
fast atomic beam experiments, 367–368
ground-state hyperfine structures, 449
Hydrogen ion
atom as optical frequency standard, 452
photodetachment, 312–313
related ions
effect of magnetic moment, 344
ground-state hyperfine structures, 449
higher order radiative corrections, 140–144
status of $1s_{1/2}$ Lamb shift calculations, 149–152
Hyperfine interaction constants, in atomic physics, 237–238
Hyperfine splitting, $^{209}Bi^{82+}$, 126
Hyperfine structure
anomalies in Cs isotopes, 355–356
atomic, and microwave standards, 446–449
early measurements, 343
Fr isotopes, 373–375
H-like ions, 344
muonium in n = 1 state, 109–117
and nuclear charge distribution, 349–352

Hyperfine structure *(continued)*
nuclear quantities for experiments, 361–363
positronium in n = 1 state, 99–104
rare earth region, 368–372
splitting for $^{209}Bi^{82+}$, 152–157

I

Indium, fast atomic beam experiments, 368
Ingvar Lindren, career in atomic physics, 1–4
Interferometry, *see* Optical interferometry
Internal quantum state, Jaynes–Cummings-type coupling
and motional state, 46–47
for trapped $^9Be^+$ ion, 47–50
Ion–atom collisions, time-dependent
correlation diagram, 274–275
future experiments, 279–281
with gas targets
on MO X-rays, 277
triple differential X-ray cross sections, 276
general problem, 274
with solid state targets
comparison to gas targets, 277–278
examples, 278–279
theoretical description, 275–276
Ionization, molecular, from high-power lasers, 222–223
Ionization potential, clusters, 245–246
Ions
atomic, *see* Atomic ions
few-electron high charged, corrections, 393–396, 399
focusing and trapping, in spectroscopy, 15–16
negative
effect of quadrupole deflectors, 317–319
monitoring with Faraday cup, 317–318
single, resonance fluorescence, 67–73
trap development, 340
trapped, as optical frequency standards, 450
ISOLDE, atomic beam experiments
atomic beam laser spectroscopy, 365–366
atomic beam magnetic resonance, 363–365
fast atomic beams, 367–368
thermal atomic beams, 363
Isotopes
Cs, hyperfine anomalies, 355–356
Fr, sequence, 373–375
radioactive, collinear laser spectroscopy, 338–339
Isotope shifts, role of nuclear charge distribution, 348–349

J

Jaynes–Cummings-type coupling
between internal and motional quantum states, 46–47
for trapped $^9Be^+$ ion, 47–50
Jellium model, metal cluster potentials, 257–258
Josephson effect, description, 81

K

Kerosene, in optical trapping, 476
Kohn–Sham equations, in density functional theory, 256–257
Krypton, cluster structure, 247

L

Lagrangian density, in supercritical QED, 197
Lamb shift
$1s_{1/2}$, calculations in H-like atoms, 149–152
Dirac energy eigenvalue as, 127–128
higher order radiative corrections, 140–144
lithium-like uranium, 125
muonium in n = 2 state, 117–119
positronium in n = 2 state, 104–106
Laser microprobe mass spectrometry, for cluster production, 242–243
Lasers
argon, for optical trapping, 472
cooling to zero-point state, 46
diode sytem, development, 455
He–Ne
as mobile system, 455

optical metrology in, 454
for optical trapping, 472
high-power
broadband X-ray generation and applications, 225–228
distributed feedback direct amplification system, 214
high harmonic generation and applications
high-order harmonic applications, 219–221
overtone generation and characterization, 215–219
molecular fragmentation and ionization, 222–223
terawatt chirped pulse amplification system, 210–214
white-light generation and applications, 223–225
X-ray laser pumping, 221
–ion beam apparatus, components, 317–318
normal pulsed, for cluster production, 244
as optical frequency standards, 450
semiconductor, specifications, 454
in spectroscopy with highly coherent radiation, 17
Ti sapphire, for optical trapping, 472
tunable radiation, generation, 288–289
X-ray, pumping, applications, 221
YAG
future role, 454
for optical trapping, 472
Laser spectroscopy
with highly coherent radiation, 17
pulsed, atomic radiative lifetimes
atomization, 290–291
error sources, 294
evaporated substances, 285
excitation, 285
experimental methods, 285
fluorescence decay, 290
short pulse generation, 286–287
transition probabilities, 284
tunable laser radiation generation, 288–289
relationship to frequency stabilization, 453
Laser tweezers
absorbing particles, 473–474
angular momentum transfer to light, 474–480
applications, 469–470
doughnut beams, 474
for optical trapping, laser types, 472–473
Laser vaporization method
for carbon cluster production, 250
for C_{60} production, 252
Lead
energy levels, 199
–uranium, in ion–atom collision experiments, 280–281
Length, measurement standards, 452–455
Light
angular momentum transfer to, optical tweezing, 474–480
associated properties of clusters, 245–246
wavelength and speed, measurement
accuracy, 458
technique, 457
white, generation and applications, 223–225
Lippman–Schwinger equation, role of nuclear mass and size, 137–138
Lithium ion
doubly excited states, 325–330
electron, cross section for photodetachment, 319–321
threshold investigations, 330–332
Logic gate, quantum, single ion, conditional dynamics, 56–58

M

Magnesium
hyperfine structure, 448
in microwave atomic clock, 449
Magnetic dipole, role in interactions in atoms, 337
Magnetic moment
clusters, 245–246
effect on H-like ion hyperfine structure, 344
effect on hyperfine structure, 349–352
Magnetic octupole
role in interactions in atoms, 337
role of ion traps, 340
Magnetization, nuclear, distribution, 352–355

Masers, in spectroscopy with highly coherent radiation, 16
Maser spectroscopy, with highly coherent radiation, 16
Mass spectrometry, carbon clusters, 250–251
Mercury
 clusters, energy dependencies, 240
 fast atomic beam experiments, 368
Metals
 –carbon clusters, production, 242–243
 cluster potentials
  Jellium model, 257–258
  3D spherical potentials, 258–259
Microwave atomic clock
 applications, 448
 Cs-based clock, 449
 Mg^+-based clock, 449
Microwave frequency standard
 and atomic hyperfine structure, 446–449
 and optical standard, chains linking, 456–457
Microwave–optical double resonance, role of ion traps, 340
Microwave spectroscopy, with highly coherent radiation, 12–13
Mirrors, coatings, for optical interferometry, 459
Models
 big bang nucleosynthesis, boron quartet system in, 302
 Fermi
  potential from nuclear charge distribution, 346–347
  two-parameter, for nuclear charge distribution, 345–346
 functions, definition, 416
 metal cluster potentials
  Jellium model, 257–258
  3D spherical potentials, 258–259
 protons and neutrons, internal nuclear degrees of freedom, 144–149
 3D, Periodic Table of Elements, 238–239
Molecular beam magnetic resonance method, in spectroscopy, 5–12
Molecular quantum theory, semi-empirical and *ab initio* methods, 422–423
Molecules
 fragmentation and ionization, from high-power lasers, 222–223
 resonance ionization spectroscopy, 339–340
Momentum, transfer in optical trapping
 angular momentum, 471, 474–480
 linear momentum, 471, 480
 macroscopic momentum, 481–484
 microscopic momentum, 480–481
 orbital angular momentum, 480
 spin angular momentum, 484–487
Motion
 coherent and Schrödinger cat states, 50–53
 driven, atoms in optical lattices, 34–35
Motional quantum state
 Jaynes–Cummings-type coupling
  and internal state, 46–47
  for trapped $^9Be^+$ ion, 47–50
 reconstruction, 53–55
Multiconfiguration Hartree–Fock calculation, boron quartet system, 301–302, 305–306
Multi-reference coupled cluster methods, cluster expansion strategy, 165–169
Muonium
 1S–2S transition, 119–120
 in n = 1 state
  hyperfine structure, 109–117
  Zeeman effect, 109–117
 in n = 2 state, Lamb shift and fine structure, 117–119

N

Nanotubes, discovery, 255
Neon, cluster structure, 247
Neutral atom detector, in laser–ion beam apparatus, 317–318
Neutrons, models, internal nuclear degrees of freedom, 144–149
Nickel, as ion–atom collision target, 278–279
Niobium, clusters, theoretical approach, 255–256
Noble gas, atoms, cluster structure, 246–247
Noble metals, clusters, periodicities, 249–250
Normal pulsed laser, for cluster production, 244
Nuclear magnetic resonance, C_{60}, 253
Nuclear tensor operators, definition, 337

Nucleus
charge distribution
Fourier–Bessel expansion, 347–348, 352
and hyperfine structure, 349–352
potential from Fermi distribution, 346–347
relationship to electron scattering, 344
role in binding energies, 348–349
shape, 357–358
two-parameter Fermi model, 345–346
effect on QED of strong fields, 137–140
–electron interaction, in atom, 336–337
finite size, in quantum electrodynamics, 90
hyperfine structure experiments, 361–363
magnetization, distribution, 352–355
properties studies with atomic spectroscopy, 356–357
rare earth region, hyperfine structure, 368
spin measurements, 237–238

O

Optical frequency standard
hydrogen atom as, 452
lasers as, 450
and microwave standard, chains linking, 456–457
trapped atoms as, 451
trapped ions as, 450
Optical interferometry
categories, 458
measurement of light wavelength
accuracy, 458
technique, 457
mirror coatings, 459
Optical lattices, atoms, 24–28
Bragg scattering, 28–34
Doppler cooling, 19–20
driven motion, 34–35
sub-Doppler cooling, 20–23
Optical pumping, in spectroscopy, 14
Optical trapping
absorbing particles, 471
energy absorption, 487–489
interaction with medium, 489–490
linear and orbital angular momentum transfer, 480
macroscopic momentum transfer, 481–484
microscopic momentum transfer, 480–481
spin angular momentum transfer, 484–487
ceramic particles, 476
components, 469
CuO particles, 476
history, 469
micron-sized reflective particles, 470–471
particles, 470
Rayleigh particles, 470
Oscillatory fields, successive, in spectroscopy, 13–14
Overtones, high, generation and characterization in laser, 215–219

P

Partial-wave renormalization
correction term evaluation, 389–390
Coulomb screened self energy, 383–385
first-order self energy, 380–383
quantum electrodynamics, 379–380
vertex and binding energy regularization correction, 388–389
wavefunction regularization correction, 385–388
Particles
absorbing, *see* Absorbing particles
ceramic, optical trapping, 476
CuO, optical trapping, 471, 476
micron-sized, reflective, trapping, 470–471
Rayleigh, optical trapping, 470
transparent, optical trapping, 470
Partioning technique, multi-dimensional, conventional, 423–429
Pauli–Villars prescription, Coulomb screened self energy, 384–385
Periodicity, noble metal clusters, 249–250
Periodic Table of Elements
clusters, 239
first layout, 236
nuclear physics research, 237
three dimensional model, 238–239, 260–263
Perturbation theory
and Bloch method, 427–428

Perturbation theory *(continued)*
for bound state QED, 145–146
connection with operators, 425–426
Photodetachment
atomic negative ions, role in physics, 312–313
cross section
electron from Li^- ion, 319–321
Li^-, threshold investigations, 330–332
partial cross sections
definition, 315–316
doubly excited states in Li^-, 325–330
Feshback resonance, 321
total cross sections
definition, 313–315
resonance structures, 315
Photoelectron affinity, clusters, 245–246
Photoionization potential, clusters, 245–246
Physics, atomic
Ingvar Lindren's career, 1–4
nuclear spins and hyperfine interaction constants, 237–238
Planck constant, in quantum electrodynamics
recent progress, 80–81
from watt-balance experiment, 81–83
Polarizability, static dipole, clusters, 245–246
Polarization, vacuum, in in quantum electrodynamics, 89
Positronium
1S–2S interval, 107–108
fine structure and Lamb shift in n = 2 state, 104–106
hyperfine structure in n = 1 state, 99–104
Potentials
Coulomb, *see* Coulomb potential
exchange, one-photon, in spinor–scalar QED, 438
from Fermi distribution, 346–347
ionization, clusters, 245–246
metal clusters
Jellium model, 257–258
3D spherical potentials, 258–259
photoionization, clusters, 245–246
role in vacuum polarization
Uehling potential, 133–134, 136–137
Wichmann–Kroll potential, 134–136
scattering, *see* Scattering potentials
Powder technology, clusters in, 260
Protons, models, internal nuclear degrees of freedom, 144–149
Pulses, successive coherent, in spectroscopy, 13–14

Q

QED, *see* Quantum electrodynamics
Quadrupoles
Fr isotopes, 373
measurements, 237–238
Quantum computation, in manipulation of trapped atomic ions, 58–59
Quantum electrodynamics
comparison of theory to experiment, 90–92
and fundamental constants
1986 adjustment, 79
1990 update, 79–80
background, 78–79
fine-structure constant, 83–85
future adjustments, 92–93
Planck constant, 81–83
recent progress, 80–81
Rydberg constant, 85–92
partial-wave renormalization, 379–380
spinor–scalar, one-photon exchange potential, 438
strong fields
$^{209}Bi^{82+}$ hyperfine splitting, 126
$^{209}Bi^{82+}$ hyperfine structure splitting, 152–157
higher order radiative corrections, 140–144
internal nuclear degrees of freedom, 144–149
Lamb shift of lithium-like uranium, 125
nuclear mass, 137–140
nuclear size, 137–140
self energy, 128–137
status of $1s_{1/2}$ Lamb shift calculations in H-like atoms, 149–152
vacuum polarization, 128–137
supercritical, status
Dirac Hamiltonian spin-orbit force, 197–198
Lagrangian density, 197
lead energy levels, 199
relativistic mean meson field theory, 196–197

Quantum Hall effect, description, 81
Quantum logic
 application to spectroscopy, 59–60
 in coherent state manipulation of trapped atomic ions
  in computation, 55–56
  conditional dynamics and single ion controlled not logic gate, 56–58
  perspectives, 58–59
  quantum computation, 58–59
  quantum registers, 58
  Schrödinger cats, 58
Quantum registers, in quantum logic, 58
Quantum states
 coherent, trapped atomic ions
  coherent and Schrödinger cat states of motion, 50–53
  conditional dynamics and single ion controlled not logic gate, 56–58
  entangled states, 45
  Jaynes–Cummings-type coupling between internal and motional states, 46–47
   for trapped $^{9}Be^{+}$ ion, 47–50
  laser cooling to zero-point state, 46
  motional state reconstruction, 53–55
  quantum computation, 58–59
  quantum logic, 55–56, 58–59
  quantum logic application to spectroscopy, 59–60
  quantum registers, 58
  Schrödinger cats, 58
  trap types, 42–45
 zero-point, laser cooling to, 46
Quantum theory, semi-empirical and *ab initio* methods, 422–423
Quasi-complete active space multi-configuration self-consistent field function, in state-specific coupled cluster theory, 169–176

R

Radiation, highly coherent, spectroscopy
 focusing and trapping of atoms and ions, 15–16
 lasers, 17
 masers, 16
 microwave spectroscopy, 12–13
 molecular beam magnetic resonance method, 5–12
 optical pumping, 14
 radio frequency spectroscopy, 12–13
 successive oscillatory fields, 13–14
Radiative corrections, higher-order, in quantum electrodynamics, 89–90
Radiative-recoil corrections, in quantum electrodynamics, 90
Radio frequency spectroscopy
 with highly coherent radiation, 12–13
 in hyperfine structure measurements, 343
Rare earth elements, nuclear region, atomic beam hyperfine structure, 368–372
Rayleigh particles, optical trapping, 470
Reaction operator, connection with perturbation theory, 425–426
Reduced-mass corrections, in quantum electrodynamics, 90
Reference functions, in state-specific coupled cluster theory
 contracted function, 176–188
 decontracted function, 169–176
Reference set, in Bloch–Lindren equation, 421
Reference state corrections, Coulomb–Breit QED
 highly charged ions, 393–396
 and nonequal energy box, 397–398
 numerical results, 399
Regularization correction
 correction term evaluation, 389–390
 vertex and binding energy terms, 388–389
Relativistic mean meson field theory, in supercritical QED, 196–197
Relativistic recoil, in quantum electrodynamics, 86–87
Renormalization, *see* Partial-wave renormalization
Resonance ionization spectroscopy
 for atom excitation state, 315–316
 atoms and molecules, 339–340
Rhenium ion
 effect of magnetic moment, 344
 ground-state hyperfine structures, 449
RSC, *see* Reference state corrections
Rydberg constant, in QED
 Dirac eigenvalue, 86
 experiment and theory comparisons, 90–92

Rydberg constant, in QED *(continued)*
  higher-order radiative corrections, 89–90
  nuclear finite size, 90
  radiative-recoil correction, 90
  recent progress, 80–81
  reduced-mass correction, 90
  relativistic recoil, 86–87
  self energy, 87–89
  vacuum polarization, 89

S

Scattering potentials, electron
  and bound states, 441–443
  one-photon exchange, 435–536
  spin-0 particle, 440
  spin-1/2 particle, 440
  two-photon exchange, 439
  two spin-0 particles, 439–440
  two spin-1/2 particles, 441
Schrödinger equation
  in density functional theory, 256
  non-linear form, 417–419
Schrödinger's cat
  motion, creation, 50–53
  in quantum logic, 58
Self energy
  correction term evaluation, 389–390
  Coulomb screened correction, 383–385
  first-order, partial-wave renormalization, 380–383
  higher order radiative corrections, 140–144
  in QED, 87–89, 128–137
  vertex and binding energy regularization correction, 388–389
  wavefunction regularization correction, 385–388
Semiconductor laser, specifications, 454
Semi-empirical methods, in quantum theory, 422–423
Single reference coupled cluster theory, non-perturbative, 165–169
Single reference many-body perturbation theory, for clusters, 165–169
Solid state material, as ion–atom collision target
  comparison to gas targets, 277–278
  examples, 278–279
Spectroscopy
  absorption, boron quartet system, 301–302
  abundance, clusters, 245–246
  atomic, for nuclear properties studies, 356–357
  atomic beam laser, experiments at ISOLDE, 365–366
  collinear fast beam laser
    development, 367
    rare earth region, 368
  collinear laser, radioactive isotopes, 338–339
  with highly coherent radiation
    focusing and trapping of atoms and ions, 15–16
    molecular beam magnetic resonance method, 5–12
    optical pumping, 14
    successive oscillatory fields, 13–14
  laser, *see* Laser spectroscopy
  maser, with highly coherent radiation, 16
  microwave, with highly coherent radiation, 12–13
  radio frequency
    with highly coherent radiation, 12–13
    in hyperfine structure measurements, 343
  resonance ionization
    for atom excitation state, 315–316
    atoms and molecules, 339–340
Stained glass, production methods, 242
State-specific coupled cluster theory
  with contracted reference function, 176–188
  with decontracted reference function, 169–176
Strong fields, quantum electrodynamics
  $^{209}Bi^{82+}$ hyperfine splitting, 126
  $^{209}Bi^{82+}$ hyperfine structure splitting, 152–157
  higher order radiative corrections, 140–144
  internal nuclear degrees of freedom, 144–149
  Lamb shift of lithium-like uranium, 125
  nuclear mass, 137–140
  nuclear size, 137–140
  status of $1s_{1/2}$ Lamb shift calculations in H-like atoms, 149–152
Sub-Doppler cooling, atoms in optical lattices, 20–23

Supersonic jets, for cluster production, 244

T

Thallium, fast atomic beam experiments, 368
Thermonuclear fusion, boron quartet system in associated research, 304
Thulium, spin measurements, 368
Titanium sapphire laser, for optical trapping, 472
Transition metals, cluster production by laser vaporization method, 250
Transition probabilities, atoms, 284
Trapping, atoms and ions, in spectroscopy, 15–16

U

Uehling potential, role in vacuum polarization, 133–134, 136–137
Ultraviolet region, atomic radiative lifetimes
  atomization, 290–291
  error sources, 294
  excitation, 285
  experimental methods, 285
  fluorescence decay, 290
  short pulse generation, 286–287
  transition probabilities, 284
  tunable laser radiation generation, 288–289
Uranium
  Lamb shift, 140–144
  –lead, in ion–atom collision experiments, 280–281
  lithium-like, Lamb shift, 125

V

Vacuum, supercritical QED status
  Dirac Hamiltonian spin-orbit force, 197–198
  Lagrangian density, 197
  lead energy levels, 199
  relativistic mean meson field theory, 196–197
Vacuum polarization
  higher order radiative corrections, 140–144
  quantum electrodynamics of strong fields, 128–137
Vacuum ultraviolet region, atomic radiative lifetimes
  atomization, 290–291
  error sources, 294
  excitation, 285
  experimental methods, 285
  fluorescence decay, 290
  short pulse generation, 286–287
  transition probabilities, 284
  tunable laser radiation generation, 288–289
Vertex term, in regularization correction, 388–389
Virtual set, in Bloch–Lindren equation, 421

W

Watt-balance experiment, Planck constant, 81–83
Wavefunction, regularization correction, 385–388
Wavelength, light, measurement
  accuracy, 458
  technique, 457
Wave operators
  connection with perturbation theory, 425–426
  definition, 416
  energy-independent operators, 417–419
Wichmann–Kroll potential, role in vacuum polarization, 134–136

X

X-ray laser pumping, applications, 221
X-rays
  broadband generation and applications, in high-power lasers, 225–228
  ion–atom collision with gas targets
    MO X-rays, 277
    triple differential cross sections, 276
  ion–atom collision with solid state targets
    comparison to gas targets, 277–278
    examples, 278–279

Y

YAG lasers
  future role, 454
  for optical trapping, 472
Ytterbium, fast atomic beam experiments, 367–368

Z

Zeeman effect, muonium in n = 1 state, 109–117
Zero-point quantum state, laser cooling to, 46
Zinc dust, optical trapping, 471

ISBN 0-12-034830-6
90065
9 780120 348305